Current Status of Fresh Water Microbiology

Ravindra Soni • Deep Chandra Suyal •
Lourdes Morales-Oyervides • Jaspal Singh
Chauhan
Editors

Current Status of Fresh Water Microbiology

Editors
Ravindra Soni
Department of Agricultural Microbiology
Indira Gandhi Krishi Vishwa Vidyalaya
Raipur, Chhattisgarh, India

Deep Chandra Suyal
Department of Science
Vidyadayini Institute of Science, Management
and Technology
Bhopal, Madhya Pradesh, India

Lourdes Morales-Oyervides
Facultad de Ciencias Quimicas
Universidad Autonoma de Coahuila
Saltillo, Mexico

Jaspal Singh Chauhan
Department of Himalayan Aquatic Biodiversity
Hemvati Nandan Garhwal University (A Central
University)
Srinagar, Uttarakhand, India

ISBN 978-981-99-5017-1   ISBN 978-981-99-5018-8   (eBook)
https://doi.org/10.1007/978-981-99-5018-8

© The Editor(s) (if applicable) and The Author(s), under exclusive license to Springer Nature Singapore Pte Ltd. 2023
This work is subject to copyright. All rights are solely and exclusively licensed by the Publisher, whether the whole or part of the material is concerned, specifically the rights of translation, reprinting, reuse of illustrations, recitation, broadcasting, reproduction on microfilms or in any other physical way, and transmission or information storage and retrieval, electronic adaptation, computer software, or by similar or dissimilar methodology now known or hereafter developed.
The use of general descriptive names, registered names, trademarks, service marks, etc. in this publication does not imply, even in the absence of a specific statement, that such names are exempt from the relevant protective laws and regulations and therefore free for general use.
The publisher, the authors, and the editors are safe to assume that the advice and information in this book are believed to be true and accurate at the date of publication. Neither the publisher nor the authors or the editors give a warranty, expressed or implied, with respect to the material contained herein or for any errors or omissions that may have been made. The publisher remains neutral with regard to jurisdictional claims in published maps and institutional affiliations.

This Springer imprint is published by the registered company Springer Nature Singapore Pte Ltd.
The registered company address is: 152 Beach Road, #21-01/04 Gateway East, Singapore 189721, Singapore

Paper in this product is recyclable.

Preface

Freshwater ecosystems are dynamic natural resources, providing sources of potable water, food, animal habitats, and recreation. Perspectives of microbial dynamics in freshwater bodies provide a comprehensive and systematic analysis of microbial ecology in these ecosystems. These microbes are at the hub of biogeochemical cycles (carbon, nitrogen, phosphorus, potassium, and other elements). Moreover, they are an integral part of the aquatic food web and control the quality of freshwater bodies. Unfortunately, our freshwater microbial diversity is under threat due to several factors including overexploitation of species, increasing population, pollution, climate change, construction of dams, and introduction of exotic species. More specifically, pollution in freshwater reservoirs is a serious concern nowadays which affects the environment as well as humans. Therefore, multidirectional efforts viz. basic and advanced research, rapid and error-free data analysis, smart data compilation, and their worldwide sharing are urgently needed to identify, characterize, and conserve freshwater microbial diversity.

Nowadays, scientific knowledge is expanding very quickly due to the voluminous research, utilization of next-generation analytical tools, and higher data generation, which need to be compiled and shared effectively among the beneficiaries. From this perspective, this book is a perfect documentation of primary and secondary data-based information on the latest research findings, case studies, experiences, and innovations. This book deals with the various aspects of freshwater microbiology including diverse habitats, associated microorganisms, their ecological interactions, and applications. Moreover, it also discusses the issue of pollution in freshwater bodies and puts forward available strategies for its eco-friendly solution.

We acknowledge the suggestions and encouragement made by colleagues and well-wishers. Moreover, we are grateful to all the authors who have contributed to this book. Suggestions for the improvement of the book will be highly appreciated and incorporated in the subsequent editions.

Raipur, Chhattisgarh, India	Ravindra Soni
Bhopal, Madhya Pradesh, India	Deep Chandra Suyal
Saltillo, Mexico	Lourdes Morales-Oyervides
Srinagar, Uttarakhand, India	Jaspal Singh Chauhan

Contents

1   Freshwater Microbiology: Recent Updates and Prospects   1
Virgilio Bocanegra-García, Humberto Martínez-Montoya,
María Concepción Tamayo-Ordoñez, Israel Estrada-Camacho,
Alejandra Aguilar-Salazar, Gloria Dhení Guaní-Sánchez,
Gerardo de Jesús Sosa-Santillán, and Erika Acosta-Cruz

**2   The Chemical Composition of the Water in the Rivers,
Lakes, and Wetlands of Uttarakhand**   29
Manjul Gondwal, Navneet Kishore, Ravindra Soni,
Rakesh Kumar Verma, and Bhanu Pratap Singh Gautam

**3   Microbial Diversity of Cold-Water Reservoirs and Their
Prospective Applications**   49
Vidhi Jain, Prashant Arya, Shivani Maheshkumar Yagnik,
Vikram Hiren Raval, and Namita Ashish Singh

**4   Overview of Microbial Associations and Their Role Under
Aquatic Ecosystems**   77
Manali Singh, Parul Chaudhary, Shivani Bhutani, Shruti Bhasin,
Anshi Mehra, and Keshawananad Tripathi

**5   Plant-Microbe Interaction in Freshwater Ecosystem for
Improving Water Quality**   117
Matta Gagan, Pant Gaurav, G. K. Dhingra, Kumar Avinash,
Nayak Anjali, and Kumar Pawan

6   Microbial Interactions with Aquatic Plants   135
N. V. T. Jayaprada, Jayani J. Wewalwela, G. A. H. Galahitigama,
and P. A. N. P. Pandipperuma

7   Status of Microplastic Pollution in the Freshwater Ecosystems   161
Vaishali Bhatt, Neha Badola, Deepti Semwal, and Jaspal Singh
Chauhan

vii

8 Heavy Metal Pollution in Water: Cause and Remediation Strategies 181
Damini Maithani, Hemant Dasila, Ritika Saxena, Anjali Tiwari, Devesh Bhatt, Komal Rawat, and Deep Chandra Suyal

9 Recent Advances in Biological Wastewater Treatment 205
José Antonio Rodríguez-de la Garza, Pedro Pérez-Rodríguez, Laura María González-Méndez, and Silvia Yudith Martínez-Amador

10 Recent Developments in Wastewater Treatments 241
Marlene Lariza Andrade-Guel, Christian Javier Cabello-Alvarado, Lucía Fabiola Cano-Salazar, Carlos Alberto Ávila-Orta, and Víctor Javier Cruz-Delgado

11 Wastewater Treatment: Perspective and Advancements 265
Divya Goel, Vineet Kumar Maurya, and Sudhir Kumar

12 Overview of Methods and Processes Used in Wastewater Treatment 289
Rewa Kulshrestha, Aakriti Sisodiya, and Soumitra Tiwari

13 Role of Microorganisms in Polluted Water Treatment 303
Inoka C. Perera, K. A. G. de Alwis, and P. I. T. Liyanage

14 Bioremediation of Polluted Water 321
U. M. Aruna Kumara, N. V. T. Jayaprada, and N. Thiruchchelvan

15 Pollution in Freshwater: Impact and Prevention 347
Nandan Singh and Maitreyie Narayan

16 Freshwater Pollution: Overview, Prevention, and Control 359
Pragati Srivastava and Manvika Sahgal

17 Iron-Degrading Bacteria in the Aquatic Environment: Current Trends and Future Directions 367
David Waikhom, Soibam Ngasotter, Laishram Soniya Devi, Soibam Khogen Singh, and Sukham Munilkumar

18 Bioactive Compounds from Aquatic Ecosystem 387
Surendra Puri, Rohit Mahar, and Gunjan Goswami

19 Freshwater Blue–Green Algae: A Potential Candidate for Sustainable Agriculture and Environment for the Welfare of Future Planet Earth 409
Arun Kumar Rai, Binu Gogoi, and Rabina Gurung

20 Factors Affecting Fish Migration 425
Uddesh Ramesh Wanjari, Anirban Goutam Mukherjee, Abilash Valsala Gopalakrishnan, Reshma Murali, Sandra Kannampuzha, and D. S. Prabakaran

Editors and Contributors

About the Editors

Ravindra Soni is an assistant professor at the *Department of Agricultural Microbiology,* I.G.K.V., Raipur, CG, India. He has more than 13 years of experience in teaching and research. He is well experienced in the field of agricultural microbiology, microbial ecology, molecular biology and biotechnology, especially with reference to higher altitudes. He has published more than 100 research articles and chapters in well-reputed international journals and books. He also edited ten books with international publishers.

Deep Chandra Suyal is Assistant Professor at the Department of Science, Vidyadayini Institute of Science, Management and Technology, Bhopal, Madhya Pradesh, India. He has been actively engaged in research for 12 years and has experience in the fields of agricultural microbiology, molecular biology and biotechnology, and microbial ecology. He is currently investigating the genomics and proteomics of cold-adapted microorganisms. He has published several research articles, book chapters, and edited books with reputed international journals and publishers.

Lourdes Morales-Oyervides obtained her PhD in Engineering from University College Cork, Ireland (2015). Later on (2015–2016), she collaborated as a postdoctoral researcher at the Dairy Processing Technology Centre (Limerick, Ireland). Currently, she is a Professor at the Autonomous University of Coahuila (México) and she is recognized as a member of the National System of Researchers (SNI) from CONACyT (México) since 2016. Her research interests are focused on the design, optimization, and scale-up of sustainable bioprocesses to produce food additives, enzymes, biopolymers, and other added-value compounds.

Jaspal Singh Chauhan is a highly accomplished environmental scientist, renowned for his expertise in the field of environmental science and his significant contributions to aquatic ecology, water pollution, plastic pollution, and climate change. He earned his Doctorate in Environmental Science from GB Pant University of Agriculture & Technology, Pantnagar, Uttarakhand, India, in 2009. He currently serves as an Assistant Professor and Head of the Department of Himalayan Aquatic Biodiversity at Hemvati Nandan Garhwal University, a prestigious central university in Uttarakhand. Dr. Chauhan possesses an impressive 15 years of teaching and research experience in various aspects of the environment. He has authored approximately 25 research articles, which have been published in reputed national and international journals and books. His dedication to sharing his findings and expertise has led him to actively participate in numerous international and national seminars and conferences, where he has presented his research work to fellow experts in the field. In addition to his academic accomplishments, Dr. Chauhan is deeply committed to raising awareness about environmental protection among the general public. He has played a pivotal role in organizing several impactful programs, such as Save Ganga, Save Water Springs, Save Himalaya, and Plastic-Free Day. Recognizing his outstanding contributions to the field of environmental science, Dr. Chauhan was honored with the prestigious "Environmentalist of the Year" Award in 2021 by the National Environmental Science Academy.

Contributors

Erika Acosta-Cruz Departamento de Biotecnología, Facultad de Ciencias Químicas, Universidad Autónoma de Coahuila, Saltillo, Mexico

Alejandra Aguilar-Salazar Laboratorio Interacción Ambiente-Microorganismo, Centro de Biotecnología Genómica, Instituto Politécnico Nacional, Reynosa, Tamaulipas, Mexico

Marlene Lariza Andrade-Guel Departamento de Química Macromolecular y Nanomateriales, Centro de Investigación en Química Aplicada, Saltillo, Coahuila, Mexico

Nayak Anjali Department of Zoology and Environmental Sciences, Gurukula Kangri (Deemed to be University), Haridwar, India

Prashant Arya Department of Microbiology and Biotechnology, University School of Sciences, Gujarat University, Ahmedabad, Gujarat, India

Carlos Alberto Ávila-Orta Departamento de Materiales Avanzados, Centro de Investigación en Química Aplicada, Saltillo, Coahuila, Mexico

Kumar Avinash Department of Zoology and Environmental Sciences, Gurukula Kangri (Deemed to be University), Haridwar, India

Neha Badola Aquatic Ecology Lab, Department of Himalayan Aquatic Biodiversity, Hemvati Nandan Bahuguna Garhwal University (A Central University), Srinagar, Uttarakhand, India

Shruti Bhasin Department of Biotechnology, Banasthali Vidyapith, Jaipur, Rajasthan, India

Devesh Bhatt Department of Chemistry, G.B. Pant University of Agriculture & Technology, Pantnagar, India

Vaishali Bhatt Aquatic Ecology Lab, Department of Himalayan Aquatic Biodiversity, Hemvati Nandan Bahuguna Garhwal University (A Central University), Srinagar, Uttarakhand, India

Shivani Bhutani Department of Biotechnology, Invertis University, Bareilly, Uttar Pradesh, India

Virgilio Bocanegra-García Laboratorio Interacción Ambiente-Microorganismo, Centro de Biotecnología Genómica, Instituto Politécnico Nacional, Reynosa, Tamaulipas, Mexico

Christian Javier Cabello-Alvarado Departamento de Materiales Avanzados, Centro de Investigación en Química Aplicada, Saltillo, Coahuila, Mexico
CONAHCYT - Centro de Investigación en Química Aplicada, Saltillo, Mexico

Lucía Fabiola Cano-Salazar Facultad de Ciencias Químicas, Universidad Autónoma de Coahuila, Saltillo, Coahuila, Mexico

Parul Chaudhary Animal Genomics Lab, Department of Animal Biotechnology, NDRI, Karnal, India

Víctor Javier Cruz-Delgado Departamento de Procesos de Transformación, Centro de Investigación en Química Aplicada, Saltillo, Coahuila, Mexico

Hemant Dasila Department of Microbiology, Akal College of Basic Sciences, Eternal University, Sirmaur, Himanchal Pradesh, India

K. A. G. de Alwis Synthetic Biology Laboratory, Department of Zoology and Environment Sciences, Faculty of Science, University of Colombo, Colombo, Sri Lanka

Gerardo de Jesús Sosa-Santillán Departamento de Biotecnología, Facultad de Ciencias Químicas, Universidad Autónoma de Coahuila, Saltillo, Mexico

Laishram Soniya Devi College of Fisheries, Central Agricultural University (I), Mohanpur, Tripura, India

G. K. Dhingra Department of Botany, Sri Dev Suman Uttarakhand University, Rishikesh Campus, Rishikesh, Uttarakhand, India

Israel Estrada-Camacho U.A.M. Reynosa Aztlán, Universidad Autónoma de Tamaulipas, Reynosa, Mexico

Matta Gagan Department of Zoology and Environmental Sciences, Gurukula Kangri (Deemed to be University), Haridwar, India

G. A. H. Galahitigama Department of Export Agriculture, Faculty of Agricultural Sciences, Sabaragamuwa University of Sri Lanka, Belihuloya, Sri Lanka

Pant Gaurav Department of Zoology and Environmental Sciences, Gurukula Kangri (Deemed to be University), Haridwar, India

Bhanu Pratap Singh Gautam Department of Chemistry, Laxman Singh Mahar Government Post Graduate College, Pithoragarh, India

Divya Goel Department of Biotechnology, H.N.B. Garhwal University (A Central University), Srinagar, Uttarakhand, India

Binu Gogoi Department of Botany, School of Life Sciences, Sikkim University, Tadong, Sikkim, India

Manjul Gondwal Department of Chemistry, Laxman Singh Mahar Government Post Graduate College, Pithoragarh, India

Laura María González-Méndez Universidad Autónoma Agraria Antonio Narro, Saltillo, Coahuila, Mexico

Abilash Valsala Gopalakrishnan Department of Biomedical Sciences, School of Bio-Sciences and Technology, Vellore Institute of Technology, Vellore, Tamil Nadu, India

Gunjan Goswami Government Inter College (GIC) Pitradhar, Augustmuni, Rudraprayag, Uttarakhand, India

Gloria Dhení Guaní-Sánchez Laboratorio Interacción Ambiente-Microorganismo, Centro de Biotecnología Genómica, Instituto Politécnico Nacional, Reynosa, Tamaulipas, Mexico

Rabina Gurung Department of Botany, School of Life Sciences, Sikkim University, Tadong, Sikkim, India

Vidhi Jain Department of Microbiology, University College of Science, Mohanlal Sukhadia University, Udaipur, Rajasthan, India

N. V. T. Jayaprada Department of Agricultural Technology, Faculty of Technology, University of Colombo, Homagama, Sri Lanka

Sandra Kannampuzha Department of Biomedical Sciences, School of Bio-Sciences and Technology, Vellore Institute of Technology, Vellore, Tamil Nadu, India

Navneet Kishore Department of Chemistry, Maitreyi College, University of Delhi, New Delhi, India

Rewa Kulshrestha Department of Food Processing and Technology, Atal Bihari Vajpayee Vishwavidyalaya, Bilaspur, Chhattisgarh, India

U. M. Aruna Kumara Department of Agricultural Technology, Faculty of Technology, University of Colombo, Homagama, Sri Lanka

Sudhir Kumar Department of Biotechnology, H.N.B. Garhwal University (A Central University), Srinagar, Uttarakhand, India

Editors and Contributors

P. I. T. Liyanage Synthetic Biology Laboratory, Department of Zoology and Environment Sciences, Faculty of Science, University of Colombo, Colombo, Sri Lanka

Rohit Mahar Department of Chemistry, H.N.B. Garhwal University (A Central University), Srinagar, Uttarakhand, India

Damini Maithani Department of Microbiology, Lovely Professional University, Phagwara, Punjab, India

Silvia Yudith Martínez-Amador Universidad Autónoma Agraria Antonio Narro, Saltillo, Coahuila, Mexico

Humberto Martínez-Montoya Departamento de Tecnología de Alimentos, U.A.M. Reynosa Aztlán, Universidad Autónoma de Tamaulipas, Reynosa, Mexico

Vineet Kumar Maurya Department of Botany and Microbiology, H.N.B. Garhwal University (A Central University), Srinagar, Uttarakhand, India

Anshi Mehra Department of Biotechnology, Invertis University, Bareilly, Uttar Pradesh, India

Anirban Goutam Mukherjee Department of Biomedical Sciences, School of Bio-Sciences and Technology, Vellore Institute of Technology, Vellore, Tamil Nadu, India

Sukham Munilkumar ICAR-Central Institute of Fisheries Education, Mumbai, India

Reshma Murali Department of Biomedical Sciences, School of Bio-Sciences and Technology, Vellore Institute of Technology, Vellore, Tamil Nadu, India

Maitreyie Narayan Department of Forestry and Environmental Science, Kumaun University, Nainital, Uttarakhand, India

Soibam Ngasotter ICAR-Central Institute of Fisheries Education, Mumbai, India

P. A. N. P. Pandipperuma Department of Agricultural Technology, Faculty of Technology, University of Colombo, Homagama, Sri Lanka

Kumar Pawan Department of Zoology and Environmental Sciences, Gurukula Kangri (Deemed to be University), Haridwar, India

Inoka C. Perera Synthetic Biology Laboratory, Department of Zoology and Environment Sciences, Faculty of Science, University of Colombo, Colombo, Sri Lanka

Pedro Pérez-Rodríguez Universidad Autónoma Agraria Antonio Narro, Saltillo, Coahuila, Mexico

D. S. Prabakaran Department of Radiation Oncology, College of Medicine, Chungbuk National University, Cheongju, Republic of Korea
Department of Biotechnology, Ayya Nadar Janaki Ammal College (Autonomous), Sivakasi, India

Surendra Puri Department of Chemistry, H.N.B. Garhwal University (A Central University), Srinagar, Uttarakhand, India

Arun Kumar Rai Department of Botany, School of Life Sciences, Sikkim University, Tadong, Sikkim, India

Vikram Hiren Raval Department of Microbiology and Biotechnology, University School of Sciences, Gujarat University, Ahmedabad, Gujarat, India

Komal Rawat Department of Chemistry, G.B. Pant University of Agriculture & Technology, Pantnagar, India

José Antonio Rodríguez-de la Garza Facultad de Ciencias Químicas, Universidad Autónoma de Coahuila, Saltillo, Coahuila, Mexico

Manvika Sahgal Department of Microbiology, G. B. Pant University of Agriculture & Technology, Pantnagar, Uttarakhand, India

Ritika Saxena School of Biotechnology, IFTM University, Moradabad, India

Deepti Semwal Aquatic Ecology Lab, Department of Himalayan Aquatic Biodiversity, Hemvati Nandan Bahuguna Garhwal University (A Central University), Srinagar, Uttarakhand, India

Manali Singh Department of Biotechnology, Invertis University, Bareilly, Uttar Pradesh, India

Namita Ashish Singh Department of Microbiology, University College of Science, Mohanlal Sukhadia University, Udaipur, Rajasthan, India

Nandan Singh Department of Forestry and Environmental Science, Kumaun University, Nainital, Uttarakhand, India

Soibam Khogen Singh College of Fisheries, Central Agricultural University (I), Mohanpur, Tripura, India

Jaspal Singh Chauhan Department of Himalayan Aquatic Biodiversity, Hemvati Nandan Bahuguna Garhwal University (A Central University), Srinagar, Uttarakhand, India

Aakriti Sisodiya Department of Food Processing and Technology, Atal Bihari Vajpayee Vishwavidyalaya, Bilaspur, Chhattisgarh, India

Ravindra Soni Department of Agricultural Microbiology, Indira Gandhi Krishi Vishwa Vidyalaya, Raipur, Chhattisgarh, India

Pragati Srivastava Department of Microbiology, G. B. Pant University of Agriculture & Technology, Pantnagar, Uttarakhand, India

Deep Chandra Suyal Department of Science, Vidyadayini Institute of Science, Management and Technology, Bhopal, Madhya Pradesh, India

María Concepción Tamayo-Ordoñez Departamento de Biotecnología, Facultad de Ciencias Químicas, Universidad Autónoma de Coahuila, Saltillo, Mexico

N. Thiruchchelvan Department of Agricultural Biology, Faculty of Agriculture, University of Jaffna, Jaffna, Sri Lanka

Anjali Tiwari Department of Environmental Science, G.B. Pant University of Agriculture & Technology, Pantnagar, Uttarakhand, India

Soumitra Tiwari Department of Food Processing and Technology, Atal Bihari Vajpayee Vishwavidyalaya, Bilaspur, Chhattisgarh, India

Keshawananad Tripathi Department of Biotechnology, Invertis University, Bareilly, Uttar Pradesh, India

Rakesh Kumar Verma Department of Zoology, Laxman Singh Mahar Government Post Graduate College, Pithoragarh, India

David Waikhom ICAR-Central Institute of Fisheries Education, Mumbai, India

Uddesh Ramesh Wanjari Department of Biomedical Sciences, School of Bio-Sciences and Technology, Vellore Institute of Technology, Vellore, Tamil Nadu, India

J. J. Wewalwela Department of Agricultural Technology, Faculty of Technology, University of Colombo, Homagama, Sri Lanka

Shivani Maheshkumar Yagnik Department of Microbiology and Biotechnology, University School of Sciences, Gujarat University, Ahmedabad, Gujarat, India

Freshwater Microbiology: Recent Updates and Prospects

1

Virgilio Bocanegra-García, Humberto Martínez-Montoya, María Concepción Tamayo-Ordoñez, Israel Estrada-Camacho, Alejandra Aguilar-Salazar, Gloria Dhení Guaní-Sánchez, Gerardo de Jesús Sosa-Santillán, and Erika Acosta-Cruz

Abstract

Along with bacteria, other microorganisms are also key components of microbiota structure of freshwater, and those include archaea, microbial eukaryotes, and viruses that together build the diverse community of microorganisms. All those microorganisms play key ecological functions and can have important impacts on animal, plant, and human health. The traditional culture-based microbiology methods have been mostly replaced, or complemented, by molecular methods, due to their accuracy, fast response, and capability to detect microorganisms that are quite difficult to grow in culture media. This chapter addresses some of the current tendencies on microbial structure and its analysis. Those methods included the "omic-based" DNA studies that allow better and more comprehensive characterization of bacterial communities, leading to metagenomic studies that help to elucidate the population structure of microbiomes in aquatic environments. Strategies for detecting water-borne pathogens are focused on detecting bacteria, protozoa, yeast, and

---

V. Bocanegra-García · A. Aguilar-Salazar · G. D. Guaní-Sánchez
Laboratorio Interacción Ambiente-Microorganismo, Centro de Biotecnología Genómica, Instituto Politécnico Nacional, Reynosa, Tamaulipas, Mexico

H. Martínez-Montoya
Departamento de Tecnología de Alimentos, U.A.M. Reynosa Aztlán, Universidad Autónoma de Tamaulipas, Reynosa, Mexico

M. C. Tamayo-Ordoñez · G. de J. Sosa-Santillán · E. Acosta-Cruz (✉)
Departamento de Biotecnología, Facultad de Ciencias Químicas, Universidad Autónoma de Coahuila, Saltillo, Mexico
e-mail: erika.acosta@uadec.edu.mx

I. Estrada-Camacho
U.A.M. Reynosa Aztlán, Universidad Autónoma de Tamaulipas, Reynosa, Mexico

© The Author(s), under exclusive license to Springer Nature Singapore Pte Ltd. 2023
R. Soni et al. (eds.), *Current Status of Fresh Water Microbiology*,
https://doi.org/10.1007/978-981-99-5018-8_1

virus in different water sources, and these methods can also be complemented by molecular tools such as molecular marker (RAPD, Microsatellites, RFLPs, and AFLPs). Those methods also allow to detect pathogenic bacteria, fungus, and viruses of public health interest, such as the enterovirus and, more recently, the coronavirus. Water bodies can also be important reservoirs of antibiotic-resistant bacteria, and the group of ESKAPE bacteria represents an important study model for antibiotic resistance and its distribution. The acronym "ESKAPE" is associated with this group due to their ability to escape antimicrobial activity and develop high levels of resistance to multiple antibiotics (multidrug resistance). Water bodies play a central role on the selective pressure and is the most important cause of the dissemination and extension of resistance and has contributed to the genetic diversification of resistance genes also affecting ecological microbial structures.

Keywords

Microbiology · Molecular markers · Metagenomics · ESKAPE · Multi-drug resistance · Microbial ecology

1.1 Introduction

The biological role of freshwater microorganisms in the carbon cycle and its association with the uptake and emission of greenhouse gases is well documented as well as the impact that disturbance of microbial communities as a consequence of anthropogenic activities have on the climate and global change (Regnier et al. 2013; Premke et al. 2022).

Bacteria are prokaryote, unicellular, taxonomically diverse, and ubiquitous microorganisms. They are the most abundant organisms on earth and their habitats include virtually every ecosystem on the planet, even extreme aquatic environments such as hot springs and ice layers in the Arctic and Antarctic regions, under high pressures or extreme acidic environments (Takai et al. 2008; Dalmasso et al. 2016; Merino et al. 2019). The adaptation capabilities of bacteria to adverse environments explain their abundance within the different ecosystems on the earth; it is estimated that the average number of bacterial cells per gram of soil is around 40 million, whereas 1 million diverse bacterial cells can be found in 1 mL of environmental freshwater and it is even higher in estuaries and open ocean waters (Whitman et al. 1998; Anas et al. 2021).

In aquatic environments, the cross-feeding and metabolite interchange represent an important class of interactions, and thus, photosynthetically fixed dissolved carbon could produce chemotactic responses (Seymour et al. 2010). Early experimental evidence suggested the existence of strong positive and negative interactions among the photoautotroph and heterotroph microorganisms (Cole 1982). Besides bacteria, there are various microorganisms in fresh water, such as fungi, algae, and viruses, although they are not microorganisms. Their importance will be addressed in this chapter. The traditional culture-based microbiology methods have been mostly replaced, or complemented, by molecular methods, due to their accuracy and fast response. Here, some of the current tendencies on microbial structure and its analysis will be addressed.

1.2 Microbiome and DNA-Based Omic Studies

As we move forward to a better understanding at microorganic system level, complexity increases as the number of microorganisms interacting do. For this reason, it is important to develop pipelines to accurately characterize the diversity of bacteria and other microorganisms in specific niches and elucidate their role within the community and the ecosystem. Water is an essential natural resource for all living organisms, and the pollution consequence of human activities has a great impact ranging from environmental effects (sometime negative), animal and human health impact to important economic losses in food production. The causes of contamination of rivers, lakes, and ponds are diverse. Still, contamination with untreated or poorly treated sewage and residual waters containing heavy metals and other xenobiotics from industry, agriculture, and livestock production are among the main drivers that serve as disturbers of native bacteria communities. In the first case, sewage serves as an important source of enteric bacteria and viruses, whereas industrial residual waters modify the physical and chemical properties of the water itself, and these changes impact the bacterial population structure affecting the carbon flux between freshwater bodies and the environment among other important aspects (Schreiber et al. 2015).

Like other ecosystems, in freshwater bodies, bacteria and other microorganisms live in a complex multispecies structure. Begon et al. (1986) defined the microbial communities as the set of microorganisms coexisting in the same space and time. However, the characterization of bacterial communities was limited to microorganisms grown in culture media, and their study was focused mainly on morphological and biochemical profiles. In 1977, Woese and Fox proposed the 16S ribosomal RNA genes as a potential marker to establish phylogenetic relationships between bacteria, and it was only after the development of Sanger sequencing technology that the 16S ribosomal RNA genes analyses was used as the benchmark to fine-tune the classification system of microorganisms.

Later in 1998, Handelsman et al. proposed the term *metagenomics* in their research, which involved the isolation of total DNA from soil bacteria without previous culture on artificial media, and their cloning using BAC vectors in order to transform competent *E. coli* cells and then sequence the genomes and express genes detected, allowing a deep understanding of the bacterial population structures. The arrival of high-throughput next-generation sequencing (NGS) techniques in 2005 with the pyrosequencing method from Roche was a milestone in the microbial communities' characterization because of its relative low cost and reliable methodology to rapid sequencing of short fragments of DNA. However, this platform is discontinued today by more efficient sequencing strategies such as synthesis sequencing, initially proposed from ion torrent, followed and enhanced then by Illumina and PacBio, all collectively called next-generation sequencing (NGS), that now represents the most accurate and used technique to characterize the metagenome of environmental samples at a different resolution of taxonomic levels; these sequencing technologies allow the rapid detection of perturbations in populations of native bacteria in freshwater deposits under different conditions that can be associated to ecological phenomena (Escobar-Zepeda et al. 2015). In the past few years, through NGS techniques, researchers worldwide have been able to elucidate the population structure of microbiomes in aquatic environments under different and extreme conditions (Table 1.1).

To this day, metagenomic characterization is mostly descriptive and without a careful approach design it does not provide insights into genes of interest or metabolic integration pathways in the analyzed environments. However, a comprehensive approach integrating every member of the microbial community transcriptome (metatranscriptomics) studies and the total set of proteins (metaproteomics) could improve the insight into the role played by every microbial member at an ecosystem level establishing accurate relationships between taxonomy and functionality profiles within microorganisms.

1.3 Current Strategies for Detection of Water-Borne Pathogens

1.3.1 Incidence of Bacterial Pathogens in Water

Water bodies, such as rivers, lakes, irrigation channels, damps, and hospital and industrial wastewaters, harbor a wide variety of microorganisms. The microorganisms commonly found in wastewater are enteric anaerobic bacteria, some yeasts and fungi, and some protists especially those resistant to harsh environments. Regarding the bacteria present in water bodies, it has been estimated that 90% of the genera and species of bacteria characterized belong to 4 phyla: Proteobacteria, Actinobacteria, Firmicutes, and Bacteroidetes. Some of the bacteria can be defined as pathogens such as *Escherichia coli* (Rice and Johnson 2000), *Pseudomonas* spp. (Pirnay et al. 2005), and *Stenotrophomonas maltophilia* (Brooke 2012). These bacteria, along with the emergence of antibiotic-resistant bacteria and its wide distribution in freshwater, is a concerning matter that will be addressed further in this chapter with the ESKAPE group.

1.3.2 Incidence of Yeast and Filamentous Fungi Pathogens in Water

The presence of these microorganisms in water has important ecological roles in organic matter transformation, and some have an additional impact, because these microorganisms can be pathogenic or display pathogenic properties and well as

1 Freshwater Microbiology: Recent Updates and Prospects

Table 1.1 Metagenomic analyses from selected freshwater vessels under diverse physicochemical conditions

Water body	Location	Water samples	Metagenome composition	Conditions	Sequencing platform
Lake Baikal[a]	Siberia, Russia	2, from 5 and 20 m depth	Verrucomicrobia	High oxygen levels.	Illumina HiSeq
			Actinobacteria		
			Proteobacteria		
			Acidobacteria		
			Bacteroidetes		
			Pseudomonadota		
Fetsui reservoir[b]	Taiwan	Samples taken from 5 m depth	Bacteroidetes	Evidence showed that typhoons disrupt the thermal stratification of the water column and change dissolved oxygen level by delivering oxygen-rich water to the reservoir.	454 Roche
			Actinobacteria		
			Cyanobacteria		
			Flavobacteria		
			Sphingobacteria		
			Alphaproteobacteria		
Lake Tanganyka[c]	DRC, Tanzania, Burundi, Zambia	Tanzanian samples from surface to 1100 m depth	Bacteroidetes	Anoxic water, stratified. Metagenomes were compared from rich oxygen surface to dark, nonoxygen, nutrient-rich depth.	Illumina HiSeq
			Ignavibacteria		
			Betaproteobacteria		
			Gammaproteobacteria		
			Alphaproteobacteria		
			Verrucomicrobia		
			Planctomycetes		
			Chloroflexi		
			Actinobacteria		
			[c]Tanganyka bacteria		
			Cyanobacteria		
Lake Tai[d]	China	10 cm depth	Proteobacteria	Water body heavily contaminated with population residual water.	Illumina HiSeq
			Firmicutes		
			Actinobacteria		
			Bacteroides		
			Acidobacteria		
El chichon[e]	Mexico	Water column and sediment samples	Found at sediment 50 °C:	Crater lake of an active volcano, formed in 1982. Geothermal activity	Illumina NextSeq
			Actinobacteria		
			Proteobacteria		
			Acidobacteria		
			Found at sediment 92 °C		

(continued)

6   V. Bocanegra-García et al.

Table 1.1 (continued)

Water body	Location	Water samples	Metagenome composition	Conditions	Sequencing platform
			Firmicutes (*Alcylobacillus* and *Sulfobacillus*)	including acidification, presence of sulfur compounds and metals, extreme water temperature.	
			Proteobacteria (*Bradtrhizobium*, *Methylobacterium*, *Sediminibacterium*)		
Lake Untersee[f]	Antarctica	80 m depth	Proteobacteria	High concentration of dissolved oxygen. High pH 9.8–12.1.	Illumina HiSeq
			Bacteroidetes		
Lonar Lake[g]	India	Surface water samples	Proteobacteria	Hyperalkaline and hypersaline basin.	Illumina HiSeq
			Firmicutes		
			Bacteroidetes		
Lake Michigan[h]	U.S.A.	Samples from river, river mouth, near-shore, and offshore, at 45 cm depth	Proteobacteria	Bacterial contamination originated from the Grand Calumet River.	Illumina
			Actinobacteria		
			Bacteroidetes		

[a] Cabello-Yeves et al. (2017)
[b] Tseng et al. (2013)
[c] Tran et al. (2021)
[d] Chen et al. (2019)
[e] Peña-Ocaña et al. (2022)
[f] Koo et al. (2018)
[g] Chakraborty et al. (2020)
[h] Nakatsu et al. (2019)

antifungal resistance (Monapathi et al. 2020). Some species of yeast that have been identified in water bodies include *Candida*, *Clavispora*, *Cyberlindnera*, *Cryptococcus*, *Hanseniaspora,* and *Yarrowia*. Yeast detected in cities water distribution systems and tap water include *Candida*, *Clavispora*, *Cryptococcus*, *Debaromyces*, *Meyerozyma,* and *Pichia*. Yeast isolates identified in sewage treatment plants are *Candida*, *Cryptococcus*, *Debaryomyces,* and *Wickerhamonyces* (Monapathi et al. 2020). Yeasts isolated from different water samples recognized as pathogenic include *Candida* spp. *Cyberlindnera*, and *Cryptococcus*. Fungal infections caused by *Candida albicans* and other *Candida* spp. are important complications in immunosuppressed patients, since frequently, candidiasis in mucosal tissues (oral, gastrointestinal, and vaginal) represent an early sign of immune system malfunction

(Panizo and Reviákina 2001). *Cryptococcus neoformans* is an encapsulated yeast, and the cause of opportunistic infections, such as meningoencephalitis in immuno-compromised patients (Pini et al. 2017).

Among filamentous fungi, some of the complications caused by some virulent species belonging to *Aspergillus, Fusarium,* and *Alternaria* may range from allergic to invasive syndromes. Immunocompetence facilitates clearance by initiating innate and adaptive host responses despite constant spore inhalation (Chotirmall et al. 2014). Regarding *Aspergillus* section Flavi includes 22 species, some of them with potentially pathogenic such as *Aspergillus felis, A. fischeri, A. fumigatiaffinis, A. fumisynnematus, A. hiratsukae, A. laciniosus, A. lentulus, A. novofumigatus, A. parafelis, A. pseudofelis, A. pseudoviridinutans, A. spinosus, A. thermomutatus,* and *A. udagawae.* The pathogenic species can display a wide range of differences in the frequency of production of toxins or toxic compounds (Tamayo-Ordóñez et al. 2021). Even though fewer pathogenic yeasts and molds have been reported in different water species compared to bacterial genera, their importance stands out, since their contact with humans with a compromised immune system could lead to serious complications and even cause death.

1.3.3 Identifying Virus in Water Samples

In raw wastewater, viruses may be ubiquitous and persistent and even after treatment, some viruses continue being distributed in treated wastewater that later arrive at water bodies. Adenovirus (HAdV), rotavirus (RoV), hepatitis A virus (HAV), and other enteric viruses, such as noroviruses (NoV), coxsackievirus, echovirus, reovirus, and astrovirus, are commonly detected in wastewater as well as some superficial water (rivers and irrigation channels) and are the principal human viral pathogens transmissible via water exposition (Corpuz et al. 2020). More dangerous viruses such as poliovirus can also be present in wastewaters. Enterovirus causing water-borne diseases such as diarrhea in children and adults are often associated with viral disease outbreaks, leading to several clinical manifestations (nausea, vomiting, and fever) and other gastrointestinal diseases, and some enteric viruses have also been related to respiratory, central nervous system (poliovirus), and other diseases (Corpuz et al. 2020).

Recently, a novel Coronavirus disease 2019 (COVID-19), which began in December 2019, that is caused by severe acute respiratory syndrome coronavirus 2 (SARS-CoV-2) (Guarner 2020) was reported. Since this virus can invade the gastrointestinal mucosa, it can be released in feces, and the presence of nucleic acid of SARS-CoV-2 has been reported in raw wastewater (Randazzo et al. 2020), sewage samples collected from hospitals (Wang et al. 2020b), and wastewater sample after secondary treatment (Randazzo et al. 2020). However, infection caused by exposition to water is not been documented, and its detection functions as an early warning system to indirectly detect SARS-Cov-2 infections in the population.

1.4 Molecular Markers for Genotyping and Identification of Bacteria in Water Bodies

1.4.1 Intraspecific and Interspecific Variability and Genotyping

Molecular markers are DNA fragments sequences associated with a physiological trait, and are used to identify a particular DNA sequence related to a function or a trait of interest (Langridge and Chalmers 2004). Different types of markers are distinguished by their ability to detect polymorphisms at single or multiple loci, or sequence patterns and can be of dominant or codominant type. Below we describe some of the molecular markers that have been most used to determine intra- and interspecific variability in bacterial genera isolated from different bodies of water.

RAPDs RAPDs (randomly amplified DNA polymorphisms) are markers that randomly amplify segments of DNA in a wide variety of species. RAPDs are based on the statistical probability of complementary sites to the oligonucleotide of ten base pairs (bp) throughout the genome. The polymorphism of the bands between individuals is due to changes in the sequence of nucleotides at oligonucleotide coupling sites and by inserting or deleting fragments at these sites. These markers are dominant; they cannot discern the dominant homozygous or heterozygous for a particular segment, so allelic frequencies must be estimated indirectly, assuming Hardy Weinberg equilibrium. Among some of the applications of these molecular markers in bacteria, it has successfully demonstrated different RAPD profiles in multidrug-resistant coliform bacteria isolated from sewage samples of Ghaziabad city, India (Raj 2012). Since RAPD-derived markers include a genome-wide analysis, they can be a very good source of physiological markers that can be obtained even if little is known about the sequence of the microorganism being studied. Their application could be considered one of the first "genomic-wide" analysis performed before the massive sequencing technology was available; however, several problems with reproducibility limited its applications out of basic research.

Microsatellites Simple repeat sequences (SSRs) or ISBPs (Insertion Site-Based Polymorphisms) are DNA sequences made up of 1–4 base pairs (mononucleotides (TT)n, dinucleotides (AT)n, or tetranucleotides (AAGG)n). These loci are found both in coding and noncoding regions of DNA, formed by breakage events that generate polymorphisms with values greater than 90%. Challagundla et al. (2018) analyzed 598 genome sequences of *Staphylococcus aureus*; this study showed that depending on the geographical area of the isolate, certain mutations in the CC5-MRSA marker are present allowing a deeply analysis of the epidemiology of these strains. A particular important application of simple repeat sequences was on *M. tuberculosis* research, because this allowed the design of a classification system of *M. tuberculosis* strains, based on a VNTR (variable number of tandem repeats) that was so specific to *M. tuberculosis* that later these were called mycobacterial interspersed repetitive units or MIRUs, allowing a great advance on mycobacterial isolates classifications, both from clinical and environmental sources. Dangerous

1 Freshwater Microbiology: Recent Updates and Prospects

strains such as the Beijing lineage were spotted with these markers and later with complementary genomic analyses on Beijing strains obtained from different clinical samples, waters, and substrates: this allowed us to know its biogeographic structure and evolutionary history of the Beijing lineage worldwide through the SNPs analysis of 4987 isolates from 99 countries (Merker et al. 2015).

RFLPs This method allows the detection of specific DNA differences recognized by restriction enzymes. Each of the endonucleases (of bacterial origin) recognizes and cuts only a specific sequence in DNA, as long as they are not protected (methylated). Therefore, any DNA that is not methylated can be recognized and cut into fragments of defined length; any mutation within those sites could change the pattern of the fragment and allow an RFLP to be detected when comparing two or more genomes. Originally, RFLP were important tools to detect and related mutations and molecular marker to key physiological functions and to design the initial molecular diagnosis strategies, since sequencing was required only to confirm the design but not to apply the technique and had very good reproducibility results. RFLPs are still currently applied to understand bacterial community dynamics in wastewater treatment systems. Wang et al. (2010) investigated how bacterial communities change in treatment plants over a year using specific PCR followed by terminal restriction fragment length polymorphism (T-RFLP) of the 16S rRNA gene. The T-RFLP results indicated lack of stability in the bacterial community structures in 2 full-scale wastewater treatment systems, with 15 days average change rates observed in the 2 systems. On the other hand, through digestion of the 16S rRNA gene with the endonuclease *Mse* I, it was shown that distinctive patterns for *Acrobacter* species are observed, indicating variation in the population structure of the genera (Figueras et al. 2012).

AFLPs AFLPs (amplified fragment length polymorphisms) is a technique that combines the digestion of two restriction enzymes, generally MseI and EcoRI, within a sequence, involving a selective amplification of restriction fragment lengths from a genome using PCR, and followed by gel electrophoresis to separate the amplified fragments and obtain a banding pattern, which can be used to identify genetic differences between individuals or populations, and can also be used for genetic mapping and phylogenetic analysis (Tamayo-Ordoñez et al. 2012).

This technique has been used to determine the biodiversity of the bacterium *Pseudomonas aeruginosa* in an aquatic environment in a study including 100 isolates of *P. aeruginosa*, where the *oprL* gene was analyzed, and a DNA-based fingerprint (AFLP) indicated a positive relationship between pollution and prevalence of *P. aeruginosa*. In the Woluwe River, *P. aeruginosa* community was almost as diverse as the global *P. aeruginosa* population. These findings illustrate the significance of river water as a stable reservoir and source for the distribution of potentially pathogenic *P. aeruginosa* strains (Pirnay et al. 2005).

In another recent study, antimicrobial tolerance to oxytetracycline and taxonomic diversity relationship in the culturable oxytetracycline-resistant (Otr) isolates of heterotrophic bacteria in two Belgian aquatic sites receiving wastewater was investigated. Profiles for ampicillin and kanamicin tolerance were detected from the taxonomic differences in the Otr bacteria detected at genera and subgenera level. In addition, *Enterobacter* sp., *Stenotrophomonas maltophilia*, and *A. veronii* resulted as potential indicator organisms to help to assess microbial tolerance in various compartments of the aquatic environment (Huys et al. 2001).

1.4.2 Identification of Species Through Sequencing of Molecular Markers

To date, depending on the purpose of each investigation, different genetic regions can be used to identify species from various samples. Among the most explored genes used for species identification are the 16S rRNA, 23S rRNA, *rpo*B, *gyr*B, *dna*K, *dsr*AB, *amo*A, *amo*B, *mip*, *hor*A, *hit*A, *rec*A, *ica, frc, oxc*, 16S–23S rDNA ISR, and IS256.

Amplification and sequencing of 16S (V3–V4 of 16S rRNA) is commonly applied to study microbial communities with third-generation sequencers (MinION from Oxford Nanopore Technologies) and this made it possible to analyze the full length of the 16S rRNA gene, and this allows a deeper identification even to species level identification and at relative low cost. The analysis of nine indigenous bacteria that can be related to food poisoning and act as opportunistic infections was carried out with due diligence. *Enterococcus faecalis* and *Enterococcus hirae* were identified at the species level with an accuracy of 96.4–97.5%. Also, using these technologies, it is possible to evaluate the antibiotic sensitivities of multiple bacteria simultaneously. Kawai et al. (2022), using these technologies, allowed rapid evaluation of antibiotic activity spectrum at the species level containing a wide variety of bacteria, such as biofilm bacteria and gut microbiota. Even though the sequencing of these genes is a powerful tool to identify bacteria, it has some limitations when it comes to related bacterial complexes that can be difficult to distinguish.

Since the sequencing and analysis of 16S rDNA ribosomal regions is the most used molecular marker, we decided to find out the abundance of bacteria identified in bodies of water and of these, which accessions have been reported in the Gen Bank from NCBI. It was possible to identify 103 sequences of 16S rDNA regions of bacteria present in different bodies of water (Fig. 1.1). After analysis, a total of 58 bacterial genera were identified.

Conclusively, we want to emphasize that the use of molecular markers allows to identify and genotype a wide variety of microorganisms from different water samples, highlighting the importance of molecular markers in the clinical and environmental areas.

1 Freshwater Microbiology: Recent Updates and Prospects

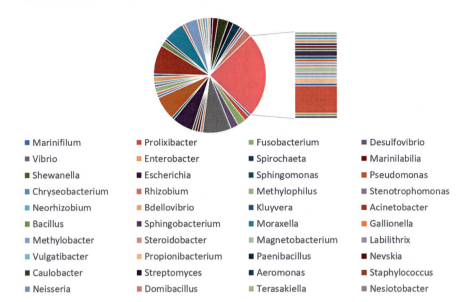

Fig. 1.1 Representation of the bacterial species identified in water bodies. The bacterial species were identified by searching in the sequences reported in the NCBI

1.4.3 Methods of Detection of Viruses in Waters

Viruses are important components of aquatic systems, since they are heavy regulators of bacteria populations, and some of them can be related to plants, animal, and human diseases, so their detection is crucial as indicated in numerous studies, including different water matrices such as surface water, treated wastewater, and irrigation water (Wang et al. 2020a; Ji et al. 2020; Rusiñol et al. 2020). Several methods to directly and indirectly detect and quantify viruses found in wastewater include epifluorescence microscopy, transmission electronic microscopy, pulsed-field gel electrophoresis, immunofluorescence assay, flow cytometry, traditional cell-culture, and molecular detection (Corpuz et al. 2020). Among the molecular techniques, PCR and especially real-time PCR are currently used.

PCR is an in vitro enzymatic reaction that amplifies or generates millions of copies of a specific DNA sequence during several repeated cycles in which the target sequence is faithfully copied. This technique uses thermostable DNA polymerase to replicate DNA strands, for which alternate high and low temperature cycles are used to separate the newly formed DNA strands after each replication phase. Among the components used in this in vitro reaction are amplification template (DNA or complementary DNA, or RNA in the case of viruses), oligonucleotides, DNA polymerase (Taq polymerase), polymerase cofactor ($MgCl_2$), PCR buffer, deoxyribonucleotide triphosphates, water, and equipment (thermocycler). Since viruses have DNA or RNA genomes, PCR is an important tool for detecting them in water samples. For the detection of RNA genome viruses, it is necessary to carry out an RT-PCR (Reverse Transcription-Polymerase Chain Reaction) that is a derived technique with a step where a RNA complementary DNA is generated that can be used directly in the PCR (Kadri 2020). An important derivative technique for virus study is the Real-time PCR or quantitative PCR that is a method that allows us to monitor the evolution (simultaneous amplification and quantification) of the PCR while it is being carried out. Cycle-by-cycle analysis of the changes in the accumulation of the PCR product, detected by a change in the fluorescent signal generated in the three steps of the PCR, is done with this technique and allows two types of determinations: the quantification absolute and quantitative comparison between samples. The quantification absolute determines the exact number of DNA or RNA molecules present in each sample. Applying this quantification method requires generating a standard curve and using the comparative Ct method. The other type of determination is the quantitative comparison of target nucleic acid, expressed in orders of magnitude concerning a calibrator.

Variants have been derived from both techniques, such as Multiplex PCR and ICC-PCR (integrated-cell culture PCR) (Corpuz et al. 2020). The application of multiplex PCR allows the simultaneous detection of multiple viruses present in a single sample using more than one set of primers in one reaction. Multiplex qPCR also enables the detection of different specific viruses using more than one fluorescent reporter. In the ICC-PCR method, the samples in which the viruses may be present are used to infect cell culture media and incubated for several days. Then the samples of cell culture are subjected to freezing to cause lysis of the cells. If viable viruses were present in the original sample and invaded the culture cells, those are from their host cells when lysis occurs. After this, PCR-based amplification is applied to the lysate (Corpuz et al. 2020), so the ICC-PCR is an effective enrichment method for virus detection.

1.4.3.1 Detection of Coronaviruses in Samples of Waters by Molecular Methods

There are extensive reviews in the literature describing the persistence of coronavirus in different types of wastewaters. The integrity of the coronavirus particle in water depends upon the wastewater characteristics, presence of suspended solids, organic matter, and temperature (Mandal et al. 2020). The methods of early detection of the virus on waters help to detect future immediate outbreaks, allow to implement measures to minimize the infections and deaths caused by COVID-19, and new molecular techniques are currently being developed that allow the effective detection of other viruses relevant for public health such as adenovirus and poliovirus,

1.5 ESKAPE Bacteria as a Model of Antibiotic-Resistant Bacteria Distribution in Water

1.5.1 Antibiotics and Drug Resistance

The use of antibiotic-producing microbes to prevent disease goes to back more than 2000 years ago from the use of moldy bread to treat open wounds (Hutchings et al. 2019). Still, it was not until the twentieth century that the microbiologist Alexander Fleming discovered the origin of Penicillin in 1928, a substance produced by mold (*Penicillum notatum*), that was lethal to several bacteria giving rise to the birth of the antibiotic era (Patel 2016), but also to the next milestone in history "antibiotic resistance" statement made by Alexander Fleming during his Nobel Prize acceptance speech (1945) (Podolsky 2018) predicting the extent and severity of antibiotic resistance over the next 10 years, statement that then became a reality, with hundreds of cases of resistance recorded throughout the world (Rosales Magallanes 2018).

Currently, antimicrobial resistance is one of the greatest threats to global health (Maillard et al. 2020). The widespread and inappropriate use of antibiotics has led to an increased prevalence of multidrug-resistant (MDR) bacteria in the environment. MDR infections led to 700,000 deaths globally in 2016 and are expected to rise to 10 million by 2050 (Brooks et al. 2018). The ESKAPE pathogens (*Enterococcus faecium, Staphylococcus aureus, Klebsiella pneumoniae, Acinetobacter baumannii, Pseudomonas aeroginosa,* and *Enterobacter* species) (Tigabu and Getaneh 2021) are a group of common opportunistic pathogens associated mainly with nosocomial infections (Santaniello et al. 2020), the acronym "ESKAPE" is associated to this group due to their ability to escape the antimicrobial activity and development of resistance to multiple antibiotics (multidrug-resistance) (Founou et al. 2018). The World Health Organization (WHO) listed in the critical priority list carbapenem-resistant *A. baumannii, P. aeruginosa, K. pneumoniae,* and *Enterobacter* spp., whereas vancomycin-resistant *E. faecium* (VRE) and methicillin/vancomycin-resistant *S. aureus* (MRSA or VRSA) in the list of high priority group (Mulani et al. 2019). The presence of the ESKAPE pathogens in the environment is likely due to contamination via sewage spills, hospital waste that has been discarded incorrectly; consequently, several antibiotic-resistant microorganisms and antibiotic-resistance genes (ARGs) have been detected in water bodies (Denissen et al. 2022). The water sources represent a reservoir of various chemical and microbiological pollutants such as chemical, heavy metal, organic pollutants, residues of pharmaceuticals (Voigt et al. 2020), including horizontal gene transfer of resistance genes, the misuse or overuse of antibiotics, and the contamination through livestock slurry and plant wastewater (Daniel et al. 2017) are factors that contribute to genetic selection pressure for the development of MDR bacterial infections in the community (Wang et al. 2018).

1.5.2 *Enterococcus*

Enterococci is a Gram-positive bacteria that commonly resides in the intestinal tracts of humans and animals but is also found in water, soil, and plants due to their high tolerance to different conditions (Cho et al. 2020), they can survive in a variety of environments, such as soil, water, food, plants, and animals (Castillo-Rojas et al. 2013), and they can survive in a wide range of temperatures (10–45 °C), pH (4.4–9.6), and in a hypersalty media with 6.5% NaCl (Ben Braïek and Smaoui 2019). Nowadays, *Enterococcus* is the leading cause of nosocomial infections, responsible for causing endocarditis, urinary tract infections, and bloodstream infections (Raza et al. 2018). There are more than 50 species of enterococci where *E. faecium* and *E. faecalis* are the most ecology and epidemiological relevant (Ramos et al. 2020). Furthermore, enterococci have intrinsic antimicrobial resistance to aminoglycosides, and cephalosporins antibiotic agents and can adapt to acquire resistance to antimicrobials such as β-lactams and vancomycin from the environment (Lee et al. 2019), by mutation and/or acquisition of genes through the plasmids and transposons (Grassotti et al. 2018). Vancomycin is an antibiotic of last resort used to treat MRSA. The vancomycin-resistance among clinical isolates is a growing health problem that involves the exchange of *van* genes between strains; where *vanA* and *vanB* are the most prevalent genotypes in *E. faecium* in clinical isolates (Melese et al. 2020), where the *vanA* gene encodes an enzyme that modifies the vancomycin binding site through the substitution of D-ala-D-ala by D-ala-D-lactate (Liu et al. 2021). Water ecosystems, in particular wastewater treatment plants, are considered one of the main hotspots of the evolution and spread of antibiotic-resistant bacteria (ARB) and ARGs into the natural environment that leads easily reach communities via aquatic environments (Ekwanzala et al. 2020). Enterococci and antimicrobials are commonly excreted in urine and feces, and most of this waste is treated in wastewater treatment plants where *E. faecalis* and *E. faecium* are dominant species identified before being discharged into surface water mainly during the dry season (Sanderson et al. 2020). *E. faecalis* and *E. faecium* are also used as an indicator of water quality criteria by fecal contamination in freshwater in the United States of America at geometric means of 33 and 35 CFU/100 mL (Wen et al. 2020) due to they are more resistant to stress than *E. coli* and other fecal coliform bacteria (Chidamba and Korsten 2018). Nevertheless, the distribution of resistant bacteria from the environment to humans may occur through contaminated food, manure, and contaminated surface water used for irrigation (Taučer-Kapteijn et al. 2016).

1.5.3 *Staphylococcus*

Staphylococcus aureus is a commensal opportunistic Gram-positive bacterium in humans (Pasachova et al. 2019) and are able to survive at a wide range of temperatures, dryness, dehydration, and low water activity (Silva et al. 2020). The genus *Staphylococcus* comprises more than 50 species, including coagulase-positive (*S. aureus*) and coagulase-negative staphylococci (*E. epidermidis*) (Kosecka-Strojek et al. 2019). *S. aureus* is one of the main pathogens in hospital and community infections associated with skin infections, pneumonia, and nosocomial bacteremia and represents public problem health (Cheung et al. 2021). The staphylococci became antibiotic resistant through genetic mutations, reducing outer membrane proteins, and acquiring resistance genes through HGT (Guo et al. 2020). The methicillin resistance in *S. aureus* (MRSA) is a significant global health concern determined by the expression of *mecA* gene, which encodes a penicillin-binding protein (PBP2a) with low affinity for β-lactam agents (Watkins et al. 2019). MRSA strains show a multidrug-resistant pattern with resistance to fluoroquinolones, macrolides, aminoglycosides, clindamycin, penicillins, cephalosporins, and carbapenem, making β-lactams ineffective (Gajdács 2019). MRSA is known as the major cause of hospital-/community-acquired infections, and morbidity infections are elevated worldwide. On the other hand, vancomycin has been one of the first-line drugs to treat MRSA infections; nowadays the intermediate and complete vancomycin-resistant *S. aureus* (VRSA) is a severe public health concern (Cong et al. 2020).

Environmental factors such as pH, temperature, nutrient content, salinity, and dissolved oxygen play important roles in biofilm development, influencing their persistence in water, especially in piping due to inefficient water treatment (Silva et al. 2022). Despite this, wastewater treatment reduces the prevalence of *S. aureus*, thus reducing its presence in surface water, a fact attributable to the fact that the effluents are diluted in the river water (Zieliński et al. 2020). Although Staphylococci are less frequent in surface water than *E. coli, Enterococcus,* and other ESKAPE pathogens, they are indicators of antimicrobial resistance in the environment mainly by the presence of *mecC* gen associated with *S. aureus* (Silva et al. 2021).

1.5.4 *Klebsiella pneumoniae*

Klebsiella spp. are Gram-negative and nonmotile bacteria resident in the environment (surface water, sewage, soil, and plants), and on mucosal surfaces such as humans and animals (Bengoechea and Sa Pessoa 2019). In humans, *K. pneumoniae* is a commensal and opportunistic pathogen found in gastrointestinal and respiratory tracts and skin of healthy people, and is responsible for causing community-acquired and nosocomial infections (urinary, respiratory tract infections, and infections of wounds and soft tissue) (Herridge et al. 2020). In contrast to other nosocomial pathogens, *K. pneumoniae* clones are divided into two phenotypic groups characterized by multidrug resistance and hypervirulence (Gonzalez-Ferrer et al. 2021). *K. pneumoniae* shows high resistance to a broad spectrum of antibiotics such as ESBL, carbapenem, quinolones, polymyxin, tigecycline, fluoroquinolones, and aminoglycosides. Recently, the WHO recognized extended-spectrum β-lactam (ESBL) producing and carbapenem-resistant *K. pneumoniae* as a critical public health threat (Wyres et al. 2020).

β-lactam antibiotics are the first-line option for the treatment of infections, but ESBL enzymes hydrolyze β-lactams (penicillins, monobactams, carbapenems, and cephalosporins) (Pillai 2022); on the other hand, the carbapenems are β-lactam antibiotics of last resort with the broadest spectrum of activity, that, in less than a decade after their use, become a global public health crisis (Cherak et al. 2021), mainly because infections caused by the ESBL/carbapenemase-producing *Enterobacteriaceae* are associated with increased mortality, length stay, and health-care costs (De angelis et al. 2020).

The environment is an important reservoir for disseminating ARB, antibiotic-resistance genes, and mobile genetic elements that interact and spread to other parts or to human and animal hosts (Samreen et al. 2021). Carbapenemase and ESBL-producing *K. pneumoniae* have been reported in aquatic environments, such as rivers, hospital effluents, and posthospital wastewater (Furlan et al. 2020). Some human health concerns associated with antibiotic resistance in the environment are the alteration of the human microbiome toward the emergence of antibiotic resistance and the potential hazard of selective pressure to create an antibiotic resistance environment (Ben et al. 2019). Nevertheless, measurements of the amounts and types of hazards (ARBs and ARGs) associated with environmental sources (water, air, soil, and food) that are related to the acquisition of resistant infections in humans are essential to establish suitable mitigation strategies (Vikesland et al. 2017).

1.5.5 *Acinetobacter baumannii*

The genus *Acinetobacter* comprises a group of Gram-negative, nonfermenting, strictly aerobic, catalase-positive and oxidase-negative coccobacillus (Torres et al. 2010). *Acinetobacter* species are widely distributed in the environment, making it possible to recover *Acinetobacter* isolates in almost 100% of the soil and water samples, as these are considered their main ecological niches (Salazar de Vegasa and Nieves 2005). Among the 32 varieties of species of the genus *Acinetobacter*, the most representative species is *Acinetobacter baumanii*, whose, while being distributed in nature, prevalence is strongly associated with the clinical-hospital environment, where some strains can survive in wet or dry inanimate objects in the hospital environment for weeks or months; in addition, it is a microorganism that is part of the normal flora of human skin and is distinguished as the microorganism that is most frequently carried persistently on the skin by hospital personnel (Phillips 2014).

Thus, all these factors increase the chances that patients will be colonized, in addition to promoting contamination of medical equipment, thus playing a determining role in the spread of infection and prolonged hospital outbreaks. The risk factors for *Acinetobacter baumannii* infection will depend on the type of infection (intrahospital, extrahospital) and its drug resistance profile, the latter being the most important. Community-acquired *Acinetobacter baumannii* infections, unlike those acquired in the hospital environment, are reported less frequently, because other risk factors are involved, such as residence in a tropical developing country, alcoholism, smoking, or diseases such as diabetes mellitus and chronic lung disease, conditions that easily allow infection to result in bacteremia or pneumonia.

It is worth mentioning that despite being rarely detected, this type of infection is of greater concern once it is acquired in the community, since it is a more severe infection than nosocomial infection and, in most cases, it is fulminant, dying in 60% of patients (Antunes et al. 2014). The *A. baumannii* genome contains several genes grouped into resistance islands, whose structure, in addition to providing high intrinsic resistance to antimicrobials, facilitates the acquisition of resistance mechanisms from other species of bacteria present in its environment, such as *Pseudomonas* spp. through mobile genetic elements (plasmids, transposons and integrons) (Agodi et al. 2013; Fournier et al. 2006), thus giving rise to an *Acinetobacter* isolate with a multidrug-resistant (MDR) phenotype with the capacity to produce extended-spectrum beta-lactamases, which confers resistance to most of the known antimicrobials, including carbapenems and colistin (Towner 2009). Therapeutic difficulties, added to the great capacity that it must survive for a long time in the hospital environment, together with the potential to develop in an acidic pH and at low temperatures, which allows it to increase its ability to invade devitalized tissues, together with the development of a biofilm ("biofilm") on human surfaces and cells, constitute this microorganism as an emerging public health problem (Gaddy and Actis 2009).

1.5.6 *Pseudomonas aeruginosa*

Bacteria of the genus *Pseudomonas* are Gram-negative microorganisms that can be classified as facultative anaerobes and are characterized as being ubiquitous and preferring humid environments (Luján Roca 2014). Specifically, *Pseudomonas aeruginosa* is a bacterium that successfully colonizes an enormous diversity of niches, as a consequence of its great nutritional versatility, it can survive in extreme environmental conditions; however, naturally, we can find it widely distributed in nature, its reservoir is the moist soil, water, wastewater, vegetation, humans and animals, humans and animals being the main hosts (CDC 2016). *P. aeruginosa* is an opportunistic pathogen responsible for a wide range of infections, mainly nosocomial, which is why it is considered highly prevalent in infections associated with health-care infections (HAIs), since it has a high adaptation to repeated changes in the microenvironment, nutrient availability, resistance mechanisms, and capacities to form biofilms (Villanueva-Ramos et al. 2019).

In humans, the species most at risk from infection is *Pseudomonas aeruginosa*, but infections by other species such as *P. paucimobilis*, *P. putida*, *P. fluorescens*, or *P. acidovorans* may also occur (Dumaru et al. 2019). Invasive infections by *P. aeruginosa* occur mainly in patients with underlying diseases that cause immunosuppression and mostly occur in the hospital setting. Community-acquired infection may occasionally occur in previously healthy patients. These infections are usually localized at the skin level (folliculitis, cellulitis, or ecthyma gangrenosum) without associated bacteremia, or present as severe clinical forms (Brady 2004; De almeida et al. 2002;). In the case of community-acquired infections, such as sepsis due to *Pseudomonas aeruginosa* in previously healthy patients, it is an infrequent clinical entity that must be considered when ecthyma gangrenosum, neutropenia, and gastrointestinal manifestations are present.

However, there is a possibility that *P. aeruginosa* can colonize parts of the human body; however, the prevalence of this colonization in healthy people is low (Rossolini and Mantengoli 2005). In case of acquiring an infection caused by *P. aeruginosa*, the vast majority of these will be related to the hospital environment, constituting a severe clinical problem, since, in most cases, there is a compromise of the host's defenses (Lyczak et al. 2000). This is because once the infection is established, *P. aeruginosa* produces a series of toxic compounds that cause not only extensive tissue damage but also interfere with the functioning of the immune system. Among the proteins involved in *P. aeruginosa* infection, we find toxins, as well as hydrolytic enzymes that degrade the membranes and connective tissue of various organs. This situation is aggravated by the difficulty in treating *P. aeruginosa* infections, since this bacterium is intrinsically resistant to multiple classes of antibiotics that are not structurally related to each other (Strateva and Yordanov 2009) due to decreased permeability of its outer membrane, the constitutive expression of several efflux pumps, and the production of enzymes that inactivate antibiotics. In addition, it can acquire new resistance mechanisms via mutations (Mesaros et al. 2007). Furthermore, in aqueous environments, this bacterium adheres to surfaces, producing a kind of aggregate called a biofilm. The formation of these accumulations of bacteria and extracellular material represents a health problem, since it contaminates devices implanted inside the body, such as intrauterine devices, catheters, or heart valves. Biofilms also represent a problem in the production process of various industries as they cause clogging and corrosion of connections and filters (Villanueva-Ramos et al. 2019). It is estimated that 65% of bacterial infections are due to the formation of biofilm and its mechanisms of tolerance to antibiotics, which represents a problem in the health field; the appearance of biofilms is strongly associated with recurrent chronic and delayed wound healing (Romeo 2020). When growing in a biofilm state, *Pseudomonas aeruginosa* increases tolerance to antibiotics up to a thousand times more than when it grows in its free form (Torres et al. 2010) and is mainly associated with the following mechanisms: quorum sensing (QS) signal, porins, efflux pumps, gene expression, membrane vesicles, extracellular DNA, and enzymes (Bolívar-Vargas et al. 2021). *P. aeruginosa* appears as an exceptional bacterium; the wide variety of virulence factors, the breadth of infections it causes, and its multiple mechanisms of resistance to antibiotics make it stand out among the pathogenic microorganisms for humans.

1.5.7 *Enterobacter* spp.

The *Enterobacter* genus is a member of the ESKAPE group, which contains the main resistant bacterial pathogens (Rice 2010). It consists of fermentative, facultative anaerobic, Gram-negative bacterial species belonging to the Enterobacteriaceae family. This genus is associated with a variety of environmental habitats, their presence in the intestinal tract as natural commensals of animal and human intestinal microbiota resulting in their wide distribution in soil, plants, water, and wastewater (Singh et al. 2018). Among this genus of bacteria, only certain subspecies/species have been associated with hospital-acquired infections and outbreaks. In humans, multiple *Enterobacter* species are known to act as opportunistic pathogens (disease-causing organisms): species, such as *E. cloacae*, *E. aerogenes*, *E. gergoviae*, and *E. agglomerans*, are associated with hospital-acquired infections and outbreaks (Akbari et al. 2016). These bacteria cause a variety of conditions, including eye and skin infections, meningitis, bacteremia (bacterial infection blood), pneumonia, and urinary tract infections.

The *Enterobacter aerogenes*, *E. cloacae*, and *E. hormaechei* species represent the most frequently described isolated species in clinical infections, especially in immunocompromised patients hospitalized in an intensive care unit (ICU), due to the adaptability of these species to antimicrobial agents and their behavior as opportunistic pathogens. These pathogens are frequently associated with a multidrug resistance (MDR) phenotype, mainly due to their adaptation to the hospital setting and their ability to readily acquire numerous mobile genetic elements containing resistance and virulence genes. These species have intrinsic resistance to ampicillin, amoxicillin, first-generation cephalosporins, and cefoxitin due to the expression of a constitutive AmpC β-lactamase. In addition, the production of extended-spectrum β-lactamases has been reported in these bacteria, making their treatment difficult (Davin-Regli et al. 2016). Antibiotic resistance, the regulation of resistance genes, and the expression of extended-spectrum beta-lactamases and carbapenemases significantly reduce therapeutic options, creating a particularly worrisome scenario with this microorganism as the protagonist of a global crisis of multiresistance to antibiotics.

Recent studies show that carbapenem-resistant enterobacteriaceae (CRE) can contaminate aquatic environments such as marine surface waters, rivers, estuaries, and contaminated drinking water (Mahon et al. 2019). CRE can reach aquatic bodies as a consequence of organic contamination from multiple sources (Mathys et al. 2019), including hospital effluents, wastewater treatment plants (PTARs), discharges from livestock and agricultural farms, water seepage, and others. Once mixed with the aquatic body, these effluents can introduce not only resistant microorganisms but also high doses of antibiotics, likely triggering the spread of resistance (Aydin et al. 2019). However, evaluating the link between the clinical and aquatic epidemiology of CRE is often challenging. While some studies have shown that clinical strains can be found in aquatic bodies, others have yet to prove such a link (Piedra-Carrasco et al. 2017).

1.5.8 Current Situation of Resistance

During the last 20 years, an increase in resistance has been observed in the 6 bacteria of the ESKAPE group; this represents a new crisis and a global public health problem (WHO 2015). Specifically, in Latin America, more than 50% of infections acquired in intensive care units (ICU) are caused by ESKAPE bacteria, with a growing tendency toward extreme drug resistance (resistance to all families of antimicrobials except 2 or 1 of them) and pan drug resistance (resistance to all families of antimicrobials) (WHO 2015).

Selective pressure is the most important cause of the dissemination and extension of resistance; in the last 70 years, the indiscriminate use of antibiotics had contributed to the genetic diversification of resistance genes, as can be seen in the current number of TEM beta-lactamases, where to date, there are at least 187 described, when in 1982 before third-generation cephalosporins were introduced into the clinic, only TEM-1 and TEM-2 were known (Corvec et al. 2013).

Antimicrobial resistance is favored by the inappropriate use of antimicrobials in human medicine, veterinary medicine, agriculture, and aquaculture. Insufficient prevention and control measures for infections associated with hospital care, incomplete treatment by patients, and lack of hygiene and sanitation, are the other factors that complicate global efforts to contain it (Bolívar-Vargas et al. 2021). The prevalence of resistance has not only affected the efforts made to hold it but has also complicated the interpretation of the phenotypic profile and addressing the appropriate treatment, since the association of different resistance mechanisms for the same is increasingly frequent.

Furthermore, the prevalence of resistance has not only increased in infection-causing bacteria. Intestinal colonization of healthy people by ESBL-producing Enterobacteriaceae has reached pandemic levels worldwide in just a few years, and it is estimated that there are 1753 million colonized people worldwide (Phillips 2014; Woerther et al. 2013). Other factors that contribute to the extension and dissemination of resistance are a consequence of the current situation in the world, where the hyperconnection among borders, for example, the food and animal trade, tourism, health and business trips, emigration, and refugees, among other events, allows resistant strains to reach anywhere (Rogers et al. 2011). In addition, wild animals can also act as a reservoir and a potential source of dissemination. An example of this is the presence of bacteria resistant to antibiotics in migratory birds, a situation that could undoubtedly favor the spread of resistance over long distances (Simoes et al. 2010). In all of the above examples, water bodies do play an important role as reservoirs and distribution nodes of antibiotic-resistant bacteria, especially those of the ESKAPE group, which, besides commonly being multidrug resistant, have an enhanced ability to transfer antibiotic-resistance genes to related bacteria present in the water bodies.

References

Agodi A, Auxilia F, Barchitta M, Brusaferro S, D'Alessandro D, Grillo OC et al (2013) Trends, risk factors and outcomes of health-care-associated infections within the Italian network SPIN-UTI. J Hosp Infect 84:52–58

Akbari M, Bakhshi B, Najar PS (2016) Particular distribution of *Enterobacter cloacae* strains isolated from urinary tract infection within clonal complexes. Iran Biomed J 20:49–55

Anas A, Tharakan B, Jasmin C, Chandran C, Vipindas PV, Narayanan S, Jaleel A (2021) Microbial community shifts along an estuary to open ocean continuum. Reg Stud Mar Sci 41:101587

Antunes LC, Visca P, Toener DJ (2014) *Acinetobacter baumannii:* evolution of a global pathogen. Pathogens Dis 71:292–301

Aydin S, Aydin ME, Ulvi A, Kilic H (2019) Antibiotics in hospital effluents: occurrence, contribution to urban wastewater, removal in a wastewater treatment plant, and environmental risk assessment. Environ Sci Pollut Res Int 26(1):544–558

Begon M, Harper JL, Townsend CR (1986) Ecology: individuals, populations and communities. Blackwell Scientific Publications, Oxford

Ben Braïek O, Smaoui S (2019) Enterococci: between emerging pathogens and potential probiotics. Biomed Res Int 2019:5938210. https://doi.org/10.1155/2019/5938210

Ben Y, Fu C, Hu M, Liu L, Wong MH, Zheng C (2019) Human health risk assessment of antibiotic resistance associated with antibiotic residues in the environment: a review. Environ Res 169: 483–493. https://doi.org/10.1016/j.envres.2018.11.040

Bengoechea JA, Sa Pessoa J (2019) Klebsiella pneumoniae infection biology: living to counteract host defenses. FEMS Microbiol Rev 43:123–144. https://doi.org/10.1093/femsre/fuy043

Bolívar-Vargas AF, Torres-Caycedo MI, Sánchez-Neira Y (2021) Biofilms de *Pseudomonas aeruginosa* como mecanismos de resistencia y tolerancia a antibióticos. Revisión narrativa. Revista de la Facultad de Ciencias de la Salud de la Universidad del Cauca 23(2):47–57. https://doi.org/10.47373/rfcs.2021.v232.1780

Brady MT (2004) Pseudomonas and related genera. In: Feigin RD, Cherry JD, Demmier GJ, Kaplan SL (eds) Textbook of pediatric infectious diseases, 5th edn. Saunders, Philadelphia, pp 1557–1573

Brooke JS (2012) *Stenotrophomonas maltophilia*: an emerging global opportunistic pathogen. Clin Microbiol Rev 25:2–41. https://doi.org/10.1128/CMR.00019-11

Brooks LE, Ul-Hasan S, Chan BK, Sistrom MJ (2018) Quantifying the evolutionary conservation of genes encoding multidrug efflux pumps in the ESKAPE pathogens to identify antimicrobial drug targets. MSystems 3:e00024-18. https://doi.org/10.1128/msystems.00024-18

Cabello-Yeves PJ, Zemskaya TI, Rosselli R, Coutinho FH, Zakharenko AS, Blinov VV, Rodriguez-Valera F (2017) Genomes of novel microbial lineages assembled from the sub-ice waters of Lake Baikal. Appl Environ Microbiol 84(1):e02132-17. https://doi.org/10.1128/AEM. 02132-17

Castillo-Rojas G, Mazari-Hiríart M, Ponce de León S, Amieva-Fernández RI, Agis-Juárez RA, Huebner J, López-Vidal Y (2013) Comparison of Enterococcus faecium and Enterococcus faecalis strains isolated from water and clinical samples: antimicrobial susceptibility and genetic relationships. PLoS ONE 8(4). https://doi.org/10.1371/journal.pone.0059491

Centers for Disease Control and Prevention (CDC) (2016) Healthy Swimming. Available online: https://www.cdc.gov/healthywater/swimming. Accessed on 4 April 2022

Chakraborty J, Sapkale V, Shah M, Rajput V, Mehetre G, Agawane S, Kamble S, Dharne M (2020) Metagenome sequencing to unveil microbial community composition and prevalence of antibiotic and metal resistance genes in hypersaline and hyperalkaline Lonar Lake, India. Ecol Indicators 110:105827. https://doi.org/10.1016/j.ecolind.2019.105827

Challagundla L, Reyes J, Rafiqullah I, Sordelli DO, Echaniz-Aviles G, Velazquez-Meza ME, Robinson DA (2018) Phylogenomic classification and the evolution of clonal complex 5 methicillin-resistant *Staphylococcus aureus* in the Western hemisphere. Front Microbiol 9:1901. https://doi.org/10.3389/fmicb.2018.01901

Chen H, Jing L, Yao Z, Meng F, Teng Y (2019) Prevalence, source and risk of antibiotic resistance genes in the sediments of Lake Tai (China) deciphered by metagenomic assembly: a comparison with other global lakes. Environ Int 127:267–275. https://doi.org/10.1016/j.envint.2019.03.048

Cherak Z, Loucif L, Moussi A, Rolain JM (2021) Carbapenemase-producing gram-negative bacteria in aquatic environments: a review. J Glob Antimicrob Resist 25:287–309. https://doi.org/10.1016/j.jgar.2021.03.024

Cheung GYC, Bae JS, Otto M (2021) Pathogenicity and virulence of *Staphylococcus aureus*. Virulence 12:547–569. https://doi.org/10.1080/21505594.2021.1878688

Chidamba L, Korsten L (2018) Relative proportions of *E. coli* and *Enterococcus* spp. may be a good indicator of potential health risks associated with the use of roof harvested rainwater stored in tanks. Environ Monit Assess 190:177. https://doi.org/10.1007/s10661-018-6554-1

Cho S, Jackson CR, Frye JG (2020) The prevalence and antimicrobial resistance phenotypes of *salmonella*, *Escherichia coli* and *enterococcus* sp. in surface water. Lett Appl Microbiol 71:3–25. https://doi.org/10.1111/lam.13301

Chotirmall SH, Mirkovic B, Lavelle GM, McElvaney NG (2014) Immunoevasive *Aspergillus* virulence factors. Mycopathologia 178:363–370. https://doi.org/10.1007/s11046-014-9768-y

Cole JJ (1982) Interactions between bacteria and algae in aquatic ecosystems. Annu Rev Ecol Syst 13:291–314

Cong Y, Yang S, Rao X (2020) Vancomycin resistant *Staphylococcus aureus* infections: a review of case updating and clinical features. J Adv Res 21:169–176. https://doi.org/10.1016/j.jare.2019.10.005

Corpuz MVA, Buonerba A, Vigliotta G, Zarra T, Ballesteros F Jr, Campiglia P, Naddeo V (2020) Viruses in wastewater: occurrence, abundance and detection methods. Sci Total Environ 745:140910. https://doi.org/10.1016/j.scitotenv.2020.140910

Corvec S, Beyrouthy R, Cremet L, Aubin GG, Robin F, Bonnet R et al (2013) TEM-187, a new extended-spectrum beta-lactamase with weak activity in a *Proteus mirabilis* clinical strain. Antimicrob Agents Chemother 57:2410–2412

Dalmasso C, Oger P, Selva G, Courtine D, L'Haridon S, Garlaschelli A et al (2016) *Thermococcus piezophilus* sp. nov., a novel hyperthermophilic and piezophilic archaeon with a broad pressure range for growth, isolated from a deepest hydrothermal vent at the Mid-Cayman Rise. Syst Appl Microbiol 39:440–444. https://doi.org/10.1016/j.syapm.2016.08.003

Daniel DS, Lee SM, Gan HM, Dykes GA, Rahman S (2017) Genetic diversity of *enterococcus faecalis* isolated from environmental, animal and clinical sources in Malaysia. J Infect Public Health 10:617–623. https://doi.org/10.1016/j.jiph.2017.02.006

Davin-Regli A, Masi M, Bialek S, Nicolas-Chanoine MH, Pagès JM (2016) Antimicrobial resistance and drug efflux pumps in *Enterobacter* and *Klebsiella*. In: Li X-Z, Elkins CA, Zgurskaya HI (eds) Efflux-mediated drug resistance in bacteria: mechanisms, regulation and clinical implications. Springer International Publishing, Basel, pp 281–306

De Almeida JF, Sztajnbok J, Trostser EJ, Vaz FA, Case report. (2002) *Pseudomonas aeruginosa* septic shock associated with ecthyma gangrenosum in an infant with agammaglobulinemia. Rev Inst Med Trop S Paulo 44:167–169

De Angelis G, Del GP, Posteraro B, Sanguinetti M, Tumbarello M (2020) Molecular mechanisms, epidemiology, and clinical importance of β-lactam resistance in enterobacteriaceae. Int J Mol Sci 21:1–22. https://doi.org/10.3390/ijms21145090

Denissen J, Reyneke B, Waso-Reyneke M, Havenga B, Barnard T, Khan S et al (2022) Prevalence of ESKAPE pathogens in the environment: antibiotic resistance status, community-acquired infection and risk to human health. Int J Hyg Environ Health 244:114006. https://doi.org/10.1016/j.ijheh.2022.114006

Dumaru R, Baral R, Shrestha LB (2019) Study of biofilm formation and antibiotic resistance pattern of gram-negative Bacilli among the clinical isolates at BPKIHS, Dharan. BMC Res Notes 12(1):38

Ekwanzala MD, Dewar JB, Kamika I, Momba MNB (2020) Comparative genomics of vancomycin-resistant enterococcus spp. revealed common resistome determinants from hospital wastewater to aquatic environments. Sci Total Environ 719:137275. https://doi.org/10.1016/j.scitotenv.2020.137275

Escobar-Zepeda A, Vera-Ponce de León A, Sanchez-Flores A (2015) The road to metagenomics: from microbiology to DNA sequencing technologies and bioinformatics. Front Genet 6:348. https://doi.org/10.3389/fgene.2015.00348

Figueras MJ, Levican A, Collado L (2012) Updated 16S rRNA-RFLP method for the identification of all currently characterised Arcobacterspp. BMC Microbiol 12:1–7. https://doi.org/10.1186/1471-2180-12-292

Founou RC, Founou LL, Essack SY (2018) Extended spectrum beta-lactamase mediated resistance in carriage and clinical gram-negative ESKAPE bacteria: a comparative study between a district and tertiary hospital in South Africa. Antimicrob Resist Infect Control 7:1–11. https://doi.org/10.1186/s13756-018-0423-0

Fournier PE, Vallenet D, Barbe V (2006) Comparative genomics of multidrug resistance in *Acinetobacter baumannii*. PLoS Genet 2:e7

Furlan JPR, Savazzi EA, Stehling EG (2020) Genomic insights into multidrug-resistant and hypervirulent Klebsiella pneumoniae co-harboring metal resistance genes in aquatic environments. Ecotoxicol Environ Saf 201:110782. https://doi.org/10.1016/j.ecoenv.2020.110782

Gaddy JA, Actis LA (2009) Regulation of *Acinetobacter baumannii* biofilm formation. Future Microbiol 4:273–278

Gajdács M (2019) The continuing threat of methicillin-resistant *Staphylococcus aureus*. Antibiotics 8:52. https://doi.org/10.3390/antibiotics8020052

Gonzalez-Ferrer S, Peñaloza HF, Budnick JA, Bain WG, Nordstrom HR, Lee JS et al (2021) Finding order in the chaos: outstanding questions in *Klebsiella pneumoniae* pathogenesis. Infect Immun 89:e00693-20. https://doi.org/10.1128/IAI.00693-20

Grassotti TT, De Angelis ZD, Da Fontoura Xavier Costa L, De Araújo AJG, Pereira RI, Soares RO et al (2018) Antimicrobial resistance profiles in enterococcus spp. isolates from fecal samples of wild and captive black capuchin monkeys (*Sapajus nigritus*) in South Brazil. Front Microbiol 9: 2366. https://doi.org/10.3389/fmicb.2018.02366

Guarner J (2020) Three emerging coronaviruses in two decades: the story of SARS, MERS, and now COVID-19. Am J Clin Pathol 153:420–421. https://doi.org/10.1093/ajcp/aqaa029

Guo Y, Song G, Sun M, Wang J, Wang Y (2020) Prevalence and therapies of antibiotic-resistance in *Staphylococcus aureus*. Front Cell Infect Microbiol 10:1–11. https://doi.org/10.3389/fcimb.2020.00107

Handelsman J, Rondon MR, Brady SF, Clardy J, Goodman RM (1998) Molecular biological access to the chemistry of unknown soil microbes: a new frontier for natural products. Chem Biol 5: R245–R249. https://doi.org/10.1016/s1074-5521(98)90108-9

Herridge WP, Shibu P, O'Shea J, Brook TC, Hoyles L (2020) Bacteriophages of *Klebsiella* spp., their diversity and potential therapeutic uses. J Med Microbiol 69:176–194. https://doi.org/10.1099/jmm.0.001141

Hutchings M, Truman A, Wilkinson B (2019) Antibiotics: past, present and future. Curr Opin Microbiol 51:72–80. https://doi.org/10.1016/j.mib.2019.10.008

Huys G, Gevers D, Temmerman R, Cnockaert M, Denys R, Rhodes G, Swings J (2001) Comparison of the antimicrobial tolerance of oxytetracycline-resistant heterotrophic bacteria isolated from hospital sewage and freshwater fishfarm water in Belgium. Syst Appl Microbiol 24:122–130. https://doi.org/10.1078/0723-2020-00008

Ji P, Aw TG, Van Bonn W, Rose JB (2020) Evaluation of a portable nanopore-based sequencer for detection of viruses in water. J Virol Methods 278:113805. https://doi.org/10.1016/j.jviromet.2019.113805

Kadri K (2020) Polymerase chain reaction (PCR): principle and applications. In: Nagpal ML, Boldura OM, Balta C, Enany S (eds) Synthetic biology—new interdisciplinary science. IntechOpen, London. https://doi.org/10.5772/intechopen.86491

Kawai Y, Ozawa N, Fukuda T, Suzuki N, Mikata K (2022) Development of an efficient antimicrobial susceptibility testing method with species identification by nanopore sequencing of 16S rRNA amplicons. PLoS One 17:e0262912. https://doi.org/10.1371/journal.pone.0262912

Koo H, Hakim JA, Morrow CD, Crowley MR, Andersen DT, Bej AK (2018) Metagenomic analysis of microbial community compositions and cold-responsive stress genes in selected Antarctic lacustrine and soil ecosystems. Life 8:29. https://doi.org/10.3390/life8030029

Kosecka-Strojek M, Sabat AJ, Akkerboom V, Becker K, van Zanten E, Wisselink G et al (2019) Development and validation of a reference data set for assigning *Staphylococcus* species based on next-generation sequencing of the 16S-23S rRNA region. Front Cell Infect Microbiol 9:1–19. https://doi.org/10.3389/fcimb.2019.00278

Langridge P, Chalmers K (2004) The principle: identification and application of molecular markers. In: Molecular marker systems in plant breeding and crop improvement. Springer, Berlin, pp 3–22. https://doi.org/10.1007/3-540-26538-4_1

Lee T, Pang S, Abraham S, Coombs GW (2019) Antimicrobial-resistant CC17 *Enterococcus faecium*: the past, the present and the future. J Glob Antimicrob Resist 16:36–47. https://doi.org/10.1016/j.jgar.2018.08.016

Liu WT, Chen EZ, Yang L, Peng C, Wang Q, Xu Z et al (2021) Emerging resistance mechanisms for 4 types of common anti-MRSA antibiotics in *Staphylococcus aureus*: a comprehensive review. Microb Pathog 156:104915. https://doi.org/10.1016/j.micpath.2021.104915

Luján Roca DA (2014) *Pseudomonas aeruginosa*: un adversario peligroso. Acta bioquím. clín. latinoam, vol 48, no 4. La Plata dic

Lyczak JB, Cannon CL, Pier GB (2000) Stablishment of *Pseudomonas aeruginosa* infection: lessons from a versatile opportunist. Microb Infect 2(9):1051–1060

Mahon BM, Brehony C, Cahill N, McGrath E, O'Connor L, Varley A et al (2019) Detection of OXA-48-like-producing Enterobacterales in Irish recreational water. Sci Total Environ 10(690): 1–6

Maillard JY, Bloomfield SF, Courvalin P, Essack SY, Gandra S, Gerba CP et al (2020) Reducing antibiotic prescribing and addressing the global problem of antibiotic resistance by targeted hygiene in the home and everyday life settings: a position paper. Am J Infect Control 48:1090–1099. https://doi.org/10.1016/j.ajic.2020.04.011

Mandal P, Gupta AK, Dubey BK (2020) A review on presence, survival, disinfection/removal methods of coronavirus in wastewater and progress of wastewater-based epidemiology. J Environ Chem Eng 8:104317. https://doi.org/10.1016/j.jece.2020.104317

Mathys DA, Mollenkopf DF, Feicht SM, Adams RJ, Albers AL, Stuever DM et al (2019) Carbapenemase-producing Enterobacteriaceae and *Aeromonas* spp present in wastewater treatment plant effluent and nearby surface waters in the US. PLoS One 14(6):e0218650

Melese A, Genet C, Andualem T (2020) Prevalence of vancomycin resistant enterococci (VRE) in Ethiopia: a systematic review and meta-analysis. BMC Infect Dis 20:124. https://doi.org/10.1186/s12879-020-4833-2

Merino N, Aronson HS, Bojanova DP, Feyhl-Buska J, Wong ML, Zhang S, Giovannelli D (2019) Living at the extremes: extremophiles and the limits of life in a planetary context. Front Microbiol 10:780. https://doi.org/10.3389/fmicb.2019.00780

Merker M, Blin C, Mona S, Duforet-Frebourg N, Lecher S, Willery E, Wirth T (2015) Evolutionary history and global spread of the *Mycobacterium tuberculosis* Beijing lineage. Nat Genet 47: 242–249. https://doi.org/10.1038/ng.3195

Mesaros N, Nordmann P, Plésiat P, Roussel-Delvallez M, Van Eldere J, Glupczynski Y et al (2007) *Pseudomonas aeruginosa*: résistance et options thérapeutiques à l'aube du deuxième millénaire. Antibiotiques 9(3):189–198

Monapathi ME, Bezuidenhout CC, James-Rhode OH (2020) Aquatic yeasts: diversity, characteristics and potential health implications. J Water Health 18:91–105. https://doi.org/10.2166/wh.2020.270

Mulani MS, Kamble EE, Kumkar SN, Tawre MS, Pardesi KR (2019) Emerging strategies to combat ESKAPE pathogens in the era of antimicrobial resistance: a review. Front Microbiol 10:539. https://doi.org/10.3389/fmicb.2019.00539

Nakatsu CH, Byappanahalli MN, Nevers MB (2019) Bacterial community 16S rRNA gene sequencing characterizes riverine microbial impact on Lake Michigan. Front Microbiol 10: 996. https://doi.org/10.3389/fmicb.2019.00996

Panizo MM, Reviákina V (2001) *Candida albicans* y su efecto patógeno sobre las mucosas. Rev Soc Venez Microbiol 21:38–45

Pasachova J, Ramirez S, Muñoz L (2019) *Staphylococcus aureus*: generalidades, mecanismos de patogenicidad y colonización celular. Ther Nova 17:25–38

Patel B (2016) Antibiotics: a savior in need of saving. City Tech Writer 11:46–49

Peña-Ocaña BA, Ovando-Ovando CI, Puente-Sánchez F, Tamames J, Servín-Garcidueñas LE, González-Toril E, Gutiérrez-Sarmiento W, Jasso-Chávez R, Ruíz-Valdiviezo VM (2022) Metagenomic and metabolic analyses of poly-extreme microbiome from an active crater volcano lake. Environ Res 203:111862. https://doi.org/10.1016/j.envres.2021.111862

Phillips M (2014) Acinetobacter species. In: Mandell GL, Dolin R, Blaser MJ (eds) Principles and practices of infectious diseases, vol 2, 8th edn. Imprint Elsevier Inc., Philadelphia, pp 2551–2558

Piedra-Carrasco N, Fabrega A, Calero-Caceres W, Cornejo-Sanchez T, Brown-Jaque M, Mir-Cros A et al (2017) Carbapenemase-producing enterobacteriaceae recovered from a Spanish river ecosystem. PLoS One 12(4):e0175246

Pillai D (2022) Genetic diversity and prevalence of extended spectrum beta-lactamase producing *Escherichia coli* and *Klebsiella pneumoniae* in aquatic environment receiving untreated hospital effluents. Res Sq:1–19

Pirnay JP, Matthijs S, Colak H, Chablain P, Bilocq F, Eldere V, Cornelis P (2005) Global *Pseudomonas aeruginosa* biodiversity as reflected in a Belgian river. Environ Microbiol 7: 969–980. https://doi.org/10.1111/j.1462-2920.2005.00776.x

Pini G, Faggi E, Bravetti E (2017) Molecular typing of clinical and environmental *Cryptococcus neoformans* strains isolated in Italy. Open J Med Microbiol 7(04):77. https://doi.org/10.4236/ojmm.2017.74007

Podolsky SH (2018) The evolving response to antibiotic resistance (1945–2018). Palgrave Commun 4:124. https://doi.org/10.1057/s41599-018-0181-x

Premke K, Wurzbacher C, Felsmann K, Fabian J, Taube R, Bodmer P, Attermeyer K, Nitzsche KN, Schroer S, Koschorreck M, Hübner E, Mahmoudinejad T, Kyba C, Monaghan MT, Hölker F (2022) Large-scale sampling of the freshwater microbiome suggests pollution-driven ecosystem changes. Environ Pollut 308:119627. Proc Natl Acad Sci U.S.A. vol 95, p 12

Raj A (2012) Antibiotic resistance, plasmid and RAPD profiles of multidrug-resistant coliform bacteria isolated from sewage samples of Ghaziabad City, India. UJERT 2:318–324

Ramos S, Silva V, de Lurdes Enes Dapkevicius M, Igrejas G, Poeta P (2020) Enterococci, from harmless bacteria to a pathogen. Microorganisms 8:1118. https://doi.org/10.3390/microorganisms8081118

Randazzo W, Truchado P, Cuevas-Ferrando E, Simón P, Allende A, Sánchez G (2020) SARS-CoV-2 RNA in wastewater anticipated COVID-19 occurrence in a low prevalence area. Water Res 181:115942. https://doi.org/10.1016/j.watres.2020.115942

Raza T, Ullah SR, Mehmood K, Andleeb S (2018) Vancomycin resistant enterococci: a brief review. J Pak Med Assoc 68:768–772

Regnier P, Friedlingstein P, Ciais P et al (2013) Anthropogenic perturbation of the carbon fluxes from land to ocean. Nat Geosci 6:597–607. https://doi.org/10.1038/ngeo1830

Rice LB (2010) Progress and challenges in implementing the research on ESKAPE pathogens. Infect Control Hosp Epidemiol 31:7–10

Rice EW, Johnson CH (2000) Survival of *Escherichia coli* O157: H7 in dairy cattle drinking water. J Dairy Sci 83:2021–2023. https://doi.org/10.3168/jds.S0022-0302(00)75081-8

Rogers BA, Aminzadeh Z, Hayashi Y, Paterson DL (2011) Country-to-country transfer of patients and the risk of multi-resistant bacterial infection. Clin Infect Dis 53:49–56

Romeo A (2020) Biofilm y Resistencia antimicrobiana. Arch Med camagüey 24(4):1–4

Rosales Magallanes GF (2018) The end of the antibiotic era, 1. https://doi.org/10.15761/cmp.1000103

Rossolini GM, Mantengoli E (2005) Treatment and control of severe infections caused by multiresistant *Pseudomonas aeruginosa*. Clin Microbiol Infect 11(Suppl 4):17–32

Rusiñol M, Martínez-Puchol S, Timoneda N, Fernández-Cassi X, Pérez-Cataluña A, Fernández-Bravo A, Moreno-Mesonero L, Moreno Y, Alonso JL, Figueras MJ, Abril JF, Bofill-Mas S, Girones R (2020) Metagenomic analysis of viruses, bacteria and protozoa in irrigation water. Int J Hyg Environ Health 224:113440. https://doi.org/10.1016/j.ijheh.2019.113440

Salazar de Vegasa EZ, Nieves B (2005) *Acinetobacter* spp: Aspectos microbiológicos, clínicos y epidemiológicos. Rev Soc Venez Microbiol 25(2):64–71

Samreen AI, Malak HA, Abulreesh HH (2021) Environmental antimicrobial resistance and its drivers: a potential threat to public health. J Glob Antimicrob Resist 27:101–111. https://doi.org/10.1016/j.jgar.2021.08.001

Sanderson H, Ortega-Polo R, Zaheer R, Goji N, Amoako KK, Brown RS et al (2020) Comparative genomics of multidrug-resistant *Enterococcus* spp. isolated from wastewater treatment plants. BMC Microbiol 20:20. https://doi.org/10.1186/s12866-019-1683-4

Santaniello A, Sansone M, Fioretti A, Menna LF (2020) Systematic review and meta-analysis of the occurrence of eskape bacteria group in dogs, and the related zoonotic risk in animal-assisted therapy, and in animal-assisted activity in the health context. Int J Environ Res Public Health 17: 3278. https://doi.org/10.3390/ijerph17093278

Schreiber C, Rechenburg A, Rind E, Kistemann T (2015) The impact of land use on microbial surface water pollution. Int J Hyg Environ Health 218(2):181–187. https://doi.org/10.1016/j.ijheh.2014.09.006

Seymour JR, Ahmed T, Durham WM, Stocker R (2010) Chemotactic response of marine bacteria to the extracellular products of *Synechococcus* and *Prochlorococcus*. Aquat Microb Ecol 59(2): 161–168

Silva V, Caniça M, Capelo JL, Igrejas G, Poeta P (2020) Diversity and genetic lineages of environmental staphylococci: a surface water overview. FEMS Microbiol Ecol 96:1–13. https://doi.org/10.1093/femsec/fiaa191

Silva V, Ferreira E, Manageiro V, Reis L, Tejedor-Junco MT, Sampaio A et al (2021) Distribution and clonal diversity of Staphylococcus aureus and other staphylococci in surface waters: detection of ST425-t742 and ST130-t843 mecC-positive MRSA strains. Antibiotics 10:1416. https://doi.org/10.3390/antibiotics10111416

Silva V, Pereira E, Igrejas G (2022) Influence of environmental factors on biofilm formation of Staphylococci isolated from wastewater and surface water. Pathogens 11:1069

Simoes RR, Poirel L, da Costa PM, Nordmann P (2010) Seagulls and beaches as reservoirs for multidrug-resistant *Escherichia coli*. Emerg Infect Dis 16:110–112

Singh NK, Bezdan D, Checinska Sielaff A, Wheeler K, Mason CE, Venkateswaran K (2018) Multidrug resistant *Enterobacter bugandensis* species isolated from the International Space Station and comparative genomic analyses with human pathogenic strains. BMC Microbiol 18:175. https://doi.org/10.1186/s12866-018-1325-2

Strateva T, Yordanov D (2009) *Pseudomonas aeruginosa* - a phenomenon of bacterial resistance. J Med Microbiol 58(9):1133–1148

Takai K, Nakamura K, Toki T, Tsunogai U, Miyazaki M, Miyazaki J et al (2008) Cell proliferation at 122 C and isotopically heavy CH4 production by a hyperthermophilic methanogen under high-pressure cultivation. Proc Natl Acad Sci U S A 105:10949–10954. https://doi.org/10.1073/pnas.0712334105

Tamayo-Ordoñez M, Huijara-Vasconselos J, Quiroz-Moreno A, Ortíz-García M, Sánchez-Teyer LF (2012) Plant tissue culture and molecular markers. In: Plant cell culture protocols. Humana Press, Totowa, pp 343–356. https://doi.org/10.1007/978-1-61779-818-4_26

Tamayo-Ordóñez MC, Contreras-Esquivel JC, Ayil-Gutiérrez BA, De la Cruz-Arguijo EA, Tamayo-Ordóñez FA, Ríos-González LJ, Tamayo-Ordóñez YJ (2021) Interspecific evolutionary relationships of alpha-glucuronidase in the genus *Aspergillus*. Fungal Biol 125:560–575. https://doi.org/10.1016/j.funbio.2021.02.005

Taučer-Kapteijn M, Hoogenboezem W, Heiliegers L, de Bolster D, Medema G (2016) Screening municipal wastewater effluent and surface water used for drinking water production for the presence of ampicillin and vancomycin resistant enterococci. Int J Hyg Environ Health 219: 437–442. https://doi.org/10.1016/j.ijheh.2016.04.007

Tigabu A, Getaneh A (2021) *Staphylococcus aureus*, ESKAPE bacteria challenging current health care and community settings: a literature review. Clin Lab 67:1539–1549. https://doi.org/10.7754/Clin.Lab.2020.200930

1 Freshwater Microbiology: Recent Updates and Prospects

Torres HA, Vázquez EG, Yagüe G, Gómez JG. (2010) Multidrug resistant *Acinetobacter baumannii*: clinical update and new highlights. Rev Esp Quimioter 23(1):12–19

Towner KJ (2009) *Acinetobacter*: an old friend, but a new enemy. J Hosp Infect 73:355–363

Tran PQ, Bachand SC, McIntyre PB et al (2021) Depth-discrete metagenomics reveals the roles of microbes in biogeochemical cycling in the tropical freshwater Lake Tanganyika. ISME J 15: 1971–1986. https://doi.org/10.1038/s41396-021-00898-x

Tseng CH, Chiang PW, Shiah FK et al (2013) Microbial and viral metagenomes of a subtropical freshwater reservoir subject to climatic disturbances. ISME J 7:2374–2386. https://doi.org/10.1038/ismej.2013.118

Vikesland PJ, Pruden A, Alvarez PJJ, Aga D, Bürgmann H, Li XD et al (2017) Toward a comprehensive strategy to mitigate dissemination of environmental sources of antibiotic resistance. Environ Sci Technol 51:13061–13069. https://doi.org/10.1021/acs.est.7b03623

Villanueva-Ramos NB, De la Mora-Fernández AR, Ríos-Burgueño ER, de Peraza-Garay FJ (2019) Detección de biopelículas en tejido de amígdalas y adenoides en pacientes con procesos infecciosos crónicos y obstructivos. Anales de Otorrinolaringología Mexicana 64:1–17

Voigt AM, Ciorba P, Döhla M, Exner M, Felder C, Lenz-Plet F et al (2020) The investigation of antibiotic residues, antibiotic resistance genes and antibiotic-resistant organisms in a drinking water reservoir system in Germany. Int J Hyg Environ Health 224:113449. https://doi.org/10.1016/j.ijheh.2020.113449

Wang X, Wen X, Criddle C, Yan H, Zhang Y, Ding K (2010) Bacterial community dynamics in two full-scale wastewater treatment systems with functional stability. J Appl Microbiol 109:1218–1226. https://doi.org/10.1111/j.1365-2672.2010.04742.x

Wang W, Arshad MI, Khurshid M, Rasool MH, Nisar MA, Aslam MA et al (2018) Antibiotic resistance: a rundown of a global crisis. Infect Drug Resist 11:1645–1658

Wang H, Kjellberg I, Sikora P, Rydberg H, Lindh M, Bergstedt O, Norder H (2020a) Hepatitis E virus genotype 3 strains and a plethora of other viruses detected in raw and still in tap water. Water Res 168:115141. https://doi.org/10.1016/j.watres.2019.115141

Wang J, Feng H, Zhang S, Ni Z, Ni L, Chen Y, Qu T (2020b) SARS-CoV-2 RNA detection of hospital isolation wards hygiene monitoring during the coronavirus disease 2019 outbreak in a Chinese hospital. Int J Infect Dis 94:103–106. https://doi.org/10.1016/j.ijid.2020.04.024

Watkins RR, Holubar M, David MZ (2019) Antimicrobial resistance in methicillin-resistant Staphylococcus aureus to newer antimicrobial agents. Antimicrob Agents Chemother 63: e01216-19. https://doi.org/10.1128/AAC.01216-19

Wen X, Chen F, Lin Y, Zhu H, Yuan F, Kuang D et al (2020) Microbial indicators and their use for monitoring drinkingwater quality—a review. Sustainability 12:1–14. https://doi.org/10.3390/su12062249

Whitman WB, Coleman DC, Wiebe WJ (1998) Prokaryotes: the unseen majority. Proc Natl Acad Sci U S A 95(12):6578–6583. https://doi.org/10.1073/pnas.95.12.6578. PMID: 9618454; PMCID: PMC33863

Woerther PL, Burdet C, Chachaty E, Andremont A (2013) Trends ins human fecal carriage of extended-spectrum beta-lactamases in the community: toward the globalization of CTX-M. Clin Microbiol Rev 26:744–758

Woese CR, Fox GE (1977) Phylogenetic structure of the prokaryotic domain: the primary kingdoms. Proc Natl Acad Sci U S A 74:5088–5090. https://doi.org/10.1073/pnas.74.11.5088

World Health Organization (2015) Worldwide country situation analysis: response to antimicrobial resistance: summary. World Health Organization. https://apps.who.int/iris/handle/10665/163473

Wyres KL, Lam MMC, Holt KE (2020) Population genomics of *Klebsiella pneumoniae*. Nat Rev Microbiol 18:344–359. https://doi.org/10.1038/s41579-019-0315-1

Zieliński W, Korzeniewska E, Harnisz M, Hubeny J, Buta M, Rolbiecki D (2020) The prevalence of drug-resistant and virulent *Staphylococcus* spp. in a municipal wastewater treatment plant and their spread in the environment. Environ Int 143:143. https://doi.org/10.1016/j.envint.2020.105914

The Chemical Composition of the Water in the Rivers, Lakes, and Wetlands of Uttarakhand

2

Manjul Gondwal, Navneet Kishore, Ravindra Soni, Rakesh Kumar Verma, and Bhanu Pratap Singh Gautam

Abstract

Ions like calcium (Ca^{2+}), magnesium (Mg^{2+}), sodium (Na^+), and potassium (K^+) predominate among the chemical components of the water in Uttarakhand's rivers, lakes, and wetlands. These waters may also contain the following significant ions: bicarbonate (HCO_3^-), sulfate (SO_4^{2-}), chloride (Cl^-), and nitrate (NO_3^-). Depending on the type of water body and the area, different ions have different concentrations. Other significant chemical components of the water in these areas include suspended particles, organic matter, and dissolved organic carbon. The concentrations of nutrients such as phosphorus (P) and nitrogen (N) vary depending on the geographical area and season. The water also contains trace amounts of elements like iron (Fe), manganese (Mn), and zinc (Zn). The water quality of the rivers, lakes, and wetlands in Uttarakhand is impacted by the presence of certain chemical components, which may have an effect on the health of aquatic life, human activities, and the environment. These water bodies need to be protected; therefore, it is important to keep an eye on their chemical makeup and take precautions to keep them healthy.

M. Gondwal · B. P. S. Gautam (✉)
Department of Chemistry, Laxman Singh Mahar Campus (SSJ University), Pithoragarh, India

N. Kishore
Department of Chemistry, Maitreyi College, University of Delhi, New Delhi, India

R. Soni
Department of Agricultural Microbiology, Indira Gandhi Krishi Vishwa Vidyalaya, Raipur, Chhattisgarh, India

R. K. Verma
Department of Zoology, Laxman Singh Mahar Campus (SSJ University), Pithoragarh, India

© The Author(s), under exclusive license to Springer Nature Singapore Pte Ltd. 2023
R. Soni et al. (eds.), *Current Status of Fresh Water Microbiology*,
https://doi.org/10.1007/978-981-99-5018-8_2

Keywords

Water body · Chemical components · Uttarakhand · Water · Wetland

2.1 Introduction

Water is an inorganic chemical substance that is clear, odorless, tasteless, and almost colorless; it is the primary component of the hydrosphere of the Earth and the fluids of all known forms of life; its chemical formula is H_2O and acts as a solvent. Despite the fact that it does not contain any calories or organic nutrients, it is necessary for all kinds of life that have ever been discovered. In accordance with its chemical formula, H_2O, each of its molecules is composed of two hydrogen atoms and one oxygen atom that are covalently bonded to one another. The angle between connected hydrogen and oxygen atoms is $104.45°$ (Brini et al. 2017). The term "water" refers to the aqueous phase of the molecule H_2O when it is at a standard pressure and temperature.

There are many different natural states that water may be found in. It results in the production of aerosols that resemble fog as well as precipitation in the form of rain. Clouds are formed by ice and water droplets that are suspended in the atmosphere. When the crystalline ice is broken up into smaller pieces, it has the ability to fall to the ground as snow. The gaseous state of water is referred to as steam or water vapors. The majority of the surface of the Earth is covered by water, the bulk of which may be found in the world's oceans and seas (Chaplin 2019). The groundwater (1.7%), glaciers and ice caps of Antarctica and Greenland (1.7%), clouds (made up of ice and liquid water suspended in air), and precipitation (0.001%) of the atmosphere all contain small quantities of water. Precipitation also contains a minimal quantity of water (Ho 1972).

Water is a very important component of the world economy. More than seventy percent of the fresh water that humans consume is put to agricultural use (Baroni et al. 2007). Water is the most vital resource necessary for the continuation of life. The pollution of water makes it unsuitable for human consumption, and the causes of this contamination include both natural processes (such as rock weathering and erosion) and human activities (such as urbanization, agriculture, industry, and population growth), among others (Oluyemi et al. 2010). Water is crucial to the continued existence of human beings. It was estimated by the WHO that 36% of Indians living in urban areas and 65% of Indians living in rural areas did not have access to safe drinking water (Kılıç 2021).

People place the most importance on the presence of naturally occurring water in the physical world that surrounds them. It is one of the most incredible, but inexplicable, combinations that can ever exist on our earth. Because of a wide range of characteristics that are unique to it, it stands out above all other substances (e.g., anomalously high values for the temperature of melting, boiling, and evaporation, and heavy dissolving capacity) (Nilsson and Pettersson 2015).

The hydrological cycle, which is a connection between the hydrosphere and the atmosphere, lithosphere, and biosphere, is responsible for the creation of the chemical composition of water on Earth. Water is a universal solvent that interacts with every component of its natural environment and is influenced by both natural and man-made factors. Water is enriched by a wide range of different substances in gaseous, solid, and liquid states that produce a huge variety of natural water types from the perspective of their chemical composition. Water is a universal solvent that interacts with every component of its natural environment and is influenced by both natural and man-made factors (Quattrini et al. 2016).

The boundaries of Uttarakhand are denoted by the presence of the Himalayas to the north, the Shivalik Hills to the south, the Ganges to the east, and the Yamuna to the west. The climate there is classified as moderate. The lowest it gets in the winter is approximately 5 °C, while the highest it gets in the summer is around 36 °C (Kumar et al. 2010).

The weathering of the rocks and improper disposal of sewage waste are the key factors that contribute to water source pollution in Uttarakhand. The mountains in this region are sloped, making these factors more likely to occur. Problems with water quality are being experienced by Uttarakhand's water sector as a consequence of turbidity and bacterial contamination of the available drinking water sources in the state. The mountainous section of the state is home to a sizeable portion of the state's population, and it is estimated that around 90% of the rural population gets the water they need for everyday life from natural water sources (Jain et al. 2010).

Fecal contamination in water is still the pollutant that most seriously affects human health. This includes the major water-borne diseases such as diarrhea, cholera, typhoid, and schistosomiasis. This is especially true in the hills of Uttarakhand and other similar states like Himachal, Jammu and Kashmir, and the North East states (Joshi et al. 2009).

2.2 The Composition of Natural Waters from a Chemical Standpoint

The dissolved inorganic components and the dissolved organic components make up the two main parts of natural water. The main inorganic elements of natural water are calcium (Ca^{2+}), magnesium (Mg^{2+}), sodium (Na^+), potassium (K^+), oxygen (O_2), hydrogen (H_2), carbon dioxide (CO_2), nitrogen (N_2), and sulphur (S). Together, these elements create a complex mixture of ions and molecules that constitutes the majority of natural water (Rittmann and McCarty 2001).

Proteins, carbohydrates, lipids, and other organic compounds make up the majority of natural water's organic constituents. Depending on the source of the water and the surrounding environment, these organic components can vary significantly. For instance, the organic components of groundwater or seawater may be different from those of river water. In addition, biological processes like photosynthesis and respiration can have a big impact on how natural water is made (Schwarzenbach et al. 2003).

Natural water's chemical makeup can vary significantly depending on its source, the surrounding environment, and the extent of chemical and biological processes. Each component's concentration can also differ significantly, with some components present in only trace amounts and others possibly in high concentrations (Ašperger 2004).

2.3 Ions Found in Nature

Conventionally, the ions, complex ions, undissociated chemicals, and colloids (in the dissolved form) that make up natural fluids are classified into macrocomponents and microcomponents. These components include: ions, complex ions, dissolved colloids, and undissociated chemicals. The macrocomponents are what make up the so-called principal ions, which define the chemical type of water and make up the majority of its natural minerals. The percentage of natural minerals in highly mineralized water can reach up to 99%, while the percentage of natural minerals in fresh water only reaches 95%. In a solution of water, hydrogen ions, nitrogen molecules, phosphorus molecules, and silicon molecules all take up positions in the center.

The quantity of different chemical elements present in the crust of the Earth and the ease with which their compounds may be dissolved in water are the two primary factors that determine the relative amounts of each mineral.

An abundance of cations such as Ca^{2+}, Na^+, Mg^{2+}, and K^+ as well as anions such as Cl^-, SO_4^{2}, HCO_3^-, and CO_3^{2-} are found in natural water.

Ions of chloride (Cl^-): This ion has a substantial capacity for migration, because sodium, magnesium, and calcium chloride compounds are very soluble in water, and chloride ions have the same property. Because of natural processes such as leaching from rocks (e.g., nephelines), minerals (such as gallite, sylvite, carnallite, and bischofite), and saline deposits, they are able to live in water. In addition to being present in the precipitation of the atmosphere, it is currently mostly associated with pollution from urban and industrial sources. Chloride ions are present in varying concentrations throughout all types of water, ranging from a few parts per million to hundreds per kilogram (in brines).

2.3.1 Ions of Sulfate (SO_4^{2-})

These are present in all surface waters, and the presence of calcium ions, with which they mix to generate the compound $CaSO_4$, which is relatively soluble, restricts their concentration. Sulfate ions may be found in all surface waters. The sedimentary rocks that are the primary contributors of sulfate to water are gypsum and anhydride, two examples of which are described below. Sulfates contribute to the enrichment of water through a variety of processes, including the oxidation of sulfide, which is abundant in the crust of the Earth, as well as the oxidation of hydrogen sulfide, which is produced during volcanic eruptions and is found in the precipitation of the atmosphere. The concentration of sulfur in water bodies may be affected by human economic activity, processes involving the breakdown and oxidation of sulfur, as well as compounds of vegetative and organic origin.

2.3.2 Ions of Carbonate and Hydrocarbonate (HCO_3^- and CO_3^{2-})

Ions of carbonate and hydrocarbonate. It is found in waters that have formed naturally in a dynamic equilibrium with carbonic acid in particular quantitative ratios and, together, the two substances produce a carbonate system of chemical equilibrium that is related to the pH of the water. When the pH of the water in a system ranges from 7 to 8.5, hydrocarbonate will predominate as the predominant ion. When the pH is lower than 5, the number of hydrocarbonate ions present is almost nonexistent. When the pH is higher than 8, the majority of ions present are carbonate ions. Various types of carbonate rocks, such as limestones, dolomites, and magnesites, are the primary contributors of HCO_3^- and CO_3^{2-}, and the decomposition of these compounds requires the presence of carbon dioxide.

The predominance of hydrocarbonate ions is seen most often in waters with considerable mineralization but sometimes occurs in fluids with low mineralization. There is some buildup of hydrocarbonate ions, but just a little. Because calcium ions are present, these ions interact with the salt HCO_3^- to generate calcium carbonate. Surface freshwaters typically have an HCO_3^- concentration that is lower than or equal to 250/mg (with the exception of soda alkaline waters in which HCO_3^- and CO_3^{2-} content can reach grams and even dozens of grams per kilogram). The sodium ion, denoted by the symbol Na^+, has a comparatively strong tendency to migrate. This may be attributed to the high solubility of all of sodium's salts. When there is very little mineralization in the water, the concentration of Na^+ is third. As the mineralization of a fluid grows, so does the concentration of sodium, and once the mineralization reaches a certain g/kg, sodium becomes the predominate ion in the fluid. The presence of chlorine ions brings a considerable portion of the sodium ions into equilibrium, which results in the formation of a stable mobile combination that moves fast through a solution.

The presence of salt deposits (rock salt), products derived from the weathering of limestone, and the removal of sodium from the absorption complex of rocks and soils by calcium and magnesium are all potential sources of sodium in rivers.

2.3.3 Potassium Ions (K^+)

Potassium is pretty similar to sodium in terms of the quantity present in the Earth's crust and the solubility of its elements. Potassium is also fairly comparable to sodium in terms of the stability of its ions. Because of its restricted potential for migration, it is only found in surface waters in very small quantities. This is the case because it plays an active role in several biological processes, such as the absorption of nutrients by living plants and other microbes.

2.3.4 Ions of Calcium (Ca^{2+})

The principal sources of calcium are carbonate rocks (limestones, dolomites), which are broken down by the carbonic acid found in water. Other sources of calcium include algae and shellfish. The reaction, on the other hand, begins to proceed in the opposite direction when the availability of carbon dioxide (with which it is in equilibrium) is low, which is then followed by the precipitation of $CaCO_3$ crystals. Gypsum, which is often found in sedimentary rocks, provides an additional source of Ca^+ in the streams that are found naturally. Calcium ions dominate the cation composition of fluids that have a relatively modest mineral content.

2.3.5 Magnesium Ions (Mg^{2+})

There is a greater concentration of calcium than magnesium in the crust of the Earth. It is possible for it to make its way into surface water as a result of the chemical weathering and disintegration of rocks such as dolomites, marls, and others. Magnesium ions are present in all naturally occurring fluids, although they very seldom ever steal the spotlight. River waters may include anything from one milligram per liter to tens of milligrams per liter of this substance. The lower biological activity of magnesium as compared to calcium, as well as the higher solubility of magnesium sulfate and magnesium hydrocarbonate in comparison to the corresponding calcium compounds, both induce an increase in the Mg^{2+} content in water. With a rise in the mineral content of the water, the proportion of magnesium to calcium begins to move toward a greater abundance of the latter.

1. **Ammonia:** Ammonia is naturally present in rivers, lakes, and wetlands due to the decomposition of organic material.
2. **Nitrate:** Nitrate is naturally present in rivers, lakes, and wetlands due to the decomposition of organic material.
3. **Phosphate:** Phosphate is naturally present in rivers, lakes, and wetlands due to the decomposition of organic material.
4. **Chloride:** Chloride is naturally present in rivers, lakes, and wetlands due to the runoff of salt from the mountain soils.
5. **Sulfate:** Sulfate is naturally present in rivers, lakes, and wetlands due to the decomposition of organic material and the runoff of sulfur-containing minerals from the mountain soils.
6. **Magnesium:** Magnesium is naturally present in rivers, lakes, and wetlands due to the dissolution of soluble salts from the mountain soils.
7. **Calcium:** Calcium is naturally present in rivers, lakes, and wetlands due to the dissolution of soluble salts from the mountain soils.
8. **Iron:** Iron is naturally present in rivers, lakes, and wetlands due to the runoff of iron-containing minerals from the mountain soils.
9. **Manganese:** Manganese is naturally present in rivers, lakes, and wetlands due to the runoff of manganese-containing minerals from the mountain soils.

10. **Zinc:** Zinc is naturally present in rivers, lakes, and wetlands due to the runoff of zinc-containing minerals from the mountain soils.
11. **Copper:** Copper is naturally present in rivers, lakes, and wetlands due to the runoff of copper-containing minerals from the mountain soils.
12. **Lead:** Lead is naturally present in rivers, lakes, and wetlands due to the runoff of lead-containing minerals from the mountain soils.
13. **Arsenic:** Arsenic is naturally present in rivers, lakes, and wetlands due to the runoff of arsenic-containing minerals from the mountain soils.
14. **Mercury:** Mercury is naturally present in rivers, lakes, and wetlands due to the runoff of mercury-containing minerals from the mountain soils.
15. **Cadmium:** Cadmium is naturally present in rivers, lakes, and wetlands due to the runoff of cadmium-containing minerals from the mountain soils.
16. **Beryllium:** Beryllium is naturally present in rivers, lakes, and wetlands due to the runoff of beryllium-containing minerals from the mountain soils.

2.3.6 Ions of Hydrogen

Despite their comparatively modest absolute number in contrast to that of other ions, hydrogen ions play a particular function in the composition of natural waters. This is because hydrogen ions have a positive charge. Hydrogen ions are always present in water, because they are formed during the electrolytic dissociation of water.

The formula for hydrogen ions is

$$H_2O \rightarrow H^+ + OH^-$$

The concentration of hydrogen ions in a water solution can be calculated using the "ionic product" of water, which is denoted by the formula

$$K_w = [H^+] \cdot [OH^-]$$

Therefore, $K_w = 10^{-14}$ is a constant value that is always equal to the product of the concentrations (in g molecules) of hydrogen ions and hydroxyl at a temperature of 22 °C. This is because $K_w = 10^{-14}$ is a product of the concentrations of hydrogen ions and hydroxyl ions concentration. Since the concentrations of hydrogen and hydroxyl ions are so negligible, it is customary to express them in the form of their logarithms with the sign switched around:

$$p^H = - \log [H^+]; p^{OH} = - \log [OH^-]$$

When describing water processes, the concentration of hydrogen ions is often utilized. When pH equals 7, the interaction between water and other substances is considered neutral. However, if pH is more than 7 or less than 7, the reaction might go in either an acidic or alkaline direction. The amount of hydrogen ions that are included within natural fluids is determined by the dissociation and hydrolysis of the combinations that are already there (Brezonik and Arnold 2012).

2.4 Gases Dissolved

Depending on the location and the surrounding environment, the percentage composition of dissolved gases in water with a focus on the biological oxygen demand (BOD) can change. In general, nitrogen, oxygen, carbon dioxide, and argon are the most prevalent dissolved gases in water. Since oxygen is necessary for aquatic organisms' respiration and metabolism, it is the most significant of these for BOD. Nitrogen, carbon dioxide, and argon make up the remaining portion of the total dissolved gas in water, which is typically between 10 and 14% oxygen. Due to higher levels of pollutants and organic matter in the water, polluted waters can have significantly lower oxygen percentages. The total amount of dissolved gas in these situations can be between 0 and 7%, with oxygen making up roughly 0–3% of the total. A significant indicator of water quality is the proportion of oxygen in the water because it can show how polluted the water is. It is also one of the key elements in determining a water body's BOD level. Overall, depending on the environment and the level of water pollution, the oxygen content of water with a focus on the BOD can range from 0 to 14% (Kumar 2020; Theodore et al. 2022).

2.5 Biological Substances

Biogenous substances are those whose origins are tied to aquatic species' critical functions, which determine whether they can survive in bodies of water. These include silicon, nitrogen, phosphorus, and iron compounds.

2.5.1 Silicon (Si)

Natural waters always include silicon, which is one of the ionogenic elements. Due to the limited solubility of silicate minerals and certain species' consumption of them, their quantity in natural waters is minimal compared to the overall salt composition (in land surface waters up to 10–20 mg/L). In the form of the meta- and ortho-silicic acids H_2SiO_3 and H_4SiO_4, as well as in colloidal form of the $x\,SiO_2 \times yH_2O$ type, it may be found in fluids in a completely dissolved condition.

2.5.2 Nitrogen (N)

It is found in natural waters as both organic molecules and a variety of inorganic ions, including ammonium NH_4^+, nitrite NO_2^-, and nitrate NO_3^- (in the amino acids and proteins of organisms, and the products of their vital activity and decomposition). These may be found in water as dissolved molecules, colloidal particles, and suspended substances.

2.5.3 Phosphorus (P)

Phosphorus is an inorganic and organic chemical that exists in water as dissolved, suspended, and colloidal entities. Being an anionogenic element, phosphorus generates the neutral phosphoric acid H_3PO_4 that dissociates into many derivative forms, including H_2PO_4, HPO_4^{2-} and PO_4^{3-}, the relationship between which is influenced by the pH of water.

2.6 High-Quality Water

Water that satisfies drinking water standards set by the Environmental Protection Agency (EPA) and other organizations and is free of contaminants is considered to be of high quality. In general, there shouldn't be a lot of bacteria, viruses, heavy metals, or other chemicals in the water. It should also look good and have a pleasing flavor and aroma.

The Environmental Protection Agency (EPA) establishes guidelines for contaminants in public drinking water systems, including bacteria, lead, arsenic, and other pollutants. Depending on the type of water system, the water source, and the population served, these standards may change. For instance, the EPA mandates that public water systems test for lead, copper, and other contaminants and disinfect water using chlorine.

Some states and localities may have their own standards or requirements for the quality of drinking water in addition to the EPA guidelines. Water filters and other items that can enhance the quality of a person's water are available for purchase by both homeowners and businesses. Many communities also have water treatment facilities that can clean up contaminants and raise the standard of the water. No matter where your water comes from, it's crucial to test it frequently to make sure it satisfies the requirements for safe drinking water. For more information if you are worried about the quality of your water, get in touch with the EPA or your local health department (EPA 2020).

2.7 River's Chemical Makeup

The chemicals in the rivers in Uttarakhand depend on where the river starts, how fast it flows, and what is around it. Most rivers in Uttarakhand have different amounts of calcium, magnesium, sodium, chloride, and sulfate ions. Ions like nitrate, phosphate, potassium, bicarbonate, iron, manganese, and copper are also often found in the rivers of Uttarakhand. Some rivers in Uttarakhand may also be affected by the runoff from factories and farms, which can add chemicals like nitrates and phosphates to the water (Seth et al. 2016; Sharma et al. 2022).

The way the rivers in Uttarakhand are made up chemically also changes with altitude. For example, rivers at higher elevations tend to have more calcium, magnesium, and bicarbonate ions, while rivers at lower elevations tend to have more sodium, chloride, and sulfate ions. The chemicals in the rivers of Uttarakhand are important for the environment because they affect the rivers' health and the health of the ecosystems around them. It can also affect the health of people, since rivers are often used to get drinking water (Ruhela et al. 2018).

2.8 Various Rivers of Uttarakhand with Their Ionic Composition

The rivers of Uttarakhand with their chemical composition are as

2.8.1 Bhagirathi River

The ions calcium, magnesium, sodium, potassium, sulfate, chloride, nitrate, and bicarbonate are found in the Bhagirathi River in Uttarakhand. By preserving the pH balance, these ions help the river remain suitable for aquatic life. As some of these ions serve as nitrogen sources for the growth of algae and other types of aquatic plants, they can also aid in reducing the eutrophication of the river (Arora et al. 2017).

High concentrations of some of these ions, including nitrate and chloride, can, however, be harmful to the river. An imbalance in the river's ecology can result from an increase in these ions, which can also promote the growth of algae and other aquatic plants. This may result in lower oxygen levels, which may have an impact on aquatic life's health. Additionally affecting the quality of drinking water and rendering it dangerous for ingestion are high quantities of nitrate and chloride (Khan et al. 2022).

2.8.2 Bhilanagna River

In general, calcium and bicarbonate predominate in the chemical makeup of the Bhilangana River in Uttarakhand, India, with minor levels of magnesium, sodium, potassium, and chloride. The amounts of dissolved organic carbon and nitrate are typically low. Aquatic life can benefit from the high calcium and bicarbonate concentrations. While bicarbonate helps to buffer the water and maintain an ideal pH for aquatic life, calcium is crucial for the skeletal development of fish and other aquatic organisms (Agarwal et al. 2018).

High calcium and bicarbonate concentrations, however, can also be harmful to aquatic life. An increase in the river's hardness brought on by too much calcium can have a detrimental effect on the health of the fish. High bicarbonate concentrations can also cause carbonates to build up on the river bottom, lowering the amount of oxygen in the water. The variety of aquatic life may suffer as a result (Amoatey and Baawain 2019).

A decrease in water clarity brought on by high quantities of dissolved organic carbon can also restrict the amount of aquatic animals' habitat. Nitrate concentrations can also be an issue since they raise the possibility of eutrophication, which can cause the water to lose oxygen (Annayat et al. 2022).

2.8.3 Alaknanda River

The Alaknanda River, which originates in the Himalayas of Uttarakhand, India, is a significant tributary of the Ganges River. Calcium, magnesium, sodium, potassium, bicarbonate, sulphate, chloride, nitrate, and silicate are the main dissolved inorganic components of the Alaknanda River. Ca^{2+} (2.68 mg/L), Mg^{2+} (1.48 mg/L), Na^+ (4.22 mg/L), K^+ (0.19 mg/L), HCO_3^- (81.41 mg/L), SO_4^{2-} (2.66 mg/L), Cl^- (2.19 mg/L), NO_3^- (8.96 mg/L), and SiO_2 (4.09 mg/L) are the elements with the highest average values in the Alaknanda River.

These inorganic substances in the Alaknanda River have both beneficial and detrimental effects on the ecology. When calcium and magnesium are present, for instance, the pH of the water is balanced and fish and other aquatic creatures are protected against environmental stresses like temperature and oxygen level variations. For aquatic plants and animals, sodium and potassium are crucial minerals for growth and reproduction. Bicarbonate, sulfate, and chloride work together to control the acidity of the water and to shield aquatic life from harmful metals. The growth of phytoplankton, the main producers in aquatic habitats, depends on nitrate and silicate.

High levels of these substances, however, can contaminate and eutrophize the water. High amounts of nutrients like nitrates and phosphates can cause excessive development of aquatic plants and algae, which lowers the amount of oxygen available to other aquatic creatures, a process known as eutrophication. Fish and other aquatic species may perish as a result of this. High levels of nitrate, chloride, and sulfate can also contaminate water, rendering it unfit for consumption or other uses. In order to safeguard the aquatic ecology of the Alaknanda River, it is crucial to maintain a healthy balance of these components (Kumar et al. 2016).

2.8.4 Mandakini River

The Mandakini River in Uttarakhand has significant amounts of the following chemical elements: calcium, magnesium, sodium, potassium, chloride, sulfate, nitrate, and phosphate. These Mandakini River elements may have a significant impact on the environment (Goswami and Singh 2018).

The presence of calcium and magnesium in the river can support aquatic life, because these ions are necessary for the growth of aquatic plants and animals. Major ions in the water like sodium and potassium can make the water more electrically conductive, which can have a big impact on the aquatic ecosystem. Pollutants that can damage the ecosystem include chloride, sulfate, nitrate, and phosphate. These contaminants have the potential to significantly affect aquatic life by causing eutrophication and oxygen depletion at high concentrations (Sain et al. 2023).

2.8.5 Ganga River

The Ganga River, commonly known as the Ganges, originates in Devprayag and flows through Biyasi Rishikesh before ending in Haridwar in the Uttarakhand region. Gangotri Glacier, Satopanth Glacier, Khatling Glacier, as well as the water melting from the snow-covered peaks of Nanda Devi, Trishul, Kedarnath, Nanda Kot, and Kamet, are the sources of the Ganga River. In Uttarakhand, the Ganga River is distinguished by a variety of chemical elements. Calcium, magnesium, sodium, potassium, chloride, sulfate, nitrate, dissolved organic carbon, and total dissolved solids are important constituents. These substances have an impact on the water's quality, which is crucial for supporting aquatic life. The majority of the dissolved solids in the Ganga River—between 15 and 25%—come from calcium. Aquatic species benefit from it, because it keeps the pH balance of water. Magnesium makes up between 10 and 15% of all dissolved solids and is the second most prevalent element. It is crucial for preserving the proper nutritional balance in water and aids in controlling the water's ionic composition. Another crucial element and crucial nutrient for aquatic life is sodium. It adds between 5 and 10% of the total dissolved solids. Potassium is found in smaller concentrations and makes up about 1–3% of all dissolved solids. For keeping the osmotic balance in water, it is crucial. Another crucial element is chloride, which accounts for 1–5% of all dissolved solids. It benefits aquatic life and aids in maintaining the ionic balance of water. Sulfate is less prevalent and makes up 0.5–2% of the total dissolved solids. It benefits aquatic life and aids in keeping the pH balance of the water. The Ganga River has a significant amount of nitrate, which makes up 0.5–2% of the total dissolved solids. Due to its role in preserving the water's nutritional balance, it is crucial for the survival of aquatic life (Mukherjee et al. 2021).

The Ganga River contains higher concentrations of dissolved organic carbon, which makes up around 5–10% of the total dissolved solids. It benefits aquatic life and aids in keeping the pH balance of the water. Total dissolved solids are a key contaminant indicator that helps determine how much contamination is present in the water. The total dissolved solids in the Ganga River range from 200 to 400 mg/L. These substances have an impact on the Ganga River's water quality, which is crucial for supporting aquatic life. Reducing the amount of pollutants and other impurities in the water is crucial for maintaining the health of the Ganga River (Kumar et al. 2015; Ahmad and Chaurasia 2019).

2.8.6 Ramganga

The Ramganga River in Uttarakhand has a complicated chemical makeup that is heavily influenced by the local geology and human activity. Bicarbonates, sulfates, chlorides, nitrates, and phosphates are some of the major ions found in water. In addition, the water contains trace amounts of iron, magnesium, calcium, zinc, copper, cadmium, lead, and arsenic (Paul 2017; Khan et al. 2015).

These chemicals have mostly had a negative impact on the Ramganga River. High levels of nitrates, phosphates, and heavy metals, for instance, can lead to eutrophication, which increases the growth of aquatic vegetation, lowers oxygen levels, and increases the frequency of hazardous algae blooms. This may lead to a decline in water quality and an increase in aquatic species mortality. In addition, exposure to heavy metals poses major health concerns for people, including neurological diseases, cancer, and renal damage. High levels of nitrates and phosphates can also result in a rise in algae, which can lower the water's oxygen content and result in a decline in fish populations (Tirth et al. 2022).

2.8.7 The Yamana River

Depending on the season, the Yamuna River in Uttarakhand has a different chemical makeup. High concentrations of calcium, magnesium, chloride, nitrate, sulfate, and potassium are typically present in the water. In addition, the water has very little dissolved oxygen and is mildly acidic (Sharma and Kansal 2011).

The ecology, especially aquatic life, may suffer as a result of the Yamuna River's chemical makeup. Eutrophication, which can result in algal blooms, lower oxygen levels, and other environmental changes, is brought on by high levels of nitrates and sulfates. High calcium and magnesium levels can produce calcium carbonate deposits as well, which can clog pipes and cause a scarcity of water. In addition, aquatic life may become hazardous at high chloride levels (Sharma et al. 2020).

2.8.8 The Kali River

A significant tributary of the Ganges is the Kali River in Uttarakhand. For the people of Uttarakhand, this river is crucial, since it supplies water for drinking, agriculture, and other uses. A wide variety of aquatic life can be found in the river.

The main ions that make up the chemical makeup of the Kali River include calcium, magnesium, sodium, potassium, sulfate, chloride, nitrate, and bicarbonate ions. Trace levels of other elements like iron, copper, zinc, and manganese are also present in the river (Trivedi 2010).

These substances are found in the Kali River, and they have both beneficial and harmful effects on the ecology. On the plus side, these components supply vital nutrients and minerals to the river's aquatic life. Additionally, they support the growth of advantageous bacteria and maintain the pH balance, which help to clean and maintain the health of the water. On the downside, certain of these elements, like iron and manganese, can contaminate water at high concentrations, which is bad for human health. In addition, sulfate and chloride levels that are too high can make the water corrosive, harming pumps, pipes, and other infrastructure (Ghosh and McBean 1998).

2.9 Wetlands' Chemical Composition

Depending on the type of wetland and its location, wetlands in Uttarakhand have different chemical compositions. Wetlands typically include high levels of organic matter, nitrogen, phosphorus, and sulfur as well as trace elements including iron, manganese, copper, and zinc. Decomposing plants, fungus, bacteria, and other microbes make up the organic matter. The trace elements may have an effect on the well-being of the wetland environment, whereas the nitrogen, phosphorus, and sulfur are vital nutrients for aquatic life. Some wetlands may have significant amounts of particular chemical substances including arsenic, mercury, and lead. In order to maintain the ecosystem's health, it is crucial to keep an eye on the chemical makeup of wetlands (Mitsch and Gosselink 2000; Sharma et al. 2018).

2.9.1 Availability of Wetland's Geotagging Data and Class of Wetlands

The Uttarakhand State Council for Science and Technology (UCOST), on the other hand, has started mapping the state's wetlands, and its findings should be made public soon. The goal of this project is to create a 1:50,000 scale map of the state's wetlands, and then categorize them according to the criteria established by the Ramsar Convention. In-depth data about the state of Uttarakhand's wetlands are available in the atlas form. Details about the wetland's location, size, type, and other attributes are included. There are many different kinds of wetlands listed in the atlas, such as marshes, ponds, lakes, reservoirs, oxbows, swamps, floodplains, and rivers (Bachheti et al. 2023).

The Uttarakhand Forest Department also keeps a database of the state's wetlands called the State Wetland Database. Details about the wetland's location, size, and composition are included. Marshes, ponds, lakes, reservoirs, oxbows, swamps, floodplains, and rivers are only some of the wetland types represented in the database.

There is a wealth of data about the state's wetlands available from the Uttarakhand State Biodiversity Board as well. The document details the wetland's precise location, as well as its kind, size, and other attributes. Marshes, ponds, lakes, reservoirs, oxbows, swamps, floodplains, and riverine wetlands are all represented in this catalogue. The wetlands of Uttarakhand are also depicted on the UK Land Cover Map. Details on the wetland's location, kind, extent, and other attributes are included. This map depicts a variety of wetland ecosystems, such as marshes, ponds, lakes, reservoirs, oxbows, swamps, floodplains, and rivers (Patakamuri 2013).

2.9.2 Wetland Coding and Classification System

Ramsar classification and the Indian Wetland Classification System serve as the foundation for Uttarakhand, India's wetland coding and classification system (IWCS). The system is made to recognize and classify wetlands based on their purpose and significance to the environment. In Uttarakhand, there are four primary categories of wetlands:

1. Natural Wetlands: These include lakes, ponds, marshes, swamps, estuaries, mangrove forests, coral reefs, lagoons, and wetlands created artificially by the construction of dams, reservoirs, and canals.
2. Human-made Wetlands: These comprise built-in wetland structures including sewage treatment plants, agricultural fields, industrial locations, and other man-made bodies of water utilized for aquaculture or irrigation.
3. Transitional Wetlands: Also known as "intermittent wetland systems," they are regions that provide as a transition between wetland and dryland ecosystems.
4. Wetlands that are particularly crucial to the preservation of biodiversity and ecosystem services are known as ecologically important wetlands.

Based on their size, hydrological properties, and other characteristics, these four categories are further subdivided (Bassi et al. 2014; Sharma and Kumar 2017).

2.10 Lake's Chemical Makeup

There are more than 500 glacial lakes in Uttarakhand. The geological and hydrological characteristics of these lakes' catchment areas determine their chemical composition. In general, the lakes have high levels of silica, magnesium, sodium, potassium, and calcium ions. They are also abundant in nutrients including nitrogen, phosphate, and sulfur as well as organic materials. Additionally, the lakes have large concentrations of suspended materials, primarily clay particles, which can change the hue and transparency of the water. The lakes may also contain significant amounts of heavy metals, including lead, arsenic, mercury, and cadmium, which are harmful to aquatic life. The amount of glacial meltwater that the Himalayan glacial lakes get also has an impact on their chemical makeup. Due to higher levels of dissolved carbon dioxide, lake water may turn acidic in the presence of glacial meltwater. Aluminum, iron, and manganese are just a few of the trace elements that the meltwater may add to the lake's water. Pollution from human activities can potentially change the lakes' chemical composition. Agricultural runoff, industrial waste, sewage, and other contaminants can contaminate the lakes. These contaminants may raise the concentrations of nutrients, organic matter, and heavy metals in the lake, which may be hazardous to aquatic life (Kumar and Sharma 2019; Babuji et al. 2023).

2.11 Effect of Chemical Constituents Present in Water of Uttarakhand on Human Beings, Marine Biota, and Environment

2.11.1 Human Beings

Uttarakhand water is generally high in minerals, which can lead to a range of health benefits. Examples of these minerals include calcium, magnesium, potassium, sodium, and iron. These minerals can help improve blood circulation, reduce blood pressure, and can improve the overall metabolism of the body. However, it can also contain pollutants and contaminants, such as pesticides, heavy metals, and industrial chemicals, which can lead to health risks if consumed in large quantities.

2.11.2 Marine Biota

The high mineral content of water in Uttarakhand can help improve the health of marine biota by providing essential nutrients. However, it can also contain pollutants, such as pesticides, heavy metals, and industrial chemicals, which can lead to toxicity and other health risks if consumed in large quantities.

2.11.3 Environment

The presence of pollutants and contaminants in the water can lead to environmental damage if not managed properly. For example, high levels of heavy metals can lead to eutrophication, which can reduce the amount of oxygen in the water and damage aquatic ecosystems. Additionally, industrial chemicals can pollute the air, land, and water and can lead to soil and water contamination (Lin et al. 2022; PR 2020).

2.12 Conclusion

The chemical composition of the water in rivers, lakes, and wetlands in Uttarakhand varies greatly depending on the water's point of origin, the time of year, and the location in which it is located. In addition to main cations like calcium, magnesium, and potassium and trace metals, the water typically contains dissolved organic and inorganic compounds including nitrates, phosphates, sulfates, and chloride. Dissolved particles and suspended sediments in the water may also be a source of toxic metals. In addition, eutrophication can lead to a rise in nutrient and contaminant levels in the water. The ecological systems of the rivers, lakes, and wetlands of

Uttarakhand are vulnerable to the effects of these substances; thus, they must be carefully monitored and maintained.

References

Agarwal NK, Singh H, Singh A, Singh G (2018) Bhilangana river regulation for Tehri hydro power project in Central Himalaya: impact on planktonic assemblages. Biojournal 13(1):1–10

Ahmad I, Chaurasia S (2019) Water quality index of Ganga River at Kanpur (U.P.). Thematics J Geogr 8(11):66–77

Amoatey P, Baawain MS (2019) Effects of pollution on freshwater aquatic organisms. Water Environ Res 91(10):1272–1287

Annayat N, Dixit M, Pani S (2022) Impact of nutrient enrichment on water quality of a tropical Ramsar wetland. Int J Agron Agric Res 20(2):47–62

Arora R, Joshi HC, Pandey IP, Tewari VK (2017) Assessment of water quality of Bhagirathi from Gangotri to Rishikesh using RS and GIS techniques. Rasayan J Chem 10(4):1167–1183

Ašperger S (2004) Chemical kinetics and reaction mechanisms. In: Chemical kinetics and inorganic reaction mechanisms, pp 3–103

Babuji P, Thirumalaisamy S, Duraisamy K, Periyasamy G (2023) Human health risks due to exposure to water pollution: a review. Water 15(14):2532

Bachheti A, Arya AK, Tyagi S, Deepti. (2023) An assessment of wetland and water-dependent avifaunal diversity in selected wetlands of Uttarakhand, Western Himalayas, India. Biol Bull 50(6):729–735

Baroni L, Cenci L, Tettamanti M, Berati M (2007) Evaluating the environmental impact of various dietary patterns combined with different food production systems. Eur J Clin Nutr 61(2): 279–286

Bassi N, Kumar MD, Sharma A, Pardha-Saradhi P (2014) Status of wetlands in India: a review of extent, ecosystem benefits, threats and management strategies. J Hydrol Reg Stud 2:1–19

Brezonik PL, Arnold WA (2012) Water chemistry: fifty years of change and progress. Environ Sci Technol 46(11):5650–5657

Brini E, Fennell CJ, Fernandez-Serra M, Hribar-Lee B, Lukšič M, Dill KA (2017) How water's properties are encoded in its molecular structure and energies. Chem Rev 117(19):12385–12414

Chaplin MF (2019) Structure and properties of water in its various states. In: Fundamentals of water, chemistry, particles, and ecology

Ghosh NC, McBean E (1998) Water quality modeling of the Kali River, India. Water Air Soil Pollut 102(1):91–103

Goswami G, Singh D (2018) Water quality and function of Mandakini River ecosystem of Central Himalaya. Int J Biosci 12(6):102–116

Ho AY (1972) Surface water. In: water: a comprehensive treatise. Plenum Press, New York, pp 35–54

Jain CK, Bandyopadhyay A, Bhadra A (2010) Assessment of ground water quality for drinking purpose, district Nainital, Uttarakhand, India. Environ Monit Assess 166:663–676

Joshi DM, Kumar A, Agrawal N (2009) Studies on physicochemical parameters to assess the water quality of river Ganga for drinking purpose in Haridwar district. Ras J Chem 2:195–203

Khan MYA, Gani KM, Chakrapani G (2015) Assessment of surface water quality and its spatial variation: a case study of Ramganga River, Ganga Basin, India. Arab J Geosci 9(1):1–9

Khan AA, Pant NC, Joshi R, Devara PCS (2022) Chemical and isotopic variability of Bhagirathi river water (Upper Ganga), Uttarakhand, India. In: Ecological significance of river ecosystems: challenges and management strategies, pp 133–146

Kılıç Z (2021) Water pollution: causes, negative effects and prevention methods. J Inst Sci Technol 3(2):129–132

Kumar A (2020) Wavelet and statistical analysis of dissolved oxygen and biological oxygen demand of Ramganga river water. Asian J Res Rev Phys:45–50

Kumar R, Sharma RC (2019) Assessment of the water quality of Glacier-fed lake Neel Tal of Garhwal Himalaya, India. Water Sci:22–28

Kumar A, Bisht BS, Joshi VD, Singh AK, Talwar A (2010) Physical, chemical and bacteriological study of water from rivers of Uttarakhand. J Hum Ecol 32(3):169–173

Kumar P, Kaushal RK, Nigam AK (2015) Assessment and management of Ganga River water quality using multivariate statistical techniques in India. Asian J Water Environ Pollut 12(4): 61–69

Kumar P, Pandey R, Singh H, Raghuvanshi D, Tripathi B, Shukla U, Shukla DN (2016) Physico-chemical characteristics of River Alaknanda. Int J Curr Res 8(12):43354–43358

Lin L, Yang H, Xu X (2022) Effects of water pollution on human health and disease heterogeneity: a review. Front Environ Sci 10:879804

Mitsch WJ, Gosselink JG (2000) Wetlands, 3rd edn. Wiley, New York

Mukherjee P, Kumar P, Gupta SK, Kumar R (2021) Seasonal variation in physicochemical parameters and suitability for various uses of Bouli pond water, Jharkhand. Water Sci:125–135

Nilsson A, Pettersson LGM (2015) The structural origin of anomalous properties of liquid water. Nat Commun 6:8998

Oluyemi EA, Adekunle AS, Adenuga AA, Makinde WO (2010) Physico-chemical properties and heavy metal content of water sources in Ife North local government area of Osun state, Nigeria. Afr J Environ Sci Technol 4(10):691–697

Patakamuri SK (2013) Land use/Land cover map of Uttarakhand for 2010. PANGAEA, Bremerhaven

Paul D (2017) Research on heavy metal pollution of river Ganga: a review. Ann Agrar Sci 15(2): 278–286

PR R (2020) Water environment pollution with its impact on human diseases in India. Int J Hydrol 4(4):152–158

Quattrini S, Pampaloni B, Brandi ML (2016) Natural mineral waters: chemical characteristics and health effects. Clin Cases Miner Bone Metab 13(3):173–180

Rittmann BE, McCarty PL (2001) Environmental biotechnology: principles and applications. McGraw-Hill, New York

Ruhela M, Paritosh K, Tyagi V, Ahamad F, Ram K (2018) Assessment of water quality of River Ganga at Haridwar with reference to water quality index. Environ Conserv J 19(3):47–58

Sain K, Mehta M, Kumar V (2023) Avalanche hazards around Kedarnath temple, Mandakini river valley, Uttarakhand-A case study. J Geol Soc India 99(2):173–176

Schwarzenbach RP, Gschwend PM, Imboden DM (2003) Environmental organic chemistry. John Wiley, New York

Seth R, Mohan M, Singh P, Singh R, Dobhal R, Singh KP, Gupta S (2016) Water quality evaluation of Himalayan Rivers of Kumaun Region, Uttarakhand, India. Appl Water Sci 6:137–147

Sharma D, Kansal A (2011) Water quality analysis of River Yamuna using water quality index in the national capital territory, India (2000–2009). Appl Water Sci 1:147–157

Sharma RC, Kumar R (2017) Water quality assessment of sacred glacial Lake Satopanth of Garhwal Himalaya, India. Appl Water Sci 7:4757–4764

Sharma RC, Chaudhary S, Kumar R, Singh S, Tiwari V, Kumari R (2018) Assessment of physico-chemical parameters of Himalayan Wetland Deoria Tal. Eur Acad Res VI(1):1–9

Sharma R, Kumar R, Satapathy SC, Al-Ansari N, Singh KK, Mahapatra RP, Agarwal AK, Le HV, Pham BT (2020) Analysis of water pollution using different physicochemical parameters: a study of Yamuna River. Front Environ Sci 8:1–13. Article 623932

Sharma MK, Kumar P, Prajapati P, Bhanot K, Wadhwa U, Tomar G, Goyal R, Prasad B, Sharma B (2022) Study of hydrochemical and geochemical characteristics and solute fluxes in Upper Ganga Basin, India. J Asian Earth Sci: X 8:100108

Tirth V, Singh RK, Tirth A, Islam S (2022) Assessment of water quality of Ramganga river in Moradabad, India. Water Qual Assess Manag India:257–272

Theodore L, Dupont RR, Reynolds J (2022) Pollution prevention solutions. Pollut Prev:147–313

Trivedi RC (2010) Water quality of the Ganga River – an overview. Aquat Ecosyst Health Manag 13(4):347–351

U.S. Environmental Protection Agency (2020) Requirements for public water systems. https://www.epa.gov/sdwa/requirements-public-water-systems

Microbial Diversity of Cold-Water Reservoirs and Their Prospective Applications

3

Vidhi Jain, Prashant Arya, Shivani Maheshkumar Yagnik, Vikram Hiren Raval, and Namita Ashish Singh

Abstract

Microorganisms including bacteria, archaea, viruses, fungi, and protists have evolved into every possible niche on the globe and now dominate the living biomass in marine environments. This chapter highlights the bacterial, fungal as well as viral diversity of microbes isolated from cold-water reservoir viz. lakes, ocean, glaciers, sea, etc. The next section focuses on the applications of microbes isolated from the cold-water reservoirs mainly in bioremediation, food industry, textile industry, and pharmaceuticals. Psychrophilic enzymes, that is, amylase, cellulase, and protease, are used in textile, detergent, cosmetics, and leather industry for different purposes. Primary and secondary metabolites produced by cold-water reservoir microbes play major role in bioremediation of hydrocarbons and development of novel drugs, which are very useful against antimicrobial resistance (AMR) pathogens. Current research on microbes that can survive in the cold and on their biomolecules has amply shown the enormous potential of psychrophiles.

Keywords

Cold-water reservoir · Diversity · Bioremediation · Metabolites · Probiotic · Drug

---

V. Jain · N. A. Singh (✉)
Department of Microbiology, University College of Science, Mohanlal Sukhadia University, Udaipur, Rajasthan, India
e-mail: namita.singh@mlsu.ac.in

P. Arya · S. M. Yagnik · V. H. Raval
Department of Microbiology and Biotechnology, University School of Sciences, Gujarat University, Ahmedabad, Gujarat, India

© The Author(s), under exclusive license to Springer Nature Singapore Pte Ltd. 2023
R. Soni et al. (eds.), *Current Status of Fresh Water Microbiology*,
https://doi.org/10.1007/978-981-99-5018-8_3

3.1 Introduction

More than 70% of the earth's surface is covered by cold-water reservoirs, including seas, oceans, lakes, rivers, ponds, and glaciers. These reservoirs support majority of the planet's biomass and play a substantial role in the global cycles of matter and energy that are necessary for life to continue on earth (Hunter-Cevera et al. 2005). The biosphere is dominated by a cold environment known as the "cryosphere" (area >33 million km^2) (Boetius et al. 2015). Microbes in the sea are most likely the source of all life on earth. They serve as the foundation of life on earth representing the richest assets and molecular diversity in nature (Sean and Jack 2015). They are referred to as multicultural tiny creatures possessing a vast range of metabolism that are cosmopolitan in nature, that is, widely distributed in air, water, soil, the sea, mountains, hot springs, and also in the bodies of living plants and animals, including humans (Onen et al. 2020). Marine microorganisms enable the storage, transfer, and turnover of essential biological materials in addition to producing the organic matter and oxygen necessary to support life. Over 50% of the oxygen produced on earth is produced by marine phototrophic microorganisms, including cyanobacteria, diatoms, and pico- and nanophytoplankton.

Ecosystem biodiversity has become a topic of rigorous study; subsequently, a great deal of information has been gathered on the distribution of microorganisms around the world. In addition, there is growing interest in the role of marine microorganisms in biogeochemical processes, biotechnology, pollution, and health (Poli et al. 2017). The immense diversity of microorganisms also gives rise to a huge amount of genetic data, bioactive substances, and biomaterials that are largely untapped but could have significant societal benefits and applications, such as improving medical treatments, probiotics and bioremediation applications, the supply of energy, and the development of industrial products and processes (Glöckner et al. 2012). Since microbes constitute the basis of life, they are vital to the habitability and sustainability of our planet and represent a largely unexplored supply of novel bioactive substances and metabolic pathways that might be utilized for new biotechnological applications and products (York 2020).

3.2 Diversity of Microbes Isolated from Cold Water Reservoirs

Microbial diversity encompasses a vast collection of variability in all kinds of microorganisms like fungi, bacteria, and viruses in the natural world vital for the continuation and conservation of global genetic resources. The structure and dynamics of food webs, global biogeochemical cycles, and the remineralization of organic matter rely on marine microbial communities (Parvathi et al. 2020; Wei et al. 2020). Psychrophiles are microorganisms capable of growing at very low temperatures ranging from -12 to 20 °C, where the optimum temperature of 15 °C shows the best growth (Sensoy 2021).

Marine bacteria, fungi, and other microorganisms build up exceptional metabolic and physiological capabilities that allow them to endure in hostile habitats and to

3 Microbial Diversity of Cold-Water Reservoirs and Their Prospective Applications 51

generate compounds that might not be formed by their earthly counterparts (Shaaban 2022). The microbial diversity of various cold-water reservoirs is listed in Table 3.1 and explained in following sections.

3.2.1 Microbial Diversity of Oceans and Seas

The abyssal plain referred to as the deep-sea is typically sited between 3000 and 6000 m deep in the global ocean. The deep sea is composed of an extensive diversity of ecosystems and is the biggest and most distant biome of the biosphere (Corinaldesi 2015). Light intensity declines with depth within this zone, and photo-synthetic yield is halted due to the absence of light. The temperatures nearing freezing and lack of organic matter at the ocean floor are the primary factors of the deep sea that control benthic productivity and biomass (Jørgensen and Boetius 2007). Microbes are the crucial members of the biota present and are believed to play a significant function in mineralization of organic matter, which in turn impacts the global biogeochemical cycles and carbon sequestration capacity of oceans. The Archaea, which degrade detrital proteins by secreting enzymes like proteases and peptidases, are one of the predominant microbial life forms found in marine sediments (Schippers et al. 2012). Bacteria and Archaea, which represent around 90% of the total benthic biomass in deep-sea sediments, contribute the highest proportion of taxonomic richness and productivity (Barnes et al. 2021).

In the sediments of Atlantic ocean and the deeper depths of the Arctic and Pacific Oceans, *Proteobacteria* is found to be dominant (Schauer et al. 2009). In the Pacific Ocean, seasonal variations have been noted, with *Bacteroidetes* being more prevalent than *Proteobacteria* in the winter (Suh et al. 2014). Sixty-four clones collected from the top surface sediments were divided into *Firmicutes* and *Gamma-proteobacteria* in a study by Khandeparker et al. (2014). *Bacillus* was the most prevalent among the *Firmicutes*, followed by *Dolosigranulum*. The dominant *Gamma-proteobacteria* was *Pseudomonas*, *Shigella*, and occasionally *Escherichia*. *Gamma-proteobacteria*, represented by *Pseudomonas* and *Enterobacteriaceae*, predominated the 58 clones isolated from deeper sediments, followed by *Beta-proteobacteria*, represented by *Limnobacter* and *Burkholderiales*. Shao et al. (2015) isolated 51 bacteria from the deep seawater column above the South-West Indian Ridge that belonged to Alpha-proteobacteria within the genera *Alterierythrobacter*, *Citricella*, *Erythrobacter*, *Kaistia*, *Lutibacterium*, *Maricaulis*, *Martelella*, *Mesorhizobium*, *Novosphingobium*, *Pseudomonas*, *Phenylobacterium*, *Roseovarius*, *Rhodobacter*, *Salipiger*, *Stappia*, *Sphingopyxis*, *Sphingomonas*, *Tistrella*, and *Thalassospira*, Gamma-proteobacteria within the genera *Alkaligens*, *Alkanovorax*, *Halomonas*, *Idiomarina*, *Marinobacter*, *Pseudoidiomarina*, and *Pseu-domonas*, *Muricauda* and *Salegentibacter* (*Bacteroides*), *Bacillus* (Firmicutes) and *Microbacterium* (Actinobacteria). A novel thermophilic sulfur-reducing bacterium, *Desulfurobacteium indicum*, was reported by Cao et al. (2017) from hydrothermal vent in the Indian Ocean. Wang et al. (2018) characterized the bacterial diversity from rare-earth-elements-rich sediment in the Indian Ocean and revealed occurrence

Table 3.1 Microbial diversity of cold-water reservoirs

Cold-water reservoir and its location	Diversity type	Abundant phyla/genera	References
Polythermal glacier, Svalbard, Norway	Bacterial	*Proteobacteria, Firmicutes, Actinobacteria*	Amato et al. (2007)
Lakes in Schirmacher Oasis, Antarctica		*Proteobacteria, Actinobacteria, Bacteroidetes, Fusobacteria, Verrucomicrobia*, and *Chlorobi*	Mojib et al. (2009)
Surface snow, glacial stream, and moraine lake in Yala glacier, Nepal		*Bacteroidetes* and *Betaproteobacteria, followed by Actinobacteria*	Liu et al. (2011)
Kafni glacier, India		*Proteobacteria, Bacteroidetes, Actinobacteria, Chloroflexi, Spirochaetae, Tenericutes, Verrucomicrobia*	Srinivas et al. (2011)
Continental Antarctica and Antarctic Peninsula		*Actinobacteria, Bacteroidetes, Proteobacteria, Firmicutes, Deinococcus-Thermus*	Peeters et al. (2012)
Austrian Alps		*Proteobacteria, Bacteroidetes, Actinobacteria, Verrucomicrobia, Nitrospira, Cyanobacteria*	Wilhelm et al. (2013)
Pacific Ocean	Fungal	*Ascomycota, Basidiomycota*	Xu et al. (2014)
Jiulong River Estuary, China	Viral	*Pelagibacter phage HTVC010P, Puniceispirillum phage HMO-2011, Thalassomonas phage BA3*	Cai et al. (2016)
Alpine glacial-fed lake, Western Tibet, Kalakuli	Bacterial	*Actinobacteria, Proteobacteria, Verrucomicrobia, Firmicutes, Planctomycetes*	Liu et al. (2017)
Tietê River, Brazil	Fungal	*Ascomycota, Basidiomycota, Chytridiomycota, Glomeromycota, Zygomycota, Neocallimastigomycota*	Ortiz-Vera et al. (2018)
East China Sea (ECS)		*Byssochlamys, Aspergillus*	Li et al. (2018)
Lake Baikal	Viral	*Myoviridae, Poxviridae, Mimiviridae, Siphoviridae, Phycodnaviridae*, and *Podoviridae*	Butina et al. (2019)
Revelva River and Lake Revvatnet, Southwestern Spitsbergen, Norway	Bacterial	*Proteobacteria, Actinobacteria, Bacteroidetes, Firmicutes, Verrucomicrobia, Tenericutes, Cyanobacteria*, and *Acidobacteria*	Kosek et al. (2019)
Antarctic freshwater	Fungal	*Ascomycota, Basidiomycota, Mortierellomycota, Mucoromycota, Chytridiomycota, Oomycota*	Ogaki et al. (2019)
Lake Magadi, Kenya		*Aspergillus, Penicillium, Acremonium, Phoma, Cladosporium, Septoriella, Talaromyces, Zasmidium, Chaetomium, Aniptodera, Pyrenochaeta, Septoria*	Orwa et al. (2020)

(continued)

3 Microbial Diversity of Cold-Water Reservoirs and Their Prospective Applications 53

Table 3.1 (continued)

Cold-water reservoir and its location	Diversity type	Abundant phyla/genera	References
Gulf of Gabès, Southern Mediterranean Sea	Bacterial	*Firmicutes, Proteobacteria, Bacteroidetes, Chloroflexi*	Jeddi et al. (2022)
Lake area of Huaibei, China		*Pseudomonadota, Bacteroidales, Chloroflexales, Acidobacteriales*, and *Firmicutes*	Shen et al. (2022)

of 49 different phyla, among which the most abundant bacteria were *Proteobacteria*, with Gamma-proteobacteria being noted in all sections of the core; this was followed by *Firmicutes, Actinobacteria, Bacteroidetes, Cyanobacteria*, and *Chloroflexi. Lactobacillus, Profundibacterium, Shigella, Escherichia, Pseudoalteromonas, Vibrio, Propionibacterium, Alteromonas, Enterobacter, Sphingomonas*, and *Staphylococcus* were the major genera isolated. Gawas et al. (2019) identified 43 heterotrophic bacteria from Central Indian Ocean Basin including members of phyla *Proteobacteria, Actinobacteria*, and *Firmicutes*. This was one of the earliest reports of *Oceanobacillus* (Firmicutes*)* and *Brachybacterium* (Actinobacteria). *Bacillus subtilis* strain (G7) was isolated by Gu et al. (2019) from deep-sea hydrothermal vent. From deep-sea sediments of the Indian Ocean, primarily from the Bay of Bengal and the Andaman Sea, Padmanaban et al. (2019) recovered 34 bacterial isolates associated with *Firmicutes, Proteobacteria*, and *Actinobacteria*, with *Firmicutes* and *Actinobacteria* being the major phyla. Bottom water samples collected from Northwest Indian Ocean reported presence of a novel planctomycete, *Gimesia benthica* (Wang et al. 2020). Qiu et al. (2021) reported a halotolerant *Halomonas sedimenti* from the deep sea sediment of the Southwest Indian Ocean. A novel hydrogen- and sulfur-oxidizing chemolithoautotroph, *Sulfovorum indicum* was reported by Xie et al. (2019) from deep sea hydrothermal plumes in the Northwestern Indian Ocean. Yaylacı (2021) isolated 13 strains of *Bacillus* species from cagewater aquaculture. In a study from different sites at varying depths of the South Indian Ocean, bacterial species belonging to *Proteobacteria, Firmicutes*, and *Chloroflexi* along with uncultured species of archaebacteria belonging to the phyla *Thaumarchaeota* and *Euryarchaeota* were identified (Zhu et al. 2022).

In all natural ecosystems, including fresh and marine waters from the ocean surface to the deep sea, eutrophic to ultra-oligotrophic lakes, lagoons, rivers, ground waters, melting water, and glacier ice, fungi are omnipresent microorganisms that are a part of the microbiota (Grum-Grzhimaylo et al. 2015; Mokhtarnejad et al. 2016). *Aspergillus sydowii*, in both nonsporulating and sporulating forms, was reported by Raghukumar et al. (2004) in the Chagos trench in the Indian Ocean. Damare et al. (2006) isolated 181 fungal cultures from 5000 m depth in the Central Indian Ocean Basin and found that most were terrestrial sporulating species belonging to *Aspergillus, Penicillium*, and *Cladosporium* genera. Various studies from deep-sea environments including hydrothermal vents and the Mariana Trench have reported fungal isolates (Nagahama et al. 2008). Singh et al. (2010) isolated 16 filamentous and 12 yeast cultures, belonging to *Ascomycota* and *Basidiomycota*, from sediments of the Central Indian Ocean Basin at depths of 5000 m. This was the first account of the isolation of filamentous fungi (*Capronia, Exophiala, Sagenomella,* and *Tilletiopsis*) from deep-sea sediments, demonstrating the unique diversity of fungi that inhabit the Indian Ocean basin. A report on fungal diversity in the Indian Ocean sediments published by Singh et al. (2011) revealed 32 fungal taxa including majorly *Ascomycota* and *Basidiomycota.* Within the *Ascomycota,* members of *Sordariomycetes, Dothideomycetes,* and *Saccharomycetes* and members belonging to *Tremellomycetes, Microbotryomycetes,* and *Ustilaginomycetes* were identified within the *Basidiomycota.* Zhang et al. (2014) conducted investigation on the Eastern Indian Ocean and found that filamentous fungi such as *Aspergillus, Penicillium, Simplicillium, Cladosporium,* and *Phoma* dominated the culturable percentage. *Ascomycota* and *Chytridiomycota* were found to be the two primary phyla of the fungal diversity in the coastal water and sediment samples from North Carolina, coupled with a significant number of novel sequences (Picard 2017). According to a study by Fotedar et al. (2018), basidiomycetous and ascomycetous yeasts were approximately equally prevalent in hypersaline Inland Sea in Qatar. The three main phyla found in the deep-sea sediments of the Indian Ocean are *Ascomycota, Basidiomycota,* and *Zygomycota* (Xu et al. 2019). In a recent study, the abyssal zones of the Indian Ocean yielded fungal phylotypes primarily from *Aspergillus, Penicillium, Ophiocordyceps,* and *Phoma* (Tang et al. 2020). The fungal diversity and composition of the Continental Solar Saltern in Añana Salt Valley was studied by Azpiazu-Muniozguren et al. (2021), and a total of 380 fungal genera were detected. *Saccharomyces* was highly prevalent in the saltern, although other halotolerant and halophilic fungi like *Wallemia, Cladosporium,* and *Trimmatostroma* were also present.

In cold seep sediments of the Gulf of Mexico, it was discovered that virus-to-prokaryote ratios and viral-like particle counts were substantially greater than in nearby sediments, indicating that these habitats may be hot spots for viruses (Kellogg 2010). Double-stranded DNA (dsDNA) viruses, particularly bacteriophages, make up the majority of virioplankton communities in the ocean, according to previous virome-based studies (Etiope et al. 2014). Novel viruses have also been identified in methane seep sediments, including a novel sister lineage to the *Microvirus* genus of *Enterobacteria* phage and a putative archaeal virus linked to an anaerobic methane-oxidizing clade (Bryson et al. 2015). In the Antarctic surface oceanic area, *Caudovirales* makes up to 72% of the total dsDNA virus community (Kennicutt 2017). Five surface and two bottom water samples from Prydz Bay were analyzed by Gong et al. (2018), and the findings showed that the majority of the DNA viruses were dsDNA viruses. The nucleocytoplasmic large DNA viruses (*Phycodnaviridae, Mimiviridae,* and *Pandoraviridae* viruses) were most prevalent in the bottom water, while *Caudovirales* (*Siphoviridae, Myoviridae,* and *Podoviridae*) phages were most prevalent in the surface seawater. *Microviridae, Inoviridae, Cellulophaga,* and *Flavobacterium* phages were found among the ssDNA viruses. Wu et al. (2020) reported that the majority of viruses they found in the surface water of the East China Sea had the characteristic head-and-tail form. These viruses belonged to *Siphoviridae, Myoviridae,* and *Podoviridae* of the

Caudovirales order. Deep sea sediments associated with cold seeps studied by Li et al. (2021) revealed the presence of *Caudovirales* order, specifically *Podoviradae*, *Myoviradae*, and *Siphoviradae*.

3.2.2 Microbial Diversity of Lakes, Rivers, and Ponds

The study of inland waterways, such as lakes, rivers, ponds, and reservoirs, is crucial for determining how human activity and climate change affect ecosystem architecture. Due to their abundance, diversity, and metabolic activity, these microprobes dominate aquatic environments and play a key role in numerous ecological processes (Savio et al. 2015). In freshwater lakes with available food, microorganisms adapt to the conditions and exist in every zone (horizontal and vertical). Microbes have a significant impact on the various biogeochemical (C, N, P, S, and other elements) cycling pathways and global energy fluxes of deep reservoirs and small lakes in freshwater environments (Savvichev et al. 2018; Liu et al. 2019b).

In comparison to Lake Urmia, which primarily contained *Proteobacteria*, *Firmicutes*, and *Actinobacteria*, the culturable microbial diversity of the Aran-Bidgol saltlake contained isolates from the genera *Halorubrum*, *Haloarcula*, *Salinibacter*, *Salicola*, and *Rhodovibrio* (Kashi et al. 2014). In their study of the bacterial diversity and abundance in pond water in Hubei Province, China, Qin et al. (2016) discovered that *Proteobacteria*, *Cyanobacteria*, *Bacteroidetes*, and *Actinobacteria* dominated the microbial communities, while potential pathogens such as *Acinetobacter*, *Aeromonas*, and some probiotics like *Comamonadaceae* unclassified and *Bacillales* unclassified were also present. After examining the microbial diversity in the hypersaline Lake Meyghan in Iran, Naghoni et al. (2017) discovered that *Haloarchaea* predominated in the high salinity brines, while bacteria dominated the low salinity brines, with *Alteromonadales* (Gammaproteobacteria) being a particularly significant taxon. Touka et al. (2018) conducted an analysis of the Ancient European Lake, The Lake Pamvotis, and found that the bacterial community comprised primarily *Proteobacteria* (β-, γ-, δ- and α-Proteobacteria), followed by phylotypes belonging to *Cyanobacteria*, *Nitrospirae*, *Acidobacteria*, *Bacteroidetes*, *Firmicutes*, *Spirochaetes*, *Planctomycetes*, *Actinobacteria*, *Gemmatimonadetes*, and most of the sequences of Archaea belonged to *Euryarchaeota*. *Bacillus shivajii* sp. nov. was discovered by Kumar et al. (2018) in a water sample from the Indian Salt Lake Sambhar. In their study of the microbial diversity in river water and sediment in Hong Kong, Deng et al. (2018) found that the major phyla in the sediments were *Proteobacteria*, *Bacteroidetes*, *Cyanobacteria*, and *Acidobacteria*. The bacterial communities contained considerable proportions of *Crenarchaeota*, *Actinobacteria*, *Chloroflexi*, *Verrucomicrobia*, and *Planctomycetes*. In two renaturalized quarry lakes in Singapore, Kumar et al. (2019a) analyzed the microbial communities in the water and sediment and discovered that *Proteobacteria*, *Actinobacteria*, *Cyanobacteria*, *Firmicutes*, and *Bacteroidetes* represented for about 90% of the total reads. *Verrucomicrobia*, *Planctomycetes*, *Nitrospirae*, *Chloroflexi*, *Spirochaetes*, *Acidobacteria*, *Gemmatimonadetes*, and

Fusobacteria were the other main phyla, while *Thaumarchaeota, Crenarchaeota,* and *Euryarchaeota* were the three principal phyla that made up the domain Archaea. Bachran et al. (2019) studied the microbial diversity from Arava Valley and found that the bacterial diversity was mainly represented by the genus *Salinimicrobium* of the order *Flavobacteriales* within the phylum *Bacteroidetes,* from the gammaproteobacterial orders *Alteromonadales* and *Oceanospirillales* as well as members from the order *Bacillales* of the phylum *Firmicutes. Euryarchaeal Halobacteria* from the orders *Halobacteriales, Haloferacales,* and *Natrialbales* dominated the Archaeal diversity. Luo et al. (2020) studied the bacterial community structure along the Lancang River in southwest China and found out that *Proteobacteria* (primarily classes of *Alphaproteobacteria, Betaproteobacteria,* and *Gammaproteobacteria*), *Bacteroidetes, Actinobacteria, Planctomycetes,* and *Firmicutes* were present in the sediment and water samples. Six archaeal phyla and a total of 28 bacterial phyla were spread throughout all samples from the three sites (Lakes Jackson and Talquin, and Pedrick Pond). *Actinobacteria, Proteobacteria, Cyanobacteria, Planctomycetes, Bacteroidetes,* and *Verrucomicrobia* are the primary phyla across all samples, according to changes in the relative abundance of dominating bacteria, while *Euryarchaeota* predominates in the archaeal phyla (Betiku et al. 2021).

According to a study by Sharma et al. (2016), 98% of the isolates found in Lonar Lake belonged to the phylum *Ascomycota,* subphylum *Pezizomycotina.* Rojas-Jimenez et al. (2017) examined the freshwater fungi that were found in ice-covered lakes in the McMurdo Dry Valleys of continental Antarctica and found sequences that, in descending order of dominance, represented taxa of the phyla *Cryptomycota, Chytridiomycota, Ascomycota, Basidiomycota,* traditional *Zygomycota,* and *Blastocladiomycota.* In their investigation on the diversity of fungi at the bottom of Lake Michigan and Lake Superior, Wahl et al. (2018) found that *Dothideomycetes, Eurotiomycetes, Leotiomycetes,* and *Sordariomycetes* accounted for about 84% of all isolates grouped by genus. A study by de Souza et al. (2021) evaluated the fungal diversity in two lakes of the South Shetland Islands and found 34 fungal taxa of the phyla *Ascomycota, Basidiomycota, Mortierellomycota, Chytridiomycota* and *Rozellomycota.* Mwirichia (2022) explored the diversity of fungi in the soda lakes of Magadi, Elmenteita, Sonachi, and Bogoria in Kenya and discovered 107 genera belonging to the phyla *Ascomycota, Basidiomycota,* and *Glomeromycota,* including *Aspergillus, Penicillium, Acremonium, Phoma, Cladosporium, Septoriella, Talaromyces, Zasmidium, Chaetomium, Aniptodera, Pyrenochaeta, Septoria, Juncaceicola, Paradendryphiella, Phaeosphaeria, Juncaceicola,* and *Biatriospora.*

According to a survey done in the Jiulong River Estuary that connects to Xiamen Sea harbour, the two most prevalent phages were HTVC010P and HMO-2011, and *Caudovirales* was the dominant viral group in the viromes (Cai et al. 2016). The viral diversity of Lake Baikal's shoreline water was researched by Butina et al. (2019), who found that the virotypes belonged to six families (*Myoviridae, Poxviridae, Mimiviridae, Siphoviridae, Phycodnaviridae* and *Podoviridae*).

3.2.3 Microbial Diversity of Glaciers

A glacier is a sizable, slowly moving deposit of hard ice that has developed over the course of thousands of years of snow accumulation. The cryosphere, which is a vital component and sensitive indicator of climatic and environmental changes, includes the glacier as one of its constituent parts. Glaciers are regarded authentic biomes (Anesio and Laybourn-Parry 2012), in which microbial life persists despite a hostile environment and plays a crucial part in the functioning of the glacial ecosystems (Hodson et al. 2008). Glaciers harbor autotrophic, chemolithotrophic, and heterotrophic bacteria because of the intense solar radiation and low oxygen levels in the atmosphere (Lutz et al. 2017). These microorganisms have been specifically reported in antarctic regions, polar arctic regions, and in high mountains (Larose et al. 2013; Hotaling et al. 2017).

Branchy bacterium sp., *Acinetobacter* sp. and *Agrococcus* sp. were among the various bacteria identified by Zhang et al. (2007) from the glacial ice of the Himalayas' East Rongbuk (ER) core. The study on bacterial diversity in a glacier foreland of high arctic on Ny-Å lesund (West Spitsbergen, Norway) by Schutte et al. (2010) found *Acidobacteria, Chlamydia, Nitrospira, Chloroflexi, Bacteroidetes, Proteobacteria, Firmicutes, Spirochaetes, Actinobacteria, Cyanobacteria, Verrucomicribia*, and *Drinococcus-Thermus* but *Spirosoma, Sphingomonas, Terromonas, Hymenobacter, Gemmatimonas, Brevundimonas*, and *Sphingopyxis* as dominated ones. *Proteobacteria, Bacteroidetes, Actinobacteria, Chloroflexi, Actinobacteria, Firmicutes, Gemmatimonadetes* and *Verucomicrobia* were found to be the most prevalent bacterial phyla in a study by Huang et al. (2013) on East Antarctica, whereas *Deinococcus-Thermus, Nitrospira, Candidate Division OP10, Planctomycetes, Candidate Division TM7*, and *Fusobacteria* showed comparatively lower abundance in snow meltwater samples. Peter and Sommaruga (2016) revealed the presence of *Sphingobacteria, Flavobacteria* and *Betaproteobacteria* as the abundant phyla on the Austrian Central Alps at glacier melting. In Livingston Island of Maritime Antarctica, Hodson et al. (2017) discovered *Proteobacteria, Bacteroidetes, Firmicutes, Acidobacteria, Cyanobacteria, Actinobacteria*, and *Verrucomicrobia* as the dominant bacterial phyla. The Garhwal Himalaya is very wealthy in terms of the occurrence of glaciers which includes one of the biggest, the Gangotri glaciers. The psychrophilic microbial diversity of glaciers in the Garhwal Himalaya, India was studied by Kumar and Sharma (2021) and they found that most of the bacteria and actinomycetes isolated from the ice samples of the glaciers (Satopanth, Bhagirathi-Kharak and Gangotri) were gram positive and most of them were in form of rods, non-spore forming, non-pigmented and with filamentous branching. *Vibrio harveyi, Pseudomonas fluorescens, Microbacterium paraoxydans, Microbacterium scheliferi, Serratia marcescens, Paenibacillus azatofixans, Ralstonia eutropha*, and *Staphylococcus cohnii* were the bacterial species isolated, the actinomycetes included *Microbacterium avium, Streptomyces rangoon, Arthrobacter sulfonivorans* and fungal species included *Aspergillus nidulans, Cladosporium cladosporioides, Verticillium nubilum, Curvularia lunata* and *Phanerochaete chrysosporium*. Glaciers represent a key linkage between coasts and their downstreaming tidewaters and are of immense significance in land-to-ocean fluxes. Garcia-Lopez et al. (2019) studied the microbial communities in coastal Glaciers and Tidewater Tongues of Svalbard Archipelago, Norway and found that the glacier microorganisms mainly corresponded to the phylum *Proteobacteria*, *Bacteroidetes* and *Cyanobacteria* and the seawater microorganisms belonged to *Bacteroidetes*, *Actinobacteria* and *Proteobacteria*. Campen et al. (2019) carried out study on Taylor Glacier (Antarctica) that revealed *Proteobacteria*, *Bacteroidetes* and *Actinobacteria* were the dominant bacterial phyla. In a glacial-fed Tibetan lake, bacteriological study conducted by Liu et al. (2019a, b) revealed *Bacteroidetes*, *Actinobacteria*, *Proteobacteria* (Alpha and Beta), *Firmicutes* and *Cyanobacteria* were dominant phyla in the samples. Sharma et al. (2020) carried out research in the Polar Regions and the Tibetan plateau and concluded that *Proteobacteria*, *Bacteroidetes*, *Cyanobacteria*, *Firmicutes*, *Verrucomicrobia*, and *Actinobacteria* were the most dominant bacterial phyla. Study of two Tibetan Plateau ice cores revealed common glacier-ice lineages including *Janthinobacterium*, *Polaromonas*, *Herminiimonas*, *Flavobacterium*, *Sphingomonas*, and *Methylobacterium* as the dominant genera (Zhong et al. 2021).

3.3 Application of Microbes Isolated from Cold Water Reservoirs

Over 90% of the oceans have temperatures of 5 °C or lower and the potential use of psychrophilic bacteria as biotechnological tools remained a promise for decades. Despite actuality, they were more well known for the damage they inflicted, such as the spoilage of refrigerated food. Certainly, psychrophiles (and the metabolites they produce) are currently regarded as sustainable and priceless resources for the development of a wide range of biotechnological processes and/or products, many of which are already protected by patents or are subject to other forms of industrial secrets (Feller 2013; Arya et al. 2022).

The main areas of research and development of new processes and/or products related with psychrophilic microorganisms are listed below:

- Psychrophilic enzymes (food technology, molecular biology, bioremediation, and medicine)
- Psychrophilic biomolecules (dietary supplements, therapeutic, cosmetic, bioremediation, nutrition, and other uses)
- Pharmaceutical and medical uses generally focused on screening of new antibiotics, anticancer drugs, and cosmeceutical products
- Biostimulation or bioaugmentation in bioremediation: degrading contaminants such as unintentional petroleum spills or old waste disposal techniques

The microbial diversity of cold-water reservoirs and its applications are depicted in Fig. 3.1 and described in following sections:

Fig. 3.1 Microbial diversity of cold-water reservoirs and their prospective applications in different fields

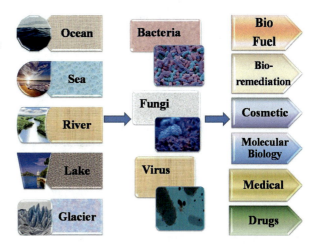

3.3.1 Bioremediation

A wide range of microbes may use hydrocarbons as their exclusive source of carbon and energy, that is, biodegradation. Regional psychrophilic microbial populations in cold habitats accomplish for low-temperature biodegradation of organic pollutants. They convert or mineralize organic contaminants into less toxic, nonhazardous chemicals (mostly through hydrocarbon bioremediation), which are subsequently incorporated into the biogeochemical cycles of the environment (Brakstad 2008; Filler et al. 2008). Cold climates present unique difficulties for hydrocarbon degraders, increased viscosity of liquid hydrocarbons, including decreased enzymatic reaction rates, decreased volatility of toxic compounds, restricted bioavailability of contaminants, low nutrient levels, and occasionally extreme salinity and pH (Margesin 2004; Aislabie et al. 2006). Various bioremediation techniques have been used to increase microbial activity during pollutant breakdown at cold temperature.

Ex-situ bioremediation and in situ bioremediation, two generally defined approaches, have been widely used for pollution remediation in cold climates, according to the implementation locations. The most popular ex situ bioremediation methods comprise biopiles, bioreactors, composting, and land farming, whereas in situ bioremediation includes natural attenuation, biostimulation, phytoremediation, bioaugmentation, and bioventing. There are some drawbacks to using psychrophilic in bioremediation in addition to their benefits. Before a bioremediation technique is implemented, a number of elements must be evaluated. Frigid temperatures, restricted access to contaminated areas in a hostile environment, and Geographic isolation make it difficult to successfully remove toxins. Therefore, while developing a soil treatment plan, it is important to take into account the primary limiting elements including pH, temperature, salinity, hydrocarbon bioavailability, nutrient availability, microbial population, soil moisture, and oxygen level (Chaudhary and Kim 2019).

A specific genetically modified organisms (GMO) psychrophile is renowned for bioremediating organic pollutants (such as hydrocarbons and toluene) from cold environments. It was developed by transforming the TOL plasmid from *Pseudomonas putida*. Examples include the recombinant expression in the Antarctic *Pseudoalteromonas haloplanktis* of the gene encoding a monooxygenase involved in the breakdown of aromatic hydrocarbons from the mesophile *Pseudomonas stutzeri* (Siani et al. 2006). Due to their capacity to break down large quantities of organic compounds in a little period of time at cold elevations, psychrophilic bacteria and fungi also function as inocula for wastewater treatment. One such cold-adapted *Arthrobacter psychrolactophilus* strain increased the biodegradability of organic molecules while completely clarifying a synthetic wastewater turbid medium by hydrolyzing proteins, carbohydrates, and lipids (Gratia et al. 2009). Further evidence is the complete phenol degradation by psychrophilic bacteria (*Rhodococcus* sp.) and yeasts (e.g., *Rhodotorula psychrophenolica*) at 10 °C in fed-batch cultivation (Margesin et al. 2005).

Owing to their physiological adaption to cold temperatures, psychrophilic enzymes are also utilized in bioremediation. Despite those challenging circumstances, cold-active enzymes, which are ten times more active than mesophilic enzymes, help to break down xenobiotics. They demonstrate the three benefits of biotechnology listed below:

1. High activity: A full reaction may be achieved with a lower dose of psychrophilic enzymes, which lowers overall experiment costs.
2. Cold activity: Since they continue to function at room temperature, processes do not require high temperatures.
3. They are heat labile and so quickly become inactive at higher temperatures.

These psychrophilic microorganisms can be utilized in the appropriate consortia throughout the winter to remove contaminants from soil or wastewater. They have the capacity to operate effectively in cold habitats. This enzyme's broad-spectrum activity makes it acceptable for use in a variety of industrial processes as well as in the bioremediation of soil that had been contaminated by effluents from food-processing plants (Kumar et al. 2019a, b).

3.3.2 As a Source of Novel Drugs

Through synthesizing and excreting secondary metabolites, many free-living microbes can colonize a certain habitat by eliminating, impeding, or distracting their rivals and/or predators (Mojib et al. 2010). The majority of actinomycetes, bacteria, and fungi are recognized for producing a variety of secondary metabolites that are significant for medicine, including antibiotics, antifungal, antiprotozoal, antiviral, and antihelminthic substances (Baltz 2007). Numerous psychrophilic microorganisms have been demonstrated to possess such antibacterial properties recently, including *Candida albicans* and methicillin-resistant *Staphylococcus aureus* that were suppressed by spore-forming and cold-adapted bacteria (Vollú et al. 2014). However, certain unprocessed extracts of various Antarctic microorganisms have antibacterial properties, such as ethanolic extracts of Antarctic fungus and organic extracts of Antarctic bacteria (Gonçalves et al. 2015; Godinho et al. 2015). Gesheva and Vasileva-Tonkova (2012) isolated glycolipids and lipopeptides (antimicrobial/antifungal compounds) from *Nocardioides* sp. A-1 (halophilic Antarctic actinomycete) that inhibit growth of *Candida tropicalis*, *Bacillus subtilis*, *S. aureus*, *Sarcinalutea*, and *Xanthomonas oryzae*. Actinomycetes bioactive metabolites prevent the growth of *Pseudomonas aeruginosa*, methicillin-resistant *S. aureus*, and *C. albicans* (Lee et al. 2012). Gram-positive and Gram-negative phytopathogens are both inhibited by the metabolites generated by *Streptomyces* species, including aromatic polyketides, glycopeptides, and polyenes (Encheva-Malinova et al. 2014; Giudice and Fani 2015). The antifungal and antimicrobial properties of Amphotericin B from *Penicillium nalgiovense* Laxa are comparable. It inhibits the growth of *C. albicans* and *S. aureus* (Svahn et al. 2015). In 2013, Bhattarai et al. isolated the pseudodepsidone metabolite, an antibacterial chemical from *Stereocaulonalpinum* lichens that inhibits the growth *of S. aureus* and *B. subtilis*. Similar to this, the cold-tolerant killer toxin from the *Mrakiafrigida* yeast has antifungal action against the fungi *Metschnikowiabicuspidata*, *C. tropicalis*, and *Candida albicans* (Hua et al. 2010; Liu et al. 2012).

Since many years ago, colored bacteria have also been known to produce antimicrobial substances. For example, *Janthinobacterium* sp., which was discovered in Antarctica in 1991, produces violacein, a violet pigment with multiple therapeutic properties, including antitumoral, antibacterial, antileishmanial, and antiviral activities (Shivaji et al. 1991; Soliev et al. 2011). According to this, *Mycobacterium smegmatis* and *M. tuberculosis* development are inhibited by *Janthinobacterium* sp. Ant5-2 (Mojib et al. 2010). This inhibitory action was linked to additional substances that may work in concert with the pigment and was not only a result of the pigment being present. Antarctic yeasts of the *Dioszegia, Cryptococcus, Exophiala, Microglossum*, and *Rhodotorula* genera produce pigments such as carotenoids and mycosporines to defend themselves from UV exposure (Vaz et al. 2011). Melanin, a different photoprotective pigment, has been isolated from the bacteria *Lysobacter oligotrophicus* (Kimura et al. 2015). An enzyme from *Chlamydomonas* sp. called photolyase is also employed to protect against light (Li et al. 2015).

Some marine isolates have also been found to produce bioactive compounds. For example, *Burkholderia cepacia* is entirely inhibited from growing by *Pseudoalteromonas* strains obtained from the Ross Sea through the production of different volatile organic compounds (VOCs) (lipophilic chemicals) (Lo Giudice and Fani 2015). Parallel to this, bacteria found in sponges produced microbial VOCs that have antibacterial effects on *S. aureus, P. aeruginosa, Xanthomonas campestris*, and *Clavibacter michiganensis* (Chávez et al. 2015). Similarly, four asterric acid nitro derivatives from *Pseudogymnoascus* sp. that were isolated by Figueroa et al. (2015) have potential uses in medicine. High and specialized antibacterial activities were exhibited by fungi isolated from different microalgae. For example, the dangerous fungi *Cladosporium sphaerospermum* and *Trypanosoma cruzitrypomastigotes* were both suppressed by *Penicillium* sp. (Godinho et al. 2013). Furthermore, a *Penicillium steckii* ethanoic extract reduced interferon alpha (IFN-) activity by 68% and yellow fever virus proliferation by 96% (Furbino et al. 2014).

Numerous research organizations have worked on the antioxidant capabilities of natural products in light from some promising findings suggesting that regular eating foods high in antioxidants might prevent or delay the development of such diseases. With similar approach, current research efforts have been concentrated on the quest for novel microbial metabolites with improved antioxidant properties from cold biological resources (Yarzábal 2016). For example, lobaric acid (depsidone), which was derived from the lichens *S. alpinum* and *Cladonia* sp., shows high antioxidant capabilities and gastroprotective effects (Bhattarai et al. 2013). Also, ramalin [γ-glutamyl-*N'*-(2-hydroxyphenyl) hydrazide] from *Ramalinaterebrata* (lichen) (Paudel et al. 2011) and β-carotene, ergosterol, torulene, torularhodin, and CoQ10 from *Sporobolomyces salmonicolor* AL1 (yeast) (Dimitrova et al. 2013) inhibit the tyrosinase enzyme activity with low cytotoxicity in human. Some microbes with their metabolites along with its applications are summarized in Table 3.2.

3.3.3 Probiotic Potential

Probiotics are defined by the Food and Agriculture Organization of the United Nations (FAO) and World Health Organization (WHO) as "live microorganisms which when administered in adequate amounts confer a health benefit on the host." According to the U.S. Food and Drug Administration (FDA), the basic guidelines for storage include a working refrigerator temperature at or below 4 °C, while the freezer compartment should be −18 °C (Mojikon et al. 2022). Therefore, all household and laboratory refrigerator temperatures should operate at 4 °C as a means to prevent or slow down the growth of foodborne pathogens such as *Salmonella, L. monocytogenes*, and *C. botulism*. Low temperatures slow down metabolism and extend the log or stationary phase of the mesophilic bacterial growth curve (Madigan et al. 2005; Mojikon et al. 2022).

As probiotics, psychrophiles are believed to have potential for use as dietary supplements in aquaculture to improve health and nutrition of livestock. While studies in this are scarce, their adaptation to low temperatures is suggested to be beneficial for a more efficient utilization in the marine habitat as compared to currently available terrestrial and/or moderate temperature adapted probiotic organisms (Collins and Margesin 2019). *Bacillus*, Lactic acid bacteria (LAB), and *Bifidobacteria* have been intensively employed as probiotic strains due to their recognition as members of the indigenous microflora of the animals, safety, and the evidence supporting their positive role (Lee et al. 2012). Marine cold-adapted bacterium, *Psychrobacternamhaensis* SO89, is used as putative dietary probiotic in Nile tilapia (*Oreochromisniloticus*) feeds (Makled et al. 2017). Biomel, GT's Organic Kombucha, KeVita Apple Cider Vinegar Tonics, and VitaCup Immunity

3 Microbial Diversity of Cold-Water Reservoirs and Their Prospective Applications 63

Table 3.2 Applications of microbes isolated from cold water reservoirs

Microorganism	Metabolite	Application	References
Streptomyces sp. LPB2019K190-4	–	Antimicrobial	Pereliaeva et al. (2022)
Rhodococcus sp. LPB2019K201-3			
Microbacterium sp. LPB2019K198-2			
Aeromonas veronii NS07	Amylase	Detergent and textile industries	Bauvois et al. (2008)
Marine Bacterium Strain YS-80-122	Protease	Peeling of leather	Wang et al. (2010)
Pseudoalteromonas haloplanktis	Cellulase	Biodegradation	Violot et al. (2005)
Rhodococcus erythropolis BZ4	Primary and secondary metabolites	Bioremediation (*n*-alkanes, phenol, anthracene, pyrene)	Margesin et al. (2013)
Rhodococcus cercidiphyllus BZ22			
Arthrobacter sulfureus BZ73			
Pimelobacter simplex BZ91			
Pseudomonas peli N3–6P	Primary and secondary metabolites	Bioremediation (petroleum hydrocarbons)	Cai et al. (2014)
Streptomyces venezuelae N3-6A			
Halomonas variabilis N3-8A			
Rhodococcus phenolicus N1–1P			
Bacillus subtilis N2–3P			
Acinetobacter oleivorans P7-1A			
Rhodococcus erythropolis N2–4P			
Rhodococcus erythropolis P6–5P			
Alcanivorax venustensis N3-7A			
Exiguobacterium antarcticum N4–1P			
Acinetobacter calcoaceticus P9-1A			
Roseovarius NS163			

Table 3.2 (continued)

Microorganism	Metabolite	Application	References
Halomoans NS159 *Glaciecola* NS168	Primary and secondary metabolites	Bioremediation (crude oil, TPH, PAH)	Chronopoulou et al. (2015)
Nocardioides sp. A-1	Glycolipids and lipopeptides (antibiotics)	Antimicrobial	Gesheva and Vasileva-Tonkova (2012)
Janthinobacterium sp. Ant5-2	Purple violet pigment	Antimicrobial	Huang et al. (2012)
Flavobacterium sp. Ant342	Flexirubin	Antimicrobial	Mojib et al. (2010)
Pseudoalteromonas sp.	Microbial volatile compounds	Antimicrobial	Maida et al. (2015)
Guehomyces pullulans	Crude and ethanolic extract	Antimicrobial	Furbino et al. (2014)
Metschnikowia australis			
Dipodascus australiensis			
Pseudogymnoascus sp.			
Penicillium steckii			
Geomyces sp.	Methanol crude extracts	Antimicrobial	Henríquez et al. (2014)
Aspergillus sydowii	Organic extracts	Antimicrobial	Godinho et al. (2015)
Penicillium allii-sativi			
Penicillium chrysogenum			
Penicillium rubens			
Stereocaulon alpinum	Lobaric acid, lobastin	Antimicrobial	Bhattarai et al. (2013)
Dioszegia, Cryptococcus, Exophiala	Carotenoids and mycosporines	UV protection	Vaz et al. (2011)
Microglossum, Rhodotorula			
Janthinobacterium sp.	Violacein (purple violet pigment)	Antimicrobial	Mojib et al. (2011, 2013)
		UV protection, cryoprotection	
Ramalinaterebrata	Usimine-like compounds	Antiaging (fibroblast cell proliferation)	Lee et al. (2010)

Coffee Pods contain *Bacillus coagulans*, which can reproduce by spore formation. The use of spores is beneficial in the manufacturing and storage process of probiotic drinks, as the *B. coagulans* spores are resistant to heat, cold temperatures, and stomach acid and remain dormant in the juice but germinate once it passes through the gastric compartment (Mojikon et al. 2022). *Lactobacillus algidus* sp. nov. was isolated from vacuum-packaged refrigerated beef (Kato et al. 2000). *L. casei, L. plantarum*, and *L. delbrueckii* were isolated and identified as probiotic culture (Yoon et al. 2006). Kasimin et al. (2020) isolated two psychrophiles strains of bacteria (*Lactobacillus* sp. strains CA1 and CA4) capable of producing antimicrobial substances from dairy products and raw milk.

3.4 Conclusion and Future Perspectives

Microbes isolated from cold reservoirs produces more novel bioactive primary and secondary metabolites compounds viz. polypeptides, macrolactones, polyketides, isocoumarins, etc., compared to terrestrial. Although majority of bacteria, fungi, and some viruses isolated from cold-water reservoirs produce metabolites with potential biotechnological applications. Nevertheless, one of the main challenges to guarantee a sustainable use of cold biological resources is still related to how to allow positive benefits from bioprospecting without incurring significant harm to the environment. Uncontrolled prospecting activities, including the logistical efforts necessary to access the natural resources, and the impact associated with them could prove in fact to be unsustainable and to severely compromise Antarctic microbial habitats. On the other hand, commercially oriented research activities and the secrecy surrounding them may compromise one of the main pillars.

The psychrophiles were formerly considered as potential resources for the advancement of several biotechnologies. Today, a significant portion of this promise has materialized, in part due to the research of psychrophilic microbes. This promise looks to be far bigger than for thermophiles when taking into consideration the broader psychrophilic biodiversity, which includes bacteria, plants, and animals, as well as the extensive regions of application. The sustainable and intelligent use of these biological and genetic resources can help other disciplines of knowledge, such as bioremediation, food, agriculture, nanotechnology, medicine, and energy production. However, the sustainable exploitation of these resources necessitates the creation of new regulations to mitigate the potential environmental damage that unregulated prospecting operations may create. Last but not least, the majority of psychrophile biotechnological applications are energy and environmentally friendly, both of which are becoming more and more important.

Acknowledgment The authors of the respective universities are grateful to their authorities for supporting this work.

References

Aislabie J, Saul DJ, Foght JM (2006) Bioremediation of hydrocarbon-contaminated polar soils. Extremophiles 10(3):171–179. https://doi.org/10.1007/s00792-005-0498-4

Amato P, Hennebelle R, Magand O, Sancelme M, Delort AM, Barbante C, Boutron C, Ferrari C (2007) Bacterial characterization of the snow cover at Spitzberg, Svalbard. FEMS Microbiol Ecol 59(2):255–264. https://doi.org/10.1111/j.1574-6941.2006.00198.x

Anesio AM, Laybourn-Parry J (2012) Glaciers and ice sheets as a biome. Trends Ecol Evol 27:219–225. https://doi.org/10.1016/j.tree.2011.09.012

Arya PS, Yagnik SM, Panchal RR, Rajput KN, Raval VH (2022) Industrial applications of enzymes from extremophiles. In: Physiology, genomics, and biotechnological applications of extremophiles. IGI Global, Philadelphia, pp 207–232. https://doi.org/10.4018/978-1-7998-9144-4.ch010

Azpiazu-Muniozguren M, Perez A, Rementeria A, Martinez-Malaxetxebarria I, Alonso R, Laorden L, Gamboa J, Bikandi J, Garaizar J, Martinez-Ballesteros I (2021) Fungal diversity and composition of the Continental Solar Saltern in Añana Salt Valley (Spain). J Fungi (Basel) 7(12):1074. https://doi.org/10.3390/jof7121074

Bachran M, Kluge S, Lopez-Fernandez M, Cherkouk A (2019) Microbial diversity in an arid, naturally saline environment. Microb Ecol 78(2):494–505. https://doi.org/10.1007/s00248-018-1301-2

Baltz RH (2007) Antimicrobials from Actinomycetas: back to the future. Microbe 2:125–133

Barnes NM, Damare SR, Shenoy BD (2021) Bacterial and fungal diversity in sediment and water column from the Abyssal Regions of the Indian Ocean. Front Mar Sci 8:687860. https://doi.org/10.3389/fmars.2021.687860

Bauvois C, Jacquamet L, Huston AL, Borel F, Feller G, Ferrer JL (2008) Crystal structure of the cold-active aminopeptidase from *Colwellia psychrerythraea*, a close structural homologue of the human bifunctional leukotriene A4 hydrolase. J Biol Chem 283(34):23315–23325. https://doi.org/10.1074/jbc.M802158200

Betiku OC, Sarjeant KC, Ngatia LW, Aghimien MO, Odewumi CO, Latinwo LM (2021) Evaluation of microbial diversity of three recreational water bodies using 16S rRNA metagenomic approach. Sci Total Environ 771:144773. https://doi.org/10.1016/j.scitotenv.2020.144773

Bhattarai HD, Kim T, Oh H, Yim JH (2013) A new pseudodepsidone from the Antarctic lichen *Stereocaulon alpinum* and its antioxidant, antibacterial activity. J Antibiot (Tokyo) 66(9): 559–561. https://doi.org/10.1038/ja.2013.41

Boetius A, Anesio AM, Deming JW, Mikucki JA, Rapp JZ (2015) Microbial ecology of the cryosphere: sea ice and glacial habitats. Nat Rev Microbiol 13(11):677–690. https://doi.org/10.1038/nrmicro3522

Brakstad OG (2008) Natural and stimulated biodegradation of petroleum in cold marine environments. In: Psychrophiles: from biodiversity to biotechnology. Springer, Berlin, pp 389–407

Bryson SJ, Thurber AR, Correa AM, Orphan VJ, Vega Thurber R (2015) A novel sister clade to the enterobacteria microviruses (family Microviridae) identified in methane seep sediments. Environ Microbiol 17:3708–3721. https://doi.org/10.1111/1462-2920.12758

Butina TV, Bukin YS, Krasnopeev AS, Belykh OI, Tupikin AE, Kabilov MR, Sakirko MV, Belikov SI (2019) Estimate of the diversity of viral and bacterial assemblage in the coastal water of Lake Baikal. FEMS Microbio Lett 366(9):fnz094. https://doi.org/10.1093/femsle/fnz094

Cai Q, Zhang B, Chen B, Zhu Z, Lin W, Cao T (2014) Screening of biosurfactant producers from petroleum hydrocarbon contaminated sources in cold marine environments. Mar Pollut Bull 86(1-2):402–410. https://doi.org/10.1016/j.marpolbul.2014.06.039

Cai L, Zhang R, He Y, Feng X, Jiao N (2016) Metagenomic analysis of Virioplankton of the Subtropical Jiulong River Estuary, China. Viruses 8(2):35. https://doi.org/10.3390/v8020035

Campen R, Kowalski J, Lyons WB, Tulaczyk S, Dachwald B, Pettit E, Welch KA, Mikucki JA (2019) Microbial diversity of an Antarctic subglacial community and high-resolution replicate sampling inform hydrological connectivity in a polar desert. Environ Microbiol 21(7): 2290–2306. https://doi.org/10.1111/1462-2920.14607

Cao J, Birien T, Gayet N, Huang Z, Shao Z, Jebbar M, Alain K (2017) Desulfurobacterium indicum sp. nov., a thermophilic sulfur-reducing bacterium from the Indian Ocean. Int J Syst Evol Microbiol 67(6):1665–1668. https://doi.org/10.1099/ijsem.0.001837

Chaudhary DK, Kim J (2019) New insights into bioremediation strategies for oil-contaminated soil in cold environments. Int Biodeterior Biodegrad 142:58–72. https://doi.org/10.1016/j.ibiod.2019.05.001

Chávez R, Fierro F, García-Rico RO, Vaca I (2015) Filamentous fungi from extreme environments as a promising source of novel bioactive secondary metabolites. Front Microbiol 6:903. https://doi.org/10.3389/fmicb.2015.00903

Chronopoulou PM, Sanni GO, Silas-Olu DI, van der Meer JR, Timmis KN, Brussaard CP, McGenity TJ (2015) Generalist hydrocarbon-degrading bacterial communities in the oil-polluted water column of the North Sea. J Microbial Biotechnol 8(3):434–447. https://doi.org/10.1111/1751-7915.12176

Collins T, Margesin R (2019) Psychrophilic lifestyles: mechanisms of adaptation and biotechnological tools. Appl Microbiol Biotechnol 103(7):2857–2871. https://doi.org/10.1007/s00253-019-09659-5

Corinaldesi C (2015) New perspectives in benthic deep-sea microbial ecology. Front Mar Sci 2:17. https://doi.org/10.3389/fmars.2015.00017

Damare S, Raghukumar C, Raghukumar S (2006) Fungi in deep-sea sediments of the Central Indian Basin. Deep Sea Res Part I Oceanogr Res Pap 53:14–27. https://doi.org/10.1016/j.dsr.2005.09.005

Deng WJ, Li N, Ying GG (2018) Antibiotic distribution, risk assessment, and microbial diversity in river water and sediment in Hong Kong. Environ Geochem Health 40(5):2191–2203. https://doi.org/10.1007/s10653-018-0092-1

Dimitrova S, Pavlova K, Lukanov L, Korotkova E, Petrova E, Zagorchev P, Kuncheva M (2013) Production of metabolites with antioxidant and emulsifying properties by Antarctic strain *Sporobolomyces salmonicolor* AL1. Appl Biochem Biotechnol 169(1):301–311. https://doi.org/10.1007/s12010-012-9983-2

Encheva-Malinova M, Stoyanova M, Avramova H, Pavlova Y, Gocheva B, Ivanova I, Moncheva P (2014) Antibacterial potential of *Streptomycete* strains from Antarctic soils. Biotechnol Biotechnol Equip 28(4):721–727. https://doi.org/10.1080/13102818.2014.947066

Etiope G, Panieri G, Fattorini D, Regoli F, Vannoli P, Italiano F, Locritani M, Carmisciano C (2014) A thermogenic hydrocarbon seep in shallow Adriatic Sea (Italy): gas origin, sediment contamination and benthic foraminifera. Mar Pet Geol 57:283–293. https://doi.org/10.1016/j.marpetgeo.2014.06.006

Feller G (2013) Psychrophilic enzymes: from folding to function and biotechnology. Scientifica (Cairo) 2013:512840. https://doi.org/10.1155/2013/512840

Figueroa L, Jiménez C, Rodríguez J, Areche C, Chávez R, Henríquez M, de la Cruz M, Díaz C, Segade Y, Vaca I (2015) 3-Nitroasterric acid derivatives from an Antarctic sponge-derived *Pseudogymnoascus* sp. fungus. J Nat Prod 78(4):919–923. https://doi.org/10.1021/np500906k

Filler DM, Snape I, Barnes DL (2008) Bioremediation of petroleum hydrocarbons in cold regions. Cambridge University Press, Cambridge. https://doi.org/10.1017/CBO9780511153595

Fotedar R, Kolecka A, Boekhout T, Fell JW, Al-Malki A, Zeyara A, Al Marri M (2018) Fungal diversity of the hypersaline Inland Sea in Qatar. Bot Mar 61(6):595–609. https://doi.org/10.1515/bot-2018-0048

Furbino LE, Godinho VM, Santiago IF, Pellizari FM, Alves TM, Zani CL, Junior PA, Romanha AJ, Carvalho AG, Gil LH, Rosa CA, Minnis AM, Rosa LH (2014) Diversity patterns, ecology and biological activities of fungal communities associated with the endemic macroalgae across the Antarctic peninsula. Microb Ecol 67(4):775–787. https://doi.org/10.1007/s00248-014-0374-9

Garcia-Lopez E, Rodriguez-Lorente I, Alcazar P, Cid C (2019) Microbial Communities in Coastal Glaciers and Tidewater Tongues of Svalbard Archipelago, Norway. Front Mar Sci 5:512. https://doi.org/10.3389/fmars.2018.00512

Gawas VS, Shivaramu MS, Damare SR, Pujitha D, Meena RM, Shenoy BD (2019) Diversity and extracellular enzyme activities of heterotrophic bacteria from sediments of the Central Indian Ocean Basin. Sci Rep 9(1):9403. https://doi.org/10.1038/s41598-019-45792-x

Gesheva V, Vasileva-Tonkova E (2012) Production of enzymes and antimicrobial compounds by halophilic Antarctic *Nocardioides* sp. grown on different carbon sources. World J Microbiol Biotechnol 28(5):2069–2076. https://doi.org/10.1007/s11274-012-1009-2

Giudice AL, Fani R (2015) Cold-adapted bacteria from a coastal area of the Ross Sea (Terra Nova Bay, Antarctica): linking microbial ecology to biotechnology. Hydrobiologia 761(1):417–441. https://doi.org/10.1007/s10750-015-2497-5

Glöckner FO, Stal LJ, Sandaa RA, Gasol JM, O'Gara F, Hernandez F, Labrenz M, Stoica E, Varela MM, Bordalo A, Pitta P (2012) Marine microbial diversity and its role in ecosystem functioning and environmental change, Marine Board Position Paper 17. Marine Board-ESF, Ostend. https://doi.org/10.13140/RG.2.1.5138.6400

Godinho VM, Furbino LE, Santiago IF, Pellizzari FM, Yokoya NS, Pupo D, Alves TM, Junior PA, Romanha AJ, Zani CL, Cantrell CL, Rosa CA, Rosa LH (2013) Diversity and bioprospecting of fungal communities associated with endemic and cold-adapted macroalgae in Antarctica. ISME J 7(7):1434–1451. https://doi.org/10.1038/ismej.2013.77

Godinho VM, Gonçalves VN, Santiago IF, Figueredo HM, Vitoreli GA, Schaefer CE, Barbosa EC, Oliveira JG, Alves TM, Zani CL, Junior PA, Murta SM, Romanha AJ, Kroon EG, Cantrell CL, Wedge DE, Duke SO, Ali A, Rosa CA, Rosa LH (2015) Diversity and bioprospection of fungal community present in oligotrophic soil of continental Antarctica. Extremophiles 19(3):585–596. https://doi.org/10.1007/s00792-015-0741-6

Gonçalves VN, Carvalho CR, Johann S, Mendes G, Alves T, Zani CL, Junior PA, Murta SM, Romanha AJ, Cantrell CL, Rosa CA (2015) Antibacterial, antifungal and antiprotozoal activities of fungal communities present in different substrates from Antarctica. Polar Biol 38(8): 1143–1152

Gong Z, Liang Y, Wang M, Jiang Y, Yang Q, Xia J, Zhou X, You S, Gao C, Wang J, He J, Shao H, McMinn A (2018) Viral diversity and its relationship with environmental factors at the surface and deep sea of Prydz Bay, Antarctica. Front Microbiol 9:2981. https://doi.org/10.3389/fmicb.2018.02981

Gratia E, Weekers F, Margesin R, D'Amico S, Thonart P, Feller G (2009) Selection of a cold-adapted bacterium for bioremediation of wastewater at low temperatures. Extremophiles 13(5): 763–768. https://doi.org/10.1007/s00792-009-0264-0

Grum-Grzhimaylo AA, Georgieva ML, Bondarenko S, Debets AJM, Bilanenko EM (2015) On the diversity of fungi from soda soils. Fungal Divers 76:27–74. https://doi.org/10.1007/s13225-015-0320-2

Gu HJ, Sun QL, Luo JC, Zhang J, Sun L (2019) A first study of the virulence potential of a *Bacillus subtilis* isolate from deep-sea hydrothermal vent. Front Cell Infect Microbiol 9:183. https://doi.org/10.3389/fcimb.2019.00183

Henríquez M, Vergara K, Norambuena J, Beiza A, Maza F, Ubilla P, Araya I, Chávez R, San-Martín A, Darias J, Darias MJ, Vaca I (2014) Diversity of cultivable fungi associated with Antarctic marine sponges and screening for their antimicrobial, antitumoral and antioxidant potential. World J Microbiol Biotechnol 30(1):65–76. https://doi.org/10.1007/s11274-013-1418-x

Hodson A, Anesio AM, Tranter M, Fountain A, Osborn M, Priscu J, Sattler B (2008) Glacial ecosystems. Ecol Monogr 78(1):41–67

Hodson AJ, Nowak A, Cook J, Sabacka M, Wharfe ES, Pearce DA, Vieira G (2017) Microbes influence the biogeochemical and optical properties of maritime Antarctic snow. Eur J Vasc Endovasc Surg 122(6):1456–1470. https://doi.org/10.1002/2016jg003694

Hotaling S, Hood E, Hamilton TL (2017) Microbial ecology of mountain glacier ecosystems: biodiversity, ecological connections and implications of a warming climate. Environ Microbiol 19:2935–2948. https://doi.org/10.1111/1462-2920.13766

3 Microbial Diversity of Cold-Water Reservoirs and Their Prospective Applications 69

Hua MX, Chi Z, Liu GL, Buzdar MA, Chi ZM (2010) Production of a novel and cold-active killer toxin by *Mrakiafrigida* 2E00797 isolated from sea sediment in Antarctica. Extremophiles 14(6): 515–521. https://doi.org/10.1007/s00792-010-0331-6

Huang JP, Mojib N, Goli RR, Watkins S, Waites KB, Ravindra R, Andersen DT, Bej AK (2012) Antimicrobial activity of PVP from an Antarctic bacterium, *Janthinobacterium* sp. Ant5-2, on multi-drug and methicillin resistant *Staphylococcus aureus*. Nat Prod Bioprospect 2(3): 104–110. https://doi.org/10.1007/s13659-012-0021-4

Huang JP, Swain AK, Thacker RW, Ravindra R, Andersen DT, Bej AK (2013) Bacterial diversity of the rock-water interface in an East Antarctic freshwater ecosystem, Lake Tawani (P). Aquat Biosyst 9(1):4. https://doi.org/10.1186/2046-9063-9-4

Hunter-Cevera J, Karl D, Buckley M (2005) Marine microbial diversity: the key to earth's habitability: this report is based on a colloquium, sponsored by the American Academy of Microbiology, held April 8–10, 2005, in San Francisco, California. American Society for Microbiology. https://doi.org/10.1128/AAMCol.8Apr.2005

Jeddi M, Karray F, Battimelli A, Danel A, Garali S, Tedetti M, Zaghden H, Mhiri N, Sousbie P, Patureau D, Sayadi S (2022) Biochemical characterization, microbial diversity and biodegradability of coastal sediments in the Gulf of Gabès, Southern Mediterranean Sea. Int J Environ Sci Technol 19:2389–2408. https://doi.org/10.1007/s13762-021-03307-0

Jørgensen BB, Boetius A (2007) Feast and famine—microbial life in the deep-sea bed. Nat Rev Microbiol 5:770–781. https://doi.org/10.1038/nrmicro1745

Kashi FJ, Owlia P, Amoozegar MA, Yakhchali B, Kazemi B (2014) Diversity of cultivable microorganisms in the eastern part of Urmia salt lake, Iran. J Microbiol Biotechnol Food Sci 4:36

Kasimin ME, Faik AA, Jani J, Abbasiliasi S, Ariff AB, Jawan R (2020) Probiotic properties of antimicrobial-producing lactic acid bacteria isolated from dairy products and raw milk of Sabah (Northern Borneo), Malaysia. Malays Appl Biol 49(3):95–106. https://doi.org/10.55230/mabjournal.v49i3.1580

Kato Y, Sakala RM, Hayashidani H, Kiuchi A, Kaneuchi C, Ogawa M (2000) *Lactobacillus algidus* sp. nov., a psychrophilic lactic acid bacterium isolated from vacuum-packaged refrigerated beef. Int J Syst Evol Microbiol 50(3):1143–1149. https://doi.org/10.1099/00207713-50-3-1143

Kellogg C (2010) Enumeration of viruses and prokaryotes in deep-sea sediments and cold seeps of the Gulf of Mexico. Deep Sea Res Part II Top Stud Oceanogr 57:2002–2007. https://doi.org/10.1016/j.dsr2.2010.05.006

Kennicutt MC (2017) Oil and gas seeps in the Gulf of Mexico. In: Ward CH (ed) Habitats and biota of the Gulf of Mexico: before the deepwater horizon oil spill: vol 1: Water quality, sediments, sediment contaminants, oil and gas seeps, coastal habitats, offshore plankton and benthos, and shellfish. Springer, New York, pp 275–358

Khandeparker R, Meena RM, Deobagkar D (2014) Bacterial diversity in deep-sea sediments from Afanasy-Nikitin seamount, equatorial Indian Ocean. Geomicrobiol J 31:942–949. https://doi.org/10.1080/01490451.2014.918214

Kimura T, Fukuda W, Sanada T, Imanaka T (2015) Characterization of water-soluble dark-brown pigment from Antarctic bacterium, *Lysobacteroligotrophicus*. J Biosci Bioeng 120(1):58–61. https://doi.org/10.1016/j.jbiosc.2014.11.020

Kosek K, Luczkiewicz A, Koziol K, Jankowska K, Ruman M, Polkowska Z (2019) Environmental characteristics of a tundra river system in Svalbard. Part 1: bacterial abundance, community structure and nutrient levels. Sci Total Environ 653:1571–1584. https://doi.org/10.1016/j.scitotenv.2018.11.378

Kumar R, Sharma RC (2021) Psychrophilic microbial diversity and physicochemical characteristics of glaciers in the Garhwal Himalaya, India. J Microbiol Biotechnol Food Sci 10:1–6. https://doi.org/10.15414/jmbfs.2096

Kumar S, Singh H, Kaur M, Kaur L, Tanuku NRS, Pinnaka AK (2018) Bacillus shivajii sp. nov., isolated from a water sample of Sambhar salt lake, India. Int J Syst Evol Microbiol 68(11): 3463–3470. https://doi.org/10.1099/ijsem.0.003008

Kumar A, Ng DHP, Wu Y, Cao B (2019a) Microbial community composition and putative biogeochemical functions in the sediment and water of tropical granite quarry lakes. Microb Ecol 77(1):1–11. https://doi.org/10.1007/s00248-018-1204-2

Kumar R, Kumar P, Giri A (2019b) Regional impact of psychrophilic bacteria on bioremediation. In: Smart bioremediation technologies. Academic Press, London, pp 119–135

Larose C, Dommergue A, Vogel TM (2013) The dynamic arctic snow pack: an unexplored environment for microbial diversity and activity. Biology (Basel) 2:317–330. https://doi.org/10.3390/biology2010317

Lee SG, Koh HY, Oh H, Han SJ, Kim IC, Lee HK, Yim JH (2010) Human dermal fibroblast proliferation activity of usimine-C from Antarctic lichen Ramalinaterebrata. Biotechnol Lett 32(4):471–475. https://doi.org/10.1007/s10529-009-0191-2

Lee LH, Cheah YK, MohdSidik S, Ab Mutalib NS, Tang YL, Lin HP, Hong K (2012) Molecular characterization of Antarctic actinobacteria and screening for antimicrobial metabolite production. World J Microbiol Biotechnol 28(5):2125–2137. https://doi.org/10.1007/s11274-012-1018-1

Li C, Ma L, Mou S, Wang Y, Zheng Z, Liu F, Qi X, An M, Chen H, Miao J (2015) Cyclobutane pyrimidine dimers photolyase from extremophilic microalga: remarkable UVB resistance and efficient DNA damage repair. Mutat Res 773:37–42. https://doi.org/10.1016/j.mrfmmm.2014.07.010

Li W, Wang M, Pan H, Burgaud G, Liang S, Guo J, Luo T, Li Z, Zhang S, Cai L (2018) Highlighting patterns of fungal diversity and composition shaped by ocean currents using the East China Sea as a model. Mol Ecol 27:564–576. https://doi.org/10.1111/mec.14440

Li Z, Pan D, Wei G, Pi W, Zhang C, Wang JH, Peng Y, Zhang L, Wang Y, Hubert CRJ, Dong X (2021) Deep sea sediments associated with cold seeps are a subsurface reservoir of viral diversity. ISME J 15(8):2366–2378. https://doi.org/10.1038/s41396-021-00932-y

Liu Y, Yao T, Jiao N, Tian L, Hu A, Yu W, Li S (2011) Microbial diversity in the snow, a moraine lake and a stream in Himalayan glacier. Extremophiles 15(3):411. https://doi.org/10.1007/s00792-011-0372-5

Liu GL, Wang K, Hua MX, Buzdar MA, Chi ZM (2012) Purification and characterization of the cold-active killer toxin from the psychrotolerant yeast Mrakiafrigida isolated from sea sediments in Antarctica. Process Biochem 47(5):822–827

Liu K, Liu Y, Jiao N, Xu B, Gu Z, Xing T, Xiong J (2017) Bacterial community composition and diversity in Kalakuli, an alpine glacial-fed lake in Muztagh Ata of the westernmost Tibetan Plateau. FEMS Microbiol Ecol 93(7). https://doi.org/10.1093/femsec/fix085

Liu K, Liu Y, Han BP, Xu B, Zhu L, Ju J, Xiong J (2019a) Bacterial community changes in a glacial-fed Tibetan lake are correlated with glacial melting. Sci Total Environ 651(pt 2):2059–2067. https://doi.org/10.1016/j.scitotenv.2018.10.104

Liu LM, Chen HH, Liu M, Yang JR, Xiao P, Wilkinson DM, Yang J (2019b) Response of the eukaryotic plankton community to the cyanobacterial biomass cycle over 6 years in two subtropical reservoirs. ISME J 13:2196–2208. https://doi.org/10.1038/s41396-019-0417-9

Luo X, Xiang X, Huang G, Song X, Wang P, Yang Y, Fu K, Che R (2020) Bacterial community structure upstream and downstream of cascade dams along the Lancang River in southwestern China. Environ Sci Pollut Res 27:42933. https://doi.org/10.1007/s11356-020-10159-7

Lutz S, Anesio AM, Edwards A, Benning LG (2017) Linking microbial diversity and functionality of arctic glacial surface habitats. Environ Microbiol 19(2):551–565. https://doi.org/10.1111/1462-2920.13494

Madigan M, Martinko J, Bender K, Buckley D, Stahl D (2005) Brock biology of microorganisms, 4th edn. Pearson Education, Edinburgh

Maida I, Bosi E, Fondi M, Perrin E, Orlandini V, Papaleo MC, Mengoni A, de Pascale D, Tutino ML, Michaud L, Lo Giudice A (2015) Antimicrobial activity of Pseudoalteromonas strains isolated from the Ross Sea (Antarctica) versus Cystic Fibrosis opportunistic pathogens. Hydrobiologia 761(1):443–457

Makled SO, Hamdan AM, El-Sayed AM, Hafez EE (2017) Evaluation of marine psychrophile, *Psychrobacternamhaensis* SO89, as a probiotic in Nile tilapia (*Oreochromisniloticus*) diets. Fish Shellfish Immunol 61:194–200. https://doi.org/10.1016/j.fsi.2017.01.001

Margesin R (2004) Bioremediation of petroleum hydrocarbon-polluted soils in extreme temperature environments. In: Applied bioremediation and phytoremediation. Springer, Berlin, pp 215–234

Margesin R, Fonteyne PA, Redl B (2005) Low-temperature biodegradation of high amounts of phenol by *Rhodococcus* spp. and basidiomycetous yeasts. Res Microbiol 156(1):68–75. https://doi.org/10.1016/j.resmic.2004.08.002

Margesin R, Moertelmaier C, Mair J (2013) Low-temperature biodegradation of petroleum hydrocarbons (n-alkanes, phenol, anthracene, pyrene) by four actinobacterial strains. Int Biodeterior Biodegrad 84:185–191. https://doi.org/10.1016/j.ibiod.2012.05.004

Mojib N, Huang J, Hoover RB, Pikuta EV, Storrie-Lombardi M, Sattler B, Andersen D, Bej AK (2009) Diversity of bacterial communities in the lakes of Schirmacher Oasis, Antarctica. In: Instruments and methods for astrobiology and planetary missions XII, vol 7441. SPIE, pp 197–207. https://doi.org/10.1117/12.831289

Mojib N, Philpott R, Huang JP, Niederweis M, Bej AK (2010) Antimycobacterial activity in vitro of pigments isolated from Antarctic bacteria. Antonie Van Leeuwenhoek 98(4):531–540. https://doi.org/10.1007/s10482-010-9470-0

Mojib N, Nasti TH, Andersen DT, Attigada VR, Hoover RB, Yusuf N, Bej AK (2011) The antiproliferative function of violacein-like purple violet pigment (PVP) from an Antarctic *Janthinobacterium* sp. Ant5-2 in UV-induced 2237 fibrosarcoma. Int J Dermatol 50(10): 1223–1233. https://doi.org/10.1111/j.1365-4632.2010.04825.x

Mojib N, Farhoomand A, Andersen DT, Bej AK (2013) UV and cold tolerance of a pigment-producing Antarctic *Janthinobacterium* sp. Ant5-2. Extremophiles 17(3):367–378. https://doi.org/10.1007/s00792-013-0525-9

Mojikon FD, Kasimin ME, Molujin AM, Gansau JA, Jawan R (2022) Probiotication of nutritious fruit and vegetable juices: an alternative to dairy-based probiotic functional products. Nutrients 14(17):3457. https://doi.org/10.3390/nu14173457

Mokhtarnejad LM, Arzanlou A, Babai-Ahari ADM, Onofri BP, Turchetti B (2016) Characterization of basidiomycetous yeasts in hypersaline soils of the Urmia Lake National Park, Iran. Extremophiles 20:915–928. https://doi.org/10.1007/s00792-016-0883-1

Mwirichia R (2022) Amplicon-based analysis of the fungal diversity across Four Kenyan Soda Lakes. Scientifica 2022:9182034. https://doi.org/10.1155/2022/9182034

Nagahama T, Abdel-Wahab M, Nogi Y, Miyazaki M, Uematsu K, Hamamoto M, Horikoshi K (2008) Dipodascus tetrasporeus sp. nov., an ascosporogenous yeast isolated from deep-sea sediments in the Japan Trench. Int J Syst Evol Microbiol 58:1040–1046. https://doi.org/10.1099/ijs.0.65471-0

Naghoni A, Emtiazi G, Amoozegar MA, Cretoiu MS, Stal LJ, Etemadifar Z, Shahzadeh Fazeli SA, Bolhuis H (2017) Microbial diversity in the hypersaline Lake Meyghan, Iran. Sci Rep 7(1): 11572. https://doi.org/10.1038/s41598-017-11585-3

Ogaki MB et al (2019) Diversity and ecology of fungal assemblages present in lakes of Antarctica. In: Rosa LH (ed) Fungi of Antarctica: diversity, ecology and biotechnological applications. Springer, Berlin, pp 69–97

Onen OI, Aboh AA, Mfam AN, Akor MO, Nweke CN, Osuagwu AN (2020) Microbial diversity: values and roles in ecosystems. Asian J Biol 9(1):10–22. https://doi.org/10.9734/ajob/2020/v9i130075

Ortiz-Vera MP, Olchanheski LR, da Silva EG, de Lima FR, Martinez LR, Sato MIZ, Jaffé R, Alves R, Ichiwaki S, Padilla G, Araújo WL (2018) Influence of water quality on diversity and composition of fungal communities in a tropical river. Sci Rep 8:14799. https://doi.org/10.1038/s41598-018-33162-y

Orwa P, Mugambi G, Wekesa V, Mwirichia R (2020) Isolation of haloalkaliphilic fungi from lake Magadi in Kenya. Heliyon 6(1):e02823. https://doi.org/10.1016/j.heliyon.2019.e02823

Padmanaban VP, Verma P, Gopal D, Sekar AK, Ramalingam K (2019) Phylogenetic identification and metabolic potential of bacteria isolated from deep sea sediments of Bay of Bengal and Andaman Sea. Indian J Exp Biol 57:561–572

Parvathi A, Jasna V, Aswathy VK, Aparna S, Nathan VK, Jyothibabu R (2020) Dominance of Wolbachia sp. in the deep-sea sediment bacterial metataxonomic sequencing analysis in the Bay of Bengal, Indian Ocean. Genomics 112:1030–1041. https://doi.org/10.1016/j.ygeno.2019.06.019

Paudel B, Bhattarai HD, Koh HY, Lee SG, Han SJ, Lee HK, Oh H, Shin HW, Yim JH (2011) Ramalin, a novel nontoxic antioxidant compound from the Antarctic lichen *Ramalinaterebrata*. Phytomedicine 18(14):1285–1290. https://doi.org/10.1016/j.phymed.2011.06.007

Peeters K, Verleyen E, Hodgson DA, Convey P, Ertz D, Vyverman W, Willems A (2012) Heterotrophic bacterial diversity in aquatic microbial mat communities from Antarctica. Polar Biol 35:543–554. https://doi.org/10.1007/s00300-011-1100-4

Pereliaeva EV, Dmitrieva ME, Morgunova MM, Belyshenko AY, Imidoeva NA, Ostyak AS, Axenov-Gribanov DV (2022) The use of Baikal psychrophilic actinobacteria for synthesis of biologically active natural products from sawdust waste. Fermentation 8(5):213

Peter H, Sommaruga R (2016) Shifts in diversity and function of lake bacterial communities upon glacier retreat. ISME J 10(7):1545–1554. https://doi.org/10.1038/ismej.2015.245

Picard K (2017) Coastal marine habitats harbor novel early-diverging fungal diversity. Fungal Ecol 25:1–13. https://doi.org/10.1016/j.funeco.2016.10.006

Poli A, Finore I, Romano I, Gioiello A, Lama L, Nicolaus B (2017) Microbial diversity in extreme marine habitats and their biomolecules. Microorganisms 5(2):25. https://doi.org/10.3390/microorganisms5020025

Qin Y, Hou J, Deng M, Liu Q, Wu C, Ji Y, He X (2016) Bacterial abundance and diversity in pond water supplied with different feeds. Sci Rep 6:35232. https://doi.org/10.1038/srep35232

Qiu X, Yu L, Cao X, Wu H, Xu G, Tang X (2021) Halomonas sedimenti sp. nov., a halotolerant bacterium isolated from deep-sea sediment of the Southwest Indian Ocean. Curr Microbiol 78(4):1662–1669. https://doi.org/10.1007/s00284-021-02425-9

Raghukumar C, Raghukumar S, Sheelu G, Gupta SM, Nagender B, Rao BR (2004) Buried in time: culturable fungi in a deep-sea sediment core from the Chagos Trench, Indian Ocean. Deep Sea Res Part I Oceanogr Res Pap 51:1759–1768. https://doi.org/10.1016/j.dsr.2004.08.002

Rojas-Jimenez K, Wurzbacher C, Bourne E, Chiuchiolo A, Priscu J, Grossart HP (2017) Early diverging lineages within Cryptomycota and Chytridiomycota dominate the fungal communities in ice-covered lakes of the McMurdo Dry Valleys, Antarctica. Sci Rep 7:15348. https://doi.org/10.1038/s41598-017-15598-w

Savio D, Sinclair L, Ijaz UZ, Parajka J, Reischer GH, Stadler P, Blaschke AP, Blöschl G, Mach RL, Kirschner AKT, Farnleitner AH, Eiler A (2015) Bacterial diversity along a 2600 km river continuum. Environ Microbiol 17:4994–5007. https://doi.org/10.1111/1462-2920.12886

Savvichev AS, Babenko VV, Lunina ON, Letarova MA, Boldyreva DI, Veslopolova EF, Demidenko NA, Kokryatskaya NM, Krasnova ED, Gaisin VA, Kostryukova ES, Gorlenko VM, Letarov AV (2018) Sharp water column stratification with an extremely dense microbial population in a small meromictic lake, Trekhtzvetnoe. Environ Microbiol 20:3784–3797. https://doi.org/10.1111/1462-2920.14384

Schauer R, Bienhold C, Ramette A, Harder J (2009) Bacterial diversity and biogeography in deep-sea surface sediments of the South Atlantic Ocean. ISME J 4:159–170. https://doi.org/10.1038/ismej.2009.106

Schippers A, Kock D, Höft C, Köweker G, Siegert M (2012) Quantification of microbial communities in subsurface marine sediments of the Black Sea and off Namibia. Front Microbiol 3:16. https://doi.org/10.3389/fmicb.2012.00016

Schutte UM, Abdo Z, Foster J, Ravel J, Bunge J, Solheim B, Forney LJ (2010) Bacterial diversity in a glacier foreland of the high Arctic. Mol Ecol 19(Suppl 1):54–66. https://doi.org/10.1111/j.1365-294X.2009.04479.x

3 Microbial Diversity of Cold-Water Reservoirs and Their Prospective Applications

Sean MG, Jack AG (2015) Microbial diversity – exploration of natural ecosystems and microbiomes. National Center for Biotechnology Information. US National Library of Medicine, Bethesda

Sensoy I (2021) A review on the food digestion in the digestive tract and the used in vitro models. Curr Res Food Sci 4:308–319. https://doi.org/10.1016/j.crfs.2021.04.004

Shaaban KA (2022) Marine microbial diversity as source of bioactive compounds. Mar Drugs 20(5):304. https://doi.org/10.3390/md20050304

Shao Z, Yuan J, Lai Q, Zheng T (2015) The diversity of PAH-degrading bacteria in a deep-sea water column above the Southwest Indian Ridge. Front Microbiol 6:853. https://doi.org/10.3389/fmicb.2015.00853

Sharma R, Prakash O, Sonawane M, Nimonkar YS, Golellu PB, Sharma R (2016) Diversity and distribution of phenol oxidase producing fungi from soda lake and description of Curvularia lonarensis sp. nov. Front Microbiol 7:1847. https://doi.org/10.3389/fmicb.2016.01847

Sharma BP, Adhikari S, Paudel G, Adhikari NP (2020) Microbial diversity in the glacial ecosystem of Antarctic, Arctic, and Tibetan Plateau: properties and response to the environmental condition. Janapriya J Interdiscip Stud 9:231–250. https://doi.org/10.3126/jjis.v9i1.35239

Shen Z, Shang Z, Wang F, Liang Y, Zou Y, Liu F (2022) Bacterial diversity in surface sediments of collapsed lakes in Huaibei, China. Sci Rep 12(1):15784. https://doi.org/10.1038/s41598-022-20148-0

Shivaji S, Ray MK, Kumar GS, Reddy GS, Saisree L, Wynn-Williams DD (1991) Identification of *Janthinobacterium lividum* from the soils of the islands of Scotia Ridge and from Antarctic peninsula. Polar Biol 11(4):267–271

Siani L, Papa R, Di Donato A, Sannia G (2006) Recombinant expression of Toluene o-Xylene Monooxygenase (ToMO) from *Pseudomonas stutzeri* OX1 in the marine Antarctic bacterium Pseudoalteromonashaloplanktis TAC125. J Biotechnol 126(3):334–341. https://doi.org/10.1016/j.jbiotec.2006.04.027

Singh P, Raghukumar C, Verma P, Shouche Y (2010) Phylogenetic diversity of culturable fungi from the deep-sea sediments of the Central Indian Basin and their growth characteristics. Fungal Divers 40:89–102. https://doi.org/10.1007/s13225-009-0009-5

Singh P, Raghukumar C, Verma P, Shouche Y (2011) Fungal community analysis in the deep-sea sediments of the Central Indian Basin by culture independent approach. Microb Ecol 61:507–517. https://doi.org/10.1007/s00248-010-9765-8

Soliev AB, Hosokawa K, Enomoto K (2011) Bioactive pigments from marine bacteria: applications and physiological roles. Evid Based Complement Alternat Med 2011:670349. https://doi.org/10.1155/2011/670349

de Souza LMD, Ogaki MB, Câmara PEAS, Pinto OHB, Convey P, Carvalho-Silva M, Rosa CA, Rosa LH (2021) Assessment of fungal diversity present in lakes of Maritime Antarctica using DNA metabarcoding: a temporal microcosm experiment. Extremophiles 25(1):77–84. https://doi.org/10.1007/s00792-020-01212-x

Srinivas T, Singh S, Pradhan S, Pratibha M, Kishore KH, Singh AK, Shivaji S (2011) Comparison of bacterial diversity in proglacial soil from Kafni Glacier, Himalayan Mountain ranges, India, with the bacterial diversity of other glaciers in the world. Extremophiles 15(6):673–690. https://doi.org/10.1007/s00792-011-0398-8

Suh SS, Park M, Hwang J, Lee S, Chung Y (2014) Distinct patterns of marine bacterial communities in the South and North Pacific Oceans. J Microbiol 52:834–841. https://doi.org/10.1007/s12275-014-4287-6

Svahn KS, Chryssanthou E, Olsen B, Bohlin L, Göransson U (2015) *Penicillium nalgiovense* Laxa isolated from Antarctica is a new source of the antifungal metabolite amphotericin B. Fungal Biol Biotechnol 2:1. https://doi.org/10.1186/s40694-014-0011-x

Tang X, Yu L, Xu W, Zhang X, Xu X, Wang Q, Wei S, Qui Y (2020) Fungal diversity of deep-sea sediments in Mid-Oceanic Ridge area of the East Pacific and the South Indian Oceans. Bot Mar 63:183–196. https://doi.org/10.1515/bot-2018-0112

Touka A, Vareli K, Igglezou M, Monokrousos N, Alivertis D, Halley J, Hadjikakou S, Frillingos S, Sainis I (2018) Ancient European lakes: reservoirs of hidden microbial diversity? The case of Lake Pamvotis (NW Greece). Open J Ecol 08:537–578. https://doi.org/10.4236/oje.2018.810033

Vaz AB, Rosa LH, Vieira ML, Garcia VD, Brandão LR, Teixeira LC, Moliné M, Libkind D, Van Broock M, Rosa CA (2011) The diversity, extracellular enzymatic activities and photoprotective compounds of yeasts isolated in Antarctica. Braz J Microbiol 42:937–947. https://doi.org/10.1590/S1517-838220110003000012

Violot S, Aghajari N, Czjzek M, Feller G, Sonan GK, Gouet P, Gerday C, Haser R, Receveur-Bréchot V (2005) Structure of a full length psychrophilic cellulase from *Pseudoalteromonas haloplanktis* revealed by X-ray diffraction and small angle X-ray scattering. J Mol Biol 348(5): 1211–1224. https://doi.org/10.1016/j.jmb.2005.03.026

Vollú RE, Jurelevicius D, Ramos LR, Peixoto RS, Rosado AS, Seldin L (2014) Aerobic endospore-forming bacteria isolated from Antarctic soils as producers of bioactive compounds of industrial interest. Polar Biol 37(8):1121–1131

Wahl HE, Raudabaugh DB, Bach EM, Bone TS, Luttenton MR, Cichewicz RH, Miller AN (2018) What lies beneath? Fungal diversity at the bottom of Lake Michigan and Lake Superior. J Great Lakes Res 44(2):263–270. https://doi.org/10.1016/j.jglr.2018.01.001

Wang F, Hao J, Yang C, Sun M (2010) Cloning, expression, and identification of a novel extracellular cold-adapted alkaline protease gene of the marine bacterium strain YS-80-122. Appl Biochem Biotechnol 162(5):1497–1505. https://doi.org/10.1007/s12010-010-8927-y

Wang S, Yu M, Wei J, Huang M, Shi X, Chen H (2018) Microbial community composition and diversity in the Indian Ocean deep sea REY-rich muds. PloS One 13:e0208230. https://doi.org/10.1371/journal.pone.0208230

Wang J, Ruan CJ, Song L, Li A, Zhu YX, Zheng XW, Wang L, Lu ZJ, Huang Y, Du W, Zhou Y, Huang L, Dai X (2020) Gimesia benthica sp. nov., a planctomycete isolated from a deep-sea water sample of the Northwest Indian Ocean. Int J Syst Evol Microbiol 70(7):4384–4389. https://doi.org/10.1099/ijsem.0.004301

Wei ZF, Li WL, Huang JM, Wang Y (2020) Metagenomic studies of SAR202 bacteria at the full-ocean depth in the Mariana Trench. Deep-Sea Res I Oceanogr Res Pap 165:103396. https://doi.org/10.1016/j.dsr.2020.103396

Wilhelm L, Singer GA, Fasching C, Battin TJ, Besemer K (2013) Microbial biodiversity in glacier-fed streams. ISME J 7(8):1651–1660. https://doi.org/10.1038/ismej.2013.44

Wu S, Zhou L, Zhou Y, Wang H, Xiao J, Yan S, Wang Y (2020) Diverse and unique viruses discovered in the surface water of the East China Sea. BMC Genomics 21:441. https://doi.org/10.1186/s12864-020-06861-y

Xie S, Wang S, Li D, Shao Z, Lai Q, Wang Y, Wei M, Han X, Jiang L (2019) Sulfurovum indicum sp. nov., a novel hydrogen- and sulfur-oxidizing chemolithoautotroph isolated from a deep-sea hydrothermal plume in the Northwestern Indian Ocean. Int J Syst Evol Microbiol 71(3). https://doi.org/10.1099/ijsem.0.004748

Xu W, Pang KL, Luo ZH (2014) High fungal diversity and abundance recovered in the deep-sea sediments of the Pacific Ocean. Microb Ecol 68:688–698. https://doi.org/10.1007/s00248-014-0448-8

Xu W, Gao YH, Gong LF, Li M, Pang KL, Luo ZH (2019) Fungal diversity in the deep-sea hadal sediments of the Yap Trench by cultivation and high throughput sequencing methods based on ITS rRNA gene. Deep Sea Res Part I Oceanogr Res Papers 145:125–136. https://doi.org/10.1016/j.dsr.2019.02.001

Yarzábal LA (2016) Antarctic psychrophilic microorganisms and biotechnology: history, current trends, applications, and challenges. In: Microbial models: from environmental to industrial sustainability. Springer, Singapore, pp 83–118

Yaylacı EU (2021) Isolation and characterization of *Bacillus* spp. from aquaculture cage water and its inhibitory effect against selected *Vibrio spp*. Arch Microbiol 204(1):26. https://doi.org/10.1007/s00203-021-02657-0

Yoon KY, Woodams EE, Hang YD (2006) Production of probiotic cabbage juice by lactic acid bacteria. Bioresour Technol 97(12):1427–1430

York A (2020) Marine microbial diversity from pole to pole. Nat Rev Microbiol 18(1):3. https://doi.org/10.1038/s41579-019-0304-4

Zhang S, Hou S, Ma X, Qin D, Chen T (2007) Culturable bacteria in Himalayan glacial ice in response to atmospheric circulation. Biogeosciences 4:1–9. https://doi.org/10.5194/bg-4-1-2007

Zhang XY, Tang GL, Xu XY, Nong XH, Qi SH (2014) Insights into deep-sea sediment fungal communities from the East Indian Ocean using targeted/ environmental sequencing combined with traditional cultivation. PloS One 9:e109118. https://doi.org/10.1371/journal.pone.0109118

Zhong ZP, Tian F, Roux S, Gazitúa MC, Solonenko NE, Li YF, Davis ME, Van Etten JL, Mosley-Thompson E, Rich VI, Sullivan MB, Thompson LG (2021) Glacier ice archives nearly 15,000-year-old microbes and phages. Microbiome 9(1):160. https://doi.org/10.1186/s40168-021-01106-w

Zhu D, Sethupathy S, Gao L, Nawaz MZ, Zhang W, Jiang J, Sun J (2022) Microbial diversity and community structure in deep-sea sediments of South Indian Ocean. Environ Sci Pollut Res Int 29(30):45793–45807. https://doi.org/10.1007/s11356-022-19157-3

Overview of Microbial Associations and Their Role Under Aquatic Ecosystems

4

Manali Singh, Parul Chaudhary, Shivani Bhutani, Shruti Bhasin, Anshi Mehra, and Keshawananad Tripathi

Abstract

Water plays a vital role in regulating the lives of living beings on earth. Life cannot be imagined without water. Aquatic plants inhabit the littoral or shallow water zones of waterbodies like river, ponds, lakes, and oceans. These plants exhibit a mutual coexistence, pathogenic infestation, commensalism, or in symbiotic association with the microbes. These plants also serve as powerful tools for removal of contaminants in the form of heavy metals in water and soil sediments. The phyllosphere consists of various microbes such as bacteria, viruses, protists, ecto- and endoparasites, nematodes, protozoa, etc. Although the zone between the microbes and roots of aquatic plants is not well defined in terms of nutrients due to their diffusion in water, still there exists a zone of interaction between them. Microbes coevolved with the plants for the fulfillment of their nutrient deficiencies in their fundamental niches. Thus, the aquatic micro biome plays a vital role in influencing and promoting the aquatic ecosystem. The plant–microbe interaction is also affected by the environment: both biotic and abiotic factors have a concomitant effect on the marine ecosystem, thereby leading to change at molecular and gene expression levels causing production of compounds by these microbial–host–environmental interactions. This chapter focuses on how the aquatic microbiomes influence the structure, growth of plants, and their diversity. It also encompasses the the molecular strategy adopted for production of

---

M. Singh (✉) · S. Bhutani · A. Mehra · K. Tripathi
Department of Biotechnology, Invertis University, Bareilly, Uttar Pradesh, India

P. Chaudhary
Department of Animal Biotechnology, Animal Genomics Lab, NDRI, Karnal, India

S. Bhasin
Department of Biotechnology, Banasthali Vidyapith, Jaipur, Rajasthan, India

© The Author(s), under exclusive license to Springer Nature Singapore Pte Ltd. 2023
R. Soni et al. (eds.), *Current Status of Fresh Water Microbiology*,
https://doi.org/10.1007/978-981-99-5018-8_4

metabolites and biofilm, siderophores, system of quorum sensing, cell to cell signaling, signal transduction, etc.

Keywords

Aquatic plants · Phyllosphere · Microbial interaction · Microbiome

4.1 Introduction

At some point of aquatic plant decomposition, plant residues launch vitamins into the water frame (Zhang et al. 2018). Some plant remember bonds with biofilm populating benthic surfaces, making submerged sediments the very last material receptor for plant residue (Wan-Lei et al. 2018). The discharge of this cloth and its succeeding migration alters benthic cloth composition (DeBusk and Reddy 2005), and drives benthic sediment network turnover (Pratiksha et al. 2020) and inhibits the performance of nutrient Elimination in sediments. Destroyed particularly by species opposition, predation, and mutualism, community turnover is pondered inside the network meeting (Yang et al. 2018). This herbal succession is Crucial to benthic geochemical balance, even though variation inside the External surroundings results in complicated, multilevel community Responses (Rocha et al. 2019). As pioneer indicators of aquatic outside pollution, benthic Prokaryotic microbial groups (archaea and microorganism) play an essential position in the elimination and stream of key vitamins (carbon, nitrogen, phosphorus, and sulfur) (Le et al. 2016) and assist in maintaining the stability of the aquatic biosphere (Mendonça et al. 2017). Long-term or sudden material entry from plant residue Release or the pollution of exogenous Escherichia coli (Gu et al. 2020) appreciably alters the composition, distribution, and biogeochemical capabilities of the benthic community. Variant of benthic microbial network in pond ecosystems, but, is confined by way of water frame length (Verpoorter et al. 2014). Scattered small ponds are broadly dispensed in agricultural areas (Holgerson and Raymond 2016). As a matter of reality, these ponds nevertheless have complicated ecological restorative capabilities, in spite of their small length (Mendonça et al. 2017). But, external pollution might result in pond surroundings dysfunction. Water bodies in agricultural regions are threatened by using wastewater from aquaculture and animal husbandry, that may be rich in excessive *E. coli* (Jiang et al. 2015). Previous researches advocate that *E. coli* can live to tell the tale and accumulate in benthic environments for a giant period following infection (Kadir and Nelson 2014). Additionally *E. coli* can reproduce in secondary habitats (Jang et al. 2015; Sadowsky and Whitman 2011), which include surface water and sediments (Bergholz et al. 2011). Under a regime in which *E. coli* invasion is constant and steady, interspecific interplay (e.g., competitive absorption of materials) with indigenous benthic microbes (e.g., Geochemical purposeful groups) happens constantly (Wanjugi and Harwood 2012). To account for this regime, indigenous microbial groups searching for noncompetitive assets (Mallon et al. 2018). The environmental tolerance of *E. coli* improves which includes the secondary succession process of the Benthic community which will increase the length and depth of aquatic surroundings damage (Xiao et al. 2016; Maal-Bared et al. 2013). Even unsuccessful exogenous invasions produce lengthy structural variation in indigenous groups (Mallon et al. 2015). Exogenous *E. coli* has a sizeable impact on bacterial community Shape and the removal of nitrogen (N) in each growing (Gu et al. 2020) and decomposing plants (Wu et al. 2021). Thus caused hinderance in aquatic plant purification and recovery functions, and releases an abundance of vitamins to the water device (Bing et al. 2019). The blended effects of continuous Nutrient enter and *E. coli* pollution triggers the strain response of The benthic prokaryotic network (Mondav et al. 2017), and results In: versions in interactions (reciprocity and opposition between bacterial and archaeal network); the stability and Turnover of the benthic microbial community; and shifts in Geo-chemical niche shape. An analysis of benthic prokaryotes at the community-degree underneath exogenous *E. coli* pollution and accelerated nutrient input elucidates the complex mechanisms of Small-scale aquatic ecosystems, and has capability ramifications for Eutrophication (Li et al. 2021).

4.2 Aquatic Plants as a Natural Source of Antimicrobial and Functional Ingredients

Marine plants survive and thrive in a competitive and hostile environment by living in complex colonies and in close proximity to others. Aquatic plants have evolved physiological adaptations, such as the production of bioactive chemicals, which provide them protection against grazers. As a reaction to ecological pressures such as rivalry for space, predation, and tidal changes, they create complex secondary metabolites. Some secondary metabolites or chemicals generated by aquatic plants may be effective in preventing harmful bacteria from growing. Anticoagulant, antiviral, antioxidant, antihelminthic, antibacterial, antifungal, anticancer, and anti-inflammatory actions have been reported for photochemical substances generated from aquatic plants, crude extracts, and their partially purified or pure components (Pérez et al. 2016).

Seaweeds are also regarded as one of the potential aquatic plants, and have been proposed as a feasible and sustainable source of biofuel production that does not disrupt the world food supply, as well as having a variety of pharmacological, industrial, and biotechnological uses. Seaweeds are also beneficial to mankind in a variety of ways, including as a resource of medications, dietary supplements, toxic products, and a possible biofuel contender (Bast 2013).

4.3 Functional Foods and Nutraceutical Products

Due to its elevated nutritional and pharmaceutical values, seaweeds have long been used as food, sea vegetables, or herbal medicine to treat and prevent a variety of diseases and disorders (Cermeno et al. 2019; Kang et al. 2019; Okolie et al. 2018).

They are also utilized as animal feed, fertilizer, fungicides, herbicides, sauces, dietary supplements, and a reservoir of agar, alginate, and carrageenan for a variety of industrial and medicinal uses. It is generally known that seaweeds have been quite successfully used as a protein source for several decades, particularly in impoverished nations. Due to the availability of critical amino acids in greater proteins, seaweeds have now become a cheaper option source of protein in recent years (Pangestuti and Kim 2015; Peng et al. 2015). On a dry weight basis, the protein content of the principal seaweed species ranges from 10 to 40%. It was also mentioned that the largest protein content in seaweeds was discovered in the winter and the lowest in the summer.

Summer's low protein content might be attributed to the degradation of phycobiliproteins, which make up the majority of the seaweed's proteins. In comparison to green and brown seaweeds, red seaweeds are said to have a higher protein content. Red seaweed's protein level is sometimes compared to that of protein-rich foods like soybeans (Kim et al. 2011a, b). The majority of seaweed proteins are divided into two functionally active categories: lectin and phycobiliproteins (Pangestuti and Kim 2011). Light-harvesting pigments found in red seaweeds (phycocyanins, allophycocyanins, and phycoerythrins) are often utilized as fluorescent probes in scientific research (Glazer 1994; Sekar and Chandramohan 2008).

Seaweed proteins' structure and biological actions are still poorly understood. However, the majority of seaweed bioactive elements, including proteins, are intracellular and protected by a very stiff and structurally complex cell wall, making effective extraction and digestion of seaweed-derived protein fractions difficult. The presence of polysaccharides in seaweed protein makes it very cohesive, according to the researchers (Adalbjörnsson and Jonsdottir 2015; Admassu et al. 2018; Fleurence et al. 2012; Harnedy and FitzGerald 2013; Wijesinghe and Jeon 2012). As a result, novel emerging technologies like microwave-assisted extraction, supercritical fluid extraction, pressurized solvent extraction, ultrasound-assisted extraction, pulsed electric field–assisted extraction, and enzyme-assisted extraction have recently been used to extract proteins with higher yields and desirable functional properties (Jiménez-Escrig et al. 2011; Samarakoon and Jeon 2012).

4.4 Bioactive Compound from Aquatic Plants

In a wide range of sectors, including food, pharmaceutical, cosmetic, nutraceutical, and biomedicine, most aquatic plants (particularly seaweeds) are frequently employed as prospective and important sources of bioactive chemicals (Peng et al. 2015). Bacterial, antifungal, antimicrobial, antiviral, as well as other biological characteristics are all present in these bioactive substances (Khalid et al. 2018). Phenolics, sulfated macromolecules, organic acids, and complex combinations of phytochemicals with antibacterial and functional characteristics are the substances responsible for these actions (Gupta and Abu-Ghannam 2011). Chemical and biological processes (gastrointestinal digestion, food processing, or fermentation) can release seaweed-derived, protein-based bioactive peptides. The most frequent

4 Overview of Microbial Associations and Their Role Under Aquatic Ecosystems

approach for releasing biocompatible hydrolysates and peptides from diverse sources is enzymatic hydrolysis (Rani and Pooja 2018; Rani et al. 2018).

The biological and highly functional actions of peptides are generally determined by the location and composition of their amino acids. Peptide bioactivity is also influenced by the protein's main sequence and the selectivity of the enzymes that discharge the peptides from the parent sequence (Pal and Suresh 2016; Rani et al. 2018). The patterns of biological and functional properties of peptides are most likely influenced by their structural features. Tyrosine, phenylalanine, tryptophan, proline, valine, leucine, lysine, isoleucine, and arginine all have a considerable impact on peptide binding to angiotensin-converting enzyme (ACE) (Pooja et al. 2017; Rani et al. 2018). Antimicrobial peptides' actions are linked to positively charged residues. Histidine, leucine, tyrosine, methionine, and cysteine amino acid residues are related to radical-scavenging activity (Pal and Suresh 2017).

The necessary amino-acid profiles of red, brown, and green seaweeds are well balanced. The glutamic acid concentration is higher in most brown seaweeds. To increase protein digestibility of seaweed fibers as well as refining process of bioactive peptides, plant, animal, and microbial enzymes can also be used. The commercialization of seaweed-derived bioactive peptides has potential applications in a variety of industries, including the nutraceutical and pharmaceutical industries (Rani et al. 2018; Samarakoon and Jeon 2012).

4.5 Categories of Macrophytes

Macrophytes are a diverse group of photosynthetic organism found in water bodies. They include bryophytes (mosses, liverwort, etc.), pteridophytes (ferns), and spermatophytes (flowering plants). An assemblage of macrophytic vegetation consists of emergent species whose vegetative parts emerge above the water surface, submerged and floating species, with each ecological group having specific features in morphology and physiological processes. A wide range of the adaptive mechanisms developed by aquatic macrophytes at the morphological, physiological, and biochemical levels enables them to inhabit various types of freshwater, brackishwater, and marine habitats. The aquatic ecosystem is composed of aquatic flora and fauna, which interact with each other in maintaining the aquatic ecosystem. The diverse and heterogeneous group of macrophytes has posed a challenge for definition and classification. Macrophytes can be loosely defined as all forms of macroscopic aquatic vegetation visible to the naked eye. This is in contrast to microphytes, that is, microscopic forms of aquatic plants, such as planktonic and periphytic algae. The fact that macrophytes live in aquatic environments, at least seasonally, makes them different from terrestrial plants that do not tolerate flooded environments. The macrophytes include taxonomically very diverse representatives: macroalgae (e.g., *Chara* and *Nitella*), mosses and liverworts (e.g., *Sphagnum* and *Riccia*), and vascular plants. Vascular plants represent the largest group of macrophytes including aquatic ferns (*Azolla, Salvinia*), Gymnosperms (rare), and Angiosperms, both monocots and dicots (Rejmánková 2011). Submerged macrophytes represent the

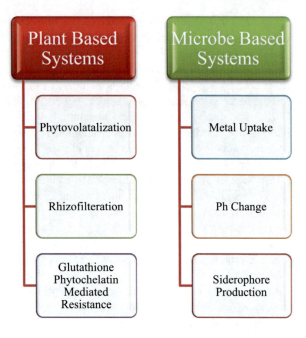

Fig. 4.1 Comparison between plant-based and microbe-based system in nutrient acquisition and accumulation

major component in aquatic ecosystems and help shape the physical and chemical environment, as well as the biota (Jeppesen and Søndergaard 1999). Aquatic macrophytes are aquatic photosynthetic organisms, large enough to see with the naked eye, that actively grow permanently or periodically submerged below, floating on, or growing up through the water surface. Aquatic macrophytes are represented by seven plant divisions: Cyanobacteria, Chlorophyta, Rhodophyta, Xanthophyta, Bryophyta, Pteridophyta, and Spermatophyta (Chambers et al. 2007). Arber (1920) and Sculthorpe (1967a, b) classified macrophytes into four different categories (Fig. 4.1) depending on their growth forms:

Emergent macrophytes: Plants rooted in soil and also emerging to a significant height above water. Submerged macrophytes: Plants grow below the surface water including few ferns, numerous mosses, and some angiosperms. Free-floating macrophytes: Plants that are not rooted to the substratum and float on the surface of the water. Floating leaf macrophytes: Plants occur in submerged sediment, and leaves are floating with long flexible petiole on the surface, which mainly include angiosperm. Besides the Arber and Sculthorpe ecological classification of macrophytes, several authors attempted to classify macrophytes into functional types (Boutin and Keddy 1993; Brock and Casanova 1997; Weiher et al. 1998). Plant functional types can be defined as sets of plants exhibiting similar responses to environmental conditions and having similar effects on the dominant ecosystem processes (Diaz and Cabido 1997; Lavorel et al. 1997). Macrophytes are important in the aquatic carbon cycle and play as primary producers a crucial role in carbon storage in aquatic systems. Macrophytes play an important role in the freshwater ecosystem functioning of many shallow water bodies: as primary producers, by providing structure in the habitat of many animal species, and provide shelter and food to invertebrates (Castella et al. 1984) and fish (Rossier 1995). Macrophytes, which are major primary producers in shallow freshwater systems, have been reported to contribute substantially to biodiversity at the ecosystem level (Zeng et al. 2012). Macrophytes are also involved in ecosystem processes such as biomineralization, transpiration, sedimentation, elemental cycling, materials transformation, and release of biogenic trace gases into the atmosphere (Carpenter and Lodge 1986). Recent studies have established the importance of aquatic macrophytes in regulating the nutrient availability in the water and enhancing the stability of lakeshores (Carpenter and Lodge 1986; Blindow et al. 2014). However, macrophytes are affected by increasing carbon concentrations (Reitsema et al. 2018). Aquatic macrophyte plants are widespread throughout the globe. Most macrophyte species are cosmopolitan, while groups of closely related species are known to replace each other in aquatic ecosystems of different parts of the world (Santamaría 2002., Zhang et al. 2019). Aquatic macrophytes are characterized by diverse growth forms and plasticity of physiological and metabolic processes, depending on changes in environmental conditions. Macrophytic vegetation includes deep-water and shallow-water species and species that actively grow in water in an emergent, submerged, or floating state. Some species are strictly aquatic, because they need to be submerged in water to complete their life cycle, while other species possess the ability to grow and reproduce in periodically inundated environments, can adapt to changes in water level, or inhabit the so-called ephemeral water bodies (e.g., floodplains, temporary springs, ponds, etc.) (Cook 1999., Jackson et al. 2009). Aquatic macrophytes play a significant role in freshwater ecosystems as they provide food and shelter to invertebrates (Rejmankova 2011) and stabilize sediments and shorelines, thus reducing turbidity of aquatic systems (Bamidele and Nyamali 2008). Submerged macrophytes affect nutrient dynamics, light attenuation, temperature regimes, hydrodynamic cycles, and substrate characteristics (Rooney et al. 2005).The macrophytes are responsible for the regulation and stabilization of mineral cycling in the water bodies, and hence, they serve as indicators for the possible degree of damage in the ecosystem (Pieczynska and Ozimek 1976). The aquatic plants are the drivers of ecosystem productivity and biogeochemical cycles, in part because they serve as a critical interface between the sediments and the overlying water column (Carpenter and Lodge 1986). Aquatic macrophytes influence metal retention indirectly by acting as traps for particulate matter, by slowing the water current, and favoring sedimentation of suspended particles (Kadlec 2000). Aquatic macrophyte also reduces sediment resuspension by offering wind protection (Brix 1997). Large aquatic macrophytes possess the ability to break down the human- and animal-derived pollutants in the water.

Macrophytes affect aquatic ecosystems in a variety of ways, especially the shallower ones where they colonize large areas. These plants change the water and physicochemical properties of sediments, influence nutrient cycling, may serve as food for aquatic invertebrates and vertebrates, both as leaves and dead biomass (detritus) and, in particular, change the spatial structure of the waterscape by increasing habitat complexity (Thomaz et al. 2004). Additionally, aquatic macrophytes are divided into several ecological groups according to their relation to environmental factors: hydatophytes (submerged species), pleistophytes (species with floating leaves), and helophytes (emergent aerial-water species) (Roshchyna 2018). Macrophytes growing at the margins of lakes or rivers and often facing challenges associated with fluctuations in water level are usually classified as a separate ecological group. These plants, commonly referred to as amphibious species, are capable of growth and reproduction in both aquatic and terrestrial environments. Amphibious plants have developed adaptations that allow them to withstand rapid emersion and submersion, including the plasticity of the leaf shape and photosynthesis process (Li et al. 2019; Maberly and Madsen 2002; Van Veen and Sasidharan 2021). Aquatic macrophytes are endowed with special adaptations to life in the submerged conditions or on the water surface. In addition to the climatic factors, such as light and temperature, inorganic carbon availability and specific factors related to aquatic environment determine the assembly and distribution of aquatic macrophytes (Barko et al. 1986). While limitations of the amount of CO_2 and sunlight for photosynthesis rarely occur in emergent macrophytes and those floating on the water surface, the growth of submerged macrophytes is strongly affected by the amount of penetrated light and the level of gas saturation (Hossain et al. 2017). Aquatic macrophytes are found mainly at depths of 0–4 m; however, their occurrence at depths exceeding 6 m has been reported, with vascular plants being found at maximum depths of about 12 m (Bornette and Puijalon 2011, Gradstein et al. 2018). In deep-water habitats, mosses frequently constitute the dominant vegetation owing to their better adaptability to lower light intensities and lower temperatures compared with vascular plants (Brönmark and Hansson 2017). An important restrictive factor for the growth of submerged macrophytes is the depth of light penetration. It is considered that the depth limit for their distribution occurs when water transparency allows <1–4% of light to reach plants (Sculthorpe 1967a, b). Temperature is an equally important factor determining the distribution of aquatic macrophytes (Barko et al. 1986; Bornette and Puijalon 2011; Santamaría 2002). Some species of macrophytes grow exclusively in tropical areas, while others are distributed only in the waters of the temperate climate zone (Chambers et al. 2008). Chemical composition of water, including salinity, strongly influences the growth and spreading of aquatic macrophytes. Macrophyte species inhabiting saltwater environments exhibit wide tolerance to osmotic stress and can grow at different salinities. For example, seagrasses, a unique plant group of approximately 60 species, which encompasses members of the families Hydrocharitaceae, Cymodoceaceae complex, and Zosteraceae, are the only angiosperms living fully submersed in the sea (Orth et al. 2006). Other plant species that have colonized saline waters (salt marsh plants, mangrove swamps, etc.) are members of several families, such as Potamogetonaceae and Ruppiaceae (Santamaría 2002). Aquatic macrophytes, similar to other plants, require a constant supply of nutrients and trace elements for growth and development (Rejmánková 2011). Nonrooted macrophytes receive ions and chemical compounds necessary for nutrition directly from water, while most aquatic plants rooted or attached to the bottom absorb them both from water and from bottom sediments (Barko et al. 1986).

4.6 Role of Macrophytes in Ecosystem

Macrophyte vegetation is an important component of various types of aquatic ecosystems. Together with phytoplankton species, these autotrophic organisms are the primary producers that provide the conversion of light energy into organic carbon compounds, thereby contributing to the formation of the trophic structure of aquatic ecosystems. During photosynthesis, macrophytes not only synthesize organic substances, but also release oxygen, which is necessary for the respiration of aquatic organisms and decomposition of organic matter. Macrophytes, especially submerged species, are important for aquatic food webs and affect the interaction among predatory, planktivorous, and benthivorous fish, as well as between fish and invertebrates (Hrivnák et al. 2006). According to available data, 37 freshwater herbivorous fish species belonging to 24 families feed on macrophytes (Opuszynski and Shireman 1995). Macrophytes, similarly to other aquatic organisms (e.g., phytoplankton, cyanobacteria), are producers and emitters of biologically active substances (allelochemicals), including low-molecular-weight volatile organic compounds, which play an important role in interspecies communication and competition. Allelochemicals released by aquatic plants perform a variety of functions, thereby affecting the composition and development of aquatic communities (Fink 2007; Li and Hu 2005; Zuo 2019). It is assumed that inhibition of phytoplankton and bacterioplankton, including cyanobacteria species, by allelochemical compounds secreted by macrophytes contributes to the stabilization of clear water states in shallow lakes (Chen et al. 2012; Mähnert et al. 2017). Additionally, macrophytes affect aquatic ecosystems through their influence on hydrological regime of water bodies (e.g., flow velocity, formation of surface waves, etc.), bottom sediment formation, and water quality (Dhote and Dixit 2009; Miler et al. 2014). Noticeable changes in pH values, dissolved gas concentrations, and ionic composition of water may result from their metabolism. Macrophytes that grow in near-shore areas can promote the stabilization of shores and contribute to reduction in erosion rates (Thomaz and da Cunha 2010).

4.7 Synergistic Action of Plant and Microbes in Aquatic Ecosystem

Microbes interact with plants in aquatic ecosystems mainly for the nutrition (organic carbon and oxygen), whereas plants developed defensive immunity and exchange of minerals. The nutritional conditions of the aquatic ecosystems depend on the distribution of the aquatic plants and microbes. Roots of the aquatic plants are directly interacting with microbial community, and provide continuous supply of oxygen, carbon, nitrogen, and nutrients. Stout (2006) reported that root-associated microbial communities of the *Lemna minor* negatively influence cadmium (Cd) acquisitions. Cd is a toxic heavy metal, which inhibits the growth and development of the *Lemna* plants (Haider et al. 2021). Several factors, such as pH, salt concentration, biological oxygen demands (BOD), chemical oxygen demands

(COD), dissolved oxygen electrolytic conductance, toxic heavy metals, organic pollutants, and available nutrients, are responsible for the plant–microbe interactions. In the aquatic ecosystems, very little information has been reported in the field of plant–microbes interactions. Moreover, some of the typical examples of plantmicrobe interaction and their roles are described (Table 4.1). The rhizospheric regions of aquatic plants are associated with maximum microbial interaction; therefore, this region exhibits different water chemistry than the other part of the aquatic plants (Stout and Nüsslein 2010). There are different types of plant–microbes interaction (terrestrial and aquatic) and these are associated with different types such as parasitic, symbiotic, mutualisms, and commensalism (Singh et al. 2019). Terrestrial plants secreted a group of chemicals, which act as signal molecules, and interact with microbes. On the other hand, an aquatic plant depends on organic carbon and oxygen (mainly rhizospheric regions) for survival of the microorganisms (Yadav et al. 2021). In general microbes make association with the plants: (1) endophytic in plant tissues and microbes provide nitrogen to host such types of microbes known as diazotropic microbes (Tripathi et al. 2013; Singh et al. 2018), (2) Ectophytic microbes (microbes associated out sides of the plants) are common example of ammonia-oxidizing microbes, methanogens, and metanotropic bacteria. The ectophytic interaction takes place in whole plants, including root, leaves, and plant trichome (Kandel et al. 2017). The transport of oxygen via diffusion through root to shoot via interconnected space. Oxygen is used as primary electron accepter for the energy generations, and this process is utilized by a group of oxidizing microbes in an oxidation processes (Morales-Olmedo et al. 2015). During digenesis process, the organic matters of sediment are degraded through anoxic microbial activities, and consumption of electron acceptor (Kristensen 2000). Under anoxic condition, bacterial cells utilized terminal electron acceptor from inorganic sources such as nitrate, sulfate, and carbon dioxides ions as terminal electron accepter to degrade the organic matters (Weber et al. 2006, Singh et al. 2018). Methane is produced by the methanogens by the reduction of carbon dioxide with hydrogen. It is poor energy-yielding process and predominant in complete combustion of all electron acceptors other than carbon dioxides (Chanton et al. 2005).

4.8 Plant-Microbe Interaction in Aquatic Plants

An aquatic environment provides a lot of opportunities for microbes to interact with the plants for their mutual survivals (Meshram and Chaugule 2018). Plant–microbe interactions are directly and indirectly involved for the mitigation of pollution known as bioremediations, and this topic has been more explored, in over of the world in the field of environmental and biological sciences (Dixit et al. 2021). Moreover, plant–microbe interactions are based on mutual benefits, whereas microbes provides minerals and metabolites to the plant in return plant provides shelter, organic carbon, and oxygen for growth and developments of the plants (Tripathi et al. 2013; Yadav et al. 2022).

Table 4.1 Sea weeds as nutritional and functional food products

S. No.	Name of seaweed	Nutritional functional food	Bioactive compound	Structure
1.	*Himanthalia elongata*	Frankfurter sausages	Fucoxanthin	
2.	*Himanthalia elongata*	Reconstructed poultry steaks	Fucoxanthin	
3.	*Undaria pinnatifida*	Beef patties	Proteins	
4.	*Laminaria japonica*	Pork patties	Polyphenols	
5.	*Saccharina latissima, Sargassum pallidum*	Vegetarian burgers	Total polysaccharides	

(continued)

Table 4.1 (continued)

S. No.	Name of seaweed	Nutritional functional food	Bioactive compound	Structure
6.	*Chondrus crispus, Eucheuma cottonii*	Ice cream	Carrageenan	
7.	*Gracilaria cornea, Gracilaria domingensis*	Marmalade and Jam	Agar	
8.	*Laminaria digitata, Laminaria hyperborean*	Beverages	Aligns/alginic acid	
9.	*Fucus vesiculosus, Ascophyllum nodosum*	Soups	Fucoidan	
10.	*Fucus vesiculosus, Laminaria hyperborean*	Caprine feed	Laminarin	

11.	Ulva lactuca, Ulva rigida	Salads	Ulvan	
12.	Saccharina latissima, Porphyra tenera	Burger	Taurine	
13.	Palmaria palmata, Digenea simplex	Medicines	Kanoids (kainic and domoic acid)	Domoic acid Kainic acid
14.	Laminaria digitata, Saccharina latissima	Medicine	PUFA (omega 3 fatty acids)	

(continued)

Table 4.1 (continued)

S. No.	Name of seaweed	Nutritional functional food	Bioactive compound	Structure
15.	*Laminaria digitata, Fucus serratus*	medicine	Carotenoids	
16.	*Laminaria japonica, Laminaria digitata*	Fertilizer	Iodine	
17.	*Porphyra tenera, Ulva lactuca*	Meat products	Calcium	
18.	*Ulva lactuca, Porphyra tenera*	Meat products	Vitamin B_{12}	

4 Overview of Microbial Associations and Their Role Under Aquatic Ecosystems

| 19. | *Monostroma nitidum* | Fresh noodles | Phenolic compounds | |
| 20. | *Undaria pinnatifida* | Pasta | Fucoxanthin | |

4.8.1 *Azolla-Anabaena* Symbiosis

Azolla is a heterosporus free floating aquatic macropyte frequently found on the water surface (Yadav et al. 2016). The macrophyte *Azolla* placed in the family *azollaceae* has been reported. In India, seven species have been reported such as *A. caroliniana, A. maxicana, A. filiculoides, A. microphylla, A. rubra, A. nilotica,* and *A. pinnata*. Among them, *Azolla pinnata* is native strain of the India, and better sporulating and temperature sensitive (Mian 2002). *Anabaena-azollae* is an example of true endo-symbiont, on the dorshal leaf cavities; *Azolla* provides carbon sources and space for the growth and development of the cyanobiont, whereas cyanobiont synthesized nitrogen source in thick wall hetrocystes, and supply to macropytes. It is this unique symbiotic association between *Azolla* and prokaryotes partner *Anabaena azollae* (Adams et al. 2006). *Azolla* cavity provides natural-ecosystem and ecological defined structure, as result of an intimate contact between host and endosymbiont embedded in a mucilaginous cavity, helping in the exchange of nutrients as well as exchange of metabolites. *Azolla-Anabaena* associations are considered a successful coevolved organism, which makes important contributions to the food, feed, biofertilizer, phytoremediation, wastewater treatments, and several biotechnological applications (Yadav et al. 2021).

4.8.2 Degradation of Organic Pollutants

The major sources of contamination in India are organic compounds such as chlorinated hydrocarbon, poly-aromatic hydrocarbon (PAH$_S$), and poly-bromide biphenyl ether (PCBs), which are common environmental pollutants. Microbes are efficient bioremediation potential due to catabolic activities, and degrade almost all types of organic compounds (Kannikka et al. 2018). Cometabolism is another mechanism that follows the microbes to degrade recalcitrant organic compounds into simple organic compounds in aquatic and terrestrial zone of macropytes (Kumwimba and Meng 2019). The biodegradations rate follows second order kinetics in aquatic ecosystems; it is directly correlated with number of microbes and amount of xenobiotic compounds (Barragán et al. 2007), whereas the microbial population depends on species of the macropytes (Chao et al. 2021). In addition to that plant supplies organic carbon to the root associate microbes and degrade recalcitrant compounds (Hoffland et al. 2020) such as pyrenes and PAH$_S$ (Mohapatra and Phale 2021). Golubev et al. (2009) found classical example of the mutualism between plant–microbe interactions whereby plants benefited in the form of plant hormone indole acetic acids (IAA) from microbes as a result microbial degradation of PAHs. Further, Golubev and his coworkers reported *Sinorhizobium meliloti* P 221 which form ectorhizospheric association with aquatic plants and mutual benefitted by the synthesizing IAA and PAHs degradations. The rhizospheric region of the aquatic plants rich sources of methanotrophs which is a groups of a and c proteobacteria used as methane as sources of energy and carbon. The methanotrophs (e.g., *Methylosinus trichosporium, Methylococcuscapsu-latus*) degrade wide range of toxic organic compounds by the synthesizing enzymes such as methane mono-oxygenase (Pandey et al. 2014). The xenobiotic compounds are mainly chlorinated ethene through cascades of the enzyme reactions; which involves the formation intermediate organic molecule formaldehyde, and finally produced terminal carbon dioxides (Yoon 2010).

4.9 Plant-Microbe Interaction Facing Environmental Challenges

Plants exist in a microbe-rich environment in nature, and they must interact with a diverse range of pathogenic, commensal, and helpful bacteria. Multiple components of the plant system are influenced by the microbiome. They influence plant health through antagonistic and synergistic interactions with the plants. Plants ability to utilize microorganisms beneficial activities while also combating microbial diseases has piqued the interest of generations of plant and microbial experts (Jones et al. 2016). Many studies looked into the fundamentals of plant-microbiota interaction, plant defense signals, and symbiotic responses to microorganisms, a microbe's genetic system for transporting signaling molecules and modulating host cell activities, the particular binary and community level conflict, as well as specific binary and community level conflict in the interaction (Hacquard et al. 2017). Environmental effect on plant and microbiome interaction is seldom studied at mechanistic and molecular levels, opening up new pathways in understanding how plants and microbes interact in nature to mediate ecological resilience.

A "holobiont" is made up of plants and microbial populations. In broad terms, a holobiont is an ecological unit formed by the assemblage or combination of a host with numerous species that surround it. Microbiota that interacts with plants has a wide range of locations, ecological functions in both the above-ground and below-ground environment, and may play a role in controlling climate tolerance and other stress responses. PGPR, fungus, actinomycetes, algae, yeasts, cyanobacteria, and others make up the microbial community in the soil system (Manzar et al. 2021). The Plant roots are home to a wide range of microfauna, both endophytic and rhizospheric (Tables 4.2 and 4.3).

4.10 Impact of Temperature on Microbial Mechanisms

Velásquez et al. (2018) have examined how pathogens are affected by environmental factors, highlighting that each pathogen has a temperature range in which it may develop and virulently. Elevated temperature has been demonstrated to suppress type IV secretion-associated pilus production and expression of virulence (vir) genes in *Agrobacterium* infection (Baron et al. 2001; Jin et al. 1993). At higher temperatures, however, the soft-rot bacteria *Pectobacterium atrosepticum* showed greater virulence, which is linked to increased production of plant cell-wall-degrading enzymes, quorum-sensing signals, and faster disease progression

Table 4.2 Common examples of the plant–microbe interactions, and its role in aquatic ecosystems

Symbiotic plants	Microbes	Role in aquatic ecosystems	References
Azolla filiculoids	*Anabaena azollae*	Nitrogen fixation	Yadav et al. (2016, 2019)
Lemna minor	*Pseudomonas sp*	Denitrification	Ying-ru et al. (2013)
Chlorella bulgaris	*Azospirillum braciliances*	Plant-growth-promoting activity	Gonzalez and Bashan (2000)
Typha latifolia	*Bacillus sp*	Nitrogen fixation	Biesboer (1984)
Scenedesmus bicellularis	*Pseudomonas diminuta*	Plant-growth-promoting activity	Mouget et al. (1995)
Nuphur sp.	*Mesorhizobiam lotti*	Nitrogen fixations	Wagner (2011)
Rooted macrophytes	*Sinorhizobium meliloti* P221	IAA production in root	Golubev et al. (2009)

(Hasegawa et al. 2005). Temperature changes also have an impact on beneficial plant-microbe interactions. It has a beneficial impact on arbuscular mycorrhizal fungus (AMF) hyphal development and plant colonisation in the majority of cases, owing to quicker plant carbon allocation to the rhizosphere, where AMF dwells (Compant et al. 2010).

4.11 Circadian Clock and Plant Immunity

External light and the internal circadian clock are involved in many areas of plant life. Interlocked transcription-translation feedback loops govern the circadian clock. Although, in addition to light, the circadian clock is a self-sustaining mechanism, certain parts of its operation may change in reaction to changes in external variables, such as temperature and humidity (Lu et al. 2017; Mwimba et al. 2018). According to Mwimba et al. (2018), humidity oscillation promotes ETI at night, implying that the host anticipates higher pathogen infection when the humidity is high at night. Effect of the circadian clock on ETI against the *Arabidopsis* oomycete pathogen Hyaloperonospora arabidopsidis (Hpa) has been studied. In order to commence infection, Hpa disperses the spores at daybreak. Through CCA1, the circadian clock has been demonstrated to modulate RPP4 (NLR)-mediated immunity against avirulent Hpa isolates—at dawn, RPP4 and RPP4-dependent genes had a modulated peak expression (Wang et al. 2011).

4.11.1 Light

Experiments including the genetic deletion of a circadian oscillator in B. cinerea, the use of continuous light to decrease fungal rhythmicity, and the use of out-of-phase light:dark cycles demonstrated that the fungal clock controls the result of the susceptible Arabidopsis-B. cinerea interaction. Light exposure may also be used

4 Overview of Microbial Associations and Their Role Under Aquatic Ecosystems 95

Table 4.3 Microbes present in phyllospheric and rhizospheric regions of different plant species

Microbial type	Microbial species	Plant species	References
Bacteria	Sphingomonads and Methylobacteria	Maize	Wallace et al. (2018)
	Pseudomonas	Apple, almond, tobacco	Aleklett et al. (2014)
	Pseudomonas, Erwinia herbicola	Sugarbeet	Thompson et al. (1993)
	Proteobacteria	Soybean, clover, rice, *Arabidopsis*	Vorholt (2012)
	Pseudomonas, Sphingomonas, Frigoribacterium	Grapewine, grape clusters	Compant et al. (2019)
	Cyanobacteria	Mangrove forests, Taiwan rain forests, tropical trees, rice	Lin et al. (2012), Venkatachalam et al. (2016), Zhu et al. (2018)
	Pseudomonas, Agrobacterium, Cupriavidus, Rhizobium	*Citrus* species	Xu et al. (2018)
	Proteobacteria, Actinobacteria, Acidobacteria, Bacteriodetes, Plantomycetes, Choloroflexi, Gemmatimonatedes	Grapevine	Faist et al. (2016), Samad et al. (2017)
	Proteobacteria, Actinobacteria	Cucumber, tomato, wheat maize	Reinhold-Hurek et al. (2015)
	Bacteriodetes, Actinobacteria	*Arabidopsis*	Bergelson et al. (2019)
	PGPR	Jerusalem artichoke	Montalbán et al. (2017)
	Acidobacteria, Alphaproteobacteria, and Gammaproteobacteria	Cottonwood	Timm et al. (2018)
	Calothrix, Scytonema	Paddy	Roger et al. (1993)
	Nostoc, Anabaena	Rice	Prasanna et al. (2009)
Fungi	*Undifilum* spp., Clavicipitaceous fungi	Legumes, Morning glory	Panaccione et al. (2014)
	Ascochyta species, *Colletotrichum gloeosporioides, Phomopsis* species	*Fagus* species, *Swida* species	Osono (2006)
	Ascomycota, Basidiomycota, Olpidiomycota	*Arabidopsis*	Bergelson et al. (2019)
	Orchidaceous m Mycorrhizal fungi	Orchids	Rudgers et al. (2020)

by pathogens to start an infection. For example, in the maize fungal pathogen *Cercospora zeae-maydis*, the blue light receptor Cercospora Regulator of Pathogenesis1 (Crp1) is required for sensing plant stomata and may mediate the biosynthesis of the light-activated toxin cercosporin, which disrupts stomatal guard cell membranes, allowing fungal infection through stomata (Kim et al. 2011a, b).

P. syringae's fitness and virulence may be influenced by light. The expression of coronatine toxin biosynthetic genes, for example, is suppressed by red light. Red light may limit bacterial entry through stomata, because coronatine is necessary for *P. syringae* to open stomata and enable bacterial entry. Light and the circadian clock, like temperature, impact both the host plant and the microbe during the creation of a plant–microbe connection (Santamaría-Hernando et al. 2018; Melotto et al. 2006).

4.11.2 Moisture

Water is necessary for life to exist on this planet. Too little water (osmotic stress) or too much water (flooding) can have a significant influence on many elements of plant and microbial life. Plants regulate the amount of the phytohormone ABA in response to a lack of water. Increased ABA activates a signaling cascade that leads to large-scale transcriptional reprogramming and physiological changes, including as stomata closing to minimize transpiration (Zhu 2016). ABA-induced stomatal closure may limit bacterial infiltration through stomata under drought stress. Increased ABA, on the other hand, can decrease the SA signalling pathway in the mesophyll cells inside the leaf, weakening SA-mediated resistance after invasion (Jiang et al. 2010).

4.11.3 Drought

Drought has an impact on plant-microbiome interactions. Drought changed the composition of microbial communities in all sampled compartments (bulk soil, rhizosphere, and root endosphere), and the closer the community is to the root, the greater the shift in composition in drought-stressed rice plants (Santos-Medellín et al. 2017). Drought increases the number of Actinobacteria and Firmicutes at the phylum level. The root endosphere shows the greatest decrease in community diversity and rise in Actinobacteria and Firmicutes transcript abundance, with specific enrichment in amino acid and carbohydrate transport activities. Drought stress produces a change in root metabolites on the host side. It needs to be seen if and how these drought-enriched metabolites "configure" root microbiome composition to boost plant stress responses.

Nonetheless, this intriguing link shows that during drought, molecular dialogues between plants and their associated microbiome may modify root microbiota in order to deal with drought stress. Deciphering this molecular conversation will help us get the essential information we need to use microbiota to improve drought tolerance in agricultural plants.

4.11.4 Humidity

Rain and/or high air humidity have long been known to be necessary conditions for many plant disease outbreaks. Plants typically die locally at the site of pathogen

4 Overview of Microbial Associations and Their Role Under Aquatic Ecosystems

infection during ETI, a process known as the hypersensitive reaction (HR). The HR is hypothesized to inhibit biotrophic pathogen development and dissemination while also activating secondary immune responses. In a variety of plant-pathogen interactions, high atmospheric humidity reduces HR cell death, which might be one of the causes for increased plant sensitivity and epidemics in humid environments (Wang et al. 2005; Wright and Beattie 2004).

4.12 Plant–Microbe Interaction in Nutrient Acquisition and Accumulation

The rhizosphere in terrestrial structures is the area of soil surrounding plant roots where there is extended microbial activity; in aquatic plants, this definition may be less clean because of diffusion of vitamins in water, but there's still a quarter Of have an effect on with the aid of plant roots on this environment (Christensen et al. 1994). Within that Area chemical conditions differ from the ones of the encompassing Surroundings resulting from a variety of techniques that Had been brought about both immediately through the interest of plant roots or by way of The interest of rhizosphere microflora. These days, there are a Range of recent research related to rhizospheres of aquatic flora And specifically their multiplied capacity for remediation of Contaminants, mainly remediation of metals through Aquatic plant–microbial interaction (Stout and Nüsslein 2010).

4.13 Application, Benefits, and Barriers of Plant-Based Remediation

Phytoremediation, the use of plants and their associated Microbial groups to eliminate or inactivate pollution. From the environment, includes any of several technology For detoxifying the surroundings with genetically Changed or wild-type vegetation (Kraemer 2005). Phytoremediation of Aquatic environments can be used as an alternative or in Addition to standard remediation methods inclusive of Ion change resins and electrodialysis, chemical precipitation, Sedimentation, microfiltration, and reverse osmosis (Rai 2009). Organic remediation strategies provide powerful opportunity treatments that are often much less luxurious and are Taken into consideration extra environmentally pleasant and publicly Acceptable than traditional technology. Various phytoremediation technology encompass phytoextraction, Phytovolatilization, phytostabilization, and rhizofiltration and are summarized by several critiques (Flathman and Lanza 1998; Prasad and Freitas 2003; Jabeen et al. 2009). Current studies have carried out a number of those technologies to phytoremediation of heavy metals in aquatic or wetland structures. As an instance, Murakami et al. (Murakami et al. 2009) used rice cultivars that collected high ranges of cadmium to take away metal from contaminated paddy fields. In those conditions in which rice was well-tailored to Boom, one rice cultivar amassed ten times as an awful lot Cd as *Thlaspi caerulescens*, a terrestrial plant well known for its potential to accumulate heavy metals. Rhizofiltration, the use of plant life to put off heavy metals From aqueous environments, has been significantly examined As a manner to take away contaminants from solutions, and may Include aquatic or terrestrial flora in hydroponic structures. Recently, sunplant and bean vegetation have been examined for their competencies to dispose of u from contaminated groundwater. In laboratory batch experiments, bean plants eliminated greater than 70% and sunplant eliminated more than 80% of u however whilst sunplant was examined in a Non-stop rhizofiltration gadget, u removal was extra Than 99% (Lee and Yang 2010). Aquatic and wetland flora which include the Water hyacinth eichhorinia crassipes (Agunbiade et al. 2009; Mishra and Tripathi 2009), the invasive Reed phragmites australis (Ghassemzadeh et al. 2008), the duckweeds spirodela Polyrrhiza (Rahman et al. 2007), lemna minor (Hou et al. 2007; Uysal and Taner 2009), and lemna gibba (Khellaf and Zerdaoui 2009; Megateli et al. 2009), the aquatic fern azolla pinnata (Rai and Tripathi 2009), and yellow Velvetleaf limnocharis flava (Abhilash et al. 2009) have recently been studied For his or her abilities to eliminate metals from aquatic Structures and show promising effects.

Pinnata become discovered to dispose of as a good deal as 94% of Hg from A solution (Rai and Tripathi 2009), whilst eichhornia crassipes was located to acquire Cr in its shoots at 223 instances the attention Inside the water (Agunbiade et al. 2009), and removed 84% of Cr from water and 95% of Zn from water (Mishra and Tripathi 2009). At the same time as metals negatively Affected increase of lemna gibba, the plant s were capable of Dispose of 90% of Cd from answer after 6–8 days (Megateli et al. 2009).

4.13.1 Metal Tolerance and Resistance: The Bacterial Aspect

Bacteria have validated expanded tolerance to metals using many diverse mechanisms. They will hold metal homeostasis, preserving concentrations of critical metals together with Zn from attaining toxic ranges inside cells (Coombs and Barkay 2005) or they'll include resistance systems, active mechanisms for getting rid of or sequestering metals (Gadd 1992). For heavy metals such as Cd, Zn, Ni, Cr, Co, and Cu, there are numerous forms of resistance systems, such as efflux Pumps to remove metals from the cellular, and sequestration Mechanisms to bind metal in the cell. The two known Sorts of efflux systems are atpases, which pump out Metals the use of ATP to pressure the reaction, and proton antiports, which use the proton gradient to pump metals across the mobile membrane (Nies 2003). Some other mechanism of bacterial metal resistance, satisfactorily known in cyanobacteria, is sequestration by way of metallothioneins. Metallothioneins bind metals to sulfhydryl businesses of cysteine residues (Huckle et al. 1993; Ybarra and Webb 1999). Recently described cadmium tolerance in Stenotrophomonas maltophilia, and discovered now not only a Cd Efflux pump but additionally accumulation of Cds particles (Pages et al. 2008).

Some microorganisms expressing metal resistance produce greater extracellular polymeric materials (eps), which bind the metal, perhaps making the microenvironment across the plant less poisonous. Manufacturing of eps has been proven to

4 Overview of Microbial Associations and Their Role Under Aquatic Ecosystems

Increase with improved metal resistance. In different lines of *Rhizobium leguminosarum*, cd tolerant Traces showed expanded tiers of glutathione, indicating that this tripeptide allows the bacterium to deal with heavy metals, in preference to an efflux system (Figueira et al. 2005). The Biomarker glutathione is a critical antioxidant that may also protect against metallic toxicity related to oxidative Strain. Any other mechanism of dealing with poisonous Metals can also involve polyphosphates, lengthy chains of orthophosphates, which might also sequester metals (Alvarez and Jerez 2004). Perrin et al. Recently said that ni exposure promoted Biofilm formation in escherichia coli cultures, which may also Serve as a defensive tolerance mechanism. Ni appeared to Be worried in adherence by using inducing transcription of Genes encoding curli, the adhesive systems vital For biofilm formation (Perrin et al. 2009). The protecting fine of eps or these other mechanisms Provided by means of root-associated microorganism shows that enriching For positive bacteria might also replace the method of Amending plant root zones with synthetic go-related Polyacrylates and hydrogels to defend roots from heavy Metal toxicity (Blaylock et al. 1997) (Fig. 4.2).

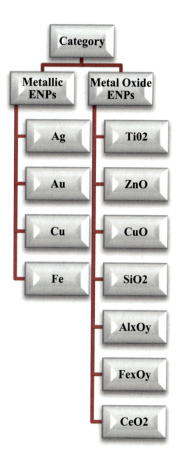

Fig. 4.2 Categories of engineered nanoparticles (Ebrahimbabaie et al. 2020)

4.13.2 Metal Tolerance and Resistance: The Plant Aspect

In plants, metal accumulation in the cells can be regulated by glutathione–phytochelatin-mediated resistance. On this system, glutathione, the identical cysteine-containing tripeptide defined above in *Rhizobium* sp., which also has several features in plant cells including coping with toxic oxygen species and amino acid delivery, Is used to synthesize phytochelatins, which chelate Heavy metals by means of formation of a thiolate. Thiolates may be transported to vacuoles for heavy metallic storage (Mendoza-Cozatl and Moreno-Sanchez 2006). Phytochelatins are activated by using heavy metals and scavenge heavy metals in plant cells (Blum et al. 2007). For a review of Plant metal tolerance mechanisms, see Jabeen et al. (2009). While some flora deal with mild degrees of toxic metals by using chelation, other plants have the potential to acquire extraordinarily high ranges of heavy metals and sequester them of their tissues. Flora that may accumulate extremely excessive concentrations of metals are termed "hyperaccumulators." There were extra than four hundred plant species recognized as such (Prasad and Freitas 2003), consisting of crop Species (Vamerali et al. 2010), and the quantity of hyperaccumulators among aquatic and wetland plants is growing (Prasad et al. 2001).

Hyperaccumulators can be defined based on bioconcentration Component (bcf), or the capacity to accumulate metals in plant tissues. As an instance, the potential to acquire more than one thousand instances the awareness of Cd (based totally on attention of metal in dry weight of Plant) than that inside the surrounding medium could be taken into consideration hyperaccumulation (Zayed et al. 1998). One of the maximum Studied hyperaccumulators is the terrestrial plant *T. caerulescens*, that's a Cd/Zn hyperaccumulator (Mijovilovich et al. 2009). Numerous aquatic vegetation have been located to have comparable abilities, along with *Salvinia minima* (Sanchez-Galvan et al. 2008), *Potamogeton natans* (Fritioff and Greger 2006), *Ceratophyllum demersum* (Robinson et al. 2006), and *S. polyrrhiza* (John et al. 2008). Metal hyperaccumulation is an adaptive manner between microbes exposed to heavy metals and flora, requiring continuous interactions a few of the cogoing on organisms. A recent proteomics look at by Farinati et al. (2009) indicated that the presence of a rhizosphere microbial Population, tailored to heavy-metallic-polluted websites, substantially enhanced the accumulation of metals in shoots Of the hyperaccumulator arabidopsis halleri. Aquatic environments consist of no longer simplest macrophytes, but also algae that may have interaction with microbes to take away contaminants from the environment. Algae can be produced in artificial structures and used to cast off contaminants. Loutseti et al. used a dried aggregate of microalgae and bacteria to dispose of cu and cd from wastewater. Munoz et al. efficaciously examined the mixture of the bacterium *Ralstonia basilensis* and the microalga *Chlorella Sorokiniana* on adsorption of Cd, Cu, Ni, and Zn. Moreover, mycorrhizal fungi associated with plant life can Beautify uptake of metals when critical metallic concentrations Are low and, vice versa, while metal quantities are too high mycorrhizae may be effective in assuaging metal Toxicity lowering plant uptake (Frey et al. 2000).

4.13.3 Mechanisms for How Aquatic Plant–Microbe Interactions Have an Effect on Phytoremediation Techniques

At the same time as many flora and microorganism have their personal mechanisms for handling heavy metal contaminants, the interplay of flora and microorganisms may additionally boom Or lower heavy metal accumulation in flora, depending on the character of the plant–microbe interaction. Due to the fact phytoremediation is an extraordinarily new era, information mechanisms of plant–microbe interactions in putting off contaminants from the surroundings is still no longer properly characterized. There have, however, been several ideas about the nature of plant–microbe interactions in metal accumulation. In their paper describing bacterial enhancement of Se and Hg uptake by wetland plants, De souza et al. (1999) proposed numerous possible mechanisms, such as bacterial stimulation of plant metal uptake compounds including siderophores; bacterial root increase promoting increasing the basis floor place; bacterial transformation of elements into greater soluble bureaucracy; or bacterial stimulation of plant transporters that could transport essential factors as well as heavy metals (within the case of selenate, The sulfate transporter. Van der lelie related the basis of this plant–microbe interaction to bacterial metal resistance, for the reason that bioavailability of metals may be altered by means of bacterial expression of resistance structures (Van der Lelie et al. 2000).

4.13.4 Plant Microbe Interaction in Perspective to Sustainable Agriculture Development

Microbes represent an important reservoir, which influence the macroorganisms. Plant health and its efficiency can be maintained by beneficial symbiotic bacteria such as *Bacillus, Pseudomonas, Rhizobium, Pantoea agglomerans, Bradyrhizobium, Actinobacteria, and Burkholderia* (Khati et al. 2018). Plant microbial interactions help in the growth of plants and increase the resistance toward stress conditions (Chaudhary et al. 2021a, b; Perreault and Laforest-Lapointe 2022). Microbes support plant rhizosphere functions via supportive expansion of roots and growth of plants (Kumar et al. 2021a, b; Chaudhary et al. 2022). In response, roots of plant secreted organic composites, which provide carbon sources and signals for microbial growth. Rhizospheric microbes adjust and affect the properties of soil/water ecosystem. Microbial activities help in transformation of nutrients, offer protection toward pathogens, and support the plant growth (Agri et al. 2021, 2022; Chaudhary et al. 2021c, d). Microbes respond rapidly to any change in the environment. Changes in microbial communities have been associated to the properties of soil under heavy metals/pesticides stress (Kumar et al. 2021a, b). Aquatic plants and their associated microbes are less studied. Application of denitrifying bacteria such as *Pseudomonas* sp. modifies the chemical conformation of root exudates in duckweed (aquatic plant) and improved excretion of stigmasterol, which is involved in N exclusion from aquatic plants via root and microbial interactions (Lu et al. 2021). *Acidobacteria* are extensively dispersed in different environments like mine water, marine sponges, and hot springs. It is also found in roots of some water plants such as *Scirpus juncoides* and *Lythrum* and increased plant growth (Tanaka et al. 2017). Coculture of *Acidobacteria* strains in aquatic duckweed species improved chlorophyll content and plant growth due to the colonization of bacteria on roots confirmed via fluorescence microscopy (Yoneda et al. 2021). It also showed the properties such as IAA and siderophore production and mineral solubilization. *Rhizobium unidicola* forms symbiotic association with aquatic plant *Neptunia oleracea* and improved their growth by production of phytohormones. Cable bacteria such as *Candidatus electrothrix* and *C. electronema* are sulfide-oxidizing and filamentous bacteria, which reduce the toxic sulfide and encourage the iron oxide development in oxygen-liberating plants, which act toward the toxic form of arsenic uptake in roots of aquatic plants (Sulu-Gambari et al. 2016; Scholz et al. 2021). These bacteria also promote the nitrogen availability to plants and present around the aquatic plant roots such as *Littorella uniflora*, *Potamogeton*, and *Equisetum* sp. (Kessler et al. 2019). Plant microbe interaction in aquatic water can be used in pollutant removal. *Bacillus* and *Microbacterium* sp. present in roots of *Lolium perenne* improved the remediation of nitrogen and phosphorus (Li et al. 2011). Presence of *Pseudomonas, Bacillus, Rhodococcus* sp., *Brachia mutica, Leptochala fusca* helps in remediation of iron, nickel, and Pb metals (Shahid et al. 2020). Remediation of benzene is increased in the presence of the aquatic plant *Syngonium podophyllum* and microbes such as *Enterobacter* and *Cronobacter* (Sriprapat and Thiravetyan 2016). Application of *Pseudomonas* sp. in the presence of *Lolium perenne* and *Arabidopsis* removed the total petroleum hydrocarbons from aquatic environment and improved the plant growth under stressful conditions (Iqbal et al. 2019). Nirophenols and phenolic compounds remediation occur with the help of microbes such as *Pseudomonas, Cupriavidus* sp., *Rhodococcus* sp., and *Pantoea* sp. and aquatic plant like *Sprirodela polyrhiza* and *Brassica napus* (Kristanti et al. 2016; Ontañon et al. 2014). Chlorpyrifos remediated by using planktonic bacterial communities with *Nymphaea alba* and *Phragmities australis* reported by Xu et al. (2018). Remediation of oil occurs by using *Ochrobactrum, Desulfovibrio* and *Halobacterium* sp. with *Halonemum strobilaceum* plant (Al-Mailem et al. 2010). Application of microbe-plant system *Bacillus* and *Pantoea* along with *Dracaena sanderiana* improved the bisphenol remediation to about 92%: this was reported by Suyamud et al. (2018). Removal of these pollutants occurs with the help of rhizofiltration and biostimulation.

4.14 Conclusion

Microbes coevolved with the plants for the fulfillment of their nutrient deficiencies in their fundamental niches. Thus, the aquatic microbiome plays a vital role in influencing and promoting the aquatic ecosystem. The plant–microbe interaction is also affected by the environment; both biotic and abiotic factors have a concomitant effect on the marine ecosystem. This leads to change at molecular and gene expression level causing production of compounds by these microbial-host-environmental interactions. Marine plants survive and thrive in a competitive and hostile environment by living in complex colonies and in close proximity to each other. Aquatic plants have evolved physiological adaptations, such as the production of bioactive chemicals, that provide them protection against grazers. Plants exist in a microbe-rich environment in nature, and they must interact with a diverse range of pathogenic, commensal, and helpful bacteria. Thus the aquatic microbiome plays a vital role in influencing and promoting the aquatic ecosystem. Aquatic phytoremediation of metals, the usage of plants to extract, contain and immobilize, or dispose of hazardous substances from aqueous environments is a completely promising vicinity, and several fairly green examples have proven the applicability of this technique to smooth business waste streams, to concentrate heavy metals, and to keep Drinking water and aquatic biodiversity. Limitations to this generation do exist and must also be taken into consideration. The utility of rhizofiltration is constrained by means of metallic availability, attention, and phytotoxicity. Environmental factors like mild, salinity, temperature, Ph, and presence of a couple of heavy metals might also have an effect on metallic uptake. Similarly limitations of phytoremediation Technology are seasonal increase of aquatic flora and contaminated biomass disposal problems. Phytoremediation centered on dissolved metals can be based totally on the software of each dead or live plant cloth or at the cultivation of aquatic plant s. No longer only aquatic macrophytes, but also algae and fungi, constitute a value-effective and green era for environmental Cleanup, a green answer regularly favored in Political decision-making. Rhizosphere microbes can lessen metal toxicity and enhance plant tolerance to dissolved metals, and may therefore be applied to deliver extended phytoprotection From dangerous results of the metals on plant life. In an immediate extension of this idea, the bacterial genes coding for metal resistance can be transplanted into the plant genome to confer elevated metallic tolerance to plants. In addition research in aquatic phytoremediation is wanted to increase knowledge of microbe–plant interactions. Such knowledge would boom the quantity of doubtlessly sizable programs and their effect together with the remedy of heavy metals from business effluents in Natural and built wetlands, or a wastewater metal stripping phase the use of rhizofiltration. A directed purposeful analysis has to inspect plant–Microbe interactions at complete organic hierarchy, starting with the genomic, transcriptomic, and proteomic analysis Of plant-associated bacteria and their extracellular enzyme activities, all of the manner to biochemical techniques And biking which might be energetic in the bacterially prompted rhizosphere. With this understanding the plant–microbe device may be carried out at discipline-scale, the use of clearly adapted indigenous microbes that have Been cultured and enriched in the laboratory. Any such multidisciplinary and included technique may also benefit aquatic metal phytoremediation the industrial significance in environmental biotechnology it deserves.

References

Abhilash PC, Pandey VC, Srivastava P, Rakesh PS, Chandran S, Singh N, Thomas AP (2009) Phytofiltration of cadmium from water by Limnocharis flava (L.) Buchenau grown in free-floating culture system. J Hazard Mater 170:791–797

Adalbjörnsson BV, Jonsdottir R (2015) Enzyme-enhanced extraction of antioxidant ingredients from algae. In: Natural products from marine algae. Springer, New York, pp 145–150

Adams DG, Bergman BIRGITTA, Nierzwicki-Bauer SA, Rai AN, Schüßler A (2006) Cyanobacterial-plant symbioses. In: The prokaryotes: a handbook on the biology of bacteria, vol 1. Springer, New York, pp 331–363

Admassu H, Gasmalla MAA, Yang R, Zhao W (2018) Bioactive peptides derived from seaweed protein and their health benefits: antihypertensive, antioxidant, and antidiabetic properties. J Food Sci 83(1):6–16

Agri U, Chaudhary P, Sharma A (2021) In vitro compatibility evaluation of agriusable nanochitosan on beneficial plant growth-promoting rhizobacteria and maize plant. Natl Acad Sci Lett 44:555–559

Agri U, Chaudhary P, Sharma A, Kukreti B (2022) Physiological response of maize plants and its rhizospheric microbiome under the influence of potential bioinoculants and nanochitosan. Plant and Soil 474:451

Agunbiade FO, Olu-Owolabi BI, Adebowale KO (2009) Phytoremediation potential of Eichornia crassipes in metal-contaminated coastal water. Bioresour Technol 100:4521–4526

Aleklett K, Hart M, Shade A (2014) The microbial ecology of flowers: an emerging frontier in phyllosphere research. Botany 92:253–266. https://doi.org/10.1139/cjb-2013-0166

Al-Mailem DM, Sorkhoh NA, Marafie M et al (2010) Oil phytoremediation potential of hypersaline coasts of the Arabian Gulf using rhizosphere technology. Bioresour Technol 101:5786–5792

Alvarez S, Jerez CA (2004) Copper ions stimulate polyphosphate degradation and phosphate efflux in Acidithiobacillus ferrooxidans. Appl Environ Microbiol 70:5177–5182

Arber A (1920) A study of aquatic angiosperms. Cambridge University Press, Cambridge, 436p

Bamidele JF, Nyamali B (2008) Ecological studies of the Ossiomo river with reference to the macrophytic vegetation. Res J Bot 3(1):29–34

Barko JW, Adams MS, Clesceri NL (1986) Environmental factors and their consideration in the management of submerged aquatic vegetation: a review. J Aquat Plant Manage 24:1–10

Baron C, Domke N, Beinhofer M, Hapfelmeier S (2001) Elevated temperature differentially affects virulence, VirB protein accumulation, and T-pilus formation in different Agrobacterium tumefaciens and Agrobacterium vitis strains. J Bacteriol 183:6852–6861

Barragán BE, Costa C, Marquez MC (2007) Biodegradation of azo dyes by bacteria inoculated on solid media. Dyes Pigm 75(1):73–81

Bast F (2013) Agronomy and cultivation methods for edible seaweeds. Int J Agric Food Sci Technol 4(7):661–666

Bergelson J, Mittelstrass J, Horton MW (2019) Characterizing both bacteria and fungi improves understanding of the Arabidopsis root microbiome. Sci Rep 9:24. https://doi.org/10.1038/s41598-018-37208-z

Bergholz PW, Noar JD, Buckley DH (2011) Environmental patterns are imposed on the population structure of Escherichia coli after fecal deposition. Appl Environ Microbiol 77(1):211–219

Biesboer DD (1984) Nitrogen fixation associated with natural and cultivated stands of Typha latifolia L. (Typhaceae). Am J Bot 71(4):505–511

Bing H, Felix GA, Xiaoying M et al (2019) Epiphytic bacterial community shift drives the nutrient cycle during Potamogeton malaianus decomposition. Chemosphere 236:124253

Blaylock MJ, Salt DE, Dushenkov S, Zakharova O, Gussman C, Kapulnik Y, Ensley BD, Raskin I (1997) Enhanced accumulation of Pb in Indian mustard by soil-applied chelating agents. Environ Sci Technol 31:860–865

Blindow A, Hargeby A, Hilt S (2014) Facilitation of clear-water conditions in shallow lakes by macrophytes: Differences between charophyte and angiosperm dominance. Hydrobiologia 737: 99–110

Blum R, Beck A, Korte A, Stengel A, Letzel T, Lendzian K, Grill E (2007) Function of phytochelatin synthase in catabolism of glutathione-conjugates. Plant J 49:740–749

Bornette G, Puijalon S (2011) Response of aquatic plants to abiotic factors: a review. Aquat Sci 73: 1–14

Boutin C, Keddy PA (1993) A functional classification of wet land plants. J Veg Sci 4:591–600

Brix H (1997) Do macrophytes play a role in constructed treatment wetlands? Water Sci Technol 35:11–17

Brock MA, Casanova MT (1997) Plant life at the edges of wetlands: ecological responses to wetting and drying patterns. In: Klomp NI, Lunt ID (eds) Frontiers in ecology: building the links. Elsevier Science, Oxford, pp 181–192

Brönmark C, Hansson LA (2017) The biology of lakes and ponds, 3rd edn. Oxford University Press, Oxford, 368p

Carpenter SR, Lodge DM (1986) Effects of submerged macrophytes on ecosystem processes. Aquat Bot 26:341–370

Castella E, Richardot-Coulet M, Roux C, Richoux P (1984) Macro-invertebrates as describers of morphological and hydrological types of aquatic ecosystems abandoned by the Rhone River. Hydrobiologia 119:219–226

Cermeno M, Stack J, Tobin P, O'Keeffe MB, Harnedy PA, Stengel D, FitzGerald RJ (2019) Peptide identification from a Porphyra dioica protein hydrolysate with antioxidant, angiotensin converting enzyme and dipeptidyl peptidase IV inhibitory activities. Food Funct 10(6): 3421–3429

Chambers PA, Lacoul P, Murphy KJ, Thomaz SM (2007) Global diversity of aquatic macrophytes in freshwater. In: Balian EV, Lévêque C, Segers H, Martens K (eds) Freshwater animal diversity assessment. Developments in hydrobiology, vol 198. Springer, Dordrecht

Chambers PA, Lacoul P, Murphy KJ, Thomaz SM (2008) Global diversity of aquatic macrophytes in freshwater. Hydrobiologia 198:9–26

Chanton J, Chaser L, Glasser P, Siegel D (2005) Carbon and hydrogen isotopic effects in microbial, methane from terrestrial environments. In: Stable isotopes and biosphere - atmosphere interactions: processes and biological controls, pp 85–112. https://doi.org/10.1016/B978-012088447-6/50006-4

Chao C, Wang L, Li Y, Yan Z, Liu H, Yu D, Liu C (2021) Response of sediment and water microbial communities to submerged vegetations restoration in a shallow eutrophic lake. Sci Total Environ 801:149701

Chaudhary P, Khati P, Chaudhary A, Gangola S, Kumar R, Sharma A (2021a) Bioinoculation using indigenous Bacillus spp improves growth and yield of Zea mays under the influence of nanozeolite. 3Biotech 11:11

Chaudhary P, Khati P, Chaudhary A, Maithani D, Kumar G, Sharma A (2021b) Cultivable and metagenomic approach to study the combined impact of nanogypsum and Pseudomonas taiwanensis on maize plant health and its rhizospheric microbiome. PloS One 16:e0250574

Chaudhary P, Khati P, Gangola S, Kumar A, Kumar R, Sharma A (2021c) Impact of nanochitosan and *Bacillus* spp on health, productivity and defence response in *Zea mays* under field condition. 3Biotech 11:237

Chaudhary P, Chaudhary A, Parveen H, Rani A, Kumar G, Kumar A, Sharma A (2021d) Impact of nanophos in agriculture to improve functional bacterial community and crop productivity. BMC Plant Biol 21:519

Chaudhary P, Chaudhary A, Bhatt P, Kumar G, Khatoon H, Rani A, Kumar S, Sharma A (2022) Assessment of soil health indicators under the influence of nanocompounds and Bacillus spp. in field condition. Front Environ Sci 9:769871

Chen J, Zhang H, Han Z, Ye J, Liu Z (2012) The influence of aquatic macrophytes on Microcystis aeruginosa growth. Ecol Eng 42:130–133

Christensen PB, Revsbech NP, Sand-Jensen K (1994) Microsensor analysis of oxygen in the rhizosphere of the aquatic macrophyte Littorella uniflora (L.) Ascherson. Plant Physiol 105: 847–852

Compant S, van der Heijden MGA, Sessitsch A (2010) Climate change effects on beneficial plant-microorganism interactions. FEMS Microbiol Ecol 73:197–214

Compant S, Samad A, Faist H, Sessitsch A (2019) A review on the plant microbiome: ecology, functions, and emerging trends in microbial application. J Adv Res 19:29–37. https://doi.org/10.1016/j.jare.2019.03.004

Cook CDK (1999) The number and kinds of embryobearing plants which have become aquatic: a survey. Perspect Plant Ecol Evol Syst 2(1):79–102

Coombs JM, Barkay T (2005) New findings on evolution of metal homeostasis genes: evidence from comparative genome analysis of bacteria and archaea. Appl Environ Microbiol 71:7083–7091

De Souza MP, Huang CPA, Chee N, Terry N (1999) Rhizosphere bacteria enhance the accumulation of selenium and mercury in wetland plants. Planta 209:259–263

DeBusk WF, Reddy KR (2005) Litter decomposition and nutrient dynamics in a phosphorus enriched everglades marsh. Biogeochemistry 75:217–240

Dhote S, Dixit S (2009) Water quality improvement through macrophytes – a review. Environ Monit Assess 152:149–153. https://doi.org/10.1007/s10661-008-0303-9. PMID: 18537050

Diaz S, Cabido M (1997) Plant functional types and ecosystem function in relation to global change. J Veg Sci 8:463–474

Dixit SP et al (2021) Genome analyses revealed genetic admixture and selection signatures in Bos indicus. Sci Rep 11(1):1–11

Ebrahimbabaie P, Meeinkuirt W, Pichtel J (2020) Phytoremediation of engineered nanoparticles using aquatic plants: mechanisms and practical feasibility. J Environ Sci 93:151

Faist H, Keller A, Hentschel U, Deeken R (2016) Grapevine (*Vitis vinifera*) crown galls host distinct microbiota. Appl Environ Microbiol 82:5542–5552

Farinati S, DalCorso G, Bona E, Corbella M, Lampis S, Cecconi D, Polati R, Berta G, Vallini G, Furini A (2009) Proteomic analysis of Arabidopsis halleri shoots in response to the heavy metals cadmium and zinc and rhizosphere microorganisms. Proteomics 9:4837–4850

Figueira EM, Lima AI, Pereira SI (2005) Cadmiumtolerance plasticity in Rhizobium leguminosarum bv. viciae: glutathione as a detoxifying agent. Can J Microbiol 51:7–14

Fink P (2007) Ecological functions of volatile organic compounds in aquatic systems. Mar Freshw Behav Physiol 40(3):155–168

Flathman PE, Lanza GR (1998) Phytoremediation: current views on an emerging green technology. J Soil Contam 7:415–432

Fleurence J, Morançais M, Dumay J, Decottignies P, Turpin V, Munier M et al (2012) What are the prospects for using seaweed in human nutrition and for marine animals raised through aquaculture? Trends Food Sci Technol 27(1):57–61

Frey B, Zierold K, Brunner I (2000) Extracellular complexation of Cd in the Hartig net and cytosolic Zn sequestration in the fungal mantle of Picea abies–Hebeloma crustuliniforme ectomycorrhizas. Plant Cell Environ 23:1257–1265

Fritioff A, Greger M (2006) Uptake and distribution of Zn, Cu, Cd, and Pb in an aquatic plant, Potamogeton natans. Chemosphere 63:220–227

Gadd GM (1992) Metals and microorganisms: a problem of definition. FEMS Microbiol Lett 79: 197–203

Ghassemzadeh F, Yousefzadeh H, Arbab-Zavar MH (2008) Removing arsenic and antimony by Phragmites australis: rhizofiltration technology. J Appl Sci 8:1668–1675

Glazer AN (1994) Phycobiliproteinsda family of valuable, widely used fluorophores. J Appl Phycol 6(2):105–112

Golubev SN, Schelud'ko AV, Muratova AY, Makarov OE, Turkovskaya OV (2009) Assessing the potential of rhizobacteria to survive under phenanthrene pollution. Water Air Soil Pollut 198(1–4):5–16

4 Overview of Microbial Associations and Their Role Under Aquatic Ecosystems

Gonzalez LE, Bashan Y (2000) Increased growth of the microalga *Chlorella vulgaris* when coimmobilized and cocultured in alginate beads with the plant-growth-promoting bacterium Azospirillum brasilense. Appl Environ Microbiol 66(4):1527–1531

Gradstein R, Vanderpoorten A, Reenen G, van Cleef A (2018) Mass occurrence of the liverwort Herbertus sendtneri in a glacial lake in the Andes of Colombia. Rev Acad Colomb Cienc Exactas Fís Nat 42:221–229

Gu L, Wu J, Hua Z et al (2020) The response of nitrogen cycling and bacterial communities to E. coli invasion in aquatic environments with submerged vegetation. J Environ Manage 261: 110204

Gupta S, Abu-Ghannam N (2011) Recent developments in the application of seaweeds or seaweed extracts as a means for enhancing the safety and quality attributes of foods. Innov Food Sci Emerg Technol 12(4):600–609

Hacquard S, Spaepen S, Garrido-Oter R, Schulze-Lefert P (2017) Interplay between innate immunity and the plant microbiota. Annu Rev Phytopathol 55:565–589

Haider FU, Liqun C, Coulter JA, Cheema SA, Wu J, Zhang R, Wenjun M, Farooq M (2021) Cadmium toxicity in plants: impacts and remediation strategies. Ecotoxicol Environ Saf 211: 1118–1187

Harnedy PA, FitzGerald RJ (2013) Extraction of protein from the macroalga Palmaria palmata. LWT Food Sci Technol 51(1):375–382

Hasegawa H, Chatterjee A, Cui Y, Chatterjee AK (2005) Elevated temperature enhances virulence of Erwinia carotovora subsp. Carotovora strain EC153 to plants and stimulates production of the quorum sensing signal, N-acyl homoserine lactone, and extracellular proteins. Appl Environ Microbiol 71:4655–4663

Hoffland E, Kuyper TW, Comans RN, Creamer RE (2020) Eco-functionality of organic matter in soils. Plant and Soil 455(1):1–22

Holgerson MA, Raymond PA (2016) Large contribution to inland water CO2 and CH4 emissions from very small ponds. Nat Geosci 9:222–226

Hossain K, Yadav S, Quaik S, Pant G, Maruthi AY, Ismail N (2017) Vulnerabilities of macrophytes distribution due to climate change. Theor Appl Climatol 129:1123–1132

Hou W, Chen X, Song G, Wang Q, Chi Chang C (2007) Effects of copper and cadmium on heavy metal polluted waterbody restoration by duckweed (Lemna minor). Plant Physiol Biochem 45: 62–69

Hrivnák R, Oťaheľová H, Jarolímek I (2006) Diversity of aquatic macrophytes in relation to environmental factors in the Slatina river (Slovakia). Biologia 61(4):413–419

Huckle JW, Morby AP, Turner JS, Robinson NJ (1993) Isolation of a prokaryotic metallothionein locus and analysis of transcriptional control by trace metal ions. Mol Microbiol 7:177–187

Iqbal A, Mukherjee M, Rashid J et al (2019) Development of plant-microbe phytoremediation system for petroleum hydrocarbon degradation: an insight from alkb gene expression and phytotoxicity analysis. Sci Total Environ 671:696–704

Jabeen R, Ahmad A, Iqbal M (2009) Phytoremediation of heavy metals: physiological and molecular mechanisms. Bot Rev 75:339–364. https://doi.org/10.1007/s12229-009-9036-x

Jackson MB, Ishizawa K, Ito O (2009) Evolution and mechanisms of plant tolerance to flooding stress. Ann Bot 103:137–142

Jang J, Di D, Han D et al (2015) Dynamic changes in the population structure of Escherichia coli in the Yeongsan River basin of South Korea. FEMS Microbiol Ecol 11:123–125

Jeppesen E, Søndergaard M (1999) Lake and catchment management in Denmark. Hydrobiologia 396:419–432

Jiang C-J, Shimono M, Sugano S, Kojima M, Yazawa K, Yoshida R, Inoue H, Hayashi N, Sakakibara H, Takatsuji H (2010) Abscisic acid interacts antagonistically with salicylic acid signaling pathway in rice-Magnaporthe grisea interaction. Mol Plant Microbe Interact 23:791–798

Jiang L, Zhu D, Chen Y et al (2015) Analysis on fecal coliform pollution in surface waters of China. Adv Sci Technol Water Resour 35(3):11–18. (In Chinese)

Jiménez-Escrig A, Gomez-Ordõnez E, Rupérez P (2011) Seaweed as a source of novel nutraceuticals: sulfated polysaccharides and peptides. In: Advances in food and nutrition research, vol 64. Elsevier, Amsterdam, pp 325–337

Jin S, Song YN, Deng WY, Gordon MP, Nester EW (1993) The regulatory VirA protein of Agrobacterium tumefaciens does not function at elevated temperatures. J Bacteriol 175:6830–6835

John R, Ahmad P, Gadgil K, Sharma S (2008) Effect of cadmium and lead on growth, biochemical parameters and uptake in Lemna polyrrhiza L. Plant Soil Environ 54:262–270

Jones JDG, Vance RE, Dangl JL (2016) Intracellular innate immune surveillance devices in plants and animals. Science 354:6395. https://doi.org/10.1126/science.aaf6395

Kadir K, Nelson KL (2014) Sunlight mediated inactivation mechanisms of Enterococcus faecalis and Escherichia coli in clear water versus waste stabilization pond water. Water Res 50(1):307–317

Kadlec RH (2000) The inadequacy of first-order removal models. Ecol Eng 15:105–119

Kandel SL, Joubert PM, Doty SL (2017) Bacterial endophyte colonization and distribution within plants. Microorganisms 5(4):77

Kang HK, Lee HH, Seo CH, Park Y (2019) Antimicrobial and immunomodulatory properties and applications of marine-derived proteins and peptides. Mar Drugs 17(6):350

Kannikka B, Seesha CP, Subhasha SN (2018) Bioremediation by microalgae. In: Tripathi K, Rathor NK, Abraham G (eds) The role of photosynthetic microbes in agriculture and industry. Nova Science, Hauppauge, pp 151–172

Kessler AJ, Wawryk M, Marzocchi U, Roberts KL, Wong WW, Risgaard Petersen N, Meysman FJ, Glud RN, Cook PL (2019) Cable bacteria promote DNRA through iron sulfide dissolution. Limnol Oceanogr 64:1228–1238

Khalid S, Abbas M, Saeed F, Bader-Ul-Ain H, Suleria HAR (2018) Therapeutic potential of seaweed bioactive compounds. In: Seaweed biomaterials. IntechOpen, London

Khati P, Parul Bhatt P, Nisha Kumar R, Sharma A (2018) Effect of nanozeolite and plant growth promoting rhizobacteria on maize. 3Biotech 8:141

Khellaf N, Zerdaoui M (2009) Phytoaccumulation of zinc by the aquatic plant, Lemna gibba L. Bioresour Technol 100:6137–6140

Kim H, Ridenour JB, Dunkle LD, Bluhm BH (2011a) Regulation of stomatal tropism and infection by light in Cercospora zeae-maydis: evidence for coordinated host/pathogen responses to photoperiod? PLoS Pathog 7:e1002113

Kim J-K, Cho ML, Karnjanapratum S, Shin I-S, You SG (2011b) In vitro and in vivo immunomodulatory activity of sulfated polysaccharides from Enteromorpha prolifera. Int J Biol Macromol 49(5):1051–1058

Kraemer U (2005) Phytoremediation: novel approaches to cleaning up polluted soils. Curr Opin Biotechnol 16:1–9. Interesting recent review discussing phytoextraction and transgenic plants, and modern genetic approaches to phytoremediation

Kristanti F, Rahayu S, Huda A (2016) The determinant of financial distress on Indonesian family firm. Procedia Soc Behav Sci 219:440–447. https://doi.org/10.1016/j.sbspro.2016.05.018

Kristensen E (2000) Organic matter diagenesis at the oxic/anoxic interface in coastal marine sediments, with emphasis on the role of burrowing animals. In: Life at interfaces and under extreme conditions. Springer, Dordrecht, pp 1–24

Kumar G, Suman A, Lal S, Ram RA, Bhatt P, Pandey G, Chaudhary P, Rajan S (2021a) Bacterial structure and dynamics in mango (Mangifera indica) orchards after long term organic and conventional treatments under subtropical ecosystem. Sci Rep 11:20554

Kumar G, Lal S, Maurya SK, Bhattacherjee AK, Chaudhary P, Gangola S et al (2021b) Exploration of Klebsiella pneumoniae M6 for paclobutrazol degradation, plant growth attributes, and biocontrol action under subtropical ecosystem. PLoS One 16(12):e0261338

Kumwimba MN, Meng F (2019) Roles of ammonia-oxidizing bacteria in improving metabolism and cometabolism of trace organic chemicals in biological wastewater treatment processes: a review. Sci Total Environ 659:419–441

Lavorel S, McIntyre S, Landsberg J, Forbes TDA (1997) Plant functional classifications: from general groups to specific groups based on response to disturbance. Trends Ecol Evol 12:474–478

Le HT, Ho CT, Trinh QH et al (2016) Responses of aquatic bacteria to terrestrial runoff: effects on community structure and key taxonomic groups. Front Microbiol 7:889

Lee M, Yang M (2010) Rhizofiltration using sunflower (Helianthus annuus L.) and bean (Phaseolus vulgaris L. var. vulgaris) to remediate uranium contaminated groundwater. J Hazard Mater 173: 589–596

Li FM, Hu HY (2005) Isolation and characterization of a novel antialgal allelochemical from Phragmites communis. Appl Environ Microbiol 71(11):6545–6553

Li H, Zhao HP, Hao HL et al (2011) Enhancement of nutrient removal from eutrophic water by a plant-microorganisms combined system. Environ Eng Sci 28:543–554

Li G, Hu S, Hou H, Kimura S (2019) Heterophylly: phenotypic plasticity of leaf shape in aquatic and amphibious plants. Plants 8:420

Li G, Jian-yi W, Hua Z-l (2021) Benthic prokaryotic microbial community assembly and biogeochemical potentials in E. coli - stressed aquatic ecosystems during plant decomposition. Environ Pollut 275:116643

Lin CS, Lin YH, Wu JT (2012) Biodiversity of the epiphyllous algae in a chamaecyparis forest of northern Taiwan. Bot Stud 53:489–499

Lu H, McClung CR, Zhang C (2017) Tick tock: circadian regulation of plant innate immunity. Annu Rev Phytopathol 55:287–311

Lu Y, Kronzucker HJ, Shi W (2021) Stigmasterol root exudation arising from Pseudomonas inoculation of the duckweed rhizosphere enhances nitrogen removal from polluted waters. Environ Pollut 287:117587

Maal-Bared R, Bartlett KH, Bowie WR et al (2013) Phenotypic antibiotic resistance of Escherichia coli and E. coli O157 isolated from water, sediment and biofilms in an agricultural watershed in British Columbia. Sci Total Environ 443:315–323

Maberly SC, Madsen TV (2002) Freshwater angiosperm carbon concentrating mechanisms: processes and patterns. Funct Plant Biol 29:393–405

Mähnert B, Schagerl M, Krenn L (2017) Allelopathic potential of stoneworts. Fottea Olomouc 17(2):137–149

Mallon CA, Elsas JDV, Salles JF (2015) Microbial invasions: the process, patterns, and mechanisms. Trends Microbiol 23(11):719–729

Mallon CA, Roux XL, Doorn GSV et al (2018) The impact of failure: unsuccessful bacterial invasions steer the soil microbial community away from the invader's niche. ISME J 12(3):728

Manzar N, Singh Y, Kashyap AS, Sahu PK, Rajawat MVS, Bhowmik A et al (2021) Biocontrol potential of native Trichoderma spp. against anthracnose of great millet (Sorghum bicolour L.) from Tarai and hill regions of India. Biol Control 152:1049–9644

Megateli S, Semsari S, Couderchet M (2009) Toxicity and removal of heavy metals (cadmium, copper, and zinc) by Lemna gibba. Ecotoxicol Environ Saf 72:1774–1780

Melotto M, Underwood W, Koczan J, Nomura K, He SY (2006) Plant stomata function in innate immunity against bacterial invasion. Cell 126:969–980

Mendonça R, Müller RA, Clow D et al (2017) Organic carbon burial in global lakes and reservoirs. Nat Commun 8:1694

Mendoza-Cozatl DG, Moreno-Sanchez R (2006) Control of glutathione and phytochelatin synthesis under cadmium stress. Pathway modeling for plants. J Theor Biol 238:919–936

Meshram BG, Chaugule BB (2018) An introduction to cyanobacteria: diversity and potential applications. In: Tripathi K, Rathor NK, Abraham G (eds) The role of photoesynthetic microbes in agriculture and industry, Nova Science Publisher, pp 1–39

Mian MH (2002) Azobiofer: a technology of production and use of Azolla as biofertiliser for irrigated rice and fish cultivation. In: Biofertilisers in action. Rural Industries Research and Development Corporation, Canberra, pp 45–54

Mijovilovich A, Leitenmaier B, Meyer-Klaucke W, Kroneck PM, Gotz B, Kupper H (2009) Complexation and toxicity of copper in higher plants. II. Different mechanisms for copper versus cadmium detoxification in the copper-sensitive cadmium/zinc hyperaccumulator Thlaspi caerulescens (Ganges Ecotype). Plant Physiol 151:715–731

Miler O, Albayrak I, Nikora V, O'Hare M (2014) Biomechanical properties and morphological characteristics of lake and river plants: implications for adaptations to flow conditions. Aquat Sci 76(4):465–481

Mishra VK, Tripathi BD (2009) Accumulation of chromium and zinc from aqueous solutions using water hyacinth (Eichhornia crassipes). J Hazard Mater 164:1059–1063

Mohapatra B, Phale PS (2021) Microbial degradation of naphthalene and substituted naphthalenes: Metabolic diversity and genomic insight for bioremediation. Front Bioeng Biotechnol 9:144

Mondav R, Mccalley CK, Hodgkins SB et al (2017) Microbial network, phylogenetic diversity and community membership in the active layer across a permafrost thaw gradient. Environ Microbiol 19(8):3201

Montalbán B, Thijs S, Lobo MC, Weyens N, Ameloot M, Vangronsveld J et al (2017) Cultivar and metal-specific effects of endophytic bacteria in *Helianthus tuberosus* exposed to Cd and Zn. Int J Mol Sci 18:2026. https://doi.org/10.3390/ijms18102026

Morales-Olmedo M, Ortiz M, Sellés G (2015) Effects of transient soil waterlogging and its importance for rootstock selection. Chil J Agric Res 75:45–56

Mouget JL, Dakhama A, Lavoie MC, de la Noüe J (1995) Algal growth enhancement by bacteria: is consumption of photosynthetic oxygen involved? FEMS Microbiol Ecol 18(1):35–43

Murakami M, Nakagawa F, Ae N, Ito M, Arao T (2009) Phytoextraction by rice capable of accumulating Cd at high levels: reduction of Cd content of rice grain. Environ Sci Technol 43:5878–5883

Mwimba M, Karapetyan S, Liu L, Marqués J, McGinnis EM, Buchler NE, Dong X (2018) Daily humidity oscillation regulates the circadian clock to influence plant physiology. Nat Commun 9: 4290

Nies DH (2003) Efflux-mediated heavy metal resistance in prokaryotes. FEMS Microbiol Rev 781: 1–27

Okolie CL, Mason B, Critchley AT (2018) Seaweeds as a source of proteins for use in pharmaceuticals and high-value applications. In: Novel proteins for food, pharmaceuticals, and agriculture: sources, applications, and advances. Wiley, Hoboken, p 217

Ontañon OM, González PS, Ambrosio LF et al (2014) Rhizoremediation of phenol and chromium by the synergistic combination of a native bacterial strain and Brassica napus hairy roots. Int Biodeter Biodegr 88:192–198

Opuszynski K, Shireman JV (1995) Herbivorous fishes: culture and use for weed management. In cooperation with James E. Weaver, Director of the United States Fish and Wildlife Service's National Fisheries Research Center. CRC Press, Boca Raton

Orth RJ, Carruthers TJB, Dennison WC, Duarte CM, Fourqurean JW, Heck KL, Hughes AR, Kendrick GA, Kenworthy WJ, Olyarnik S, Short FT, Waycott M, Williams SL (2006) A global crisis for seagrass ecosystems. BioScience 56(12):987–996

Osono T (2006) Role of phyllosphere fungi of forest trees in the development of decomposer fungal communities and decomposition processes of leaf litter. Can J Microbiol 52:701–716. https://doi.org/10.1139/w06-023

Pages D, Rose J, Conrod S, Cuine S, Carrier P, Heulin T, Achouak W (2008) Heavy metal tolerance in Stenotrophomonas maltophilia. PLoS One 3:e1539

Pal GK, Suresh PV (2016) Sustainable valorisation of seafood by-products: recovery of collagen and development of collagen-based novel functional food ingredients. Innov Food Sci Emerg Technol 37(part B):201–215. https://doi.org/10.1016/j.ifset.2016.03.015

Pal GK, Suresh PV (2017) Comparative assessment of physico-chemical characteristics and fibril formation capacity of thermostable carp scales collagen. Mater Sci Eng C 70:32–40

Panaccione DG, Beaulieu WT, Cook D (2014) Bioactive alkaloids in vertically transmitted fungal endophytes. Funct Ecol 28:299–314

Pandey VC, Singh JS, Singh DP, Singh RP (2014) Methanotrophs: promising bacteria for environmental remediation. Int J Environ Sci Technol 11(1):241–250

Pangestuti R, Kim S-K (2011) Biological activities and health benefit effects of natural pigments derived from marine algae. J Funct Foods 3(4):255–266

Pangestuti R, Kim S-K (2015) Seaweed proteins, peptides, and amino acids. In: Seaweed sustainability. Elsevier, Amsterdam, pp 125–140

Peng Y, Hu J, Yang B, Lin X-P, Zhou X-F, Yang X-W, Liu Y (2015) Chemical composition of seaweeds. In: Seaweed sustainability. Elsevier, Amsterdam, pp 79–124

Pérez MJ, Falqué E, Domínguez H (2016) Antimicrobial action of compounds from marine seaweed. Mar Drugs 14(3):52

Perreault R, Laforest-Lapointe I (2022) Plant-microbe interactions in the phyllosphere: facing challenges of the anthropocene. ISME J 16:339–345

Perrin C, Briandet R, Jubelin G, Lejeune P, Mandrand- Berthelot MA, Rodrigue A, Dorel C (2009) Nickel promotes biofilm formation by Escherichia coli K-12 strains that produce curli. Appl Environ Microbiol 75:1723–1733

Pieczynska E, Ozimek T (1976) Ecological significance of macrophytes. Int J Ecol Environ Sci 2: 115–128

Pooja K, Rani S, Prakash B (2017) In silico approaches towards exploration of rice bran proteins derived angiotensin-I-converting enzyme inhibitory peptides. Int J Food Prop 20:2178–2191

Prasad MNV, Freitas HMD (2003) Metal hyperaccumulation in plants — biodiversity prospecting for phytoremediation technology. Electron J Biotechnol 6:285–321. Review of phytoremediation technologies that includes list of known bacteria, algae, lichens, fungi, and plants that show resistance to metals and have potential applications for metal removal from the environment

Prasad MNV, Malec P, Waloszek A, Bojko M, Strzalka K (2001) Physiological responses of Lemna triscula L. (duckweed) to cadmium and copper bioaccumulation. Plant Sci 161:881–889

Prasanna R, Jaiswal P, Nayak S, Sood A, Kaushik BD (2009) Cyanobacterial diversity in the rhizosphere of rice and its ecological significance. Indian J Microbiol 49:89–97

Pratiksha B, Madhusmita M, Kim JY et al (2020) Benthic archaeal community structure and carbon metabolic profiling of heterotrophic microbial communities in brackish sediments. Sci Total Environ 706:135709

Rahman MA, Hasegawa H, Ueda K, Maki T, Okumura C, Rahman MM (2007) Arsenic accumulation in duckweed (Spirodela polyrhiza L.): a good option for phytoremediation. Chemosphere 69:493–499

Rai PK (2009) Heavy metal phytoremediation from aquatic ecosystems with special reference to aquatic macrophytes. Crit Rev Env Sci Technol 39:697–753. Detailed review of phytoremediation focusing on heavy metals and aquatic plants. Highlights advantages of phytoremediation for use in aquatic systems over conventional technologies and several hypotheses as to why plants hyperaccumulate metals

Rai PK, Tripathi BD (2009) Comparative assessment of Azolla pinnata and Vallisneria spiralis in Hg removal from G.B. Pant Sagar of Singrauli industrial region, India. Environ Monit Assess 148:75–84

Rani S, Pooja K (2018) Elucidation of structural and functional characteristics of collagenase from Pseudomonas aeruginosa. Process Biochem 64:116–123

Rani S, Pooja K, Pal GK (2018) Exploration of rice protein hydrolysates and peptides with special reference to antioxidant potential: computational derived approaches for bioactivity determination. Trends Food Sci Technol 80:61–70

Reinhold-Hurek B, Bunger W, Burbano CS, Sabale M, Hurek T (2015) Roots shaping their microbiome: global hotspots for microbial activity. Annu Rev Phytopathol 53:403–424

Reitsema RE, Meire P, Schoelynck J (2018) The future of freshwater macrophytes in a changing world: dissolved organic carbon quantity and quality and its interactions with macrophytes. Front Plant Sci 9:629

Rejmánková E (2011) The role of macrophytes in wetland ecosystems. J Ecol Field Biol 34(4): 333–345

Rejmankova E (2011) The role of macrophytes in wetland ecosystems. J Ecol Field Biol 34(4): 333–345

Robinson B, Kim N, Marchetti M, Moni C, Schroeter L, van den Dijssel C, Milne G, Clothier B (2006) Arsenic hyperaccumulation by aquatic macrophytes in the Taupo Volcanic Zone, New Zealand. Environ Exp Bot 58:206–215

Rocha MIA, Recknagel F, Minoti RT et al (2019) Assessing the effect of abiotic variables and zooplankton on picocyanobacterial dominance in two tropical mesotrophic reservoirs by means of evolutionary computation. Water Res 149:120–129

Roger PA, Zimmerman WJ, Lumpkin TA (1993) Microbiological management of wetland rice fields. In: Soil microbial ecology: applications in agricultural and environmental management. Marcel Dekker, New York, pp 417–455

Rooney VJN, Girwat MW, Savin MC (2005) Links between phytoplankton and bacterial community dynamics in a coastal marine environment. Microb Ecol 49:163–175

Roshchyna NO (2018) Modern condition and analysis of anthropogenous-climatic transformation of vegetation of lakes of the northern Steppe land. Ecol Noospherol 29(2):142–148

Rossier O (1995) Spatial and temporal separation of littoral zone fishes of Lake Geneva (Switzerland–France). Hydrobiologia 300(301):321–327

Rudgers JA, Afkhami ME, Bell-Dereske L, Chung YA, Crawford KM, Kivlin SN et al (2020) Climate disruption of plant-microbe. Annu Rev Ecol Evol Syst 51:561–586. https://doi.org/10.1146/annurev-ecolsys-011720-090819

Sadowsky MJ, Whitman RL (2011) Conclusions and future use of fecal indicator bacteria for monitoring water quality and protecting human health

Samad A, Trognitz F, Compant S, Antonielli L, Sessitsch A (2017) Shared and host specific microbiome diversity and functioning of grapevine and accompanying weed plants. Environ Microbiol 19:1407–1424

Samarakoon K, Jeon Y-J (2012) Bio-functionalities of proteins derived from marine algae a review. Food Res Int 48(2):948–960

Sanchez-Galvan G, Monroy O, Gomez J, Olguin EJ (2008) Assessment of the hyperaccumulating lead capacity of Salvinia minima using bioadsorption and intracellular accumulation factors. Water Air Soil Pollut 194:77–90

Santamaría L (2002) Why are most aquatic plants widely distributed? Dispersal, clonal growth and small-scale heterogeneity in a stressful environment. Acta Oecol 23:137–154

Santamaría-Hernando S, Rodríguez-Herva JJ, Martínez-García PM, Río-Álvarez I, González-Melendi P, Zamorano J, Tapia C, Rodríguez-Palenzuela P, López-Solanilla E (2018) Pseudomonas syringae pv. Tomato exploits light signals to optimize virulence and colonization of leaves. Environ Microbiol 20:4261–4280

Santos-Medellín C, Edwards J, Liechty Z, Nguyen B, Sundaresan V (2017) Drought stress results in a compartment-specific restructuring of the rice root-associated microbiomes. mBio 8:e00764

Scholz VV, Martin BC, Meyer R et al (2021) Cable bacteria at oxygen-releasing roots of aquatic plants: a widespread and diverse plant-microbe association. New Phytol 232(5):2138–2151

Sculthorpe CD (1967a) The biology of aquatic vascular plants. Edward Arnold, London, 610p

Sculthorpe CD (1967b) The biology of aquatic vascular plants. St. Martin's Press, New York, 610p

Sekar S, Chandramohan M (2008) Phycobiliproteins as a commodity: trends in applied research, patents and commercialization. J Appl Phycol 20(2):113–136

Shahid MJ, Ali S, Shabir G et al (2020) Comparing the performance of four macrophytes in bacterial assisted floating treatment wetlands for the removal of trace metals (Fe, Mn, Ni, Pb, and Cr) from polluted river water. Chemosphere 243:125353

Singh SK, Singh SK, Kannaujiya VK, Rahman MA, Dixit K, Kapur AS, Sundaram S (2018) Algal based CO_2 sequestration: a sustainable route for CO_2 mitigation. In: Tripathi K, Rathor NK, Abraham G (eds) The role of photeosynthetic microbes in agriculture and industry. Nova Science, Hauppauge, pp 209–118

Singh PP, Kujur A, Yadav A, Kumar A, Singh SK, Prakash B (2019) Mechanisms of plant-microbe interactions and its significance for sustainable agriculture. In: PGPR amelioration in sustainable agriculture. Woodhead Publishing, Sawston, pp 17–39

Sriprapat W, Thiravetyan P (2016) Efficacy of ornamental plants for benzene removal from contaminated air and water: effect of plant associated bacteria. Int Biodeter Biodegr 113:262–268

Stout L, Nüsslein K (2010) Biotechnological potential of aquatic plant-microbe interactions. Curr Opin Biotechnol 21(3):339–345

Stout LM (2006) "Influence of plant-associated microbial communities on heavy metal uptake by the aquatic plant Lemna minor" Doctoral Dissertations Available from Proquest. AAI3242356. https://scholarworks.umass.edu/dissertations/AAI3242356

Sulu-Gambari F, Seitaj D, Meysman FJ, Schauer R, Polerecky L, Slomp CP (2016) Cable bacteria control iron–phosphorus dynamics in sediments of a coastal hypoxic basin. Environ Sci Technol 50:1227–1233

Suyamud B, Thiravetyan P, Panyapinyopol B, Inthorn D (2018) Dracaena sanderiana endophytic bacteria interactions: effect of endophyte inoculation on bisphenol a removal. Ecotoxicol Environ Saf 157:318–326

Tanaka Y, Matsuzawa H, Tamaki H, Tagawa M, Toyama T, Kamagata Y, Mori K (2017) Isolation of novel bacteria including rarely cultivated phyla, Acidobacteria and Verrucomicrobia, from the roots of emergent plants by simple culturing method. Microbes Environ 32:288–292

Thomaz SM, da Cunha ER (2010) The role of macrophytes in habitat structuring in aquatic ecosystems: methods of measurement, causes and consequences on animal assemblages' composition and biodiversity. Acta Limnol Bras 22(2):218–236

Thomaz SM, Bini LM, Pagioro TA, Murphy KJ, dos Santos AM, Souza DC (2004) The Upper Paraná River and its floodplain. In: Thomaz SM, Agostinho AA, Hahn NS (eds) The Upper Paraná River and its floodplain: physical aspects, ecology and conservation. Backhuys, Leiden, pp 331–352

Thompson IP, Bailey MJ, Fenlon JS, Fermor TR, Lilley AK, Lynch JM et al (1993) Quantitative and qualitative seasonal changes in the microbial community from the phyllosphere of sugar beet (Beta vulgaris). Plant and Soil 150:177–191

Timm CM, Carter KR, Carrell AA, Jun S-R, Jawdy SS, Vélez JM et al (2018) Abiotic stresses shift belowground populus-associated bacteria toward a core stress microbiome. mSystems 3: e00070–e00017

Tripathi K, Sharma NK, Rai V, Rai AK (2013) Low cellular P-quota and poor metabolic adaptations of the freshwater cyanobacterium Anabaena fertilissima Rao during Pi-limitation. Antonie Van Leeuwenhoek 103(2):277–291

Uysal Y, Taner F (2009) Effect of pH, temperature, and lead concentration on the bioremoval of lead from water using Lemna minor. Int J Phytoremediation 11:591–608

Vamerali T, Bandiera M, Mosca G (2010) Field crops for phytoremediation of metal-contaminated land. A review. Environ Chem Lett 8:1–17

Van der Lelie D, Corbisier P, Diels L, Gills A, Lodewyckx C, Mergeay M, Taghavi S, Spelmans N, Vangronsveld J (2000) The role of bacteria in the phytoremediation of heavy metals. In: Terry N, Banuelos GS (eds) Phytoremediation of contaminated soil and water. Lewis Publishers, Boca Raton, pp 265–281

Van Veen H, Sasidharan R (2021) Shape shifting by amphibious plants in dynamic hydrological niches. New Phytol 229:79

Velásquez AC, Castroverde CDM, He SY (2018) Plant-pathogen warfare under changing climate conditions. Curr Biol 28:R619–R634

Venkatachalam S, Ranjan K, Prasanna R, Ramakrishnan B, Thapa S, Kanchan A et al (2016) Diversity and functional traits of culturable microbiome members, including cyanobactera in the rice phyllosphere. Plant Biol 18:627–637

Verpoorter C, Kutser T, Seekell DA et al (2014) A global inventory of lakes based on high-resolution satellite imagery. Geophys Res Lett 41(18):6396–6402

Vorholt JA (2012) Microbial life in the phyllosphere. Nat Rev Microbiol 10:828–840. https://doi.org/10.1038/nrmicro2910

Wagner SC (2011) Biological nitrogen fixation. Nat Educ Knowl 3(10):15

Wallace J, Kremling KA, Kovar LL, Buckler ES (2018) Quantitative genetics of the maize leaf microbiome. Phytobiomes J 2:208–224

Wang C, Cai X, Zheng Z (2005) High humidity represses Cf-4/Avr4- and Cf-9/Avr9-dependent hypersensitive cell death and defense gene expression. Planta 222:947–956

Wang W, Barnaby JY, Tada Y, Li H, Tör M, Caldelari D, Lee DU, Fu X-D, Dong X (2011) Timing of plant immune responses by a central circadian regulator. Nature 470:110–114

Wanjugi P, Harwood VJ (2012) The influence of predation and competition on the survival of commensal and pathogenic fecal bacteria in aquatic habitats. Environ Microbiol 15(2):517–526

Wan-Lei X, Wei P, Qi L et al (2018) Aquatic plant debris changes sediment enzymatic activity and microbial community structure. Environ Sci Pollut Res Int 22:21801–21810

Weber KA, Achenbach LA, Coates JD (2006) Microorganisms pumping iron: anaerobic microbial iron oxidation and reduction. Nat Rev Microbiol 4(10):752–764

Weiher E, Clarke GDP, Keddy PA (1998) Community assembly rules, morphological dispersion, and the coexistence of plant species. Oikos 81:309–322

Wijesinghe W, Jeon Y-J (2012) Enzyme-assistant extraction (EAE) of bioactive components: a useful approach for recovery of industrially important metabolites from seaweeds: a review. Fitoterapia 83(1):6–12

Wright CA, Beattie GA (2004) Pseudomonas syringae pv. tomato cells encounter inhibitory levels of water stress during the hypersensitive response of Arabidopsis thaliana. Proc Natl Acad Sci U S A 101:3269–3274

Wu J, Gu L, Hua Z et al (2021) Effects of Escherichia coli pollution on decomposition of aquatic plants: variation due to microbial community composition and the release and cycling of nutrients. J Hazard Mater 401:123252

Xiao L, Chunyu L, Thompson ML et al (2016) E. coli surface properties differ between stream water and sediment environments. Front Microbiol 7:1732

Xu L, Naylor D, Dong Z, Simmons T, Pierroz G, Hixson KK et al (2018) Drought delays development of the sorghum root microbiome and enriches for monoderm bacteria. Proc Natl Acad Sci 115:E4284–E4293. https://doi.org/10.1073/pnas.1717308115

Yadav RK, Tripathi K, Ramteke PW, Varghese E, Abraham G (2016) Salinity induced physiological and biochemical changes in the freshly separated cyanobionts of Azolla microphylla and Azolla caroliniana. Plant Physiol Biochem 106:39–45

Yadav RK, Tripathi K, Mishra V, Ramteke PW, Singh PK, Abraham G (2019) Proteomic evaluation of the freshly isolated cyanobionts from *Azolla microphylla* exposed to salinity stress. Symbiosis 77(3):249–256

Yadav RK, Chatrath A, Tripathi K, Gerard M, Ahmad A, Mishra V, Abraham G (2021) Salinity tolerance mechanism in the aquatic nitrogen fixing pteridophyte Azolla: a review. Symbiosis 83(2):129–142

Yadav RK, Ramteke PW, Tripathi K, Varghese E, Abraham G (2022) Salinity induced alterations in the growth and cellular ion content of Azolla caroliniana and Azolla microphylla. J Plant Growth Regul 42:867

Yang JW, Wu W, Chung CC et al (2018) Predator and prey biodiversity relationship and its consequences on marine ecosystem functioning interplay between nanoflagellates and bacterioplankton. ISME J 12(6):1532–1542

Ybarra GR, Webb R (1999) Effects of divalent metal cations and resistance mechanisms of the cyanobacterium Synechococcus sp. strain PCC 7942. J Hazard Subst Res 2:1–9

Ying-ru Z, Yu-fang L, Hai-lin Z, Wei-ming S (2013) Aerobic denitrifying characteristics of duckweed rhizosphere bacterium RWX31. Afr J Microbiol Res 7(3):211–219

Yoneda Y, Yamamoto K, Makino A, Tanaka Y, Meng X-Y, Hashimoto J, Shin-ya K, Satoh N, Fujie M, Toyama T (2021) Novel plant-associated Acidobacteria promotes growth of common floating aquatic plants, Duckweeds. Microorganisms 9:1133

Yoon S (2010) Towards practical application of methanotrophic metabolism in chlorinated hydrocarbon degradation, greenhouse gas removal, and immobilization of heavy metals. Doctoral dissertation, University of Michigan

Zayed A, Gowthaman S, Terry N (1998) Phytoaccumulation of trace elements by wetland plants: I. Duckweed. J Environ Qual 27:715–721

Zeng J, Bian Y, Xing P, Wu QL (2012) Macrophyte species drive the variation of bacterioplankton community composition in a shallow freshwater lake. Appl Environ Microbiol 78(1):177–184

Zhang L, Zhang S, Lv X et al (2018) Dissolved organic matter release in overlying water and bacterial community shifts in biofilm during the decomposition of, Myriophyllum verticillatum. Sci Total Environ 633:929–937

Zhang M, Molinos JG, Su G, Zhang H, Xu J (2019) Spatially structured environmental variation plays a prominent role on the biodiversity of freshwater macrophytes across China. Front Plant Sci 10:161. https://doi.org/10.3389/fpls.2019.00161. PMID: 30853965

Zhu J-K (2016) Abiotic stress signaling and responses in plants. Cell 167:313–324

Zhu H, Li S, Hu Z, Liu G (2018) Molecular characterization of eukaryotic algal communities in the tropical phyllosphere based on real-time sequencing of the 18S rDNA gene. BMC Plant Biol 18:365

Zuo Z (2019) Why algae release volatile organic compounds – the emission and roles. Front Microbiol 10:491

Plant-Microbe Interaction in Freshwater Ecosystem for Improving Water Quality

5

Matta Gagan, Pant Gaurav, G. K. Dhingra, Kumar Avinash, Nayak Anjali, and Kumar Pawan

Abstract

In an aqueous environment, microorganisms and plants interact primarily for organic oxygen and carbon, while plants acquire resistance mechanisms and exchange of minerals. This interaction results in a strong link between the two that is heavily dependent on the mutual nutrient supply. Nowadays, water quality of freshwater resources is heavily polluted due to industrial waste, agricultural runoff, oil spillage, and several other factors. In addition to their mutual advantages, plant-microbe interactions have an impact on the water quality, particularly in the rhizosphere, which gives aquatic systems the innate capacity to reduce pollution from the water column. This chapter provides in-depth information and recent advances in the study of plant-microbe interactions, ecological and biochemical components, their role in fresh water ecosystems, and their potential to enhance water quality.

Keywords

Aquatic plants · Aquatic pollution · Plant-microbe interaction · Biofilm · Water quality

---

M. Gagan (✉) · P. Gaurav · K. Avinash · N. Anjali · K. Pawan
Department of Zoology and Environmental Sciences, Gurukula Kangri (Deemed to be University), Haridwar, India
e-mail: drgaganmatta@gkv.ac.in

G. K. Dhingra
Department of Botany, Sri Dev Suman Uttarakhand University, Rishikesh Campus, Rishikesh, Uttarakhand, India

© The Author(s), under exclusive license to Springer Nature Singapore Pte Ltd. 2023
R. Soni et al. (eds.), *Current Status of Fresh Water Microbiology*,
https://doi.org/10.1007/978-981-99-5018-8_5

5.1 Introduction

Wetlands are the transitional areas between land and water and are distinguished by soil surfaces that are waterlogged above them and support a wide variety of plant and animal species. The macrophytes and microphytes, including phytoplankton, diatoms, and other dominant algae, that make up the floral richness and diversity of aquatic ecosystems are incredibly diverse. Large plants known as aquatic macrophytes grow in water and in areas where land and water converge. The primary chemical components of surface water necessary for the correct development of macro- and microphytes are the optimal concentrations of key nutrients like Nitrogen and Phosphorus as well as organic Carbon and other nutrient elements. In addition to macro- and microphytes, microbial consortia exist at different community levels and are typically seen as detrital microbial mats, biofilms, and cluster of different microorganisms (planktons, bacteria, and microalgae) (Xie et al. 2015). They significantly contribute to nutrient cycling (nitrification, denitrification, sulfate reduction, methanogenesis, and metal ion (Noori et al. 2021). Consortia of microbes, especially those present in the rhizosphere and rhizoplane, and on the solid surfaces of sediments, form biofilms on submerged plants' leaves. Several environmental factors impact biofilm formation and its structure, including an abundance of nutrients (eutrophication), their supply, and the concentration of hazardous materials in the water (Giaramida et al. 2013; Calheiros et al. 2009). The eutrophication process, natural deterioration processes, and effects of human activity all affect the freshwater aquatic systems' water quality. Numerous academic studies have been conducted on water quality and their remediation by microorganisms (McGrane 2016; Ismail et al. 2019; Oliver et al. 2019; Bhatt et al. 2022) the remediation processes (Jalal et al. 2011; Kurniawan et al. 2021; Swaleh et al. 2022).

In addition, previous scientific studies appear to show that the majority of water quality improvement studies have focused on environmental contaminants and showed that these contaminants can be removed only by aquatic plant or by different microorganisms. Small number of publications are available implying direct effect of plant (macrophytes) and microbe interactions (Dhote and Dixit 2009; Sood et al. 2012; Kochi et al. 2020; Lu et al. 2014) and its potential benefits on water quality improvement (Stottmeister et al. 2003; Rodrigues 2007; Toyama et al. 2011; Chakraborty et al. 2013; Prakash et al. 2022). The researches on the interaction of microbes with aquatic macrophytes were discussed in this chapter. With an emphasis on their combined effect on the quality of freshwaters, several facets of microorganisms, microbial communities, and their involvement in aquatic ecosystems have been reviewed.

5.2 The Function of Microbial Communities (Biofilms) in Freshwater Ecosystems

Based on the nutrients rich and prevailing environmental scenarios, microorganisms account for the majority of all inland water habitats quantitatively and biochemically (Hahn 2006). This high microbial diversity supports the efficient functioning of freshwater ecosystems (Zehr 2010). The culturable bacterial group, which includes Actinobacteria, alpha-, beta-, and gamma-proteobacteria, firmicutes, and Bacteroides, constitutes the majority of the microbial biodiversity in freshwater ecosystems and archaea (Wang et al. 2008; Calheiros et al. 2009; Wei et al. 2011; Dash et al. 2021).

Microbial populations are present on solid substrates and plant surfaces as biofilm (Gagnon et al. 2007). The predominant bacterial groups frequently found in the assemblage, primarily in freshwaters, are depicted in Fig. 5.1. Extracellular polymeric material (EPS) is a porous slime matrix made of lipids, proteins, nucleic acids, and polysaccharides, which is used to produce biofilms, which are then filled with microbe cells (Wotton 2011; Branda et al. 2005). In a biofilm, microbial cells are confined to a specific microniche inside a complex, homeostatically stable community that exhibits strong metabolic coordination. This gives the microbes ecologically different characteristics (Costerton et al. 1995). When habitats and environmental factors change, the strong microbial assemblage in a biofilm is prone to significant change (Yannarell and Triplett 2004; Witteveen et al. 2020). According to Crump and Koch (2008), various plant types support various microbial species.

Additionally, molecular methods like terminal restriction fragment length polymorphism (TRFLP) fingerprints of PCR amplified 16S rDNA fragments and denaturing gradient gel electrophoresis (DGGE) can quickly provide knowledge about the general trend of the microbial population of Biofilm (Truu et al. 2009; Kumar et al. 2022). According to metagenomics research, consortia of microbes are good for their growth (Singh et al. 2021; Suyal and Soni 2022). Bacterial cells in a biofilm have numerous possibilities to share genetic info via horizontal gene transfer (HGT), which confers tolerance, resistance, and capability of chemical degradation. Bacterial survival in natural habitats is commonly attributed to HGT (Ventura et al. 2007). The genetically stable populations of microorganisms produce diverse responses and sensitivities to different anthropogenic disturbances in a biofilm (Mcclellan et al. 2008). The vulnerability of the bacterial population in Biofilm to toxicants is mainly influenced by PO_4^{3-} ions (Kamaya et al. 2004; Guasch et al. 2007; Tlili et al. 2010). In particular, when there is food scarcity, Tlili et al. (2010) showed how the microbiota changes in reaction to toxicants like Cu and diuran (herbicide).

Water quality is deteriorating nowadays due to increasing anthropogenic activities such as commercial as well as industrial. Freshwater resources are sinking and the condition is very harsh as far as the water quality is concerned. Different studies have been conducted on the water quality of freshwater resources, that is, rivers, lakes, groundwater, etc., globally: these studies have exposed the increasing

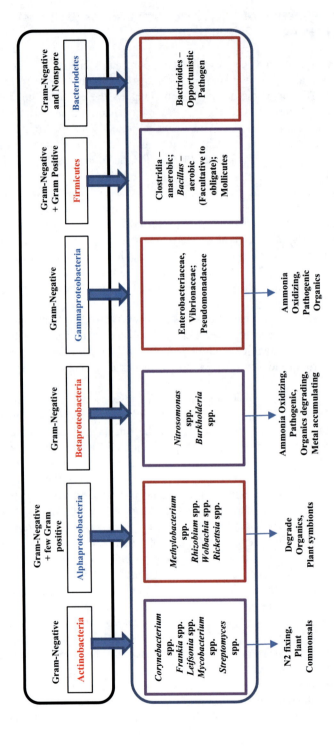

Fig. 5.1 Common bacterial group in aquatic environment

levels of contamination in the water in terms of change in physicochemical parameters (Matta et al. 2018d, 2020b, c, 2022a, d; Gjyli et al. 2020; Kumar et al. 2021; Bensoltane et al. 2021) and heavy metal toxicity (Matta et al. 2018b, c, 2020d, c) due to industrial and agricultural runoff and other sources. It was also observed in several studies that decreasing water quality has negative impact on the planktons diversity (Matta et al. 2018a, 2020a) and limnological condition including species diversity in the water (Matta and Uniyal 2017). However, in the UN agenda of 2030 of Sustainable development Goals (SDGs), water, health, and hygiene practice is very important for making a healthy environment (Matta et al. 2022b).

5.3 The Importance of Aquatic Plant-Microbe Interactions in Freshwater Ecosystems

The macrophytes in the freshwater ecosystem include members of 4 distinct groups: (a) free-floating (e.g., *Pistia stratiotes*) (b) Floating (only leaves) (e.g., *Hydrilla* spp.), (c) submerged macrophytes (e.g., *Chara* spp.), and (d) Emergent (e.g., *Phragmites australis*) and are only found in the macroscopic flora. According to various studies, it was revealed that the distribution of aquatic plants and microbiological species is significantly influenced by the freshwater's nutritional quality (Buosi et al. 2011; Wu et al. 2007; Debbarma et al. 2021) in the increasing order from oligotrophic to hypertrophic. Many microbial communities in water and soil directly interact with macrophytes in the rhizoplane. There is no negative effect on micriobial community's structure by the macrophytes. This has been proved by several studies even in sediments too (Munch et al. 2007; Ahn et al. 2007). The prolonged surface that aquatic plants' roots give for the benthic microbial community to rest and the unique niches they provide as for individual microbes ensure a constant supply of oxygen, nutrients, and organic carbon (Stottmeister et al. 2003). Microbes provide mineral nutrients and protective immunity to aquatic plants in exchange, building solid interactions between the two. To prevent the entry of this hazardous metal into the plants, Stout (2006) showed how the interaction between plants and microbes on Lemna minor negatively affects the uptake of Cd metal ions. Water chemistry (salt concentrations, pH, dissolved oxygen, electrical conductivity, dissolved organic matter, and hazardous organic pollutants), redox conditions, and nutrients availability are a few elements that affect how plants and microbes interact in freshwater bodies (Buosi et al. 2011; Ahn et al. 2007).

However, less studies are available on the importance of plant-microbe interactions in aquatic ecosystems, some typical examples of these interactions and their function in the aquatic system are shown in Table 5.1. The connection between aquatic macrophytes and microbes is also seen in Table 5.1, with the nitrogen cycle being the primary beneficiary. A zone of influence for aquatic plants, the rhizoplane has distinct water chemistry from the rest of the water column due to vigorous microbial activity (Stout and Nüsslein 2010). According to this, regardless of the plant and microbe species, aquatic plants and microorganisms interact in most aquatic regimes, including artificial wetlands, in parasitic and symbiotic ways.

Table 5.1 Environmental perspectives of plant-microbe interaction in aquatic ecosystem

Microbe	Plant	Interaction type	Environmental significance	References
Methylosinus trichosporium	Rooted macrophytes	Ectorhizospheric	Degrade trichloroethylene	Tsien et al. (1989)
Mycobacterium gilvum	*Phragmites australis*	Ectorhizospheric	Degradation of bezo[a]pyrine	Toyama et al. (2011)
Hydrogeno phaga S1; *Agrobacterium radiobacter S2*	*Phragmites australis*	Ectorhizospheric	Degradation of acid Orange-7	Davies et al. (2006)
Bacillus cereus GXBC1	*Pistia stratiotes*	Ectorhizospheric	Enhanced Cr (VI) uptake	Chakraborty et al. (2013)
a, b, c proteobacteria	*Typha latifolia* (L.)	Endorhizospheric	Reduction of Fe (III) into	Ye et al. (2001), Carranza-Alvarez and Alonso-Castro (2008)
AMF	*Ipomea aquatic*	Endorhizospheric	Enhanced cd uptake	Bhaduri and Fulekar (2012)
Microbacterium sp.	*Phragmites communis*	Endorhizospheric	Degrade Chloropyriphos	Chen et al. (2012)
Rhizoplane bacteria	*Eichhornia crassipes (Mart.)*	Rhizospheric	Reduce the toxicity of heavy metals	Duraivadivel et al. (2020)

Unlike aquatic plants, which rely more on resources like organic carbon and oxygen (particularly at the rhizoplane) that are needed largely by microbes to thrive, upland plants also produce a variety of chemical messages to link with other species (Badri et al. 2009). Commonly, there are two different kinds of symbiotic relationships that microbes can have with plants: (1) endophytic, which involves the colonization of a plant's internal tissues, including nitrogen fixing diazotrophs and other nutrient assimilation AMF and (2) ectophytic, for example, ammonia-oxidizing bacteria and methanotropic bacteria (Nielsen et al. 2001; Sorrell et al. 2002; Sřaj-Krzˇicˇ et al. 2006; Wei et al. 2011). An essential plant association is an ectophytic, which involves roots and leaves. Ectophytic interaction affects the elemental cycles in aquatic ecosystems by influencing many biochemical events that take place at the interacting surface (Laanbroek 2010).

The oxygen is transferred from stem to root via interlinked lacunae in the plant-microbe interaction within ectophytic zone of influence (ectorhizosphere) (Sand-Jensen et al. 2005), with some of it being expelled from the roots either through humidity-induced pressured flow by air venture method, generally considered as radial oxygen loss (ROL). The ROL is defined by vegetation types and the redox potential of water, with the maximum amount of oxygen stimulating the growth of aerobic nitrifying bacteria (Reddy et al. 1989; Wiessner et al. 2002; Stottmeister et al. 2003; Inoue and Tsuchiya 2008; Soda et al. 2007) and heterotrophic bacteria aerobically decomposing organic materials present as plant extracts. Oxygen is primarily used as a principal electron acceptor in energy production (Bodelier 2003) and in various beneficial oxidation activities (Laanbroek 2010). Furthermore, organic matter diagenesis in sediments occurs via oxic and anoxic microbiological activities, with the use of electron acceptors such as oxygen resulting in an oxygen deficiency region. To disintegrate organic material under this kind of anaerobic condition, bacterial cells (facultative anaerobes) that can use NO^{3-}, SO_4^{2-}, and CO_2 as terminal acceptors of electrons become more effective (Steenberg et al. 1993). This increases the sediments' electron transport system (ETS) activity (Germ and Simc˘ic˘ 2011). By reducing CO_2 with H_2, methanogens release methane gas (CH_4). Following the total utilization of all other electron acceptors besides CO_2, the lowest energy-yielding activity, CH_4 generation, dominates in freshwater environments (Rejmankova and Post 1996; Conrad 2008).With radial oxygen loss, which involves oxygen being released from the roots and traveling through interconnected lacunae, aquatic macrophytes replace the oxygen in the deep waters. By employing O_2 as an electron acceptor and organic secretions as carbon, ROL and root organic secretions trigger aerobic microorganisms to execute metabolic action for their survival. These essential conditions form the basis of most plant-microbe interactions (endorhizospheric and ectorhizospheric). Plants receive nutritional minerals in exchange and defenses against infections and hazardous contaminants. Anaerobic living forms use terminal electron acceptors like NO -, CO -, CH -, and SO_2- to break down organic matter in the lowest oxygen region or in sediments, producing minerals and gases. The rhizoplane, which includes both the interior and external rhizospheric zones, is made into a zone with a high electron transport system (ETS) and energy usage by bacteria.

5.4 Improvement of Water Quality Due to the Interaction of Plants and Microbes

As noted in the earlier segment, the interaction of plants and bacteria in the environment is extremely visible and impacts the general quality of media. The interaction between plants and microorganisms in freshwater ecosystems is abundant and necessary for their existence. Environmental pollution prevention, generally referred to as bioremediation and the most extensively studied area in environmental and biological sciences worldwide, is the combined impact of plant-microbe interactions in a larger sense (Pilon-Simts and Freeman 2006). Generally, the relationship between plants and microorganisms is based on mutually beneficial exchanges. For example, plants give microbes the minerals and metabolites they need to develop, while microbes give oxygen and organic carbon to plants. Table 5.1 gives some studies showing the plant microbe interaction in improving water quality.

5.4.1 Organic Pollutants Degradation

Freshwater ecosystems are the most sensitive of all ecosystems because of the widespread field application of organic substances like chlorinated organic compounds, poly-aromatic hydrocarbons (PAHs), poly-brominated biphenyls ethers (PBEs), and poly-chlorinated biphenyls (PCBs). There are well-known bioremediators due to their catabolic activity, which can break down almost all kinds of organic compounds (Hiraishi 2008; Fennell et al. 2011).

One of the primary tools used by bacteria to catabolically degrade resistant organic compounds and obtain organic carbon and electron acceptors from the rich rhizosphere of land and aquatic macrophytes is cometabolism (Stottmeister et al. 2003; Kour et al. 2021). While the microbial community relies primarily on the macrophytic species, the rate of microbial degradation is of second-order kinetics and related to the number of bacteria and amount of xenobiotics in water sources (Paris et al. 1981). (Calheiros et al. 2009). Additionally, the organic carbon supplied by plants to rhizospheric bacteria aids in degrading complex resistant organic substances, including PAHs and pyrenes (Mordukhova et al. 2000; Mori et al. 2005; Jouanneau et al. 2005; Debbarma et al. 2017).

A typical instance of this concerted mutual gain was reported by Golubev et al. (2009), in which plants receive the indole acetic acid (IAA), harmone responisble for growth, as a product of the rhizospheric microbial breakdown of PAHs. Other researchers have previously found similar findings on other aquatic species and sediments (Huang et al. 2004; Escalante-Espinosa et al. 2005). The microbe *Sinorhizobium meliloti* P 221, which forms an ectorhizospheric connection with aquatic plants and is capable of synthesizing IAA by degrading PAHs, was isolated and identified by Golubev and colleagues. Additionally, prior studies by Gasol and Duarte (2000) suggest that bacteria have the best chances of surviving in the productive aquatic environment of algae, where they can make efficient use of the carbon generated from the algae to grow and reproduce.

Problems with taste and odor are brought on by an upsurge in bacteria in freshwater (Okabe et al. 2002). The dissolved organic matter (DOM) (Tranvik 1998), which includes PCBs (polychlorinated biphenyls) (Ghosh et al. 1999) and atrazine, can be degraded by aquatic-plant-associated Biofilm, which primarily contains amines, aliphatic aldehydes, and phenolics (Guasch et al. 2007). A group of a and c proteobacteria known as universal methanotrophs, which use methane as a carbon source, are also abundant in the rhizoplane of aquatic plant species (Semrau et al. 2010). Many toxic organic compounds are degraded by Particulate methane monooxygenase (pMMO) generated by the methanotrophs (Pandey et al. 2014; Yoon 2010), especially chlorinated ethenes (Tsien et al. 1989; Yoon 2010) by a series of enzymatic reactions that first produce formaldehydes and then the terminal compound CO_2.

5.4.2 Inorganic Pollutants Removal

Aquatic biota are unaffected by the modest concentrations of metal ions that normally exist in aqueous ecosystems as a result of gradual leaching from rocks and soil (Zhou et al. 2008). In many regions worldwide, increased metal ions in water are mostly caused by industrial, agricultural, and municipal wastes. Numerous bio/chemical parameters, such as the pH and Eh (redox potential) of the water, the occurrence of hydrated iron oxides, metal carbonates, and plant-microbe interaction as a biofilm on the rhizosphere of macrophytes, affect the movement of metal ions in the water (Hansel et al. 2001; Carranza-Alvarez and Alonso-Castro 2008). Most metals in water occur in the form of cations, which stick to the negatively charged EPS of the biofilm matrix and stop metal ions from entering it and the plants.

Most freshwater macrophytes absorb metal ions from water and have iron plaque around their roots and submerged sections (King and Garey 1999; Hansel et al. 2001). When iron is oxidized by molecular oxygen or by bacteria that can oxidize iron, such as *Ferroplasma* sp. and *Leptospirillum ferroxidans*, a coating of iron (hydroxide) precipitate forms surrounding the plant components (King and Garey 1999). The plant's root porosity increases the oxygen level at the rhizoplane, which influences radial oxygen loss (Li et al. 2011). More iron plaques may grow as a result of iron-oxidizing bacteria. The process of root porosity, plaque development, ROL, and the toxic response to As (arsenic) metalloid have also been shown by Li et al. (2011); however, the latter was found to have significantly diminished.

Rhizoplanes are where interactions between plants and microbes are most active. The biofilm in the figure, which is assumed to be present on the rhizosphere of an aquatic plant, is involved in a number of bio−/physicochemical processes, including nitrification (1) denitrification, (2) sulfate reduction, (3) iron oxidation, and (4) methane oxidation by the appropriate microbes. The chemical ions produced as a result of these reactions enter the water column, where they trigger additional physicochemical metal sulfate (MS) precipitates, which are seen on the surface of water bodies and these are mostly formed when hydrogen sulfide combines with metal cations.

Iron plaques are formed on the surface of plants when Fe^3+ and water interact to generate Fe (OH) 3. Most facultative anaerobes break down organic materials and plant matter to obtain carbon in anoxic environments, and they use CO_2, SO_4^{2-}, and NO_3- as terminal electron acceptors at enhanced plaque formation. Sulfate reduction, which occurs when sulfate-reducing bacteria are connected with aquatic macrophytes as biofilm, is another significant metal-separating process in the aquatic environment after the oxidation of iron (Machemer and Wildeman 1992). This process lowers the pH, which is necessary for the microorganism cell to biosorbent the metal ions from the water column (Han and Gu 2010).

Additionally, metal ions sequester metal ions from the water column by reacting with hydrogen sulfide in water (as a result of sulfate reduction) to generate metal sulfide, which precipitates in acidogenic circumstances (Webb et al. 1998; Kumar et al. 2019). In order to remove contaminants from water, macrophytes and algae may work together. For instance, Munoz et al. (2006) observed that the microalga *Chlorella sorokiniana*, which is associated with the bacterium *Ralstonia basilensis*, enhanced the adsorption of toxic metals like cadmium, copper, nickel, and zinc, particularly for Cu (II) adsorption due to the presence of more copper binders. Most aquatic plants also associate with mycorrhizae as endophytic symbionts, which improves P absorption and nutrient transfer in the plants (Thingstrup et al. 2000; Sr˘aj-Krz˘ic˘ et al. 2006). Toxic contaminants like heavy metals are kept away from plants by mycorrhizal interactions. By obstructing the phosphorus membrane transportation system, a chemical counterpart of As, the mycorrhizal interaction in vetiver grass (a widespread wetland species of the Indian subcontinent, South East Asia, and Australia) protects from the As (III) (Meharg and Hartley-Whitaker 2002).

The growth of plants and microorganisms in an aquatic environment heavily relies on the presence of nutritional ions like different mineral elements, P, and N. Water bodies get eutrophicated as a result of too many nutrient ions, which is followed by a cyanobacterial bloom and the production of toxins (Giaramida et al. 2013). Overabundant nutrient ions from the water are taken up by aquatic macrophytes, which prevent algal growth. *Pistia stratiotes, Ipomea aquatica, Eichhornia crassipes*, and *Spirodela polyrhiza* are free-floating macrophytes that play a significant role in the removal of nutritional ions like dissolved inorganic nitrogen such as ammonium NH_4. *Nitrosomonas* and *Nitrobacter*, two aerobic chemoautotrophic bacteria, work together in the rhizosphere to oxidize ammonium to NH_4NO_2.

5.5 Future Perspectives

Characterizing the microbial population associated with the rhizoplane of aquatic plants using metagenomics and other technological advances will be crucial for future studies on plant-microbe interaction and its function in environmental cleanup and restoration. The behavior of plant-microbe interactions at the rhizoplane of free-floating aquatic macrophytes under increased CO_2 in the atmosphere and at increased ambient temperature, as well as the formation of new interactive combinations in freshwater regimes, would be fascinating to study, according to the second point. It is important to investigate further the content and role of the microbial community in a biofilm dealing with a specific plant species, including the effects of nearby harmful chemicals on microbial assemblages, how the microbial community changes in response to climate change, and the environmental implications of newly identified transgenic plant-microbe interaction.

5.6 Conclusion

Particularly on the rhizoplane, plant-microbe interaction is frequent in freshwater environments. Plants exude a variety of organic molecules (plant exudates) that include lipids, phenolic compounds, amino acids, polysaccharides, and nucleic acids to preserve the developing soft tissues, to absorb minerals according to the local electrochemical situation, and to attract bacteria that form a community of distinctive

5 Plant-Microbe Interaction in Freshwater Ecosystem for Improving Water Quality

features that carry out specific functions. Based on the connections, these encounters might range from favorable to negative in character. Actinobacterial species, a, b, c, and d, proteobacterial species, bacteriodetes, firmicutes, and archaea species interact and coexist in a complex system of exopolymeric substance (EPS), which creates a matrix of microbial network (biofilm). The composition of the consortia of microbes changes among different species of plants depending on the type and amount of organic carbon as well as the Oxygen level at the rhizoplane. Known as radial oxygen loss, rooted macrophytes in deep seas continuously replace the oxygen lost by microbial and chemical intake by delivering through the plant's linked lacunae from the stem to the root where the O_2 is discharged (ROL). However, anaerobic life forms in the absence of or in low oxygen levels. The ROL at the rhizoplane delivers it a high electron transport (ETS) area where O_2 functions as electron acceptor requisite for the existence of aerobic life forms.

Each freshwater species of plants have a specific and predetermined pattern of microorganisms in addition to the bio- or physicochemical characteristics. While both species depend on one another for life, the interaction between flora and fauna makes the resources in the area available and, inadvertently, contributes to a greater degree to the removal of environmental pollutants. Examples include mycorrhizal interaction, which improves nutrient absorption and shields plants from hazardous metals by preventing their immediate access, and bacterial species, which break down PAHs to produce indole acetic acid (IAA), a hormone that promotes plant life.

References

Ahn C, Gillevet PM, Sikaroodi M (2007) Molecular characterization of microbial communities in treatment microcosm wetlands as influenced by macrophytes and phosphorus loading. Ecol Indic 7(4):852–863

Badri DV, Weir TL, van der Lelie D, Vivanco JM (2009) Rhizosphere chemical dialogues: plant–microbe interactions. Curr Opin Biotechnol 20(6):642–650

Bensoltane MA, Zeghadnia L, Bordji N, Gagan M, Bouranen S (2021) Drinking water quality assessment using principal component analysis: case study of the town of Souk Ahras, Algeria. Egypt J Chem 64(6):3069–3075. https://doi.org/10.21608/EJCHEM.2021.53654.3112

Bhaduri AM, Fulekar MH (2012) Assessment of arbuscular mycorrhizal fungi on the phytoremediation potential of Ipomoea aquatica on cadmium uptake. 3Biotech 2(3):193–198

Bhatt K, Suyal DC, Kumar S, Singh K, Goswami P (2022) New insights into engineered plant-microbe interactions for pesticide removal. Chemosphere 309(2):136635

Bodelier PLE (2003) Interactions between oxygen-releasing roots and microbial processes in flooded soils and sediments. In: Root ecology. Springer, Berlin, Heidelberg, pp 331–362

Branda SS, Vik Å, Friedman L, Kolter R (2005) Biofilms: the matrix revisited. Trends Microbiol 13(1):20–26

Buosi PRB, Pauleto GM, Lansac-Tôha FA, Velho LFM (2011) Ciliate community associated with aquatic macrophyte roots: effects of nutrient enrichment on the community composition and species richness. Eur J Protistol 47(2):86–102

Calheiros CS, Duque AF, Moura A, Henriques IS, Correia A, Rangel AO, Castro PM (2009) Changes in the bacterial community structure in two-stage constructed wetlands with different plants for industrial wastewater treatment. Bioresour Technol 100(13):3228–3235

Carranza-Alvarez C, Alonso-Castro AJ (2008) Accumulation and distribution of heavy metals in Scirpus americanus and Typha latifolia from an artificial lagoon in San Luis Potosí, México. Water Air Soil Pollut 188(1):297–309

Chakraborty R, Mitra AK, Mukherjee S (2013) Synergistic chromium bioremediation by water lettuce (Pistia) and bacteria (Bacillus cereus GXBC-1) interaction. J Biol Chem Res 30(2): 421–431

Chen WM, Tang YQ, Mori K, Wu XL (2012) Distribution of culturable endophytic bacteria in aquatic plants and their potential for bioremediation in polluted waters. Aquat Biol 15(2): 99–110

Conrad R (2008) Methanogenic microbial communities associated with aquatic plants. In: Plant surface microbiology. Springer, Berlin, Heidelberg, pp 35–50

Costerton JW, Lewandowski Z, Caldwell DE, Korber DR, Lappin-Scott HM (1995) Microbial biofilms. Annu Rev Microbiol 49(1):711–745

Crump BC, Koch EW (2008) Attached bacterial populations shared by four species of aquatic angiosperms. Appl Environ Microbiol 74(19):5948–5957

Dash B, Sahu N, Singh AK, Gupta SB, Soni R (2021) Arsenic efflux in enterobacter cloacae RSN3 isolated from arsenic-rich soil. Folia Microbiol 66:189–196

Davies LC, Pedro IS, Novais JM, Martins-Dias S (2006) Aerobic degradation of acid orange 7 in a vertical-flow constructed wetland. Water Res 40(10):2055–2063

Debbarma P, Raghuwanshi S, Singh J, Suyal DC, Zaidi MGH, Goel R (2017) Comparative in situ biodegradation studies of polyhydroxybutyrate film composites. 3Biotech 7(178):1–9

Debbarma P, Joshi D, Maithani D, Dasila H, Suyal DC, Kumar S, Soni R (2021) Sustainable bioremediation strategies to manage environmental pollutants. In: Shah MP (ed) Removal of refractory pollutants from wastewater treatment plants. CRC Press, Boca Raton, FL. https://doi.org/10.1201/9781003204442-14

Dhote S, Dixit S (2009) Water quality improvement through macrophytes—a review. Environ Monit Assess 152(1):149–153

Duraivadivel P, Gowtham HG, Hariprasad P (2020) Co-occurrence of functionally diverse bacterial community as biofilm on the root surface of Eichhornia crassipes (Mart.) Solms-Laub. Sci Total Environ 714:136683

Escalante-Espinosa E, Gallegos-Martínez ME, Favela-Torres E, Gutiérrez-Rojas M (2005) Improvement of the hydrocarbon phytoremediation rate by Cyperus laxus lam. Inoculated with a microbial consortium in a model system. Chemosphere 59(3):405–413

Fennell DE, Du S, Liu H, Liu F, Haggblom MM (2011) Dehalo- genation of polychlorinated dibenzo-p-dioxins and dibenzofu- rans, polychlorinated biphenyls and brominated flame retardants and potential as a bioremediation strategy. In: Moo-Young M, Butler M, Webb C, Moreira A, Grodzinski B, Cui ZF, Agathos S (eds) Comprehensive biotechnology, 2nd edn. Pergamon Press, Oxford, pp 136–149

Gagnon V, Chazarenc F, Comeau Y, Brisson J (2007) Influence of macrophyte species on microbial density and activity in constructed wetlands. Water Sci Technol 56(3):249–254

Gasol JM, Duarte CM (2000) Comparative analyses in aquatic microbial ecology: how far do they go? FEMS Microbiol Ecol 31(2):99–106

Germ M, Simc̆ic̆ T (2011) Vitality of aquatic plants and microbial activity of sediment in an oligotrophic Lake (lake Bohinj, Slovenia). J Limnol 70(2):305–312

Ghosh U, Weber AS, Jensen JN, Smith JR (1999) Granular activated carbon and biological activated carbon treatment of dissolved and sorbed polychlorinated biphenyls. Water Environ Res 71(2):232–240

Giaramida L, Manage PM, Edwards C, Singh BK, Lawton LA (2013) Bacterial communities' response to microcystins exposure and nutrient availability—linking degradation capacity to community structure. Int Biodeterior Biodegrad 84:111–117

Gjyli L, Vlachogianni T, Kolitari J, Matta G, Metalla O, Gjyli S (2020) Marine litter on the Albanian coastline: baseline information for improved management. Ocean Coast Manag 187: 105108. https://doi.org/10.1016/j.ocecoaman.2020.105108

5 Plant-Microbe Interaction in Freshwater Ecosystem for Improving Water Quality

Golubev SN, Schelud'ko AV, Muratova AY, Makarov OE, Turkovskaya OV (2009) Assessing the potential of rhizobacteria to survive under phenanthrene pollution. Water Air Soil Pollut 198(1–4):5–16

Guasch H, Lehmann V, Van-Beusekom B, Sabater S, Admiraal W (2007) Influence of phosphate on the response of periphyton to atrazine exposure. Arch Environ Contam Toxicol 52(1):32–37

Hahn MW (2006) The microbial diversity of inland waters. Curr Opin Biotechnol 17:256–261

Han X, Gu J (2010) Sorption and transformation of toxic metals by microorganisms. In: Mitchell R, Gu J (eds) Environmental microbiology

Hansel CM, Fendorf S, Sutton S, Newville M (2001) Characterization of Fe plaque and associated metals on the roots of mine-waste impacted aquatic plants. Environ Sci Technol 35(19): 3863–3868

Hiraishi A (2008) Biodiversity of dehalorespiring bacteria with special emphasis on polychlorinated biphenyl/dioxin dechlori- nators. Microbes Environ 23:1–12

Huang X-D, El-Alawi Y, Penrose DM, Glick BR, Greenberg BM (2004) A multi-process phytoremediation system for removal of polycyclic aromatic hydrocarbons from contaminated soils. Environ Pollut 130:465–476

Inoue TM, Tsuchiya T (2008) Interspecific differences in radial oxygen loss from the roots of three Typha species. Limnol 9:207–211

Ismail M, Akhtar K, Khan MI, Kamal T, Khan MA, Asiri M, Khan SB (2019) Pollution, toxicity and carcinogenicity of organic dyes and their catalytic bio-remediation. Curr Pharm Des 25(34): 3645–3663

Jalal KCA, Alam MZ, Matin WA, Kamaruzzaman BY, Akbar J, Hossain T (2011) Removal of nitrate and phosphate from municipal wastewater sludge by Chlorella vulgaris, Spirulina platensis and Scenedesmus quadricauda. IIUM Engin J 12(4)

Jouanneau Y, Willison JC, Meyer C, Krivobok S, Chevron N, Besombes J-L, Blake G (2005) Stimulation of pyrene mineral- ization in freshwater sediments by bacterial and plant bioaugmentation. Environ Sci Technol 39:5729–5735

Kamaya Y, Takada T, Suzuki K (2004) Effect of medium phosphate levels on the sensitivity of Selenastrum capricornutum to chemicals. Bull Environ Contam Toxicol 73:995–1000

King GM, Garey MA (1999) Ferric iron reduction by bacteria associated with the roots of freshwater and marine macrophytes. Appl Environ Microbiol 65(10):4393–4398

Kochi LY, Freitas PL, Maranho LT, Juneau P, Gomes MP (2020) Aquatic macrophytes in constructed wetlands: a fight against water pollution. Sustainability 12(21):9202

Kour D, Kaur T, Devi R, Yadav A, Singh M, Joshi D, Singh J, Suyal DC et al (2021) Beneficial microbiomes for bioremediation of diverse contaminated environments for environmental sustainability: present status and future challenges. Environ Sci Pollut Res 28:24917–24939

Kumar P, Gupta SB, Anurag RS (2019) Bioremediation of cadmium by mixed indigenous isolates Serratia liquefaciens BSWC3 and Klebsiella pneumoniae RpSWC3 isolated from industrial and mining affected water samples. Pollution 5(2):351–360

Kumar A, Matta G, Bhatnagar S (2021) A coherent approach of water quality indices and multivariate statistical models to estimate the water quality and pollution source apportionment of river ganga system in Himalayan region, Uttarakhand, India. Environ Sci Pollut Res Int 28(31):42837. https://doi.org/10.1007/s11356-021-13711-1

Kumar S, Joshi D, Debbarma P, Singh M, Yadav AN, Singh N, Suyal DC, Soni R, Goel R (2022) Denaturing gradient gel electrophoresis (DGGE) analysis of the fungi involved in biodegradation. In: Udayanga D, Bhatt P, Manamgoda D, Saez JM (eds) Mycoremediation protocols. Springer Protocols Handbooks, Humana, NY, pp 93–100

Kurniawan SB, Ahmad A, Said NSM, Imron MF, Abdullah SRS, Othman AR, Hasan HA (2021) Macrophytes as wastewater treatment agents: nutrient uptake and potential of produced biomass utilization toward circular economy initiatives. Sci Total Environ 790:148219

Laanbroek HJ (2010) Methane emission from natural wetlands: interplay between emergent macrophytes and soil microbial processes: a mini-review. Ann Bot 105:141–153

Li H, Ye ZH, Wei ZJ, Wong MH (2011) Root porosity and radial oxygen loss related to arsenic tolerance and uptake in wetland plants. Environ Pollut 159(1):30–37

Lu Y, Zhou Y, Nakai S, Hosomi M, Zhang H, Kronzucker HJ, Shi W (2014) Stimulation of nitrogen removal in the rhizosphere of aquatic duckweed by root exudate components. Planta 239(3):591–603. https://doi.org/10.1007/s00425-013-1998-6

Machemer SD, Wildeman TR (1992) Adsorption compared with sulphide precipitation as metal removal processes from acid mine drainage in a constructed wetland. Contam Hydrol 9:115–131

Matta G, Uniyal DP (2017) Assessment of species diversity and impact of pollution on limnological conditions of river ganga. Int J Water 11(2):87–102. https://doi.org/10.1504/IJW.2017.083759

Matta G, Kumar A, Naik PK, Tiwari AK, Berndtsson R (2018a) Ecological analysis of nutrient dynamics and phytoplankton assemblage in the Ganga River system, Uttarakhand. Taiwan Water Conser 66(1):1–12; (ISSN: 0492-1505)

Matta G, Kumar A, Kumar A, Naik PK, Kumar A (2018b) Assessment of heavy metal indexing on Ganga River system assessing heavy metals toxicity and ecological impact on river water quality. INAE letters, an official journal of the Indian National Academy of Engineering. (ISSN: 2366-326X: EISSN: 2366-3278), vol 3, p 123. https://doi.org/10.1007/s41403-018-0041-4

Matta G, Kumar A, Tiwari AK, Naik PK, Berndtsson R (2018c) HPI appraisal of concentrations of heavy metals in dynamic and static flow of Ganga River System. Environment, development and sustainability, Springer Nature. (ISSN: 1387-585X; EISSN: 1573–2975). https://doi.org/10.1007/s10668-018-01182-3

Matta G, Naik P, Kumar A, Gjyli L, Tiwari AK, Machell J (2018d) Comparative study on seasonal variation in hydro-chemical parameters of Ganga River water using comprehensive pollution index (CPI) at Rishikesh (Uttarakhand) India. Desalin Water Treat 118:87–95. https://doi.org/10.5004/dwt.2018.22487

Matta G, Kumar A, Nayak A et al (2020a) Water quality and planktonic composition of river henwal (India) using comprehensive pollution index and biotic-indices. Trans Indian Natl Acad Eng 5:541–553. https://doi.org/10.1007/s41403-020-00094-x

Matta G, Nayak A, Kumar A et al (2020b) Water quality assessment using NSFWQI, OIP and multivariate techniques of Ganga River system, Uttarakhand, India. Appl Water Sci 10:206. https://doi.org/10.1007/s13201-020-01288-y

Matta G, Kumar A, Nayak A et al (2020c) Determination of water quality of Ganga River system in Himalayan region, referencing indexing techniques. Arab J Geosci 13:1027. https://doi.org/10.1007/s12517-020-05999-z

Matta G, Nayak A, Kumar A, Kumar P, Kumar A, Tiwari AK, Naik PK (2020d) Evaluation of heavy metals contamination with calculating the pollution index for Ganga River system. Taiwan Water Conserv 68(3):10. https://doi.org/10.6937/TWC.202009/PP_68(3).0005

Matta G, Kumar A, Nayak A et al (2022a) Appraisal of spatial–temporal variation and pollution source estimation of Ganga River system through pollution indices and environmetrics in upper ganga basin. Appl Water Sci 12:33. https://doi.org/10.1007/s13201-021-01552-9

Matta G, Kumar P, Uniyal DP, Joshi DU (2022b) Communicating water, sanitation, and hygiene under sustainable development goals 3, 4, and 6 as the panacea for epidemics and pandemics referencing the succession of COVID-19 surges. ACS ES T Water 2:667. https://doi.org/10.1021/acsestwater.1c00366

Matta G, Kumar A, Nayak A, Kumar P, Kumar A, Naik PK, Singh SK (2022c) Assessing heavy metal index referencing health risk in Ganga River system. Int J River Basin Manag 1:1–11

Matta G, Kumar A, Nayak A, Kumar P, Pant G (2022d) Pollution complexity quantification using NPI and HPI of river ganga system in Himalayan region. Proc Indian Natl Sci Acad 1-13:651. https://doi.org/10.1007/s43538-022-00111-3

Mcclellan K, Altenburger R, Schmitt-Jansen M (2008) Pollution induced community tolerance as a measure of species interaction in toxicity assessment. J Appl Ecol 45:1514–1522

McGrane SJ (2016) Impacts of urbanisation on hydrological and water quality dynamics, and urban water management: a review. Hydrol Sci J 61(13):2295–2311

Meharg AA, Hartley-Whitaker J (2002) Arsenic uptake and metabolism in arsenic resistant and nonresistant plant species. New Phytol 154:29–43

Mordukhova EA, Sokolov SL, Kochetkov VV, Kosheleva IA, Zelenkova NF, Boronin AM (2000) Involvement of naphthalene dioxygenase in indole-3-acetic acid biosynthesis by Pseu-domonas putida. FEMS Microbiol Lett 190:279–285

Mori K, Toyama T, Sei K (2005) Surfactants degrading activities in the rhizosphere of giant duckweed ("Spirodela polyrrhiza"). Jpn J Water Treat Biol 41:129–140

Munch C, Neu T, Kuschk P, Roske I (2007) The root surface as the definitive detail for microbial transformation processes in constructed wetlands—a biofilm characteristic. Water Sci Technol 56(3):271–276

Munoz R, Alvarez MT, Munoz A, Terrazas E, Guieysse B, Mattiasson B (2006) Sequential removal of heavy metals ions and organic pollutants using an algal-bacterium consortium. Chemosphere 63(6):903–911

Nielsen LB, Finstek K, Welsh DT, Donelly A, Herbert RA, de Wit R, Lomstein BA (2001) Sulphate reduction and nitrogen fixation rates associated with roots, rhizomes and sediments from Zostera noltli and Spartina maritime meadows. Environ Microbiol 3:63–71

Noori R, Ansari E, Bhattarai R, Tang Q, Aradpour S, Maghrebi M, Kløve B (2021) Complex dynamics of water quality mixing in a warm mono-mictic reservoir. Sci Total Environ 777: 146097

Okabe S, Kokazi T, Watanabe Y (2002) Biofilm formation potentials in drinking waters treated by different advanced treatment processes. Water Sci Technol 2(4):97–104

Oliver S, Corburn J, Ribeiro H (2019) Challenges regarding water quality of eutrophic reservoirs in urban landscapes: a mapping literature review. Int J Environ Res Public Health 16(1):40

Pandey VC, Singh JS, Singh DP, Singh RP (2014) Methanotrophs: promising bacteria for environmental remediation. Int J Environ Sci Technol 11(1):241–250

Paris DF, Steen WC, Baughman GL, Barnett-Jr JT (1981) Second-order model to predict microbial degradation of organic compounds in natural waters. Appl Environ Microbiol 41(3):603–609

Pilon-Simts EAH, Freeman JL (2006) Environmental cleanup using plants: biotechnological advances and ecological considerations. Front Ecol Environ 4(4):203–210

Prakash O, Pathak A, Kuamr A, Juyal VK, Joshi HC, Gangola S, Patni K, Bhandari G, Suyal DC, Nand V (2022) Spectroscopy and its advancements for environmental sustainability. In: Suyal DC, Soni R (eds) Bioremediation of environmental pollutants. Springer, Cham, pp 317–338

Reddy LG, Patrick WH, Lindau CW (1989) Nitrification-denitrifica- tion at the plant root-sediment interface in wetlands. Limnol Ocean 34:1004–1013

Rejmankova E, Post R (1996) Methane in sulfate—rich and sulfate—poor wetland sediments. Biogeochemistry 54:57–70

Rodrigues BF (2007) Arbuscular mycorrhizae in association with aquatic and marshy plant species in Goa, India. Aquat Bot 86(3):291–294

Sand-Jensen K, Pedersen O, Binzer T, Borum J (2005) Contrasting oxygen dynamics in the freshwater isoetid Lobelia dortmanna and the marine seagrass Zostera marina. Ann Bot 96: 613–623

Semrau JD, DiSpirito AA, Yoon S (2010) Methanotrophs and copper. FEMS Microbiol Rev 34: 496–531

Singh D, Bhasin S, Madan N, Singh M, Suyal N, Singh N, Soni R, Suyal DC (2021) Metagenomics: insights into microbial removal of the contaminants. In: Bhatt P, Gangola S, Udayanga D, Kumar G (eds) Microbial technology for sustainable environment. Springer, Singapore, pp 293–306

Soda S, Ike M, Ogasawara Y, Yoshinaka M, Mishima D, Fujita M (2007) Effects of light intensity and water temperature on oxygen release from roots into water lettuce rhizosphere. Water Res 41:487–491

Sood A, Uniyal PL, Prasanna R, Ahluwalia AS (2012) Phytoremediation potential of aquatic macrophyte, Azolla. Ambio 41(2):122–137

Sorrell BK, Downes MT, Stanger CL (2002) Methanotrophic bacteria and their activity on submerged aquatic macrophytes. Aquat Bot 72(2):107–119

Srˇaj-Krzˇicˇ N, Pongrac P, Klemenc M, Kladnik A, Regvar M, Gabersˇcˇik A (2006) Mycorrhizal colonization in plants from intermittent aquatic habitats. Aquat Bot 85:331–336

Steenberg C, Sweerts JP, Cappenberg T (1993) Microbial biogeo- chemical activities in lakes: stratification and eutrophication. In: Ford T (ed) Aquatic microbiology—an ecological approach. Blackwell Scientific Publications, Boston, pp 69–100

Stottmeister U, Wießner A, Kuschk P, Kappelmeyer U, Kastner M, Bederski O, Muller RA, Moormann H (2003) Effects of plants and microorganisms in constructed wetlands for waste-water treatment. Biotechnol Adv 22:93–117

Stout LM (2006) Influence of plant associated microbial communities on heavy metal uptake by the aquatic plant Lemna minor. Doctrol dissertation for UMass Amherst. Paper AAI3242356. http://scholarworks.umass.edu/dissertations/AAI3242356

Stout L, Nüsslein K (2010) Biotechnological potential of aquatic plant–microbe interactions. Curr Opin Biotechnol 21(3):339–345

Suyal DC, Soni R (2022) Bioremediation of environmental pollutants. Springer Nature, Switzerland; Hardcover ISBN: 9783030861681; eBook ISBN: 9783030861698

Swaleh M, Abubakar L, Mwaguni S, Munga D, Okuku E (2022) The potential of marine micro-algae grown in wastewater to remove nutrients and produce biomass. A Scientific Journal of Kenya Marine and Fisheries Research Institute

Thingstrup I, Kahiluoto H, Jakoben I (2000) Phosphate transport by hyphae of field communities of arbuscular mycorrhizal fungi at two levels of P fertilization. Plant Soil 221:181–187

Tlili A, Be'rardb A, Rouliera J, Volata B, Montuellea B (2010) PO 3- dependence of the tolerance of autotrophic and heterotrophic biofilm communities to copper and diuron. Aquat Toxicol 98: 165–177

Toyama T, Furukawa T, Maeda N, Inoue D, Sei K, Mori K, Kikuchi S, Ike M (2011) Accelerated biodegradation of pyrene and bezo[a]pyrene in the Phragmites australis rhizosphere by bacteria—root exudates interactions. Water Res 45:1629–1638

Tranvik LJ (1998) Degradation of dissolved organic matter in humic waters by bacteria. In: Hessen DO, Tranvik LJ (eds) Aquatic humic substances—book series, ecological studies. Springer, Berlin, Heidelberg, pp 259–283

Truu M, Juhanson J, Truu J (2009) Microbial biomass activity and community composition in constructed wetlands. Sci Total Environ 407:3958–3971

Tsien HC, Brusseau GA, Hanson RS, Waclett LP (1989) Biodegra- dation of trichloroethylene by Methylosinus trichosporium OB3b. Appl Environ Microbiol 55(12):3155–3161

Ventura M, Canchaya C, Tauch A, Chandra G, Fitzgerald GF, Chater KF, vanSinderen D (2007) Genomics of actinobacteria: tracing the evolutionary history of an ancient phylum. Microbiol Mol Biol Rev 71(3):495–548

Wang Y, Inamori R, Kong H, Xu K, Inamori Y, Kondo T, Zhang J (2008) Influence of plant species and wastewater strength on constructed wetland methane emissions and associated microbial populations. Ecol Eng 32:22–29

Webb JS, McGinness S, Lappin-Scott HM (1998) Metal removal by sulphate-reducing bacteria from natural and constructed wet- lands. J Appl Microbiol 84:240–248

Wei B, Yu X, Zhang S, Gu L (2011) Comparison of the community structures of ammonia-oxidizing bacteria and archaea in rhizoplanes of floating aquatic macrophytes. Microbiol Res 166:468–474

Wiessner A, Kuschk P, Kastner M, Stottmeister U (2002) Abilities of helophyte species to release oxygen into rhizospheres with varying redox conditions in laboratory-scale hydroponic systems. Int J Phytoremediation 4:1–15

Witteveen NH, Freixa A, Sabater S (2020) Local and regional environmental factors drive the spatial distribution of phototrophic biofilm assemblages in Mediterranean streams. Hydrobiologia 847(10):2321–2336

Wotton RS (2011) EPS (extracellular polymeric substances), silk, and chitin: vitally important exudates in aquatic ecosystems. J N Am Benthol Soc 30(3):762–769

Wu X, Xi W, Ye W, Yang H (2007) Bacterial community composition of a shallow hypertrophic freshwater lake in China revealed by 16S rRNA gene sequences. FEMS Microbiol Ecol 61:85–96

Xie X, Borjigin T, Zhang Q, Zhang Z, Qin J, Bian L, Volkman JK (2015) Intact microbial fossils in the permian lucaogou formation oil shale, Junggar basin, NW China. Int J Coal Geol 146:166–178

Yannarell AC, Triplett EW (2004) Within and between lake variability in the composition of bacterioplankton communities: investigations using multiple spatial scales. Appl Environ Microbiol 70:214–223

Ye ZH, Cheng KC, Wong MH (2001) Copper uptake in *Typha latifolia* as affected by iron and manganese plaque on the root surface. Can J Bot 79(3):314–320

Yoon S (2010) Towards practical application of methanotrophic metabolism in chlorinated hydrocarbon degradation greenhouse gas removal and immobilization of heavy metals. Doctoral Dissertation. Department of Environmental Engineering. University of Michigan, USA

Zehr JP (2010) Microbes in earth's aqueous environments. Front Microbiol 1(4):1–2

Zhou Q, Zhang J, Fu J, Shi J, Jiang G (2008) Biomonitoring: an appealing tool for assessment of metal pollution in the aquatic ecosystem. Analyt Chim Acta 606:135–150

Microbial Interactions with Aquatic Plants

6

N. V. T. Jayaprada, Jayani J. Wewalwela, G. A. H. Galahitigama, and P. A. N. P. Pandipperuma

Abstract

A fresh water ecosystem consists of rich floral diversity, which includes macrophytes, microphytes, diatoms, and many other algae spp. In addition, a fresh water ecosystem provides best niche for various levels of microbial communities such as biofilms and planktonic microalgal-bacterial consortia. Microbial accumulation as biofilms on aquatic plant surfaces is robust, and biofilms have the ability to change its structure with habitat and environmental changes. Adhesion of biofilms to plant surfaces is the main way bacteria interact with plant tissues. Through these interactions, aquatic plants provide nutrients, organic carbon, and oxygen to microorganisms and in return aquatic plants mainly receive mineral nutrients and defensive immunity. In the process, these interactions significantly contribute to nutrient recycling, energy flow in aquatic ecosystems, and removal of environmental pollutants. Thereby, also in terms of environmental perspective, aquatic plant-microbe interaction has substantially granted water quality.

Keywords

Bioremediation · Aquatic ecosystem · Biofilm · Quorum sensing

---

N. V. T. Jayaprada · J. J. Wewalwela (✉) · P. A. N. P. Pandipperuma
Department of Agricultural Technology, Faculty of Technology, University of Colombo, Homagama, Sri Lanka
e-mail: jayani@at.cmb.ac.lk

G. A. H. Galahitigama
Department of Export Agriculture, Faculty of Agricultural Sciences, Sabaragamuwa University of Sri Lanka, Belihuloya, Sri Lanka

© The Author(s), under exclusive license to Springer Nature Singapore Pte Ltd. 2023
R. Soni et al. (eds.), *Current Status of Fresh Water Microbiology*, https://doi.org/10.1007/978-981-99-5018-8_6

6.1 Microbial and Floral Diversity in Aquatic Ecosystems

Microorganisms dominate all inland water habitats by both numerically and biochemically and proper functioning of an aquatic ecosystem is aided by the rich microbial diversity and it depends on the available nutrient content and prevailing environmental conditions (Baker and Orr 1986).

Bacteria are abundant on the surface of freshwater plants, ranging in density from 100 to 107 cm^{-2} (Fry and Humphrey 1978; Hossell and Baker 1979a, b; Kudryavtsev 1984). Epiphytic bacteria form an integral part of the leaf-surface community (Cummins and Klug 1979; Bhatt et al. 2022). Epiphytic bacteria are metabolically more active than planktonic bacteria present in the surrounding water (Fry and Humphrey 1978). Epiphytic bacteria serve passively as food for higher trophic levels, and as active members of the microbial community. However, bacteria are not distributed evenly over plant surfaces (Hossell and Baker 1979a, b; Ramsay 1974).

Wetlands are the transitional zones between land and water bodies. Shallow overlying water-logged soils harboring rich floral and faunal diversity are the characteristic features of wetlands. Floral diversity of freshwater ecosystem is rich in diversity of macrophytes and microphytes such as phytoplankton, diatoms, and other algae. The large plants growing in the water and wetlands are aquatic macrophytes. Microphytes are microscopic algae found in freshwater and marine water systems (Srivastava et al. 2017). Macroscopic flora of aquatic macrophytes includes four different groups: (1) emergent (e.g., *Phragmites australis*), (2) floating leaved (e.g., *Hydrilla* spp.), (3) free floating (e.g., *Pistia stratiotes*) and (4) submerged macrophytes (e.g., *Chara* spp.) (Srivastava et al. 2008).

The part of root remaining in contact with water or soil of all macrophytes is rhizoplane, and it is the most active zone due to the presence of various microbial communities (Davies et al. 2006; Munch et al. 2007). Extended surface for microbial community is provided by roots of aquatic plants ensuring continuous supply of nutrients, organic C, and O (Stottmeister et al. 2003). And at the same time aquatic plants are provided with mineral nutrients and defensive immunity in return from microbial community and this way both are forming an interrelationship (Srivastava et al. 2017).

The status of freshwater is one of the major dependent factors for the distribution of aquatic plants and microbial species in the following order: oligotrophic > mesotrophic > eutrophic > hypertrophic (Wu et al. 2007; Buosi et al. 2011). For the proper growth of both macrophytes and microphytes, they require optimum concentrations of major nutrients such as N (>45 mgL^{-1}) and P (>0.25 mgL^{-1}) along with organic C and other elements (Srivastava et al. 2008).

Microbial association exists at various levels of microbial community and can be observed as microbial mat, biofilms, and planktonic microalgal-bacterial assemblages and contribute to the nutrient cycling in nitrification, denitrification, sulfate reduction, methonogenesis, and metal ion reduction. And also serve in energy flow in aquatic ecosystem as a feed to zooplanktons. They serve in degrading environmental pollutants and alter the water quality (Paerl and Pinckney 1996;

Cotner and Biddanda 2002; Battin et al. 2003; Hahn 2006). Microbial biofilms generally occur on the leaves of submerged plants and rhizosphere. Some environmental conditions (excessive nutrients and their availability) and the toxic substances on the water affect biofilms and their structure (Giaramida et al. 2013; Calheiros et al. 2009).

6.2 Microbial Biofilms and Quorum Sensing in Aquatic Ecosystems

In aquatic environments, in order to survive in harsh environments, bacteria prefer to form microbial communities, which are attached to different surfaces and are embedded in an extracellular polymeric matrix. These specific microbial communities are called microbial biofilms (Nazir et al. 2019; Afonso et al. 2021; Jayaprada et al. 2020). In order to maintain a high density of bacterial cells in a biofilm, bacteria produce a matrix of substances called extracellular polymeric substances (EPSs) (Afonso et al. 2021). Polysaccharides, proteins, lipids, and extracellular DNA are the main components of EPS (Afonso et al. 2021). Biofilm formation occurs following several main steps, which are attachment to a surface, formation of microcolonies, colony maturation, and detachment phases (Davey and O'toole 2000; Gong et al. 2019). Biofilm formation is started by the attachment of planktonic cells to surfaces like aquatic plants, which are called adhesive biofilms. Some planktonic cells gather at an air-liquid interface and form floating biofilms. Next step is the propagation of cells followed by the aggregation of these newly formed cells to form microcolonies. EPSs formation occurs at this stage, which helps in the attachment and aggregation of cells with scaffolds gradually forming mature biofilms (Dang and Lovell 2016). Lumps of biofilms can be separated due to friction, pressure, and rapid water flow returning to their planktonic state, moving to find new niches. It is mentioned that biofilm formation and composition are affected by variables like temperature, salinity, pH, nutrient concentration of marine and freshwater along with other factors like geographical location, season, availability of light, depth of water, and availability of tides (Ghannoum et al. 2020).

Microbial biofilm is a consortium of both prokaryotes (bacteria and archaea) and eukaryotes (algae and fungi) (Zhang et al. 2019). This leads to a rich microbial community, creating a microenvironment where thousands of different species can exist together in a complex makeup, maintaining remarkable features like gene expression heterogeneity and division of functions in the community. These functions are carried out by cell-to-cell communication called quorum sensing (Antunes et al. 2019). In a nutshell, quorum sensing coordinates population-density-dependent changes in microbial behavior. During the process of biofilm formation, microorganisms communicate with each other through quorum sensing. It is reported that induction of biofilm formation is manifested through quorum-sensing-regulated metabolic activity of planktonic cells (Hmelo 2017; Lami 2019). Microbial biofilms are also controlled by factors like interactions with the environment, interactions with the geography of the location, cycling of nutrient and organic matter, and photosynthesis, which are harnessed through quorum sensing. As a result, biofilms are capable of adapting to dynamic environments influencing ecosystem processes and functioning (Bauer and Mathesius 2004; Joint et al. 2007; Stien et al. 2016). Quorum sensing and biofilm formation involve the biosynthesis of and response to diffusible signal molecules, which differ among different types of bacteria (Mooney et al. 2018; Saxena et al. 2019).

These signal-molecule-based quorum sensing is reported to be the basis behind microbial interactions with aquatic plants defined as interkingdom communication (Bauer and Mathesius 2004; Joint et al. 2007; Stien et al. 2016; Wijewardene et al. 2022). Hence, in both fresh and marine water environments, biofilms have been the subject of studies in terms of microbial-aquatic plants interactions. Therefore, in the following chapters, microbial-aquatic plants interactions will be discussed in terms of biofilms-mediated quorum sensing.

6.3 Aquatic Plant-Microbe Interactions and Its Role

6.3.1 Role of an Aquatic Plant in Plant-Microbe Interaction

Aquatic environment consists of numerous plants species with higher diversity. Aquatic plants existing in a particular water reservoir provide significant impact on microbial growth and development through forming symbiotic relationship (Yoneda et al. 2021). In an aquatic ecosystem, macrophytes are categorized into four different groups such as emergent, floating leaved, free floating, and submerged macrophytes (Srivastava et al. 2008). The surrounding area of macrophytes root is known as "rhizoplane," which is a highly active zone where most of biochemical reactions take place (Munch et al. 2007). Aquatic plants play a major role in structuring microbial assemblages (De Wolf et al. 2022). Such interactions are well described in terrestrial systems as well as in aquatic systems (Srivastava et al. 2017). Aquatic macrophytes directly influence water chemistry by altering pH, dissolved oxygen, and dissolved organic carbon concentration by releasing bioactive compounds (Vilas et al. 2017; Shahid et al. 2020). The neutral pH induced by vegetation encourages the settlement of dissolved contaminants (Borne et al. 2014; Debbarma et al. 2017). Therefore, microbial composition in the aquatic environment would slightly influence due to this alternated chemical nature, thus ultimately impacting its physicochemical condition of water (Collins et al. 2004). In addition, aquatic plants have significantly influenced nutrient uptake and storing of numerous trace metal(loid)s, thus controlling the competition for nutrients by algae and other phytoplankton (Liu et al. 2016). For instance, plants might remove phosphorous by direct uptake and inhibit growth of algae communities (Urakawa et al. 2017). Therefore, aquatic plants maintain the water quality and overwhelm algae blooms in water reservoirs, ensuring macrophyte-dominated clear aquatic ecosystem (Bakker et al. 2010).

Apart from chemical and microbial composition of adjacent water, plants' roots would influence on microbial biogeography in large scale, through providing surface for microbes to form "microbial biofilms" on the surfaces of roots and shoot parts

(Srivastava et al. 2017). Biofilm formation affects the special distribution of microbes under small-scale water reservoirs like lakes (Souffreau et al. 2018). Furthermore, the composition of plant-associated biofilms is species specific and varies with the surrounding water (Souffreau et al. 2018; Kumar et al. 2019; He et al. 2021). Root exudates secrete by macrophytes and deliver essential nutrients for microbes' growth (Ashraf et al. 2018). Furthermore, roots facilitate air circulation in rhizosphere through supplying of oxygen, thus accelerating the aerobic degradation of organic matter by rhizosphere bacteria. As a result of this degradation, nutrients are being accumulated in rhizosphere, which uptake by plants (Shahid et al. 2020). Simultaneously, Ijaz et al. 2016 reported roots assist in the reduction of biological oxygen demand (BOD) and chemical oxygen demand (COD) by enabling microbial populations to ingest the carbon molecules. Aquatic macrophytes have been widely used in constructed wetlands and floating treatment wetland (FTW) to remediate contaminants from polluted water artificially (Shahid et al. 2020). The selection of an appropriate plant species for these systems depends upon their availability, contaminant type, and climatic region. Moreover, the pollutant remediation potential of selected macrophyte should be good and ensure the ecosystem sustainability (Shahid et al. 2020). Furthermore, to develop an effective FTW, macrophyte with robust shoot growth, widely distributed root system, tolerance to toxic compounds, higher nitrogen removal capability, and availability of large aerenchyma in their roots and rizomes characteristics are highly concerned (Li et al. 2011; Lamers et al. 2013).

6.3.2   Role of Microbes in Plant-Microbe Interaction

Microbial communities in aquatic environment have a significant role in pollutant removal progression and plant growth promotion (Shahid et al. 2020). The microbial diversity depends on the nutrient and existing environmental circumstances (Zehr 2010). Freshwater microbial diversity includes actinobacteria, α-proteobacteria, β-proteobacteria, γ-proteobacteria, archea, bacteriodetes, cyanobacteria, fungi, etc. (Wei et al. 2011; Shahid et al. 2020). Among them, several bacteria have a distinct ability to form biofilms, which are known as epiphytic microbes; thus, microbial presence in the aquatic bodies can be categorized into biofilm-forming bacteria and water column bacteria (Shahid et al. 2020).

These biofilms have significant impact on continuation of biochemical cycles and other ecosystem processes in aquatic environment (Battin et al. 2003). A biofilm is mainly composed of aquatic plant and related bacteria communities. Biofilms are composed of extracellular polymeric substances (EPSs), proteins, comprised of polysaccharides, lipids, and nucleic acids, which embrace the cell together (Branda et al. 2005). The further development and maintenance of biofilms depend upon minor molecules secretion by microbes such as secondary metabolic compounds, antibiotics, and proteins (Lasa and Penadés 2006; López et al. 2010). Furthermore, the nature of biofilm is affected by growth conditions (e.g., water flow, light availability, nutrient conditions) and type of substrate (Shahid et al. 2020).

In addition, endophytes also contribute to strengthen the plant-microbe relationship through protecting plant from biotic stresses, relieving abiotic stresses, and facilitating metal uptake from contaminated water (Bacon and White 2016). Endophytic microbes hasten the organic and inorganic degradation process, which eventually produces nutrients essential for plant uptake (Wei et al. 2014; Peiris et al. 2022). For instance, rhizobacteria stimulate the growth of plants and restrict trace metal uptake, hence protecting plants from metal-induced toxicity (Ma et al. 2015). Moreover, endophytes involved in decontamination of other pollutants amplify metal translocation process (Babu et al. 2015). For example, endophytes in a FTWs system, vegetated with P. *australis* effectively, removed most of trace metal (loid)s from the polluted river (Shahid et al. 2019). According to previous research findings, it has been proven that prior inoculated plants with endophytes would reduce the disease incidents from microbial infections and biological agents such as insects and nematodes (Rasche et al. 2006). Endophytic response to stresses mainly occurs through production of siderophore, antimicrobial metabolics, phosphate solubilizing compounds, and phytohormones (Glick 2012; Pinheiro et al. 2013; Matsuoka et al. 2013; Jasim et al. 2014). Endophytes enable to mitigate stress conditions by accelerating of photosynthetic process, enhancing translocation, activating of antioxidative enzyme functions, etc. (Shahid et al. 2020).

6.3.3 Aquatic Plant-Microbe Interactions in Aquatic Ecosystem

Aquatic macrophyte-microbe interaction is a crucial aspect in maintaining of ecosystem equilibrium. However, very limited studies have been conducted to examine the significance of this relationship (Srivastava et al. 2017). However, symbiotic relationship between plant and microbe leads to positive outcomes for both organisms. Table 6.1 illustrates the several examples for plant-microbe interactions found in aquatic environment.

The nitrogen fixation by microbes occurs mainly in rhizoplane with the aid of numerous nitrogen-fixing bacteria (Suyal et al. 2018; Shahid et al. 2020). The plant root would supply oxygen and organic matter for nitrogen-metabolizing bacteria (Jun-Xing et al. 2012; Rawat et al. 2019; Wewalwela et al. 2020), whereas aquatic plant receives nutrients (e.g., Ammonium) from microbial degradation (Ormeño-Orrillo et al. 2013). Moreover, roots also reduce nitrogen concentration from the rhizosphere and regulate the pH and redox potential in the rhizosphere (Husson 2013). Scientists have identified significant N-fixing bacterial genera, namely, *Azospirillum, Pseudomonas, Enterobacter, Vibrio*, and *Klebsiella* (Lamers et al. 2012; López-Guerrero et al. 2012). Several aquatic plants such as P. *australis*, *J. effuses, J. balticus, Sagittaria triflolia*, and *Zostera marina* associated with nitrogen-fixing bacteria effectively fix nitrogen in aquatic reservoirs (Hiraishi 2008; Neori and Agami 2017). Similar types of interaction have been recorded by Stirk and vanStaden (2003), Rivas et al. (2003), Hay et al. (2004), and Wagner (2012) with different nitrogen-fixing species (see Table 6.1). Further, symbiotic relationship between nodulating bacteria of water legume (Ghosh et al. 2015) and

6 Microbial Interactions with Aquatic Plants

Table 6.1 Aquatic ecosystem interactions between plants and microbes

Plant species	Microbial species	Major function in ecosystem	References
P. australis	Nitrogen-fixing bacteria	Nitrogen fixation	Hiraishi (2008), Neori and Agami (2017)
J. effuses			
J. balticus			
Sagittaria triflolia			
Zostera marina			
Azolla filiculoides	Arthrobacter spp.		Stirk and vanStaden (2003)
Neptunia natans nov.	Devosia neptuniae sp.		Rivas et al. (2003)
Hemiaulus hauki	Richelia intracelluaris		Hay et al. (2004)
Nuphur spp.	Mesorhizobium loti		Wagner (2012)
Water legume	Root-nodulating bacteria		Ghosh et al. (2015)
Water fern	Cyanobacteria		Zheng et al. (2009)
Eichhornia crassipes	Bacteria	Denitrification	Gao et al. (2014)
Phragmites australis	Nitrosomonas	Ammonia oxidation	Okabe et al. (2012)
Ulva australis	Pseudoalteromonas tunicate	Allelopathic impact on algae	Rao et al. (2006)
Vitiveria ziznioides	Arbuscularmycorrhiza	Allelopathic impact on bacteria	Nautiyal et al. (2013)
Lemna aequinoctalis	Acidobacteria F-183	IAA production in roots	Yoneda et al. (2021)
	Acidobacteria TBR-22		
Chlorella vulgaris	Azospirillum brasilense	Plant growth promoting	Gonzalez and Bashan (2000)

cyanobacteria of water fern (Zheng et al. 2009) enhanced the nitrogen fixation in plants.

In addition, nitrogen fixation of some epiphytic bacteria is useful in denitrification and ammonia oxidation (Gao et al. 2014). For instance, *Eichhornia crassipes* remediates eutrophic water through generating of gaseous nitrogen (Gao et al. 2014). Moreover, ammonia oxidation by *Nitrosomonas*-associated *Phragmites australis* is also reported by Okabe et al. 2012. Some specific epiphytic microorganisms associated with aquatic macrophytes are involved with secreting with allelopathic compounds, which retard other algae growth (Rao et al. 2006; Srivastava et al. 2007; Hempel et al. 2008; Table 6.1).

Growth and development of aquatic macrophytes would enhance due to beneficial symbiotic bacteria; these are also identified as plant-growth-promoting bacteria (PGPB) (Rajwar et al. 2018; Pole et al. 2022). These organisms serve for host plant by supplying nutrients and plant hormones (ex: indole acetic acid-IAA). Furthermore, bacteria would provide nutrients through nitrogen fixation and phosphate solubilization (Shahid et al. 2020). For instance, Yoneda et al. 2021 revealed that *Lemna aequinoctalis* (Duckweed) associated with *Acidobacteria* strains F-183 and TBR-22 promoted healthy growth of duckweeds through IAA production. Simultaneously, tested *Acidobacteria* strains enhanced the chlorophyll content in the fronds of duckweeds. Similar of type results have been discussed by Gonzalez and Bashan 2000 (Table 6.1).

6.4 Environment Perspectives of Aquatic Plant-Microbe Interactions

6.4.1 Remediation of Pollutants by Microorganism-Plant Association

Aquatic ecosystem is adversely affected globally by enhanced anthropogenic activities during last few decades (Hill 2020). Industrialization and urbanization processes have accelerated the water pollution, which led to increase in numerous health hazardous resulting in deaths due to consumption of contaminated water (Xu et al. 2018a; Supreeth 2022). For instance, bioaccumulation of toxic pollutants exists in water, encourage trace metal(loid)s accretion in human body, and later change the physiological functions of the body. Simultaneously, when aquatic pollution increases, it negatively affects the ecological sustainability of the particular area (Supreeth 2022). Interestingly, due to epidemic COVID-19 pandemic situation prevailed in last 2 years has reduced the total environmental pollution (total environment including water, soil, and air ecosystem) significantly compared to previous years (Muhammad et al. 2020). However, consider to aquatic ecosystem, application of viable, economical and eco-friendly strategy would provide sustainable solution to mitigate water pollution from contaminants. Among the various remediation strategies use in contaminants removal in water; biological remediation or bioremediation has been widely used. Biological remediation usually relies with microorganisms and phytoremediation processes. However, according to previous research studies related to remediation suggest that microorganisms accompanied with plants would accelerate the remediation process in advance (Giri et al. 2017; Supreeth 2022). The plant-microbe combined system provides several advantages over each of these methods (Soni et al. 2022; Supreeth 2022). Through this combination, pollutants would be degraded and uptake by plants more effectively (Nasr 2019).

6.4.2 Remediation of Contaminants

Due to the phytotoxic impact, phytoremediation of pollutants by macrophytes and biodegradation of contaminants by microorganisms alone frequently lead to

6 Microbial Interactions with Aquatic Plants

inadequate metal removal in aquatic ecosystems (Supreeth 2022). However, combination of these two methods showed significant impact on contaminants removal from the polluted water (Mishra et al. 2020). Nonetheless, the micro-environmental condition in polluted water quite differs compared to noncontaminated water. Mainly, plant required nutrient concentration and dissolved oxygen content are low in toxic water (Supreeth 2022); thus, both organisms might face unfavorable conditions to survive. Therefore, plants initiate to interrelate with root associated microbes through releasing several chemical substances which leads to create interactions (Supreeth 2022). Macrophytes would provide micro-aerophilic conditions and accelerate the degradation of contaminants through producing of numerous root exudates. In addition, plant roots stimulate microbe activities on contaminants through providing the substrates and then changing of media pH (Hussain et al. 2018). Simultaneously, microbes presence in rhizosphere or/and epiphytic bacteria involve in nutrient cycling and ensuring the vigorous plant growth (Clark et al. 2018; Kumar et al. 2021; Debbarma et al. 2021; Suyal and Soni 2022). Moreover, endophytic bacteria reside inside the plant tissues increase the plant's capability for cope with various biotic and abiotic stresses (O'Brien et al. 2020).

Presently, microbe-plant combined remediation system is widely used to remediate pollutants such as excess nutrients, heavy metals, petroleum-hydrocarbon, pharmaceuticals, and personal care products (PPCP) through artificially prepared systems, namely, constructed wetlands (CW) and floating treatment wetlands (FTW) (Zhang and Shao 2013; Shahid et al. 2020; Dash et al. 2021).

6.4.3 Removal of Excess Nutrients in Polluted Water

Nutrient accumulation (especially nitrates and phosphates) in water reservoirs result in severe environmental issues like "eutrophication" (Padedda et al. 2017). Several anthropogenic activities such as intensive catchment agriculture and industrialization could be reason for this situation and indirectly affect human health (Jiang et al. 2019). Hence, removal of these compounds is an essential process. However, efficiency of plant-microbe strategy depends on pollutant concentration, previous research experiments have proved beyond doubt that plant-microbe interaction in aquatic rhizoplane would enhance the cleaning process (Chen et al. 2012; Xu et al. 2018b). For instance, Organophosphorous compounds present in polluted water was removed by the plant species *Nymphaea alba*, *Phragmites australis*, and *Myriophyllum verticillatum* with the assistance of bacterial colonies in the plankton (Xu et al. 2018b). Furthermore, *Lolium perenne* plant combined with microbial species such as *Bacillus* sp. MOE1 and *Microbacterium* sp. MOE2 increased the remediation of ammonium nitrogen and phosphorous available in the eutrophic water (Li et al. 2011). Similarly, nitrogen presence in polluted water mitigated by immobilized nitrogen cycling bacteria (INCB) associated with water plants such as *Eichhornia crassipes* and *Elodea muttallii* (Chang et al. 2006). Similarly, Wu et al. 2016 reported several plant species (*T. augustifolia*, *P. australis*, and *A. calamus* L.) have removed nitrogen from water associated with microbial community.

Submerged plant species *Vallisneria natans* inoculated by *Pseudomonas putida* KT2440 enhanced carbon availability in water (Gan et al. 2018). Plant-microbes interaction effectively used in floating treatment wetlands technology to reduce nitrogen and phosphorous from polluted water. For instance, aquatic macrophyte, *Festuca arundinacea* combined with denitrifying polyphosphate accumulating microbes diminished the total nitrogen significantly; since improved the plant growth and biomass production (Zhao et al. 2011). Another study under FTW system revealed that aquatic plant named *Oenanthe javanica* interact with *Archaea* and Anaerobic ammonium oxidation bacteria resulted removal of NH_4^+-N, NO_3- N, and total nitrogen relatively higher in percentage (Wang et al. 2018). In addition, *Typha domingensis* and *Phragmites australis* plants associated with inoculated rhizospheric and endophytic bacteria improved the remediation performances of polluted river water through reducing nitrate and total nitrogen content (Shahid et al. 2019). Concurrently, some plants species such as *Cymbidium faberi, Ipomoea aquatica, Corbicula fluninea*, and *Thalia dealbata* also effectively used to remove excess nutrients from polluted water (Li et al. 2010; Zhao et al. 2011; Zhang and Shao 2013) and summarized results of these experiments shown in Table 6.2.

6.4.4 Plant-Microbe Interaction Relevant to Trace Metal(Loid)S Removal

In an aquatic ecosystem, phytoremediation process mainly occurs through phytofiltration technique (Islam et al. 2015). However, phytofiltration mechanism is encouraged by bacteria, thus enhancing the bioavailability of trace metal(loid)s (Khan et al. 2015). Several rhizospheric and endophytic bacteria promote trace metal (loid)s removal from contaminated water through absorbing metal(loid)s ions into their cell walls. Releasing of specific metal-binding proteins and peptides, enable microorganisms to accumulate toxic metals and this facilitates plant's hormone and redox signaling process due to toxic metal exposure (Cobbett and Goldsbrough 2002). Especially, endophytic bacteria involve with trace metal bioaccumulation and detoxification (Shin et al. 2015; Ijaz et al. 2016). For instance, *Pseudomonas fluorescence* G10 and *Micobacterium* sp. G16 on *Brassica napus* enhanced the Pd accretion in plant shoots (Sheng et al. 2008). On the other hand, some bacteria release extracellular polymeric substances (EPSs) to reduce metal(loid)s bioavailability (Rajkumar et al. 2009). For example, *Azobacter* sp. produced complexes with Cd through secreting of numerous EPS and decrease metal uptake by *Triticum aestivum* (Joshi and Juwarkar 2009). Furthermore, Shahid et al. 2019 revealed that few plants, namely, *B. mutica, T. domingensis, P. australis*, and *L. fusca*, remediated trace metal(loid)s (e.g., Fe, Ni, Pd, Cr, and Mn) associated with endophytic bacteria from contaminated river water by floating treatment wetlands. In addition, plant root and biofilm interaction enhanced remediation of Cu and Ni (Tanner and Headley 2011; Ladislas et al. 2015), Cr (Chakraborty et al. 2013), Cu (Munoz et al. 2006), Cr and Zn (Abou-Shanab et al. 2007) etc. displayed in Table 6.2.

6 Microbial Interactions with Aquatic Plants

Table 6.2 Plant-microbe interaction systems that improve the removal of pollutants and nutrients from aquatic ecosystems

Xenobiotics	Plant species	Relevant microorganisms	References
Organophosphorous compounds (ex: Chlorpyrifos)	*Nymphaea alba; Phragmites australis; Myriophyllum verticillatum*	Planktonic bacterial populations	Xu et al. (2018b)
	Phragmites communis	*Microbacterium* spp.	Chen et al. (2012)
	Nymphaea spp.	*Pseudomonas* spp.	
	Najas spp.	*Paenibacillus* spp.	
Ammonium nitrogen and phosphorous	*Lolium perenne*	*Bacillus* sp. MOE1 and *Microbacterium* sp. MOE2	Li et al. (2011)
	Eichhornia crassipes and Elodea muttallii	Immobilized nitrogen cycling bacteria (INCB)	Chang et al. (2006)
Nitrogen	*Typha augustifolia; Phragmites australis; Acorus calamus L.*	Microbial community	Wu et al. (2016)
	Festuca arundinacea	Denitrifying bacteria	Zhao et al. (2011)
Carbon dioxide	*Vallisneria natans*	*Pseudomonas putida* KT2440	Gan et al. (2018)
Ammonium nitrogen and nitrate nitrogen	*Oenanthe javanica*	*Archaea* and anaerobic ammonium oxidation bacteria	Wang et al. (2018)
Total nitrogen and total nitrogen	*Typha domingensis* and *P. australis*	Rhizospheric and endophytic bacteria	Shahid et al. (2019)
Excess nutrients	*Cymbidium faberi; Ipomoea aquatica; Corbicula fluninea; Thalia dealbata*	Bacterial biofilm	Li et al. (2011), Zhao et al. (2011), Zhang and Shao (2013)
Trace metal (Pd)	*Brassica napus*	*Pseudomonas fluorescence* G10 and *Micobacterium* sp. G16	Sheng et al. (2008)
Trace metal (Cd)	*Triticum aestivum*	*Azobacter* sp.	Joshi and Juwarkar (2009)
Trace metals (ex: Fe, Ni, Pd, Cr, and Mn)	*Brachia mutica; Typha domingensis; Phragmites australis; Leptochala fusca*	Endophytic bacteria	Shahid et al. (2019)
Trace metals (ex: Cu and Ni)	*Carex virgate; Cyperus ustulatus; Juncus edgariae; Schoenoplectus tabemaemontani*	Biofilms	Tanner and Headley (2011), Ladislas et al. (2015)
Trace metal (Cr)	*Pistia stratiotes*	*Bacillus cereus* GXBC1	Chakraborty et al. (2013)

(continued)

Table 6.2 (continued)

Xenobiotics	Plant species	Relevant microorganisms	References
Trace metal (Cu)	*Chlorella sorokiniana*	*Ralstonia basilensis*	Munoz et al. (2006)
Trace metals (Cr and Zn)	*Eichhornia crassipes*	*Ochrobactrum anthropi* *Bacillus cereus*	Abou-Shanab et al. (2007)
Amines, phenolics, aliphatic aldehydes, polychlorinated biphenyles and atrazine	Aquatic macrophyte	Biofilms	Ghosh et al. (1999), Tranvik (1998), Guasch et al. (2007)
Methane	Aquatic macrophyte	*Proteobacteria*	Yoon (2010), Pandey et al. (2014)
Benzene	*Euphorbia milii, Hedera helix, Chlorophytum comosum*	Endophytic and Epiphytic bacteria	Sriprapat and Thiravetyan (2016)
Phenolic compounds	*Brassica napus*	*Pantoea* sp.	Ontañon et al. (2014)
Phenol	*Phragmite australis*	*Acinetobacter, Bacillus cereus, Pseudomonas* sp.	Saleem et al. (2018a, b)
	Lemna aoukikusa	*Acinetobacter, Calcoaceticus*	Yamaga et al. (2010)
	Chlorella sorokiniana	*Pseudomonas migulae, Sphingomonas, Yanoikuyae*	Borde et al. (2003)
Nitro-phenols	*Spirodela polyrhiza*	*Nitrophenols degrading bacteria*	Kristanti et al. (2014)
Decaclorobiphenyl PCB-209	*Ocimum basilicum L.*	*Acinetobacter, Bacillus, Lysinibacillus, Novosphingobium, Pseudomonas, Rhizobium* etc.	Sánchez-Pérez et al. (2020)
Bisphenol A	*Dracaena sanderiana*	*Bacillus thuringiensis* and *Pantoea dispersa*	Suyamud et al. (2018)
Carbamazepine	*Armoracia rusticana*	*Rhizobium radiobactoer* and *Diaphorobacter nitroreducens*	Sauvêtre et al. (2018)
Indusrial dye	*Phragmites australis*	*A. Juniistrain, Rhodococcus* sp. and *P. indoloxydans*	Tara et al. (2019)
Petroleum hydrocarbon	*Lolium perenne L.*	*Pseudomonas* sp.	Iqbal et al. (2019)
	A. thaliana		
	Halonemum strobilaceum	*Ochrobactrum* sp., *Desulfovibrio* sp., *Halobacterium* sp., etc.	Al-Mailem et al. (2010)

6 Microbial Interactions with Aquatic Plants

Fungi also have a potential to remediate toxic metals from the polluted water. Root exudates invite fungi to the rhizosphere (Shahid et al. 2020; Kumar et al. 2022). Fungi genera of *Penicillium, Aspergillus,* and *Rhizopus* are widely used in trace metal(loid)s removal (Ahalya et al. 2003). Fungi closely associate with wetlands plant roots and form symbiotic relationship, which leads to release of metal-chelating siderophores and concomitant detoxification of metal(loid)s (Liu et al. 2015; Saha et al. 2016).

6.4.5 Degradation of Organic Pollutants Through Plant-Microbe Interaction

Microbes act as excellent bioremediators, since they enable to digest all organic contaminant classes effectively (Hiraishi 2008; Fenchel et al. 2012). Microbes degrade these pollutants available in the rhizosphere of the aquatic ecosystem by a cometabolism process; whereas complex carbon-based compounds break down to organic carbon and electron acceptors (Shahid et al. 2020). However, success of degradation process depends on microbial population, concentration of organic pollutants/xenobiotics, and macrophyte species in natural aquatic ecosystem (Calheiros et al. 2009). In this plant-microbe relationship, plants provide organic carbon to microbes, which assist them to perform the degradation process and bacteria release indole acetic acid (IAA) to improve plants' development (Mori et al. 2005; Golubev et al. 2009). Interestingly, the biofilms attached to aquatic macrophytes are capable of digesting of amines, phenolics, aliphatic aldehydes, polychlorinated biphenyles, and atrazine (Tranvik 1998; Ghosh et al. 1999; Guasch et al. 2007). Moreover, methanotrophs containing a collection of proteobacteria in aquatic plant rhizosphere degrade various harmful organic complexes and produce methane that may be utilized for generating carbon and energy by microbes (Yoon 2010; Pandey et al. 2014). Sriprapat and Thiravetyan (2016) revealed that several macrophytes (*Euphorbia milii, Hedera helix, Chlorophytum comosum,* etc.) associated with both endophytic and epiphytic bacteria removed the benzene in hydroponic condition effectively. The bacteria strain *Pantoea* sp. inoculated into rhizosphere of *Brassica napus* showed phenol digestion is accompanied by Cr (V) reduction into Cr (III); hence, both parties receive benefits from the relationship (Ontañon et al. 2014). Another study relevant to phenol degradation through plant-bacteria synergism showed *Phragmite australis* associated with *Acinetobacter, Bacillus cereus,* and *Pseudomonas* sp. microbes accelerated the process (Saleem et al. 2018a, b). Similar type of study conducted by Saleem et al. 2018a, b, indicated above mentioned phenol-degrading bacteria associated with *Typha domingensis* successfully removed the phenols from the water (Saleem et al. 2018a, b). Moreover, Yamaga et al. 2010 and Borde et al. 2003 also provide evidences for phenol degradation through plant-microbe interaction (Table 6.1). Besides, water contaminated with various nitro-phenols (NPs) rhizoaugmented with NP-degrading bacteria allied with *Spirodela polyrhiza* roots completely removed available NPs from polluted water (Kristanti et al. 2014). A research conducted by

Sánchez-Pérez et al. 2020 described that Decaclorobiphenyl PCB-209 remediated by *Ocimum basilicum L.* associated microorganisms.

6.4.6 Degradation of Pharmaceuticals and Personal Care Products(PPCPS) Through Plant-Microbe Interaction

According to a study conducted by Suyamud et al. 2018 resulted microorganism-plant combined system increased the removal of Bisphenol A (BPA), which is considered as a toxic compound to human endocrine system. Here, *Bacillus thuringiensis* and *Pantoea dispersa* associated with *Dracaena sanderiana* plant root enhanced the BPA more than 90%. A disobedient contaminant Carbamazepine (CBZ) present in aquatic atmosphere was removed by a plant species *A. rusticana* in combination with *R. radiobactoer* and *D. nitroreducens* (Sauvêtre et al. 2018). The results of these experiments are given in Table 6.2.

6.4.7 Degradation of Industrial Waste Through Plant–Microbe Interaction

The inoculated bacteria species (e.g., A. *juniistrain*, *Rhodococcus* sp. and P. *indoloxydans*) associated with *Phragmites australis* degrade dye substances in FTWs system (Tara et al. 2019). Similar study revealed that parallel to the contaminant degradation, trace metal concentration was also removed by 87% in the treated water. Correspondingly, Iqbal et al. 2019 discussed that the inoculation of some bacteria would mitigate the stress conditions, which affect plant growth and development. For example, together with *Pseudomonas* sp., *L. perenne and A. thaliana* reduced petroleum hydrocarbon; simultaneously enhancing the plant growth (Iqbal et al. 2019). Furthermore, *H. strobilaceum* allied with several microorganisms to remediate oil compounds present in contaminated water (Al-Mailem et al. 2010).

6.5 Biotechnological Applications of Aquatic Plant–Microbe Interactions

Many studies have taken place in the context of aquatic plant microbiomes targeting many approaches. This has relied on development of various microbial biotechnological approaches to enhance our understanding of both aquatic plants as well as plant microbiomes.

6.5.1 Plant-Microbe Symbiosis in Transgenics

It is well known that the endophytic bacterial strains showing both pollutant removal and plant growth-promoting functions have been more efficient in phytoremediation.

This plant microbiome association has been successfully used in treating aquatic contamination of crude oil. *Pseudoarthrobacter phenanthrenivorans* (MS2) and *Azospirillum oryzae* have been successful in both crude oil removal and plant-growth-promoting function in terms of chlorophyll content, water potential, proline, amino acids, and antioxidant enzymes (Saeed et al. 2021).

However, developing transgenics has become another promising approach, which has been extensively used in plant-microbe symbiosis. In order to biodegrade toxic substances, many microorganisms and plants have been genetically engineered, which has been difficult to degrade by naturally occurring species.

One example depicts the expression of metal-tolerant genes in bacterial strains to increase Cd binding, and engineered bacteria strain associated reduced Cd toxicity in aquatic plant environment roots. Yeast has been genetically engineered by inserting the arsenic removal gene *WaarsM* from a soil fungi *Westerdykella aurantiaca*, and has been successful in rice-associated arsenic bioremediation (Verma et al. 2016). Recent efforts have also been taken to understand the association between aquatic rhyzosphere microbial communities and their associated root exudates using a systems biology approach involving metabolomics and metagenomics (Singh et al. 2021). Metabolomics and Metagenomics tools have been used in this study, which has been demonstrated using a system biology approach. This study reports of developing a RhizoFlowCell (RFC) system in order to check the dynamics of root exudation patterns of *Pandanus amaryllifolius* when it was exposed to naphthalene pollutant. The developed system was able to capture the complexity of root exudates in the aquatic rhizosphere and results were further analyzed using LC-qTOF-MS. The obtained metabolomic profile has significantly helped them to understand the response of roots to changing levels of naphthalene. They further have observed the formation of an active microbial biofilm during the process. Fluorescence in situ hybridization and Illumina Miseq Next-Generation Sequencing of metagenomic DNA experiments have revealed the ability of aquatic plant roots to attract bacterial communities. This has been facilitated through the metabolic compounds secreted by the root system (Lee et al. 2013).

6.5.2 Biorefineries: Cultivation Systems

Certain bacterial communities, especially plant-growth-promoting bacteria (PGPB), play a significant role in algae under phototrophic conditions (Lian et al. 2018): 10–70% increase in algae growth rate has been found in many instances in the association of PGPB. Therefore, many attempts have been taken on cultivation of algae and growth-enhancing bacteria together for better productivity. Further concern has been taken to maintain suitable levels of bacterial communities to achieve mass cultivation of microalgae with stagnant higher growth rates (Lian et al. 2018; Fuentes et al. 2016). In addition to that, several incidences have been reported regarding recycling of the harvested algal-bacterial community into High Rate Algal Ponds in order to safeguard the stability of the community (Lian et al. 2018; Makut et al. 2019). Cocultivation of microalgae and certain bacteria have been undertaken to enhance algal lipid production. This has been achieved by cocultivating microalgae *Chlorella sacchrarophila* and *Cellvibrio pealriver* together in a bioreactor. *Cellvibrio pealriver* was allowed to grow in an inner bag, which has been embedded in the bioreactor. This strategy has increased the lipid production by 825.34–929.79 mg·L^{-1} (Xie et al. 2021). Similar strategy between bacterium *Idiomarina loihiensis* and microalgae *Chlorella variabilis* has witnessed significant increase in both protein and lipid contents in algae (Rajapitamahuni et al. 2019).

Microalgae carry a significant potential in contributing to achieve clean energy by production of Hydrogen. Fakhimi N. and Tavakoli suggest the possibility of improving hydrogen production through the algae-bacteria cocultivation. Cocultivation of *Escherichia coli*, *Pseudomonas stutzeri*, and *Pseudomonas putida* bacterial strains along with *Chlamydomonas reinhardtii* enhanced hydrogen production significantly (Fakhimi and Tavakoli 2019).

Furthermore, certain studies report that during microalgae harvesting, the algal-bacterial interactions have caused a significant increase in algae flocculant size, thereby facilitating easy settlement, even in lesser flocculation contributive condition. The formation of the settleable flocs has been achieved by the bacteria surrounding the phycosphere of algae (Lee et al. 2013). Extracellular Polysaccharide Substances of bacterium, and the cell wall and secretory proteins of algae has contributed to this formation of settleable flocs. Quorum sensing between the two components also most likely causes this phenomenon. However, this is significantly important in algae biorefineries, since 20–30% of biomass production cost is taken by algae harvesting.

6.5.3 Other Biotechnological Applications

Development of light microbial solar/fuel cells is another application of the usage of synergism between electricity producing bacteria and algae (Feng et al. 2018). Periodic current generation with a significant energy conversion efficiency has been reported when green alga, *Chlamydomonas reinhardtii*, and an iron-reducing bacterium, *Geobacter sulfurreducens,* are cocultured together. *G. sulfurreducens* has yielded current by oxidizing formate, which has been produced by *C. reinhardtii*. This is a demonstration of the electricity generation through syntrophic interactions between phototrophs and electricity-generating bacteria (Nishio et al. 2013).

Biogas production is another biotechnological use of the algae-bacteria interactions. A study reports the successful production of both biohydrogen and biogas using mutulistic cocultivation of algae (*Chlamydomonas* sp. and *Scenedesmus* sp.) and *Rhizobium* sp. The study further shows that these mutually living bacteria have the innate ability to quench O_2, which led to the further utilization of spent algal-bacterial biomass for biogas production (Cantera et al. 2021).

Another study shows the importance in algal and prokaryotic community structure in an outdoors pilot High Rate Algal Pond (HRAP), which has been also used for the wastewater treatment and biogas upgrading. According to the study, the presence of NH_4^+ and CO_2 tolerant algae *Chlorella vulgaris,* with certain prokaryotic bacterial strains, resulted in the production of biomethane with significant amounts of CH_4, CO_2, and no H_2S (Wirth et al. 2015).

An enriching advance of algal-bacterial coexistence is bioethanol production, which has also gained significant attention. A study shows the ability of certain marine bacteria to convert algae starch granules to ethanol (De Maia et al. 2020). Polikovsky et al. show that the consortium of genetically altered *Maribacter* sp. and *Roseovarius* sp. significantly affects the *Ulva mutabilis* growth rate and photosynthetic constituents and their contents. This has contributed to bioethanol production by causing significant increase in glucose and glycerol and decrease in xylose and glucuronic acid (Polikovsky et al. 2020).

Another high potential area of the use of bacterial-aquatic plant such as algae interactions is for the production of industrially important chemicals and fuels in biorefifineries. For example, Liu et al. demonstrate the possibility of isoprene production using *Synechococcus elongates* and *Escherichia coli* coculture (Liu et al. 2021). In addition to that since algae-based food products are already a good option for food scarcity associated with increasing population and reducing fertile land area, scientists suggest the importance of using aquatic plant-bacteria interactions to produce high value food products such as nutraceuticals and also low value food products for aquaculture. The potential of using aquatic plant-microbes interactions for our benefit is enormous. With advancement of certain biotechnological tools and areas such as CRISPR-Cas, genomics, transcriptomics, metagenomics, proteomics, and metabolomics, the scientific community has taken more interest in unraveling more and more possibilities.

6.6   Conclusion

Microbial interactions with aquatic plants have immense importance. Microbial accumulation as biofilms in aquatic and marine water using quorum sensing led to many interactions such as to remediate pollutants by microorganism, removal of excess nutrient in pollutant water, trace element metal removal, organic pollutant removal, and pharmaceuticals industrial waste removal. Moreover, there are vast varieties of biotechnological applications using the plant-microbial interactions in aquatic and marine biosystems.

References

Abou-Shanab RAI, Angle JS, Van-Berkum P (2007) Chromate tolerant bacteria for enhanced metal uptake by Eichhornia crassipes (Mart.). Int J Phytoremediation 9:91–105

Afonso AC, Gomes IB, Saavedra MJ, Giaouris E, Simões LC, Simões M (2021) Bacterial coaggregation in aquatic systems. Water Res 196:117037

Ahalya N, Ramachandra T, Kanamadi R (2003) Biosorption of heavy metals. Res J Chem Environ 7:71–79

Al-Mailem DM, Sorkhoh NA, Al-Awadhi H, Eliyas M, Radwan SS (2010) Biodegradation of crude oil and pure hydrocarbons by extreme halophilic archaea from hypersaline coasts of the Arabian Gulf. Extremophiles 14:321–328

Antunes J, Leão P, Vasconcelos V (2019) Marine biofilms: diversity of communities and of chemical cues. Environ Microbiol Rep 11(3):287–305

Ashraf S, Afzal M, Naveed M, Shahid M, Ahmad Zahir Z (2018) Endophytic bacteria enhance remediation of tannery effluent in constructed wetlands vegetated with Leptochloa fusca. Int J Phytoremediation 20:121–128

Babu AG, Shea PJ, Sudhakar D, Jung I-B, Oh B-T (2015) Potential use of Pseudomonas koreensis AGB-1 in association with Miscanthus sinensis to remediate heavy metal (loid)-contaminated mining site soil. J Environ Manag 151:160–166

Bacon CW, White JF (2016) Functions, mechanisms and regulation of endophytic and epiphytic microbial communities of plants. Symbiosis 68:87–98

Baker JH, Orr DR (1986) Distribution of epiphytic bacteria on freshwater plants. J Ecol 74:155–165

Bakker ES, Van Donk E, Declerck SAJ, Helmsing NR, Hidding B, Nolet BA (2010) Effect of macrophyte community composition and nutrient enrichment on plant biomass and algal blooms. Basic Appl Ecol 11:432–439. https://doi.org/10.1016/j.baae.2010.06.005

Battin TJ, Kaplan LA, Newbold JD, Hensen CME (2003) Contributions of microbial biofilms to ecosystem processes in stream mesocosm. Nature 426:439–442

Bauer WD, Mathesius U (2004) Plant responses to bacterial quorum sensing signals. Curr Opin Plant Biol 7(4):429–433

Bhatt K, Suyal DC, Kumar S, Singh K, Goswami P (2022) New insights into engineered plant-microbe interactions for pesticide removal. Chemosphere 309(2):136635

Borde X, Guieysse B, Delgado O, Munoz R, Hatti-Kaul R, NugierChauvin C, Patin H, Mattiasson B (2003) Synergistic relationships in algal-bacterial microcosms for the treatment of aromatic pollutants. Bioresour Technol 86(3):293–300

Borne KE, Fassman-Beck EA, Tanner CC (2014) Floating treatment wetland influences on the fate of metals in road runoff retention ponds. Water Res 48:430–442

Branda SS, Vik Å, Friedman L, Kolter R (2005) Biofilms: the matrix revisited. Trends Microbiol 13:20–26

Buosi PRB, Pauleto GM, Lansac-To'ha FA, Velho LFM (2011) Ciliate community associated with aquatic macrophyte roots: effects of nutrient enrichment on the community composition and species richness. Eur J Protistol 47(2):86–102

Calheiros CSC, Duque AE, Moura A, Henriques IS, Correia A, Rangel AOSS, Castro PML (2009) Changes in the bacterial community structure in two-stage constructed wetlands with different plants for industrial wastewater treatment. Bioresour Technol 100(13):3228–3235

Cantera S, Fischer PQ, Sánchez-Andrea I, Marín D, Sousa DZ, Muñoz R (2021) Impact of the algal-bacterial community structure, physio-types and biological and environmental interactions on the performance of a high rate algal pond treating biogas and wastewater. Fuel 302:121148

Chakraborty R, Mitra AK, Mukherjee S (2013) Synergistic chromium bioremediation by water lettuce (Pistia) and bacteria (Bacillus cereus GXBC-1) interaction. J Biol Chem Res 30(2): 421–431

Chang HQ, Yang XE, Fang YY et al (2006) In-situ nitrogen removal from the eutrophic water by microbial-plant integrated system. J Zhejiang Univ Sci B 7:521–531. https://doi.org/10.1631/jzus.2006.B0521

Chen W-M, Tang Y-Q, Mori K, Wu X-L (2012) Distribution of culturable endophytic bacteria in aquatic plants and their potential for bioremediation in polluted waters. Aquat Biol 15:99–110

Clark DR, Ferguson RMW, Harris DN, Matthews Nicholass KJ, Prentice HJ, Randall KC et al (2018) Streams of data from drops of water: 21st century molecular microbial ecology. WIREs Water 5:e1280. https://doi.org/10.1002/wat2.1280

Cobbett C, Goldsbrough P (2002) Phytochelatins and metallothioneins: roles in heavy metal detoxification and homeostasis. Annu Rev Plant Biol 53:159–182

Collins B, McArthur JV, Sharitz RR (2004) Plant effects on microbial assemblages and remediation of acidic coal pile runoff in mesocosm treatment wetlands. Ecol Eng 23:107–115. https://doi.org/10.1016/j.ecoleng.2004.07.005

Cotner JB, Biddanda BA (2002) Small players, large role: microbial influence on biogeochemical processes in pelagic aquatic ecosystems. Ecosystems 5:105–121

Cummins KW, Klug MJ (1979) Feeding ecology of stream invertebrates. Annu Rev Ecol Syst 10: 147–172

Dang H, Lovell CR (2016) Microbial surface colonization and biofilm development in marine environments. Microbiol Mol Biol Rev 80(1):91–138

Dash B, Sahu N, Singh AK, Gupta SB, Soni R (2021) Arsenic efflux in Enterobacter cloacae RSN3 isolated from arsenic-rich soil. Folia Microbiol 66:189–196

Davey ME, O'toole GA (2000) Microbial biofilms: from ecology to molecular genetics. Microbiol Mol Biol Rev 64(4):847–867

Davies LC, Pedro IS, Novais JM, Martins-Dias S (2006) Aerobic degradation of acid orange 7 in a vertical flow constructed wetlands. Water Res 40(10):2055–2063

De Maia JL, Cardoso JS, da Silveira Mastrantonio DJ, Bierhals CK, Moreira JB, Costa JAV, de Morais MG (2020) Microalgae starch: a promising raw material for the bioethanol production. Int J Biol Macromol 165:2739–2749

De Wolf EI, Calder WJ, Harrison JG, Randolph GD, Noren BE, Weinig C (2022) Aquatic Macrophytes are associated with variation in biogeochemistry and bacterial assemblages of Mountain Lakes. Front Microbiol 12:777084. https://doi.org/10.3389/fmicb.2021.777084

Debbarma P, Raghuwanshi S, Singh J, Suyal DC, Zaidi MGH, Goel R (2017) Comparative in situ biodegradation studies of polyhydroxybutyrate film composites. 3Biotech 7(178):1–9

Debbarma P, Joshi D, Maithani D, Dasila H, Suyal DC, Kumar S, Soni R (2021) Sustainable bioremediation strategies to manage environmental pollutants. In: Shah MP (ed) Removal of refractory pollutants from wastewater treatment plants. CRC Press, Boca Raton, FL. https://doi.org/10.1201/9781003204442-14

Fakhimi N, Tavakoli O (2019) Improving hydrogen production using co-cultivation of bacteria with Chlamydomonas reinhardtii microalga. Mat Sci Energy Technol 2(1):1–7

Fenchel T, Blackburn H, King GM, Blackburn TH (2012) Bacterial biogeochemistry: the ecophysiology of mineral cycling. Academic Press, Cambridge, MA

Feng H, Tang C, Wang Q, Liang Y, Shen D, Guo K, He Q, Jayaprada T, Zhou Y, Chen T, Ying X, Wang M (2018) A novel photoactive and three-dimensional stainless steel anode dramatically enhances the current density of bioelectrochemical systems. Chemosphere 196:476–481

Fry JC, Humphrey NCB (1978) Techniques for the study of bacteria epiphytic on aquatic macrophytes. In: Lovelock DW, Davies R (eds) Techniques for the study of mixed populations, pp 1–29

Fuentes JL, Garbayo I, Cuaresma M, Montero Z, González-del-Valle M, Vílchez C (2016) Impact of microalgae-bacteria interactions on the production of algal biomass and associated compounds. Mar Drugs 14(5):100

Gan L, Zhao H, Wang A et al (2018) Pseudomonas putida inoculation promotes submerged plant Vallisneria natans growth by carbon conversion in a plant-microbe interaction. Mar Freshw Res 69:851–858. https://doi.org/10.1071/MF17117

Gao Y, Yi N, Wang Y, Ma T, Zhou Q, Zhang Z, Yan S (2014) Effect of Eichhornia crassipes on production of N_2 by denitrification in eutrophic water. Ecol Eng 68:14–24

Ghannoum M, Parsek M, Whiteley M, Mukherjee PK (eds) (2020) Microbial biofilms. John Wiley & Sons, Hoboken, NJ

Ghosh U, Weber AS, Jensen JN, Smith JR (1999) Granular activated carbon and biological activated carbon treatment of dissolved and sorbed polychlorinated biphenyls. Water Environ Res 71:232–240

Ghosh PK, Kumar De T, Maiti TK (2015) Production and metabolism of indole acetic acid in root nodules and symbiont (rhizobium undicola) isolated from root nodule of aquatic medicinal legume Neptunia Oleracea Lour. J Bot 2015:1–11

Giaramida L, Manage PM, Edwards C, Singh BK, Lawton LA (2013) Bacterial communities' response to microcystins exposure and nutrient availability—linking degradation capacity to community structure. Int Biodeterior Biodegrad 84:111–117

Giri K, Rai JPN, Pandey S, Mishra G, Kumar R, Suyal DC (2017) Performance evaluation of isoproturon-degrading indigenous bacterial isolates in soil microcosm. Chem Ecol 33(9): 817–825

Glick BR (2012) Plant growth-promoting bacteria: mechanisms and applications. Scientifica 2012: 1–15

Golubev SN, Schelud'ko AV, Muratova AY, Makarov OE, Turkovskaya OV (2009) Assessing the potential of rhizobacteria to survive under phenanthrene pollution. Water Air Soil Pollut 198:5–16

Gong M, Yang G, Zhuang L, Zeng EY (2019) Microbial biofilm formation and community structure on low-density polyethylene microparticles in lake water microcosms. Environ Pollut 252:94–102

Gonzalez LE, Bashan Y (2000) Increased growth of the microalga Chlorella vulgaris when coimmobilized and cocultured in alginate beads with the plant growth promoting bacterium Azospirillum brasilense. Appl Environ Microbiol 66(4):1527–1531

Guasch H, Lehmann V, Van Beusekom B, Sabater S, Admiraal W (2007) Influence of phosphate on the response of periphyton to atrazine exposure. Arch Environ Contam Toxicol 52:32–37

Hahn MW (2006) The microbial diversity of inland waters. Curr Opin Biotechnol 17:256–261

Hay ME, Parker JD, Burkepile DE, Caudill CC, Wilson AE, Hallinan ZP, Chequer AD (2004) Mutualisms and aquatic community structure: the enemy of my enemy is my friend. Ann Rev Ecol Evol Syst 35:175–197

He D, Zheng J, Ren L, Wu QL (2021) Substrate type and plant phenolicsinfluence epiphytic bacterial assembly during short-term succession. Sci Total Environ 792:148410. https://doi.org/10.1016/j.scitotenv.2021.148410

Hempel M, Blume M, Blindow I, Gross EM (2008) Epiphytic bacterial community composition on two common submerged macrophytes in brackish waterand fresh water. BMC Microbiol 8:58; http://www.biomedcentral.com/1471-2180/8/58

Hill MK (2020) Understanding environmental pollution. Cambridge University Press, Cambridge

Hiraishi A (2008) Biodiversity of dehalorespiring bacteria with special emphasis on polychlorinated biphenyl/dioxin dechlorinators. Microbes Environ 23:1–12

Hmelo LR (2017) Quorum sensing in marine microbial environments. Annu Rev Mar Sci 9:257–281

Hossell JC, Baker JH (1979a) A note on the enumeration of epiphytic bacteria by microscopic methods with particular reference to two freshwater plants. J Appl Bacteriol 46:87–92

Hossell JC, Baker JH (1979b) Epiphytic bacteria of the freshwater plant Ranunculus penicillatus: enumeration, distribution and identification. Archiv Fur Hydrobiol 86:322–337

Hussain I, Aleti G, Naidu R et al (2018) Microbe and plant assistedremediation of organic xenobiotics and its enhancement by genetically modified organisms and recombinant technology: a review. Sci Total Environ 628–629:1582–1599. https://doi.org/10.1016/j.scitotenv.2018.02.037

Husson O (2013) Redox potential (eh) and pH as drivers of soil/plant/microorganism systems: a transdisciplinary overview pointing to integrative opportunities for agronomy. Plant Soil 362: 389–417

Ijaz A, Iqbal Z, Afzal M (2016) Remediation of sewage and industrial effluent using bacterially assisted floating treatment wetlands vegetated with typha domingensis. Water Sci Technol 74: 2192–2201

Iqbal A, Mukherjee M, Rashid J et al (2019) Development of plantmicrobe phytoremediation system for petroleum hydrocarbon degradation: an insight from alkb gene expression and phytotoxicity analysis. Sci Total Environ 671:696–704. https://doi.org/10.1016/j.scitotenv.2019.03.33

Islam MS, Saito T, Kurasaki M (2015) Phytofiltration of arsenic and cadmium by using an aquatic plant, micranthemum umbrosum: phytotoxicity, uptake kinetics, and mechanism. Ecotoxicol Environ Saf 112:193–200

Jasim B, Jimtha John C, Shimil V, Jyothis M, Radhakrishnan E (2014) Studies on the factors modulating indole-3-acetic acid production in endophytic bacterial isolates from P iper nigrum and molecular analysis of ipdc gene. J Appl Microbiol 117:786–799

Jayaprada T, Hu J, Zhang Y, Feng H, Shen D, Geekiyanage S, Yao Y, Wang M (2020) The interference of nonylphenol with bacterial cell-to-cell communication. Environ Pollut 257:113352

Jiang Q, Song X, Liu J et al (2019) Enhanced nutrients enrichment and removal from eutrophic water using a self-sustaining in situ photomicrobial nutrients recovery cell (PNRC). Water Res 167:115097. https://doi.org/10.1016/j.watres.2019.115097

Joint I, Tait K, Wheeler G (2007) Cross-kingdom signalling: exploitation of bacterial quorum sensing molecules by the green seaweed Ulva. Philos Trans R Soc Lond B Biol Sci 362(1483): 1223–1233

Joshi PM, Juwarkar AA (2009) In vivo studies to elucidate the role of extracellular polymeric substances from azotobacter in immobilization of heavy metals. Environ Sci Technol 43:5884–5889

Jun-Xing Y, Yong L, Zhi-Hong Y (2012) Root-induced changes of pH, eh, Fe (II) and fractions of Pb and Zn in rhizosphere soils of four wetland plants with different radial oxygen losses. Pedosphere 22:518–527

Khan MU, Sessitsch A, Harris M, Fatima K, Imran A, Arslan M, Shabir G, Khan QM, Afzal M (2015) Cr-resistant rhizo-and endophytic bacteria associated with Prosopis juliflora and their potential as phytoremediation enhancing agents in metal-degraded soils. Front Plant Sci 5:755

Kristanti RA, Toyama T, Hadibarata T et al (2014) Sustainable removal of nitrophenols by rhizoremediation using four strains of bacteria and giant duckweed (Spirodela polyrhiza). Water Air Soil Pollut 225:1. https://doi.org/10.1007/s11270-014-1928-7

Kudryavtsev VM (1984) Bacteria on vascular aquatic plants. Hydrobiol J 19:50–55

Kumar P, Gupta SB, Anurag, Soni R (2019) Bioremediation of cadmium by mixed indigenous isolates Serratia liquefaciens BSWC3 and Klebsiella pneumoniae RpSWC3 isolated from industrial and mining affected water samples. Pollution 5(2):351–360

Kumar P, Dash B, Suyal DC, Gupta SB, Singh AK, Chowdhury T, Soni R (2021) Characterization of arsenic-resistant *Klebsiella pneumoniae* RnASA11 from contaminated soil and water samples and its bioremediation potential. Curr Microbiol 78:3258–3267

Kumar S, Joshi D, Debbarma P, Singh M, Yadav AN, Singh N, Suyal DC, Soni R, Goel R (2022) Denaturing gradient gel electrophoresis (DGGE) analysis of the fungi involved in biodegradation. In: Udayanga D, Bhatt P, Manamgoda D, Saez JM (eds) Mycoremediation protocols. Springer Protocols Handbooks, Humana, NY, pp 93–100

Ladislas S, Gerente C, Chazarenc F, Brisson J, Andres Y (2015) Floating treatment wetlands for heavy metal removal in highway stormwater ponds. Ecol Eng 80:85–91

Lamers LP, Van Diggelen JM, Op Den Camp HJ, Visser EJ, Lucassen EC, Vile MA, Jetten MS, Smolders AJ, Roelofs JG (2012) Microbial transformations of nitrogen, sulfur, and iron dictate vegetation composition in wetlands: a review. Front Microbiol 3:156

Lamers LP, Govers LL, Janssen IC, Geurts JJ, Van der Welle ME, Van Katwijk MM, Van der Heide T, Roelofs JG, Smolders AJ (2013) Sulfide as a soil phytotoxin—a review. Front Plant Sci 4:268

Lami R (2019) Quorum sensing in marine biofilms and environments. Quorum Sens 1:55–96

Lasa I, Penadés JR (2006) Bap: a family of surface proteins involved in biofilm formation. Res Microbiol 157:99–107

Lee YJ, Mynampati K, Drautz D, Arumugam K, Williams R, Schuster S, Kjelleberg S, Swarup S (2013) Understanding aquatic rhizosphere processes through metabolomics and metagenomics approach. In: EGU general assembly conference abstracts, pp EGU2013–EGU7810

Li X-N, Song H-L, Li W, Lu X-W, Nishimura O (2010) An integrated ecological floating-bed employing plant, freshwater clam and biofilm carrier for purification of eutrophic water. Ecol Eng 36:382–390

Li H, Zhao H-P, Hao H-L, Liang J, Zhao F-L, Xiang L-C, Yang X-E, He Z-L, Stoffella PJ (2011) Enhancement of nutrient removal from eutrophic water by a plant–microorganisms combined system. Environ Eng Sci 28:543–554

Lian J, Wijffels RH, Smidt H, Sipkema D (2018) The effect of the algal microbiome on industrial production of microalgae. Microb Biotechnol 11(5):806–818

Liu W-L, Guan M, Liu S-Y, Wang J, Chang J, Ge Y, Zhang C-B (2015) Fungal denitrification potential in vertical flow microcosm wetlands as impacted by depth stratification and plant species. Ecol Eng 77:163–171

Liu JT, Sun JJ, Fang SW, Han L, Feng Q, Hu F (2016) Nutrient removal capabilities of four submerged macrophytes in the Poyang Lake Basin. Appl Ecol Environ Res 14:107–124. https://doi.org/10.15666/aeer/1402_107124

Liu H, Cao Y, Guo J, Xu X, Long Q, Song L, Xian M (2021) Study on the isoprene-producing co-culture system of Synechococcus elongates–Escherichia coli through omics analysis. Microb Cell Factories 20(1):1–18

López D, Vlamakis H, Kolter R (2010) Biofilms. Cold Spring Harbor Perspect Biol 2:a000398

López-Guerrero MG, Ormeño-Orrillo E, Acosta JL, Mendoza-Vargas A, Rogel MA, Ramírez MA, Rosenblueth M, Martínez-Romero J, Martínez-Romero E (2012) Rhizobial extrachromosomal replicon variability, stability and expression in natural niches. Plasmid 68:149–158

Ma Y, Rajkumar M, Rocha I, Oliveira RS, Freitas H (2015) Serpentine bacteria influence metal translocation and bioconcentration of Brassica juncea and Ricinus communis grown in multi-metal polluted soils. Front Plant Sci 5:757

Makut BB, Das D, Goswami G (2019) Production of microbial biomass feedstock via co-cultivation of microalgae-bacteria consortium coupled with effective wastewater treatment: a sustainable approach. Algal Res 37:228–239

Matsuoka H, Akiyama M, Kobayashi K, Yamaji K (2013) Fe and P solubilization under limiting conditions by bacteria isolated from Carex kobomugi roots at the Hasaki coast. Curr Microbiol 66:314–321

Mishra A, Mishra SP, Arshi A et al (2020) Plant-microbe interactions for bioremediation and phytoremediation of environmental pollutants and agro-ecosystem development. Bioremediation Ind Waste Environ Saf. https://doi.org/10.1007/978-981-13-3426-9_17

Mooney JA, Pridgen EM, Manasherob R, Suh G, Blackwell HE, Barron AE, Bollyky PL, Goodman SB, Amanatullah DF (2018) Periprosthetic bacterial biofilm and quorum sensing. J Orthop Res 36(9):2331–2339

Mori K, Toyama T, Sei K (2005) Surfactants degrading activities in the rhizosphere of giant duckweed (Spirodela polyrrhiza). Jpn J Water Treat Biol 41:129–140

Muhammad S, Long X, Salman M (2020) COVID-19 pandemic and environmental pollution: a blessing in disguise? Sci Total Environ 728:138820. https://doi.org/10.1016/j.scitotenv.2020.138820

Munch C, Neu T, Kuschk P, Roske I (2007) The root surface as the definitive detail for microbial transformation processes in constructed wetlands—a biofilm characteristic. Water Sci Technol 56(3):271–276

Munoz R, Alvarez MT, Munoz A, Terrazas E, Guieysse B, Mattiasson B (2006) Sequential removal of heavy metals ions and organic pollutants using an algal-bacterium consortium. Chemosphere 63(6):903–911

Nasr M (2019) Environmental perspectives of plant-microbe nexus for soil and water remediation. In: Kumar V, Prasad R, Kumar M, Choudhary D (eds) Microbiome in plant health and disease. Springer, Singapore. https://doi.org/10.1007/978-981-13-8495-0_18

Nautiyal CS, Srivastava S, Chauhan PS, Seem K, Mishra A, Sopory SK (2013) Plant growth-promoting bacteria Bacillus amyloliquefaciens NBRISN13 modulates gene expression profile of leaf and rhizosphere community in rice during salt stress. Plant Physiol Biochem 66:1–9

Nazir R, Zaffar MR, Amin I (2019) Bacterial biofilms: the remarkable heterogeneous biological communities and nitrogen fixing microorganisms in lakes. Freshwater Microbiol 1:307–340

Neori A, Agami M (2017) The functioning of rhizosphere biota in wetlands–a review. Wetlands 37: 615–633

Nishio K, Hashimoto K, Watanabe K (2013) Light/electricity conversion by defined cocultures of chlamydomonas and geobacter. J Biosci Bioeng 115(4):412–417

O'Brien AM, Yu ZH, Luo D, Laurich J, Passeport E, Frederickson ME (2020) Resilience to multiple stressors in an aquatic plant and its microbiome. Am J Bot 107(2):273–285. https://doi.org/10.1002/ajb2.1404

Okabe S, Nakamura Y, Satoh H (2012) Community structure and in situ activity of nitrifying bacteria in *Phragmites* root associated biofilms. Microbes Environ 27(3):288–292

Ontañon OM, González PS, Ambrosio LF et al (2014) Rhizoremediation of phenol and chromium by the synergistic combination of a native bacterial strain and Brassica napus hairy roots. Int Biodeterior Biodegrad 88:192–198. https://doi.org/10.1016/j.ibiod.2013.10.017

Ormeño-Orrillo E, Hungria M, Martinez-Romero E (2013) Dinitrogen-fixing prokaryotes. Prokaryotes prokaryotic physiol. Biochemist 1:427–451

Padedda BM, Sechi N, Lai GG, Mariani MA, Pulina S, Sarria M, Satta CT, Virdis T, Buscarinu P, Lugliè A (2017) Consequences of eutrophication in the management of water resources in mediterranean reservoirs: a case study of Lake Cedrino (Sardinia, Italy). Global Ecol Conserv 12:21–35. https://doi.org/10.1016/j.gecco.2017.08.004

Paerl HW, Pinckney JL (1996) A mini review of microbial consortia: their roles in aquatic production and biogeochemical cycling. Microb Ecol 31(3):225–247

Pandey VC, Singh J, Singh D, Singh RP (2014) Methanotrophs: promising bacteria for environmental remediation. Int J Environ Sci Technol 11:241–250

Peiris C, Wathudura PD, Gunatilake SR, Gajanayake B, Wewalwela JJ, Abeysundara S, Vithanage M (2022) Effect of acid modified tea-waste biochar on crop productivity of red onion (Allium cepa L.). Chemosphere 288:132551

Pinheiro EA, Carvalho JM, dos Santos DC, Feitosa AO, Marinho PS, Guilhon GMS, Santos LS, de Souza AL, Marinho AM (2013) Chemical constituents of aspergillus sp EJC08 isolated as endophyte from Bauhinia guianensis and their antimicrobial activity. An Acad Brasi Ciênc 85: 1247–1253

Pole A, Srivastava A, Zakeel MCM, Sharma VK, Suyal DC, Singh AK, Soni R (2022) Role of microbial biotechnology for strain improvement for agricultural sustainability. In: Soni R, Suyal DC, Yadav AN, Goel R (eds) Trends of applied microbiology for sustainable economy. Academic Press, Elsevier, London, pp 285–317

Polikovsky M, Califano G, Dunger N, Wichard T, Golberg A (2020) Engineering bacteria-seaweed symbioses for modulating the photosynthate content of Ulva (Chlorophyta): significant for the feedstock of bioethanol production. Algal Res 49:101945

Rajapitamahuni S, Bachani P, Sardar RK, Mishra S (2019) Co-cultivation of siderophore-producing bacteria Idiomarina loihiensis RS14 with Chlorella variabilis ATCC 12198, evaluation of microalgal growth, lipid, and protein content under iron starvation. J Appl Phycol 31(1):29–39

Rajkumar M, Ae N, Freitas H (2009) Endophytic bacteria and their potential to enhance heavy metal phytoextraction. Chemosphere 77:153–160

Rajwar J, Chandra R, Suyal DC, Tomer S, Kumar S, Goel R (2018) Comparative phosphate solubilizing efficiency of psychrotolerant *Pseudomonas jesenii* MP1 and *Acinetobacter* sp. ST02 against chickpea for sustainable hill agriculture. Biologia 73(8):793–802

Rasche F, Velvis H, Zachow C, Berg G, Van Elsas JD, Sessitsch A (2006) Impact of transgenic potatoes expressing anti-bacterial agents on bacterial endophytes is comparable with the effects of plant genotype. Soil type and pathogen infection. J Appl Ecol 43(3):555–566

Ramsay AJ (1974) The use of autoradiography to determine the proportion of bacteria metabolising in an aquatic habitat. J Gen Microbiol 80:363–373

Rao D, Webb JS, Kjelleberg S (2006) Microbial colonization and competition on the marine alga Ulva australis. Appl Environ Microbiol 72:5547–5555

Rawat N, Sharma M, Suyal DC, Singh DK, Joshi D, Singh P, Goel R (2019) Psyhcrotolerant bio-inoculants and their co-inoculation to improve *Cicer arietinum* growth and soil nutrient status for sustainable mountain agriculture. J Soil Sci Plant Nutr 19(3):639–647

Rivas R, Willems A, Subba-Rao NS, Mateos PF, Dazzo FB, Kroppenstedt RM, Martlnez-Molina E, Gillis M, Vela'zquez E (2003) Description of Devosia neptuniae sp—nov. that nodulates and fixes nitrogen in symbiosis with Neptunia natans, an aquatic legume from India. System Appl Microbiol 26(1):47–53

Saeed M, Ilyas N, Arshad M, Sheeraz M, Ahmed I, Bhattacharya A (2021) Development of a plant microbiome bioremediation system for crude oil contamination. J Environ Chem Eng 9(4): 105401

Saha M, Sarkar S, Sarkar B, Sharma BK, Bhattacharjee S, Tribedi P (2016) Microbial siderophores and their potential applications: a review. Environ Sci Pollut Res 23:3984–3999

Saleem H, Arslan M, Rehman K, Tahseen R, Afzal M (2018a) *Phragmites australis*—a helophytic grass—can establish successful partnership with phenol-degrading bacteria in a floating treatment wetland. Saudi J Biol Sci 26:1179–1186

Saleem H, Rehman K, Arslan M, Afzal M (2018b) Enhanced degradation of phenol in floating treatment wetlands by plant-bacterial synergism. Int J Phytoremediation 20:692–698

Sánchez-Pérez BN, Zenteno-Rojas A, Rincón-Molina CI et al (2020) Rhizosphere and endophytic bacteria associated to Ocimum basilicum L with decaclorobiphenyl removal potential. Water Air Soil Pollut 231:1. https://doi.org/10.1007/s11270-020-04481-6

Sauvêtre A, May R, Harpaintner R et al (2018) Metabolism of carbamazepine in plant roots and endophytic rhizobacteria isolated from *Phragmites australis*. J Hazard Mater 342:85–95. https://doi.org/10.1016/j.jhazmat.2017.08.006

Saxena P, Joshi Y, Rawat K, Bisht R (2019) Biofilms: architecture, resistance, quorum sensing and control mechanisms. Indian J Microbiol 59(1):3–12

Shahid MJ, Tahseen R, Siddique M, Ali S, Iqbal S, Afzal M (2019) Remediation of polluted river water by floating treatment wetlands. Water Supply 19:967–977

Shahid JM, Al-surhanee AA, Kouadri F, Ali S, Nawaz N, Afzal M, Rizwan M, Ali B, Soliman MH (2020) Role of microorganisms in the remediation of wastewater in floating treatment wetlands: a review. Sustainability 12:5559. https://doi.org/10.3390/su12145559

Sheng X-F, Xia J-J, Jiang C-Y, He L-Y, Qian M (2008) Characterization of heavy metal-resistant endophytic bacteria from rape (Brassica napus) roots and their potential in promoting the growth and lead accumulation of rape. Environ Pollut 156:1164–1170

Shin CJ, Nam JM, Kim JG (2015) Floating mat as a habitat of Cicuta virosa, a vulnerable hydrophyte. Landsc Ecol Eng 11:111–117

Singh D, Bhasin S, Madan N, Singh M, Suyal N, Singh N, Soni R, Suyal DC (2021) Metagenomics: insights into microbial removal of the contaminants. In: Bhatt P, Gangola S, Udayanga D, Kumar G (eds) Microbial technology for sustainable environment. Springer, Singapore, pp 293–306

Soni R, Suyal DC, Yadav AN, Goel R (2022) Trends of applied microbiology for sustainable economy. Academic Press, Elsevier, London; Paperback ISBN: 9780323915953; eBook ISBN: 9780323915960

Souffreau C, Busschaert P, Denis C, Van Wichelen J, Lievens B, Vyverman W et al (2018) A comparative hierarchical analysis of bacterioplankton and biofilm metacommunity structure in an interconnected pond system: hierarchical metacommunity study of pond bacteria. Environ Microbiol 20:1271–1282. https://doi.org/10.1111/1462-2920.14073

Sriprapat W, Thiravetyan P (2016) Efficacy of ornamental plants for benzene removal from contaminated air and water: effect of plant associated bacteria. Int Biodeterior Biodegradation 113:262–268

Srivastava J, Chandra H, Singh N (2007) Allelopathic response of *Vetiveria zizanioides* (L.) Nash on members of the family Enterobacteriaceae and pseudomonas spp. Environmentalist 27:253–260

Srivastava J, Gupta A, Chandra H (2008) Managing water quality with aquatic macrophytes. Rev Environ Sci Biotechnol 7:255–266

Srivastava JK, Chandra H, Kalra SJS, Mishra P, Khan H, Yadav P (2017) Plant–microbe interaction in aquatic system and their role in the management of water quality: a review. Appl Water Sci 7: 1079–1090. https://doi.org/10.1007/s13201-016-0415-2

Stien D, Sanchez-Ferandin S, Lami R (2016) Quorum sensing and quorum quenching in the phycosphere of phytoplankton: a case of chemical interactions in ecology. J Chem Ecol 42(12):1201–1211

Stirk WA, vanStaden J (2003) Occurrence of cytokinins likecompounds in two aquatic ferns and their exudates. Environ Exp Bot 49:77–85

Stottmeister U, Wießner A, Kuschk P, Kappelmeyer U, Kastner M, Bederski O, Muller RA, Moormann H (2003) Effects of plants and microorganisms in constructed wetlands for waste-water treatment. Biotechnol Adv 22:93–117

Supreeth M (2022) Enhanced remediation of pollutants by microorganisms–plant combination. Int J Environ Sci Technol 19:4587–4598. https://doi.org/10.1007/s13762-021-03354-7

Suyal DC, Soni R (2022) Bioremediation of environmental pollutants. Springer Nature, London; Hardcover ISBN: 9783030861681; eBook ISBN: 9783030861698

Suyal DC, Kumar S, Joshi D, Soni R, Goel R (2018) Quantitative proteomics of psychotrophic diazotroph in response to nitrogen deficiency and cold stress. J Proteome 187:235–242

Suyamud B, Thiravetyan P, Panyapinyopol B, Inthorn D (2018) Dracaena sanderiana endophytic bacteria interactions: effect of endophyte inoculation on bisphenol a removal. Ecotoxicol Environ Saf 157:318–326. https://doi.org/10.1016/j.ecoenv.2018.03.066

Tanner CC, Headley TR (2011) Components of floating emergent macrophyte treatment wetlands influencing removal of storm water pollutants. Ecol Eng 37:474–486

Tara N, Arslan M, Hussain Z, Iqbal M, Khan QM, Afzal M (2019) On-site performance of floating treatment wetland macrocosms augmented with dye-degrading bacteria for the remediation of textile industry wastewater. J Clean Prod 217:541–548

Tranvik LJ (1998) Degradation of dissolved organic matter in humic waters by bacteria. In: Aquatic humic substances. Springer, Berlin/Heidelberg, Germany, pp 259–283

Urakawa H, Dettmar DL, Thomas J (2017) The uniqueness and biogeochemical cycling of plant root microbial communities in a floating treatment wetland. Ecol Eng 108:573–580

Verma S, Verma PK, Meher AK, Dwivedi S, Bansiwal AK, Pande V, Srivastava PK, Verma PC, Tripathi RD, Chakrabarty D (2016) A novel arsenic methyltransferase gene of Westerdykella aurantiaca isolated from arsenic contaminated soil: phylogenetic, physiological, and biochemi-cal studies and its role in arsenic bioremediation. Metallomics 8(3):344–353

Vilas MP, Marti CL, Adams MP, Oldham CE, Hipsey MR (2017) Invasive macrophytes control the spatial and temporal patterns of temperature and dissolved oxygen in a shallow lake: a proposed feedback mechanism of macrophyte loss. Front Plant Sci 8:2097. https://doi.org/10.3389/fpls.2017.02097

Wagner SC (2012) Biological nitrogen fixation. Nat Educ Knowl 3(10):15

Wang P, Jeelani N, Zuo J, Zhang H, Zhao D, Zhu Z, Leng X, An S (2018) Nitrogen removal during the cold season by constructed floating wetlands planted with Oenanthe javanica. Mar Freshw Res 69:635–647

Wei B, Yu X, Zhang S, Gu L (2011) Comparison of the community structures of ammonia-oxidizing bacteria and archaea in rhizoplanes of floating aquatic macrophytes. Microbiol Res 166:468–474

Wei Y, Hou H, ShangGuan Y, Li J, Li F (2014) Genetic diversity of endophytic bacteria of the manganese-hyperaccumulating plant phytolacca americana growing at a manganese mine. Eur J Soil Biol 62:15–21

Wewalwela JJ, Tian Y, Donaldson JR, Baldwin BS, Varco JJ, Rushing B, Lu H, Williams MA (2020) Associative nitrogen fixation linked with three perennial bioenergy grasses in field and greenhouse experiments. GCB Bioenergy 12:1104–1117

Wijewardene L, Wu N, Fohrer N, Riis T (2022) Epiphytic biofilms in freshwater and interactions with macrophytes: current understanding and future directions. Aquat Bot 176:103467

Wirth R, Lakatos G, Maróti G, Bagi Z, Minárovics J, Nagy K, Kondorosi É, Rákhely G, Kovács KL (2015) Exploitation of algal-bacterial associations in a two-stage biohydrogen and biogas generation process. Biotechnol Biofuels 8(1):1–14

Wu X, Xi W, Ye W, Yang H (2007) Bacterial community composition of a shallow hypertrophic freshwater lake in China revealed by 16S rRNA gene sequences. FEMS Microbiol Ecol 61:85–96

Wu H, Xu K, He X, Wang X (2016) Removal of nitrogen by three plant species in hydroponic culture: plant uptake and microbial degradation. Water Air Soil Pollut 227:1. https://doi.org/10.1007/s11270-016-3036-3

Xie Z, Lin W, Luo J (2021) Co-cultivation of microalga and xylanolytic bacterium by a continuous two-step strategy to enhance algal lipid production. Bioresour Technol 330:124953

Xu X, Nie S, Ding H, Hou FF (2018a) Environmental pollution and kidney diseases. Nat Rev Nephrol 14:313–324. https://doi.org/10.1038/nrneph.2018.11

Xu XJ, Lai GL, Chi CQ et al (2018b) Purification of eutrophic water containing chlorpyrifos by aquatic plants and its effects on planktonic bacteria. Chemosphere 193:178–188. https://doi.org/10.1016/j.chemosphere.2017.10.171

Yamaga F, Washio K, Morikawa M (2010) Sustainable biodegradation of phenol by Acinetobacter calcoaceticus P23 isolated from the rhizosphere of duckweed Lemna aoukikusa. Environ Sci Technol 44(16):6470–6474

Yoneda Y, Yamamoto K, Makino A, Tanaka Y, Meng X-Y, Hashimoto J, Shin-ya K, Satoh N, Fujie M, Toyama T et al (2021) Novel plant-associated Acidobacteria promotes growth of common floating aquatic plants, Duckweeds. Microorganisms 9:1133. https://doi.org/10.3390/microorganisms9061133

Yoon S (2010) Towards practical application of methanotrophic metabolism in chlorinated hydrocarbon degradation, greenhouse gas removal, and immobilization of heavy metals. Ph.D. Thesis, University of Michigan, Ann Arbor, MI

Zehr JP (2010) Microbes in earth's aqueous environments. Front Microbiol 1(4):1–2

Zhang L, Shao H (2013) Heavy metal pollution in sediments from aquatic ecosystems in China. Clean Soil Air Water 41:878–882

Zhang W, Ding W, Li YX, Tam C, Bougouffa S, Wang R, Pei B, Chiang H, Leung P, Lu Y, Sun J (2019) Marine biofilms constitute a bank of hidden microbial diversity and functional potential. Nat Commun 10(1):1–10

Zhao T, Fan P, Yao L, Yan G, Li D, Zhang W (2011) Ammonifying bacteria in plant floating Island of constructed wetland for strengthening decomposition of organic nitrogen. Trans Chin Soc Agric Eng 27:223–226

Zheng W, Rang L, Bergman B (2009) Structural characteristics of the cyanobacterium–azolla symbioses. In: Pawlowski K (ed) Prokaryotic symbionts in plants. Springer, Berlin/Heidelberg, Germany, pp 235–263

Status of Microplastic Pollution in the Freshwater Ecosystems

7

Vaishali Bhatt, Neha Badola, Deepti Semwal, and Jaspal Singh Chauhan

Abstract

Only 7% of the total plastic produced worldwide is recycled, allowing the majority of plastics to build in the environment and pose a tremendous threat to living and nonliving things of the planet. These plastics in the form of fine particles [microplastics (MPs)] have nowadays emerged as a significant water pollutant. The MPs in the freshwater system are still a subject of research that has to be thoroughly investigated, because it has not yet been fully understood. The sources, circulation, and final disposition of MPs in the freshwater environments are the main topics of this chapter. Further, the movement and dispersion of MPs in the freshwater bodies are impacted by a number of variables, which are also discussed. For scientific communities all across the world, it is advised that a uniform technique be created for identifying MPs. The current chapter will help to better understand the origin, movement, fate, and distribution of MPs in the freshwater ecosystem.

Keywords

Pollution · Microplastic · Water · Ecosystem

V. Bhatt · N. Badola · D. Semwal · J. Singh Chauhan (✉)
Aquatic Ecology Lab, Department of Himalayan Aquatic Biodiversity, Hemvati Nandan Bahuguna Garhwal University (A Central University), Srinagar, Uttarakhand, India
e-mail: jaspal.chauhan@hnbgu.ac.in

© The Author(s), under exclusive license to Springer Nature Singapore Pte Ltd. 2023
R. Soni et al. (eds.), *Current Status of Fresh Water Microbiology*,
https://doi.org/10.1007/978-981-99-5018-8_7

7.1 Introduction

Microplastic contamination is a problem that is getting worse and is widespread in the world's soil, air, and water. Because MPs are present in both natural and artificial water cycles, MP pollution is observed almost throughout the whole water cycle (hydrologic cycle) on the earth. The different aquatic ecosystems have been affected by MPs, with the oceans serving as catch basins and rivers as transporters of the material. The presence of small plastic objects (5 mm) accumulating in aquatic ecosystems was first noted in marine areas in the 1970s (Hays and Cormons 1974). Thompson et al. used the term MP to describe the tiny plastic pieces, and there has been an exponential increase in the number of scientific research on the topic of MPs since then (Thompson et al. 2004). There was not much research on MPs in the literature on freshwater ecosystems before the twenty-first century; however, Hays and Cormons (1974) reported MPs in the rivers of North America (Hays and Cormons 1974). The substantial artificial polymer contamination of rivers was previously unnoticed (or disregarded), but since these early papers, a lot more research has been done, and MPs have now been reported in freshwater systems, including lakes and rivers, all around the world. Researches show that microplastic is building up in freshwaters just as it is in the oceans. In a solitary waterbody, such as a lake or pond in a mountain range, the microplastic will persist forever, breaking down further into smaller MPs and nanoplastics and harming the ecology as a whole. Rivers and lakes provide water supplies to humans and act as habitats to aquatic species having significant ecological and economic importance as well as providing opportunities for recreation, and the production of aquatic goods. Therefore, understanding the prevalence, distribution, and impacts of MPs in freshwaters is crucial. This chapter describes the latest evidence on the occurrence, outcome, and influence of MPs in aquatic habitats to better understand the current state and fate of MPs in the freshwater ecosystems.

7.2 Sources and Fate of MPS in Freshwater Ecosystem

MPs can enter the environment through a variety of routes, and a route that is crucial in one area may not be so crucial in another. Particles from the diverse sources have a variety of morphologies, compositions, sizes, and other characteristics. The qualities of particles produced under carefully monitored industrial circumstances are often more uniform and homogenous. Intentional industrial manufacturing is one of the possible sources and entry points for designed MPs. For instance, plastic pellets used as feedstock in the production of plastic or microbeads purposefully made for cosmetic products. In 2012, it is projected that over 4000 tonnes of MP beads were utilized in cosmetics across Europe (Gouin et al. 2015). Although being only a minor portion of the anticipated total load of environmental MPs, primary MPs are a comparatively simple problem to manage and eliminate. However, unmanaged secondary sources of human origin are the primary causes of MP pollution. These sources include inadequately collected and disposed plastic garbage that is dumped directly into the environment or that is incorrectly collected and dumped in landfills and later on dispersed into the environment via wind or water. Additionally, synthetic paints, automobile tires, and industrial abrasion processes are anticipated to make contributions to the productions of MPs (Lassen et al. 2012). Other causes include pollutant from factories and building sites, which can enter aquatic environments through the wind and surface runoff. The use of plastic sheets for crop cultivation is one of the most significant causes of plastic pollution of agricultural soils and is regarded as a key agricultural emission (Xu et al. 2006). Synthetic textiles are a significant additional source, since they wash up with high amounts of MP fibers in waste-water (Napper and Thompson 2016). The fibers were always present in the several environmental compartments that had previously been studied, but the fragments were mostly found in urban runoff. These fibers may originate from a variety of places, such as landfills, trash incineration, degradation of macroplastics, synthetic fibers used in clothing and home furnishings, and landfills. The characterization indicates that the primary source of these fibers being clothing is the most likely theory. The fibers in the air, including MPs, may be carried by the wind and dumped on the land surfaces or transported to aquatic environments. These fibers, which include MPs, could be carried by the wind to the aquatic environment or left behind in terrestrial ecosystem. It is not feasible to determine whether the fibers are coming from nearby or far off sources when they are detected. MPs that have been discovered in the remote lakes imply that long-distance transfer may also be possible (Free et al. 2014; Zhang et al. 2016; Liu et al. 2021b). When looking into MPs in freshwater, it appears that atmospheric fallout may also be a substantial source for its availability. Other sources, such as fibers that may detach directly from persons strolling down the street while wearing their clothing, must also be taken into account in addition to atmospheric fallout (Fig. 7.1).

After entering the environment, MP does not stay in one place; instead, it is moved to different environments compartments, where it spends varied amounts of time. The flow of MPs from the land to the river system is influenced by the local weather, distance from the river, type of land use overland runoff, and dispersion into roadside ditches. The flow conditions, daily discharge, and the morphology at each individual site will control the migration of bulk plastics and MPs within the riverine system (Balas et al. 2001). Since MPs will be transported and dispersed to diverse environmental compartments more quickly than macroplastics, they are also vulnerable to varying rates of degradation. While the movement to sediments and the development of biofilms on MPs surfaces may also slow down the rate of deterioration by reducing light exposure. Most of the knowledge we presently have about the degradation of plastics comes from laboratory studies that recurrently concentrate on a particular mechanism, such as photo, thermal, or biodegradation, which restricts our knowledge of the degradation of plastics under environmental conditions where multiple mechanisms occur simultaneously. In contrast, no research has been done on MP formation rates. This is significant, because some polymers tend to depolymerize slowly and break down into smaller pieces (Lambert and Wagner 2016) and, as these fragments continue to fragmentate, MPs are eventually formed.

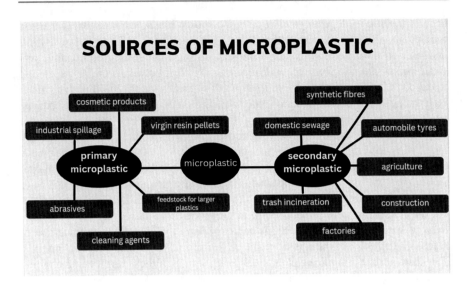

Fig. 7.1 Sources of MPs in freshwater ecosystems

Environmental factors, polymer characteristics, chemical additive type and quantity, and polymer properties are key factors that impact MP degradation and fragmentation. Crystallinity, a molecular characteristic that affects density and permeability of MPs due to more organized and tightly structured polymer chains, is a significant polymer attribute. This has an impact on how MPs behave when they are wet and their swelling, impacting bacterial binding to surfaces. While stabilizers like antioxidants and antimicrobials extend the lifespan of plastics, which leads to MP fragmentation into ever smaller particles, including nanoplastic formation, transformation of fragments, degradation into nonpolymer organic molecules, and further transformations/degradation into other chemicals (Lambert et al. 2013). Pollutants such as DDT, PCBs, metals, and dioxins interact with MPs and get adsorbed on its surface, affecting their mobility and bioavailability (Mato et al. 2001; Ashton et al. 2010). Pharmaceuticals, personal care items, and other industrial chemicals that enter the environment are likely to co-occur with MPs in the freshwater environment via physical, chemical, or pore-filling adsorption. It would be fascinating to know how much of the sorbed chemicals become accessible in the water column as a result of the MP's continuing deterioration as well as alterations in the environment (Fig. 7.2).

7.3 MPS in Freshwater Systems

Over the past 10 years, there have been numerous studies on MPs pollution in the marine environment; however, there have been few studies on MPs contamination in the freshwater ecosystem. There are around six times as many articles reported for

7 Status of Microplastic Pollution in the Freshwater Ecosystems

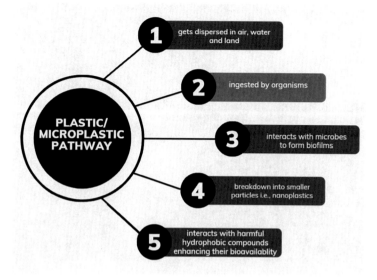

Fig. 7.2 Pathway of MPs in freshwater ecosystems

MP in the marine environment as there are in freshwater. Since freshwater systems, whether directly or indirectly, are the main source of drinking water, appropriate investigations must be conducted to look into the MPs pollution in freshwater too. Additionally, the release of land based as well as aerial MPs into freshwater ecosystems results in their transit to the marine waters. The study of MPs in freshwater systems is advancing, and MPs have now been found in freshwater systems on several different continents. Researchers have found MPs all over the world, from Asia to Antarctica, in rivers, lakes, ponds, wetlands as well as reservoirs. Despite having a very uneven distribution, MPs have been found in every compartment of the freshwater ecosystem. MPs are reported from fish, birds, amphibians, and other biota to river bottom sediments, beach sediments, surface water, and drinking water (Fig. 7.3).

7.4 MPs in River

MPs have been a growing subject in limnological research. Researchers have focused on MP pollution in the last decade in freshwater, beginning with lakes and later shifting to rivers, reservoirs, and wetlands. Among all the research on rivers, most were published in last 5 years. The geographical cover of these researches is not homogeneous and most of the researches are carried out in Asia. To summarize the research on rivers, more than 50 rivers along the globe are investigated for MPs. Most of these studies are carried out in China, among which the Yangtze River basin is the most studied worldwide (Xiong et al. 2019; Hu et al. 2018). Many European

Fig. 7.3 Status: Studies of MPs around the world

rivers are also reported to be polluted with MPs, among which the Rhine River has received most attention (Mani et al. 2019). In North America, St. Lawrence River system has been extensively studied (Castañeda et al. 2014; Crew et al. 2020). MP concentration in rivers is far less than marine and estuarine systems. MPs have been found in all the compartments of river system as surface water, bottom sediment, shore sediment, and biota reporting varying levels of contamination. The maximum number of MPs was reported from the Sinos river in Brazil (Ferraz et al. 2020), that is, 330.2 particles per liter and yellow river in China (Liu et al. 2021a), that is, 5358–654,000 particles per m^3. In river sediments, the most polluted rivers were river Yangtze (Hu et al. 2018) and river Fuhe (Zhou et al. 2021) in China and River Ganga (Sarkar et al. 2019) in India, that is, 35.76 to 3185.33 items/kg, 1049 ± 462 items/kg and 99.27–409.86 items/kg, respectively. In most of the rivers, fibers were the dominant type of MPs, while in some rivers like Han river in South Korea, Thames river in Canada, Ravi river in Pakistan, the Nakdong river in South Korea, Cisadane river in Indonesia, and Ottawa river in Canada, fragments were the most dominant type of MPs (Table 7.1).

MP concentration at water surface is highly dynamic and altered by flow regimes, precipitation, seasons, and proximity to point of entry. For example, the number of MPs in the surface water of 9 rivers in USA doubled after introduction of effluent from waste water treatment plants (Mccormick et al. 2016). Hydrological conditions and sampling season also play an important role in the abundance of MPs reported in the rivers. Seasonal changes can alter the occurrence of MPs in the rivers. For instance, MPs concentration reduced in the 10 urban, suburban, and rural river bed sediments in the United Kingdom and reached to 2812/kg from 6350/kg (Hurley et al. 2018). While in another study in a South African river, mean MP concentration

7 Status of Microplastic Pollution in the Freshwater Ecosystems

Table 7.1 MPs in rivers

Reference	Country	River	Sample type	Mean particle	Dominant Ttype
	Vietnam	Saigon river			
Ferraz et al. (2020)	Brazil	Sinos river	Surface water	330.2 particles L^{-1}	Fiber
Chauhan et al. (2021)	India	Alaknanda river	Water/ sediment	566/389 particles	Fiber
Dris et al. (2014)	Paris	Seine river	Surface water	62 and 101 fibers/L	Fiber
Vermaire et al. (2017)	Canada	Ottawa river	Surface water	0.05 and 0.24/L	Fragment
Hou et al. (2021)	USA	Illinois river	Fish	MPs present	Fiber
Martínez Silva and Nanny (2020)	USA	Magdalena river	Sediment	25.5 to 102.4 fibers/kg 10.4–12.7 fragments/kg	
Luo et al. (2019)	China	Suzhou river	Surface water	1.8–2.4 items/L	
Luo et al. (2019)	China	Huangpu river	Surface water	1.8–2.4 items/L	
Alam et al. (2019)	Indonesia	Ciwalengke river	Surface water	5.85 ± 3.28/L	Fiber
Sulistyowati et al. (2022)	Indonesia	Cisadane river	Surface water	13.33 and 113.33 particles m^{-3}	Fragment
Mani et al. (2019)	Europe	Rhine river	Bed sediment	0.26 ± 0.01 and 11.07 ± 0.6 × 10^3/ kg^{-1}	
Deocaris et al. (2019)	Phillepines	Pasig river	Surface water	34 MP fragments	
Hu et al. (2018)	China	Yangtze river	Surface water	0.48 to 21.52 items L^{-1}	Fiber
Zhang et al. (2020)	China	Qin river	Sediment	0–97 items·kg^{-1}	
Chen et al. (2021)	Malaysia	Langat river	Surface water	4.39 particles/L	Fiber
Wong et al. (2020)	Taiwan	Tamsui river	Surface water	2.5 ± 1.8/m^3 and 83.7 ± 70.8/m^3	
Wong et al. (2020)	Taiwan	Dahan river	Surface water	83.7 ± 70.8/m^3	
Wong et al. (2020)	Taiwan	Keelung river	Surface water	2.5 ± 1.8/m^3 and 83.7 ± 70.8/m^3	
Wong et al. (2020)	Taiwan	Xindian river	Surface water	2.5 ± 1.8/m^3	
Napper et al. (2021)	India	Ganges river	Surface water	140 MP particles	Fiber
Eo et al. (2019)	South Korea	Nakdong river	Surface water	293 ± 83 to 4760 ± 5242 particles/m^3	Fragment

(continued)

Table 7.1 (continued)

Reference	Country	River	Sample type	Mean particle	Dominant Ttype
Xu et al. (2021)	China	Fenghua river	Surface water	300 N/m^3 to 4000 N/m^3	Fiber
Sarkar et al. (2019)	India	Ganga river	Sediment	99.27–409.86 items/kg	
Liu et al. (2021a)	China	Yellow river	Surface water	5358–654,000 N/m^3	Fiber
Yin et al. (2022)	China	Xiangjiang river	Sediment	288 ± 60 items/kg	
Sekudewicz et al. (2021)	Poland	Vistula river	Surface water	1.6 to 2.55 items L^{-1}	Fiber
Hu et al. (2018)	China	Yangtze river	Sediment	35.76 to 3185.33 items kg^{-1}	Fragment
Jiang et al. (2019)	Tibet	5 Tibetan rivers	Surface water	483 to 967 items/m^3	Fiber
Campanale et al. (2020)	Italy	Ofanto river	Surface water	0.9 ± 0.4 p/m^3 to 13 ± 5 p/m^3	
Irfan et al. (2020)	Pakistan	Ravi river	Surface water	190 ± 141 MPs/m^3 to 16,150 ± 80 MPs/m^3	Fragment
Constant et al. (2020)	France	Rhone river		6 kg km^{-2} year^{-1}	Fiber
Constant et al. (2020)	France	Tet river		6 kg km^{-2} year^{-1}	Fiber
Liu et al. (2021b)	China	Dafeng river	Surface water	3 × 10^{-4}-2.5 × 10^{-3} items/L and 4 × 10^{-5}-9 × 10^{-4} items/L	
Zhou et al. (2020)	China	Tuojiang river basin	Surface water	911.57 ± 199.73 to 3395.27 ± 707.22 items/m^3	Fiber
Zhou et al. (2021)	China	Fuhe river	Sediment	1049 ± 462 items/kg	
Simon-Sánchez et al. (2019)	China	Ebro river	Surface water	3.5 ± 1.4 MPs·m^{-3}	Fiber
Lechner (2020)	China	Danube river		316.8 ± 4664.6 items Per 1000 m^3	
Corcoran et al. (2019)	Canada	Thames river	Sediment	6 to 2444/kg	Fragment
Xiong et al. (2019)	China	Yangtze river		4.92 × 105 items/km^2	
Park et al. (2020a)	South Korea	Han river	Surface water	7.0 ± 12.9 particles/m^3	Fragment
de Carvalho et al. (2021)	France	Garonne river	Surface water	0.15 particles m^{-3}	
Wu et al. (2020)	Macao	Maozhou river	Surface water	4.0 ± 1.0 to 25.5 ± 3.5 items·L^{-1}	

(continued)

7 Status of Microplastic Pollution in the Freshwater Ecosystems

Table 7.1 (continued)

Reference	Country	River	Sample type	Mean particle	Dominant Ttype
Barrows et al. (2018)	USA	Gallatin river	Surface water	1.2 particles L^{-1}	Fiber
Dai et al. (2022)	China	Qiantang river	Surface water	1.5–9.4 items L^{-1}	Fiber
Patel et al. (2020)	India	Sabarmati river	Sediment	134.53 mg/kg to 581.706 mg/kg	
Lin et al. (2018)	China	Pearl river		379 to 7924 items·m^{-3}	Fiber
Wang et al. (2020)	China	Manas river	Surface water	21 ± 3–49 ± 3 items/L	Fiber
Huang et al. (2021a)	China	West river	Surface water	2.99 to 9.87 items/L	Fiber

changed from 6.3 ± 4.3/kg in summer to 160.1 ± 139.5/kg in winter (Nel et al. 2018). Most of the studies have reported MPs in the surface water of rivers, which may not be a reliable representation of entire river. For example, MP concentrations in six South Korean bays were four times higher at the surface than in the rest of the water column (Song et al. 2018). MPs have also been ingested by a wide range of aquatic organisms via feeding, respiration, drinking, or some other routes. MPs have been reported from biota in rivers from many rivers in the world. These organisms may act like reserves of MPs in the rivers. These particles can be egested back in the rivers, continuing its cycle in environment, or can be accumulated in the biota and transferred through the food web (Bhatt and Chauhan 2022), affecting the biological processes of organisms as well as communities and ecosystem.

7.5 MPs in Lakes

MPs have been found in lakes throughout Africa, Asia, Europe, and North America. MPs have been documented in North America from the Great Lakes (Zbyszewski et al. 2014), which have enormous watershed populations. In Europe, MPs have been found from heavily populated Swiss Lakes (Faure et al. 2015) to underpopulated lakes such as Lakes Bolsena (Fischer et al. 2016), and Lake Garda (Klein et al. 2018). MPs have been documented in Asia from the isolated Mongolian lake Hovsgol (Free et al. 2014) and the remote Chinese lakes on the Tibetan Plateau (Zhang et al. 2016). They have also reportedly been seen in China's Lake Taihu (Su et al. 2016), Qinghai (Xiong et al. 2018), Poyang (Jian et al. 2020), and Indian lake, Vembanad (Devi et al. 2020). The finding of MPs in fish taken from Lake Victoria in and Lake Ziway (Merga et al. 2020) was evidence of their presence in African lakes (Biginagwa et al. 2016). MPs have been found in all the compartments of Lakes as surface water, sediment, and biota reporting different levels of contamination. The maximum number of MPs was reported from the Lake Ziway, Ethiopia

(Merga et al. 2020), that is, 30,000 particles/m^3 with a range of 400 to 124,000 particles/m^3. In lake sediments, the most polluted lakes were Lake Onego, in Russia (Zobkov et al. 2020), that is, 2188.7 ± 1164.4 and lake St. Clair, Canada (Zbyszewski et al. 2014). Fiber is the dominant type of MP in most of the lakes in surface water as well as sediments, except for Lakes Michigan, Maggiore, Iseo, and Garda where fragments were the most dominant (Table 7.2).

7.6 MPs in Groundwater

MP pollution in groundwater has attracted less researches than that in other types of natural ecosystems, such as the ocean, rivers, lakes, and soil. In addition to serving as a wall between groundwater MP contamination, soil is also the most likely location for MPs to invade the groundwater system, which can have serious repercussions when groundwater is utilized for drinking purposes. Groundwater systems may get contaminated with MPs as a result of the transfer of MPs through soil migration, runoffs from garbage, industry, and urbanization. Groundwater is difficult to examine and maintain, since it is unseen. Some research papers on groundwater pollution with MPs have been published. In some of these researches, the issue of soil pollution was discussed, and it was found that the soil may serve as a pathway for MPs to reach groundwater systems. These investigations constitute the most recent contributions to the knowledge on groundwater contamination with MPs, and distribution comments and remarks have been made and published for those investigations. In a karst aquifer system in the United States, MPs (microfibers) were found at median and maximal concentrations of 6.4 and 15.2 n/L, respectively (Panno et al. 2019). In a similar study in China, mean concentration of MPs was 4.50 particles/L (An et al. 2022). While in another Karst system in Italy, 28 MPs were found in every liter of water on average (Balestra et al. 2023). MPs were absent in treated potable drinking water from taps in Germany (Weber et al. 2021), while a low concentration (0.48 particles/L) of MPs was reported in an aquifer from Iran (Esfandiari et al. 2022). Contrastingly, an outrageous amount (2103 particles/L) of MPs was found in Chinese groundwater (Khant and Kim 2022), which was higher than other observed concentrations of MPs in groundwater around the globe. The most prevalent MP polymer in groundwater systems are PE and PET (Huang et al. 2021b), while the most prevalent shapes are fragments and fibers (Table 7.3). The MP contamination of surface water is far higher than that of groundwater: it may be because anthropogenic activity has directly affected and contaminated it. When surface water sources are adjacent to WWTP and STP sites, these facilities act as entry points for MPs into the water (Park et al. 2020b). In soil as well as groundwater, MPs may operate as a carrier of these dangerous compounds by absorbing persistent organic contaminants and metals (Huang et al. 2021b). The above studies have provided data indicating that research on groundwater should receive international attention.

7 Status of Microplastic Pollution in the Freshwater Ecosystems 171

Table 7.2 MPs in freshwater lakes

Reference	Country	Lake	Sample type	Mean particle	Dominant type
Zhang et al. (2016)	Northern Tibet	4 lakes, Siling co basin	Shore sediment	8 ± 14 to 563 ± 1219 items/m	
Xiong et al. (2018)	China	Qinghai lake	Surface water	0.05105 to 7.58105 items km^2	Sheet and fibers
Xiong et al. (2018)			Shore sediment	50 to 1292 items m^2	
Xiong et al. (2018)			Fish	2–15/Ind	
Wang et al. (2017)	China	20 urban lakes	Surface water	1660.0 ± 639.1 to 8925 ± 1591 N/m^3	Fiber
Vaughan et al. (2017)	Uk	Edgbaston pool	Sediment	25–30/100 g	Fiber/film
Su et al. (2016)	China	Taihu	Surface water	3.4–25.8 items/L	Fiber
Su et al. (2016)			Sediment	11.0–234.6 items/kg Dw	
Su et al. (2016)			Asian clams	0.2–12.5 items/g	
Sighicelli et al. (2018)	Italy	Maggiore, Iseo, and Garda	Surface water	4000 ± 2700 to $57,000 \pm 36,000/$km^2	Fragments
Mason et al. (2016)	USA	Michigan	Surface water	17,000 particles/km^2	Fragments
Zobkov et al. (2020)	Russia	Lake Onego	Sediment	2188.7 ± 1164.4	Fiber
Mao et al. (2020)	China	Wuliangsuhai Lake	Surface water	3.12 to 11.25 N/L	Fiber
Li et al. (2019)	China	18 lakes (Yangtze river basin)	Surface water	240 items/m^3 to 1800 items/m^3	Fiber
Li et al. (2019)			Sediment	90 items/kg to 580 items/kg	Fiber
Jian et al. (2020)	China	Poyang lake	Surface water	1064 ± 90 Mp/m^3	Fiber
Jian et al. (2020)			Sediment	1936 ± 121 Mp/kg	Fragments
Klein et al. (2018)	Germany	Lake Garda	Sediment	108–1108	
Hu et al. (2020)	China	Dongting lake	Sediment	21–52 items/100 g	Fiber
Hu et al. (2020)			Surface water	0.62–4.31 items/m^2	Fiber
Fischer et al. (2016)	Italy	Lake Bolsena	Surface water	2.68 to 3.36 particles/m^3	Fiber

(continued)

Table 7.2 (continued)

Reference	Country	Lake	Sample type	Mean particle	Dominant type
Fischer et al. (2016)			Sediment	112 particles/kg dry weight	Fiber
Fischer et al. (2016)	Italy	Lake Chiusi	Surface water	0.82 to 4.42 particles/m^3	Fiber
Fischer et al. (2016)			Sediment	234 particles/kg dry weight	Fiber
Costello and Ebert (2020)	USA	Erie	Sediment	1 to 12 per m^2	Fiber
Costello and Ebert (2020)	USA	Ontario	Sediment	1 to 12 per m^2	Fiber
Biginagwa et al. (2016)	Tanzania	Victoria	Fish	MPs in 20% fish	
Anderson et al. (2017)	USA	Lake Winnipeg	Surface water	MPs present	
Devi et al. (2020)	India	Vembanad	Fish	MPs in 26% fish	Fiber
Grbić et al. (2020)	Canada	Ontario	Surface water	0.8/L	
Minor et al. (2020)	USA/Canada	Superior	Surface water	9000 to 40,000 particles/km^2	Fiber
Minor et al. (2020)			Sediment	0 to 55 particles/ kg dry weight	Fiber
Merga et al. (2020)	Ethiopia	Lake Ziway	Surface water	30,000 (400–124,000) particles/m^3	
Merga et al. (2020)			Fish	MPs in 35% fish	
Free et al. (2014)	Mongolia	Lake Hovsgol	Surface water	20,264/km^2	
Zbyszewski et al. (2014)	Canada	St. Clair	Sediment	1.726×10^6	

7.7   Conclusion

With regard to MPs, freshwater habitats face similar challenges as marine ones. According to analysis, freshwaters are overwhelmingly probable to become contaminated with MPs. However, there have only been a few studies till date focusing on the presence of MPs in freshwater. Studies that are now available have shown that MPs are present in lakes, rivers, and ground water. Freshwaters are affected by MP pollution everywhere from remote, rural, and suburban regions with little to no population and urban areas with soaring populations. To lessen the concerns with MP contamination, waste management systems must be improved, because they are a generator as well as supplier of freshwater MPs. Since MPs are

7 Status of Microplastic Pollution in the Freshwater Ecosystems

Table 7.3 MPs in groundwater

Reference	Country	Sample type	Mean particle	Dominant type
Połeć et al. (2018)	Poland	Deep well (untreated potable water)	MPs present	Fragment
Panno et al. (2019)	USA	Karst system		Fiber
Ganesan et al. (2019)	India	Not mentioned	66 particle	Fragment and fiber
Samandra et al. (2022)	Australia	Drinking water		Fragment and fiber
Mintenig et al. (2019)	Germany	Well		Fragment
Selvam et al. (2021)	India	Wells and borewells		Fiber
Weber et al. (2021)	Germany	Tap (treated potable water)	No MPs	Fragment and fiber
Strand et al. (2018)	Denmark	Tap		Fiber, fragment and film
Shruti et al. (2020)	Mexico	Public drinking water fountains		Fragment and fiber
Kirstein et al. (2021)	Sweden	Drinking water		Fragment and fiber
Oni and Sanni (2022)	Nigeria	Borehole drinking water	206 to 1691 items m^{-3}	Fragment
Ledieu et al. (2022)	France	Landfill leachates	10.3 to 106.7 particles/L	
Alvarado-Zambrano et al. (2022)	Mexico	Borehole	12.3 particles/L	
Balestra et al. (2023)	Italy	Karst system	28 items/L	Fiber
Patterson et al. (2023)	India	Borewell and open wells	29.73 ± 3.27	Fiber
An et al. (2022)	China	Karst system	4.50 items$\cdot L^{-1}$	Film and fiber
Wu et al. (2022)	China	Groundwater	17.0 ± 2.16 to 44.0 ± 1.63 n/L	Fiber
Cha et al. (2023)	Korea	Groundwater	0.02 to 3.48 particles/L	Fragment
Manh et al. (2022)	Vietnam	Well	2 to 21 particles/L	Fragment
Shi et al. (2022)	China	Groundwater	29 n/L	
Esfandiari et al. (2022)	Iran	Aquifer	0.48/L	Fiber
Khant and Kim (2022)	China	Groundwater	2103 particles/L	

transported to freshwater bodies such as groundwater, rivers, and lakes by soil migration, WWTP, air, surface water, landfill leachate, and other means, attention should be paid to the entire ecosystem as a single unit rather than to individual compartments when focusing on their reduction and removal. MP samples' characteristics, such as their forms, sizes, colors, texture, and polymer kinds, can be utilized to determine the origins and behaviors of the MPs in addition to their abundance and, hence, should be fairly analyzed. This chapter urges additional investigation to learn more about the origins and dispersal of MPs in freshwater bodies. Both biological and ecological consequences of MP exposure should be evaluated, particularly in situations where it is crucial. For the upcoming monitoring initiatives, protocols for sampling, pretreatment, and reporting of MPs should be harmonized and standardized.

References

Alam FC, Sembiring E, Muntalif BS, Suendo VJC (2019) Microplastic distribution in surface water and sediment river around slum and industrial area (case study: Ciwalengke River, Majalaya district, Indonesia). Chemosphere 224:637–645

Alvarado-Zambrano D, Rivera-Hernández JR, Green-Ruiz C (2022) First insight into microplastic groundwater pollution in Latin America: the case of a coastal aquifer in Northwest Mexico. Environ Sci Pollut Res Int 30(29):73600–73611

An X, Li W, Lan J, Adnan M (2022) Preliminary study on the distribution, source, and ecological risk of typical microplastics in karst groundwater in Guizhou Province, China. Int J Environ Res Public Health 19:14751

Anderson PJ, Warrack S, Langen V, Challis JK, Hanson ML, Rennie MD (2017) Microplastic contamination in Lake Winnipeg, Canada. Environ Pollut 225:223–231

Ashton K, Holmes L, Turner A (2010) Association of metals with plastic production pellets in the marine environment. Mar Pollut Bull 60:2050–2055

Balas CE, Williams AT, Simmons SL, Ergin A (2001) A statistical riverine litter propagation model. Mar Pollut Bull 42:1169–1176

Balestra V, Vigna B, De Costanzo S, Bellopede R (2023) Preliminary investigations of microplastic pollution in karst systems, from surface watercourses to cave waters. J Contam Hydrol 252:104117

Barrows APW, Christiansen KS, Bode ET, Hoellein TJ (2018) A watershed-scale, citizen science approach to quantifying microplastic concentration in a mixed land-use river. Water Res 147:382–392

Bhatt V, Chauhan JS (2022) Microplastic in freshwater ecosystem: bioaccumulation, trophic transfer, and biomagnification. Environ Sci Pollut Res Int 30:1–12

Biginagwa FJ, Mayoma BS, Shashoua Y, Syberg K, Khan FR (2016) First evidence of microplastics in the African Great Lakes: recovery from Lake Victoria Nile Perch and Nile Tilapia. J Great Lakes Res 42:146–149

Campanale C, Stock F, Massarelli C, Kochleus C, Bagnuolo G, Reifferscheid G, Uricchio VF (2020) Microplastics and their possible sources: the example of Ofanto river in Southeast Italy. Environ Pollut 258:113284

Castañeda RA, Avlijas S, Simard MA, Ricciardi A (2014) Microplastic pollution in St. Lawrence river sediments. Can J Fish Aquat Sci 71:1767–1771

Cha J, Lee J-Y, Chia RW (2023) Microplastics contamination and characteristics of agricultural groundwater in Haean Basin of Korea. Sci Total Environ 864:161027

7 Status of Microplastic Pollution in the Freshwater Ecosystems

Chauhan JS, Semwal D, Nainwal M, Badola N, Thapliyal P (2021) Investigation of microplastic pollution in river Alaknanda stretch of Uttarakhand. Environ Dev Sustain 1:1–15

Chen HL, Gibbins CN, Selvam SB, Ting KN (2021) Spatio-temporal variation of microplastic along a rural to urban transition in a tropical river. Environ Pollut 289:117895

Constant M, Ludwig W, Kerhervé P, Sola J, Charrière B, Sanchez-Vidal A, Canals M, Heussner S (2020) Microplastic fluxes in a large and a small Mediterranean river catchments: the Têt and the Rhône, Northwestern Mediterranean Sea. Sci Total Environ 716:136984

Corcoran PL, Belontz SL, Ryan K, Walzak MJ (2019) Factors controlling the distribution of microplastic particles in benthic sediment of the Thames River, Canada. Environ Sci Technol 54:818–825

Costello JD, Ebert JR (2020) Microplastic pollutants in the coastal dunes of Lake Erie and Lake Ontario. J Great Lakes Res 46:1754–1760

Crew A, Gregory-Eaves I, Ricciardi A (2020) Distribution, abundance, and diversity of microplastics in the upper St. Lawrence River. Environ Pollut 260:113994

Dai L, Wang Z, Guo T, Hu L, Chen Y, Chen C, Yu G, Ma LQ, Chen J (2022) Pollution characteristics and source analysis of microplastics in the Qiantang River in southeastern China. Chemosphere 293:133576

de Carvalho AR, Garcia F, Riem-Galliano L, Tudesque L, Albignac M, Ter Halle A, Cucherousset J (2021) Urbanization and hydrological conditions drive the spatial and temporal variability of microplastic pollution in the Garonne River. Sci Total Environ 769:144479

Deocaris CC, Allosada JO, Ardiente LT, Bitang LGG, Dulohan CL, Lapuz JKI, Padilla LM, Ramos VP, Padolina JB (2019) Occurrence of microplastic fragments in the Pasig River. H2Open J 2: 92–100

Devi SS, Sreedevi AV, Kumar AB (2020) First report of microplastic ingestion by the alien fish Pirapitinga (Piaractus brachypomus) in the Ramsar site Vembanad Lake, south India. Mar Pollut Bull 160:111637

Dris R, Gasperi J, Tassin B (2014) Assessing the microplastics in urban effluents and in the Seine River (Paris). In: Fate and impacts of microplastics in marine ecosystems

Eo S, Hong SH, Song YK, Han GM, Shim WJ (2019) Spatiotemporal distribution and annual load of microplastics in the Nakdong River, South Korea. Water Res 160:228–237

Esfandiari A, Abbasi S, Peely AB, Mowla D, Ghanbarian MA, Oleszczuk P, Turner A (2022) Distribution and transport of microplastics in groundwater (shiraz aquifer, Southwest Iran). Water Res 220:118622

Faure F, Demars C, Wieser O, Kunz M, de Alencastro LF (2015) Plastic pollution in Swiss surface waters: nature and concentrations, interaction with pollutants. Environ Chem 12:582–591

Ferraz M, Bauer AL, Valiati VH, Schulz UH (2020) Microplastic concentrations in raw and drinking water in the Sinos River, Southern Brazil. Water 12:3115

Fischer EK, Paglialonga L, Czech E, Tamminga M (2016) Microplastic pollution in lakes and Lake shoreline sediments–a case study on Lake Bolsena and Lake Chiusi (central Italy). Environ Pollut 213:648–657

Free CM, Jensen OP, Mason SA, Eriksen M, Williamson NJ, Boldgiv B (2014) High-levels of microplastic pollution in a large, remote, mountain lake. Mar Pollut Bull 85:156–163

Ganesan M, Nallathambi G, Srinivasalu S (2019) Fate and transport of microplastics from water sources. Curr Sci 117:1879–1885

Gouin T, Avalos J, Brunning I, Brzuska K, de Graaf J, Kaumanns J, Koning T, Meyberg M, Rettinger K, Schlatter H (2015) Use of micro-plastic beads in cosmetic products in Europe and their estimated emissions to the North Sea environment. SOFW J 141:40–46

Grbić J, Helm P, Athey S, Rochman CM (2020) Microplastics entering northwestern Lake Ontario are diverse and linked to urban sources. Water Res 174:115623

Hays H, Cormons G (1974) Plastic particles found in tern pellets, on coastal beaches and at factory sites. Mar Pollut Environ 5:44–46

Hou L, Mcmahan CD, Mcneish RE, Munno K, Rochman CM, Hoellein TJ (2021) A fish tale: a century of museum specimens reveal increasing microplastic concentrations in freshwater fish. Ecol Appl 31:e02320

Hu L, Chernick M, Hinton DE, Shi H (2018) Microplastics in small waterbodies and tadpoles from Yangtze River Delta, China. Environ Sci Technol 52:8885–8893

Hu D, Zhang Y, Shen M (2020) Investigation on microplastic pollution of Dongting Lake and its affiliated rivers. Mar Pollut Bull 160:111555

Huang D, Li X, Ouyang Z, Zhao X, Wu R, Zhang C, Lin C, Li Y, Guo X (2021a) The occurrence and abundance of microplastics in surface water and sediment of the West River downstream, in the south of China. Sci Total Environ 756:143857

Huang J, Chen H, Zheng Y, Yang Y, Zhang Y, Gao B (2021b) Microplastic pollution in soils and groundwater: characteristics, analytical methods and impacts. Chem Engin J 425:131870

Hurley R, Woodward J, Rothwell JJ (2018) Microplastic contamination of river beds significantly reduced by catchment-wide flooding. Nat Geosci 11:251–257

Irfan M, Qadir A, Mumtaz M, Ahmad SR (2020) An unintended challenge of microplastic pollution in the urban surface water system of Lahore, Pakistan. Environ Sci Pollut Res 27:16718–16730

Jian M, Zhang Y, Yang W, Zhou L, Liu S, Xu EG (2020) Occurrence and distribution of microplastics in China's largest freshwater lake system. Chemosphere 261:128186

Jiang C, Yin L, Li Z, Wen X, Luo X, Hu S, Yang H, Long Y, Deng B, Huang L (2019) Microplastic pollution in the rivers of the Tibet Plateau. Environ Pollut 249:91–98

Khant NA, Kim H (2022) Review of current issues and management strategies of microplastics in groundwater environments. Water 14:1020

Kirstein IV, Hensel F, Gomiero A, Iordachescu L, Vianello A, Wittgren HB, Vollertsen J (2021) Drinking plastics?–Quantification and qualification of microplastics in drinking water distribution systems by µFTIR and Py-GCMS. Water Res 188:116519

Klein S, Dimzon IK, Eubeler J, Knepper TP (2018) Analysis, occurrence, and degradation of microplastics in the aqueous environment. In: Freshwater microplastics. Springer, Cham

Lambert S, Wagner M (2016) Formation of microscopic particles during the degradation of different polymers. Chemosphere 161:510–517

Lambert S, Sinclair CJ, Bradley EL, Boxall AB (2013) Effects of environmental conditions on latex degradation in aquatic systems. Sci Total Environ 447:225–234

Lassen C, Hansen SF, Magnusson K, Norén F, Hartmann NIB, Jensen PR, Nielsen TG, Brinch AJS (2012) Microplastics-occurrence, effects and sources of releases to the environment in Denmark, p 2

Lechner A (2020) "Down by the river":(micro-) plastic pollution of running freshwaters with special emphasis on the Austrian Danube. In: Mare plasticum-the plastic sea. Springer

Ledieu L, Phuong N-N, Flahaut B, Radigois P, Papin J, le Guern C, Bechet B, Gasperi J (2022) Impact of a closed municipal solid waste landfill to microplastic groundwater pollution. In: MICRO 2022, Online Atlas Edition: plastic pollution from MACRO to nano, p 1

Li L, Geng S, Wu C, Song K, Sun F, Visvanathan C, Xie F, Wang Q (2019) Microplastics contamination in different trophic state lakes along the middle and lower reaches of Yangtze River Basin. Environ Pollut 254:112951

Lin L, Zuo L-Z, Peng J-P, Cai L-Q, Fok L, Yan Y, Li H-X, Xu X-R (2018) Occurrence and distribution of microplastics in an urban river: a case study in the Pearl River along Guangzhou City, China. Sci Total Environ 644:375–381

Liu RP, Li ZZ, Liu F, Dong Y, Jiao JG, Sun PP, El-Wardany RM (2021a) Microplastic pollution in Yellow River, China: current status and research progress of biotoxicological effects. China Geol 4:585–592

Liu S, Chen H, Wang J, Su L, Wang X, Zhu J, Lan W (2021b) The distribution of microplastics in water, sediment, and fish of the Dafeng River, a remote river in China. Ecotoxicol Environ saf 228:113009

Luo W, Su L, Craig NJ, Du F, Wu C, Shi H (2019) Comparison of microplastic pollution in different water bodies from urban creeks to coastal waters. Environ Pollut 246:174–182

Manh DT, Tuan PM, Thom DT, van Manh D (2022) Microplastics occurrence in groundwater of Da Nang City, Viet Nam. Vietnam J Sci Technol 60:39–49

Mani T, Primpke S, Lorenz C, Gerdts G, Burkhardt-Holm P (2019) Microplastic pollution in benthic midstream sediments of the Rhine River. Environ Sci Technol 53:6053–6062

Mao R, Hu Y, Zhang S, Wu R, Guo X (2020) Microplastics in the surface water of Wuliangsuhai Lake, northern China. Environment 723:137820

Martínez Silva P, Nanny MA (2020) Impact of microplastic fibers from the degradation of nonwoven synthetic textiles to the Magdalena River water column and river sediments by the City of Neiva, Huila (Colombia). Water 12:1210

Mason SA, Kammin L, Eriksen M, Aleid G, Wilson S, Box C, Williamson N, Riley A (2016) Pelagic plastic pollution within the surface waters of Lake Michigan, USA. J Great Lakes Res 42:753–759

Mato Y, Isobe T, Takada H, Kanehiro H, Ohtake C, Kaminuma T (2001) Plastic resin pellets as a transport medium for toxic chemicals in the marine environment. Environ Sci Technol 35:318–324

Mccormick AR, Hoellein TJ, London MG, Hittie J, Scott JW, Kelly JJ (2016) Microplastic in surface waters of urban rivers: concentration, sources, and associated bacterial assemblages. Ecosphere 7:e01556

Merga LB, Redondo-Hasselerharm PE, Van Den Brink PJ, Koelmans AA (2020) Distribution of microplastic and small macroplastic particles across four fish species and sediment in an African lake. Sci Total Environ 741:140527

Minor EC, Lin R, Burrows A, Cooney EM, Grosshuesch S, Lafrancois B (2020) An analysis of microlitter and microplastics from Lake Superior beach sand and surface-water. Sci Total Environ 744:140824

Mintenig S, Löder M, Primpke S, Gerdts G (2019) Low numbers of microplastics detected in drinking water from ground water sources. Sci Total Environ 648:631–635

Napper IE, Thompson RC (2016) Release of synthetic microplastic plastic fibres from domestic washing machines: effects of fabric type and washing conditions. Mar Pollut Bull 112:39–45

Napper IE, Baroth A, Barrett AC, Bhola S, Chowdhury GW, Davies BF, Duncan EM, Kumar S, Nelms SE, Niloy MNH (2021) The abundance and characteristics of microplastics in surface water in the transboundary Ganges River. Environ Pollut 274:116348

Nel HA, Dalu T, Wasserman RJ (2018) Sinks and sources: assessing microplastic abundance in river sediment and deposit feeders in an Austral temperate urban river system. Sci Total Environ 612:950–956

Oni BA, Sanni SE (2022) Occurrence of microplastics in borehole drinking water and sediments in Lagos, Nigeria. Environ Toxicol Chem 41:1721–1731

Panno SV, Kelly WR, Scott J, Zheng W, Mcneish RE, Holm N, Hoellein TJ, Baranski EL (2019) Microplastic contamination in karst groundwater systems. Groundwater 57:189–196

Park T-J, Lee S-H, Lee M-S, Lee J-K, Lee S-H, Zoh K-D (2020a) Occurrence of microplastics in the Han River and riverine fish in South Korea. Sci Total Environ 708:134535

Park T-J, Lee S-H, Lee M-S, Lee J-K, Park J-H, Zoh K-D (2020b) Distributions of microplastics in surface water, fish, and sediment in the vicinity of a sewage treatment plant. Water 12:3333

Patel AK, Bhagat C, Taki K, Kumar M (2020) Microplastic vulnerability in the sediments of the Sabarmati river of India. In: Resilience, response, and risk in water systems. Springer

Patterson J, Laju R, Jeyasanta KI, SHELCIYA S, Esmeralda VG, Asir NGG, Narmatha M, Booth AM (2023) Hydrochemical quality and microplastic levels of the groundwaters of Tuticorin, southeast coast of India. Hydrogeol J 31:1–18

Połeć M, Aleksander-Kwaterczak U, Wątor K, Kmiecik E (2018) The occurrence of microplastics in freshwater systems–preliminary results from Krakow (Poland). Geology Geophys Environ 44:391–400

Samandra S, Johnston JM, Jaeger JE, Symons B, Xie S, Currell M, Ellis AV, Clarke BO (2022) Microplastic contamination of an unconfined groundwater aquifer in Victoria, Australia. Sci Total Environ 802:149727

Sarkar DJ, Sarkar SD, Das BK, Manna RK, Behera BK, Samanta S (2019) Spatial distribution of meso and microplastics in the sediments of river Ganga at eastern India. Sci Total Environ 694: 133712

Sekudewicz I, Dąbrowska AM, Syczewski M (2021) Microplastic pollution in surface water and sediments in the urban section of the Vistula River (Poland). Sci Total Environ 762:143111

Selvam S, Jesuraja K, Venkatramanan S, Roy PD, Kumari V (2021) Hazardous microplastic characteristics and its role as a vector of heavy metal in groundwater and surface water of coastal south India. J Hazard Mater 402:123786

Shi J, Dong Y, Shi Y, Yin T, He W, An T, Tang Y, Hou X, Chong S, Chen D (2022) Groundwater antibiotics and microplastics in a drinking-water source area, Northern China: occurrence, spatial distribution, risk assessment, and correlation. Environ Res 210:112855

Shruti V, Pérez-Guevara F, Kutralam-Muniasamy G (2020) Metro station free drinking water fountain-A potential "microplastics hotspot" for human consumption. Environ Pollut 261: 114227

Sighicelli M, Pietrelli L, Lecce F, Iannilli V, Falconieri M, Coscia L, di Vito S, Nuglio S, Zampetti G (2018) Microplastic pollution in the surface waters of Italian Subalpine Lakes. Environ Pollut 236:645–651

Simon-Sánchez L, Grelaud M, Garcia-Orellana J, Ziveri P (2019) River Deltas as hotspots of microplastic accumulation: the case study of the Ebro River (NW Mediterranean). Sci Total Environ 687:1186–1196

Song YK, Hong SH, Eo S, Jang M, Han GM, Isobe A, Shim W (2018) Horizontal and vertical distribution of microplastics in Korean coastal waters. Environ Sci Technol 52:12188–12197

Strand J, Feld L, Murphy F, Mackevica A, Hartmann NB (2018) Analysis of microplastic particles in Danish drinking water, DCE-Danish Centre for Environment and Energy

Su L, Xue Y, Li L, Yang D, Kolandhasamy P, Li D, Shi H (2016) Microplastics in taihu lake, China. Environ Pollut 216:711–719

Sulistyowati L, Riani E, Cordova MR (2022) The occurrence and abundance of microplastics in surface water of the midstream and downstream of the Cisadane River, Indonesia. Chemosphere 291:133071

Thompson RC, Olsen Y, Mitchell RP, Davis A, Rowland SJ, John AW, Mcgonigle D, Russell AE (2004) Lost at sea: where is all the plastic? Science 304:838–838

Vaughan R, Turner SD, Rose NL (2017) Microplastics in the sediments of a UK urban lake. Environ Pollut 229:10–18

Vermaire JC, Pomeroy C, Herczegh SM, Haggart O, Murphy M (2017) Microplastic abundance and distribution in the open water and sediment of the Ottawa River, Canada, and its tributaries. Facets 2:301–314

Wang W, Ndungu AW, Li Z, Wang J (2017) Microplastics pollution in inland freshwaters of China: a case study in urban surface waters of Wuhan, China. Sci Total Environ 575:1369–1374

Wang G, Lu J, Tong Y, Liu Z, Zhou H, Xiayihazi N (2020) Occurrence and pollution characteristics of microplastics in surface water of the Manas River basin, China. Sci Total Environ 710: 136099

Weber F, Kerpen J, Wolff S, Langer R, Eschweiler V (2021) Investigation of microplastics contamination in drinking water of a German City. Sci Total Environ 755:143421

Wong G, Löwemark L, Kunz A (2020) Microplastic pollution of the Tamsui River and its tributaries in Northern Taiwan: spatial heterogeneity and correlation with precipitation. Environ Pollut 260:113935

Wu P, Tang Y, Dang M, Wang S, Jin H, Liu Y, Jing H, Zheng C, Yi S, Cai Z (2020) Spatial-temporal distribution of microplastics in surface water and sediments of Maozhou River within Guangdong-Hong Kong-Macao Greater Bay Area. Sci Total Environ 717:135187

Wu B, Li LW, Zu YX, Nan J, Chen XQ, Sun K, Li ZL (2022) Microplastics contamination in groundwater of a drinking-water source area, Northern China. Environ Res 214:114048

Xiong X, Zhang K, Chen X, Shi H, Luo Z, Wu C (2018) Sources and distribution of microplastics in China's largest inland lake–Qinghai Lake. Environ Pollut 235:899–906

Xiong X, Wu C, Elser JJ, Mei Z, Hao Y (2019) Occurrence and fate of microplastic debris in middle and lower reaches of the Yangtze River–from inland to the sea. Sci Total Environ 659:66–73

Xu G, Wang Q, Gu Q, Cao Y, Du X, Li F (2006) Contamination characteristics and degradation behavior of low-density polyethylene film residues in typical farmland soils of China. J Environ Sci Health 41:189–199

Xu Y, Chan FKS, Johnson M, Stanton T, He J, Jia T, Wang J, Wang Z, Yao Y, Yang J (2021) Microplastic pollution in Chinese urban rivers: the influence of urban factors. Resour Conserv Recycl 173:105686

Yin L, Wen X, Huang D, Zhou Z, Xiao R, Du L, Su H, Wang K, Tian Q, Tang Z (2022) Abundance, characteristics, and distribution of microplastics in the Xiangjiang river, China. Gondwana Res 107:123–133

Zbyszewski M, Corcoran PL, Hockin A (2014) Comparison of the distribution and degradation of plastic debris along shorelines of the Great Lakes, North America. J Great Lakes Res 40:288–299

Zhang K, Su J, Xiong X, Wu X, Wu C, Liu J (2016) Microplastic pollution of lakeshore sediments from remote lakes in Tibet plateau, China. Environ Pollut 219:450–455

Zhang L, Liu J, Xie Y, Zhong S, Yang B, Lu D, Zhong Q (2020) Distribution of microplastics in surface water and sediments of Qin river in Beibu gulf, China. Sci Total Environ 708:135176

Zhou G, Wang Q, Zhang J, Li Q, Wang Y, Wang M, Huang X (2020) Distribution and characteristics of microplastics in urban waters of seven cities in the Tuojiang River basin, China. Environ Res 189:109893

Zhou Z, Zhang P, Zhang G, Wang S, Cai Y, Wang H (2021) Vertical microplastic distribution in sediments of Fuhe River estuary to Baiyangdian Wetland in Northern China. Chemosphere 280:130800

Zobkov M, Belkina N, Kovalevski V, Zobkova M, Efremova T, Galakhina N (2020) Microplastic abundance and accumulation behavior in Lake Onego sediments: a journey from the river mouth to pelagic waters of the large boreal lake. J Environ Chem Eng 8:104367

Heavy Metal Pollution in Water: Cause and Remediation Strategies

8

Damini Maithani, Hemant Dasila, Ritika Saxena, Anjali Tiwari, Devesh Bhatt, Komal Rawat, and Deep Chandra Suyal

Abstract

Heavy metals are naturally present in earth's crust, and some of them are essential to living organisms for carrying out life processes. Due to their high persistence and nonbiodegradable nature, heavy metal accumulation beyond recommended concentrations may lead to hazardous effect on various life forms and environment. Contamination of water bodies may be due to natural and anthropogenic sources. Unchecked discharge from industrial sites and agricultural runoff in to adjoining water bodies makes the water unfit for human consumption. Escalating levels of these pollutants pose a threat to aquatic life forms and surrounding environment. Heavy metals can execute various health problems that may range from mild to severe. They can be toxic to living organisms at very low levels of exposure. Excessive usage of heavy metals has raised concerns over time, and

---

D. Maithani (✉)
Department of Microbiology, Lovely Professional University, Phagwara, Punjab, India

H. Dasila
Department of Microbiology, Akal College of Basic Sciences, Eternal University, Sirmaur, Himanchal Pradesh, India

R. Saxena
School of Biotechnology, IFTM University, Moradabad, India

A. Tiwari
Department of Environmental Science, G.B. Pant University of Agriculture & Technology, Pantnagar, Uttarakhand, India

D. Bhatt · K. Rawat
Department of Chemistry, G.B. Pant University of Agriculture & Technology, Pantnagar, India

D. C. Suyal
Department of Science, Vidyadayini Institute of Science, Management and Technology, Bhopal, Madhya Pradesh, India

© The Author(s), under exclusive license to Springer Nature Singapore Pte Ltd. 2023
R. Soni et al. (eds.), *Current Status of Fresh Water Microbiology*,
https://doi.org/10.1007/978-981-99-5018-8_8

consequently, their impact on the overall environment is being studied by researchers extensively. To safeguard human health and environment, proper management and greener technologies for removal of heavy metal from water bodies is required. This chapter will discuss the source, toxicity, and permitted concentrations of some of the major heavy metals in water bodies. Remediation approaches for mitigation of these toxic compounds have also been described. Physical and chemical remediation processes for heavy metal cleanup are highly expensive and sometimes generate a significant amount of secondary pollutants; therefore, the focus has now shifted toward eco-friendly approaches such as bioremediation and phytoremediation. Further research needs to be carried out to maximize the applicability of the existing techniques and developing highly efficient technologies for heavy metal removal from water bodies.

Keywords

Bioremediation · Heavy metal toxicity · Phytoremediation · Water pollution

8.1 Introduction

Metals are elements with high atomic weight and density ranging from 3.5 to 7 g cm^{-3}. These are found to be deposited on earth's crust naturally as minerals in the form of sulfates, phosphates, oxides, etc. (Singh et al. 2022). In aquatic system, these are naturally present in low amounts, but a variety of natural and anthropogenic activities have unavoidably escalated metal concentrations in water bodies (Ansari et al. 2003). Expansion of industries that extensively utilize metals and metalloids, mining operations, improper e-waste disposal, transportation, and fossil fuel burning are some major human activities that significantly increase concentration of heavy metals in aquatic systems. Animals and plants use some heavy metals such as Fe, Cu, Zn, and Mn at low concentrations for carrying out their physiological processes, but they become hazardous at higher concentrations (Pratush et al. 2018). Some of them are important for activity of enzymes as cofactors and also help to maintain osmotic balance. However, their accumulation over time can harm human health and other life forms. Escalating levels of heavy metals in aquatic systems is one of the most serious global concerns. About 40% of the world's lakes and rivers have been contaminated with heavy metals (Zhou et al. 2020; Bhatt et al. 2022).

Due to their toxic nature, heavy metals are considered as potent pollutants in water bodies and soil (Duffus 2002). Drinking water contaminated with heavy metals is a potential threat to public health. Consumption of polluted water may lead to cardiovascular disorders and renal failure and in severe cases may also lead to life-threatening disease like Parkinson's disease, Alzheimer's disease, and cancer (Singh et al. 2022). These metals come into touch with human bodies via digestion and respiration. People who work or live close to industrial regions associated with these heavy metals and their equivalent compounds have a significant risk of exposure and cause increased mortality rates globally (Rehman et al. 2018).

Biodiversity in aquatic ecosystem is also hampered due to toxic effects of these metals, and reports indicate their accumulation and deposition within tissues of fishes and other aquatic life forms.

With increasing industrialization, generation of heavy metals also increases, and thus, their disposal is of paramount importance. Unlike organic pollutants, inorganic metal ions are resistant to degradation. Their persistence in environment and bio-accumulation in different organisms through food chain is major problem associated with heavy metals (Wuana and Okieimen 2011). Bioaccumulation of heavy metals and their toxicity level in food chain increases with time as their separation and purification is not easy. Although metal ions are resistant to degradation, their bioavailability and chemical forms can be changed. Heavy metals can easily run into water bodies via industrial effluents, agriculture runoff, and household, but there are many technologies and strategies that are employed to remediate heavy metals. Treatment of heavy metal-laden wastewater is an arduous job. The prominent strategies that are employed include thermal treatment, chlorination, electrokinetic, bioleaching, precipitation and coagulation, ion exchange, membrane filtration, bioremediation, heterogenous photocatalyst, and adsorption and are some commonly used methods for heavy metal removal from wastewater (Selvi et al. 2019). However, most of these methods have certain disadvantages and sometimes may generate toxic compounds; therefore, integrated approaches and safer technologies need to be introduced. This chapter highlights toxicity and remediation strategies for heavy metal removal from water bodies.

8.2 Sources of Heavy Metal Pollution in Water Bodies

In water bodies, main source of heavy metal contamination includes landfill leaches, petrochemical spillage, urban runoff, industrial and mining wastewaters particularly from the electronic, metal finishing, and electroplating industries (Khan et al. 2008). Mining and metallurgical operations are prominent cause of increased concentration of these hazardous metals. Heavy metals often reach the aquatic environment through natural physical sources such as volcanic eruptions, air deposition, forest fire, runoffs, and geological matrix erosion. Occurrence of metals in lakes, rivers, and ponds is related to the type of soil and waterflow through sewage and surface runoffs from soils. Additional natural sources of heavy metal contamination of water include the wet and dry deposition of atmospheric salts, water-rock contact, and water-soil interaction (Gautam et al. 2014).

Anthropogenic activities account for vast accumulation of heavy metals in water. Rapid urbanization and industrialization are the main anthropogenic factors that pollute water. The primary causes are mining, metallurgical activities, coal and oil combustion, and agricultural runoff entering waterways (Wang et al. 2004). Other anthropogenic sources include alloy production, atmospheric deposition, battery production, coating, explosive manufacturing, improper stacking of industrial solid waste, leather tanning, mining, pesticides, phosphate fertilizer, photographic materials, printing pigments, sewage, smelting, steel and electroplating industries, textiles, and dyes and wood preservation (Dixit et al. 2015). Untreated wastewater from municipal, domestic sewage, and industry directly discharged into the natural water system leads to water contamination, industrial effluents, water tank leakages, dumping beside water bodies, and atmospheric deposition, which are some major sources through which these heavy metals entry into aquatic ecosystem. Electroplating is a major contributor to pollution because it releases heavy metals via water, air emissions, and solid waste in an environment that has been reported to contain high levels of heavy metals such as nickel, iron, lead, zinc, chromium, cadmium, and copper (Baby et al. 2010). Mining and ore processing are important sources of heavy metal contamination in the soil, and the recovery of ecosystems from mining operations might take decades. These activities generated a vast number of piles and dumps, which are usually discarded without treatment, and these abandoned mines contaminate water system through chemical runoff (Adler et al. 2007). Pesticide overuse and misuse in response to rising food demand have resulted in greater pollutant burdens in the environment, including rivers, lakes, aquifers, and coastal waterways. Anthropogenic sources of heavy metal have been observed to be more ahead than natural sources. Besides, lack of awareness for proper disposal of metal waste and failure of implementation of strict government policies and not following the recommended guidelines also contribute to the problem. Table 8.1 enlist major sources of some common heavy metal pollutants.

8.3 Potential Risks Associated with Heavy Metals

Heavy metals generally have negative impact on the ecosystem and aquatic life, and even trace amounts of heavy metals in water can be hazardous to aquatic life and human health. Ayangbenro and Babalola (2017) defined toxicity of metal as its capability to cause undesirable effects on organisms. Metal toxicity is determined by the environmental factors such as pH, temperature, salinity and dissolved oxygen, the presence of other toxicants, the condition of the test organism, the kinetics of toxic reactions, etc. (Ansari et al. 2003). Metal ions frequently penetrate cells, get accumulated and interact with various elements of cells, and target molecules (Chiarelli and Roccheri 2014). Heavy metals have a negative impact on the health of humans and other living organisms in both the terrestrial and aquatic environments (Das et al. 2013). Toxicity of metal ions on any living system depends upon the duration of exposure and its dose. These cause oxidative damage to biological macromolecules and may end up in damaging DNA and even halt metabolic machinery of exposed life forms. High toxicity of heavy metals can be seen in fetus and newborn babies of industrial workers that are constantly exposed to its high concentrations. Metal ion exposure in newborn babies can harm brain memory and central nervous system, disrupt the function of red blood cells, and cause physiological and behavioral problems, and its severe toxicity may cause cancer. Plants exposed to heavy metals may undergo morphological and physiological changes, which can reduce the photosynthesis rate and may trigger mutagenic changes in several plant species. Heavy metal exposures also obstruct the microbial

Table 8.1 Sources of some heavy metal pollutants

Metal	Sources	References
Iron (Fe)	• Suspended sediment • Aeolian dust transport • Hydrothermal activity • Recycling from shelf sediments • Combustion of coal, petroleum, biofuel, fossil fuel, and biomass • Wet and dry deposition of atmospheric aerosols, vertical mixing, and upwelling	Raiswell and Canfield (2012), Wang et al. (2015), Wells et al. (1995)
Zinc (Zn)	• Pigments and paints • Coal burning, metal smelting, steel works, and waste incineration • Pesticides (Zn-based fungicides) and phosphatic fertilizers • Alloys and solders • Sewage sludge • Mining work and industrial effluents (e.g., from smelting and refining) • Urban runoff from abrasion of galvanized roofs and tire rubber	Shah (2021), Rieuwerts (2015)
Lead (Pb)	• Drainage from mining sites • Smelting, manufacturing, and recycling activities • Lead batteries • Pigments and paints • Alloys and solders • Lead plumbing system • Drainage from leaded roofs and gutters • Fossil fuel and waste combustion (PVC products)	WHO (2017), Jaishankar et al. (2014)
Cadmium (Cd)	• Phosphate mining and fertilizers • Detergents and refined petroleum products • Runoff from agricultural and mining activities	Wuana and Okieimen (2011), Gautam et al. (2014), Reichelt-Brushett (2012), Rieuwerts (2015)
Nickel (Ni)	• Domestic wastewater (detergents) • Combustion of oil and coal and incineration of waste and sewage • Opencast mining of Ni laterite ore and metal smelting • Phosphatic fertilizers and sewage sludge • Leakage from plumbing	Gillmore et al. (2020), Cempel and Nikel (2006)
Copper (Cu)	• Antifouling paints • Pesticides (algicides) • Copper polishing, mining, smelting, discharge of wastewater • The use of wood preservatives • Dumping of sewage sludge	Brown and Eaton (2001), Rieuwerts (2015)

(continued)

Table 8.1 (continued)

Metal	Sources	References
Chromium (Cr)	• Metal processing • Industrial effluents (tannery sludge, textile dyes) • Chromate production and smelting • Sewage sludge and phosphate fertilizers	Geisler and Schmidt (1991), Bielicka et al. (2005), Shah (2021)
Arsenic (As)	• Smelting process of nickel, copper, zinc, and lead ores • Paints and wood preservatives • Mining, pesticides, fertilizers	Rieuwerts (2015)
Silver (Ag)	• Emissions from smelting operations • Manufacture and disposal of photographic and electrical supplies • Combustion • Industrial and municipal wastewater outfalls	Shah (2021)
Mercury (Hg)	• Agricultural practices • Municipal and industrial wastewater (chlor-alkali industry) discharge • Mining activities (artisanal gold mining) and combustion of fossil fuels • Natural degassing and direct atmospheric deposition • Sewage sludge and phosphate rock fertilizer	Jaishankar et al. (2014), UNEP (2013)
Tin (Sn)	• Mining • Coal and oil combustion • Atmospheric deposition, riverine input, and sediment resuspension • Antifoulant compounds (tributyltin)	Duan et al. (2012)

growth (Wase and Forster 1997). Presence of these heavy metals even at low concentration is toxic for humans and animals, for example, Pb exposure to human cause dysfunction in nervous system, reproductive system and kidneys. Cd is another heavy metals that get accumulated in aquatic environment from metal ore refining, alloy decomposition, and fertilizer input, which causes renal dysfunction, bone degenerations, and liver damage in humans (Dojlido and Best 1993). Cd is also associated with itai-itai disease in Japan. Hg is also one of common heavy metals that is associated with many human dysfunctions like Minamata disease. Hg toxicity has also been reported to cause abortion and physiological stress. The high concentration of even Fe and Mn in water bodies also reported to effect animal system although both of these metals are required by the body in consistent amount but in case of water contamination level of these metals tend to get increase, which poses threat to human health. Countries rapidly undergoing industrialization are found to have a high number of heavy metal-related disorders (Esslemont 1998). A correlation between historical events and the metal toxicity along with responsible metal is presented in Table 8.2.

8 Heavy Metal Pollution in Water: Cause and Remediation Strategies

Table 8.2 Historical events due to heavy metal toxicity

Year	Historical events	References
Late 1800s	Drainage water from Summerford Bing industrial area containing high concentration of chromium causing Cr toxicity in Scotland	Esslemont (1998)
1932	Minamata Bay in Japan is contaminated with sewage containing Hg, leading Hg poisoning	Harada (1995), Lenntech (2006)
1910s–1940s	Jinzu river in Japan contaminated with waste sludge containing Cd from Kamioka zinc mines, which gradually reached to drinking water and groundwater	Rieuwerts (2015)
1950s	Cd toxicity in Japan (Itai-itai disease)	Esslemont (1998)
1952	Minamata Syndrome (mass human mortality in Japan as a result of consuming Hg-polluted fish)	Lenntech (2006)
1986	At Sandoz in Germany, water used to put out a large fire releases 30 tons of fungicide containing mercury into the Upper Rhine, killing a huge number of fish	Giga (2009)
1998	A Spanish natural reserve has been contaminated by toxic chemicals from a burst dam	Lenntech (2006)
2010	Chemical spill into a river from a Cu mining-smelting complex resulting significant fish kill in China	Rieuwerts (2015)
2011	Illegal dumping of thousands of tons of waste tailings from the production of the tanning chemical chromium sulphate resulting Hexavalent Cr poisoning of drinking water in China	Rieuwerts (2015)
2014	8×10^4 tons of coal spilled into the Dan River in North Carolina, USA, causing As and Cr toxicity	Rieuwerts (2015)

Toxic effects of some heavy metals have been discussed under this section in brief.

8.3.1 Cadmium

Cadmium is considered to be one of the most toxic heavy metals. Its minute concentrations in food chain has been found to cause *itai-itai* disease. Cadmium is not an essential element for biological system unlike other heavy metals. Cadmium metal is used in the manufacturing of plastics, pigments, and nickel-cadmium batteries. It enters into the environment mostly through waste dumping and phosphate fertilizers. Due to its potential to be hazardous for both humans and animals even at minute concentrations, there is an increased concern over the function and toxicity of cadmium in the environment. The Restriction of Hazardous Chemicals (RoHS) directive of the European Union forbids the use of six substances, including cadmium, in electrical and electronic equipment, while allowing for some exceptions and exclusions from the rule (European Commission 2006). Numerous disorders, such as early atherosclerosis, hypertension, osteomalacia, cardiovascular conditions, and renal dysfunction, are all linked to cadmium exposure, including lung cancer (ARL: Cadmium toxicity 2016; Medinews direct 2009). In the 1950s, a large population near the Jinzu River in Japan is found to have kidney dysfunction and osteomalacia, resulting in deformed and fragile bones, which is caused by the ingestion of Cd for a long time. This disease is called as itai-itai named after the excruciating pain that people who had consumed Cd-contaminated rice had experienced. Klaassen et al. (2009) reported Cd to be a metallothioneins (metal-binding proteins) interfering metal that disrupts the homeostasis of an organism. Once ingested by an organism, Cd shows high bio-persistence without biomagnifying properties (Wuana and Okieimen 2011; Kumar et al. 2019).

8.3.2 Chromium

Chromium is a highly toxic metal that poses varying levels of harm to the ecosystem. Cr is mainly introduced to the environment through discharge of untreated or inadequately treated residues from tanning and leather industries, rubber manufacturing, and pulp and paper industries (Chiarelli and Roccheri 2014). Among various oxidation states of Cr, only +3 and +6 are chemically and biologically stable (Ducros 1992). Hexavalent Cr is the most toxic form of chromium for animal and human health, which is considered to be mutagenic and carcinogenic (Chiarelli and Roccheri 2014). High exposure to Cr may cause liver, kidney, and central nervous system damage. Hematological problems can be observed in fishes of freshwater contaminated with Cr.

8.3.3 Copper

Copper enters the water bodies mostly via mining, dumping of sewage sludge, and use of Cu-based pesticides in agriculture. Cu has been used by human civilization since pre-historic times. It is well known that Cu is an essential element for all living organisms. However, its high concentration may cause toxic effects (Bielmyer-Fraser et al. 2018). Excess Cu may alter enzyme functions, acid/base balance, iono-regulation, and endocrine disruption in aquatic organisms (Rieuwerts 2015). Cu shows bioaccumulation in organisms like plankton and oysters but does not magnify in the food chain (Rieuwerts 2015). At higher concentrations in human, it may damage kidney and stomach and cause diarrhea, vomiting, and loss of strength.

8.3.4 Lead

Leaded gasoline had been a significant contributor to the environmental dispersion of Pb. Tetraethyl lead was added in gasoline to boost its octane levels since the 1920s (ATSDR 2017). However, lead use has been banned in gasoline in the United States since January 1996, and USEPA has encouraged all the countries to do so (EPA 2017). Pb is a nonessential element that could be hazardous to most of the living forms. Lead can be present in water bodies in four different forms: ionic (mobile and

8 Heavy Metal Pollution in Water: Cause and Remediation Strategies

bioavailable), organic complex (bound with limited mobility and bioavailability), strongly bound (attached to solid particles like iron oxides with limited mobility), and very strongly bound (attached to clay, dead remains, or solid particle with very low mobility) (ILA 2018). Exposure of high concentrations of Pb to human may cause severe toxicity to the central nervous system, damage the fetus, alter hemoglobin synthesis, and harm kidney and reproductive system. Poisoning of agricultural food may be caused by airborne Pb by its deposition on soils, water, and fruits (WHO 1984). Bio-magnification of Pb has not been observed in the food chain (De Pooter 2013).

8.3.5 Mercury

Mercury (Hg) is a nonessential, persistent and highly toxic element, which is present in the environment in elemental, organic, and inorganic forms that are all interconvertible and toxic (Jaishankar et al. 2014). Elemental Hg^0 is a volatile liquid at room temperature, inorganic Hg is present in mercuric (Hg^{2+}: HgS, $HgCl_2$) and mercurous (Hg^{1+}: Hg_2Cl_2) forms, and organic Hg is present in the form of methylated Hg (mono and dimethyl mercury) and phenyl mercury. Toxic effects related to Hg gained global attention due to the outbreak of Minamata disease in Japan in the year 1956 (Ye et al. 2016). Methylmercury containing waste from a fertilizer industry as a byproduct of acetaldehyde discharged into the Minamata Bay for a long time polluted the marine ecosystem harming organisms. Contaminated seafood consumption caused injury to the nervous system, brain, heart, eyes, kidney, and lungs. Methylmercury also shows bio-magnification, and it dissolves well in water and can persist fatty tissues by crossing biological membrane (Solomon 2008).

8.3.6 Nickel

Opencast mining for the extraction of nickel (Ni) ore Ni laterite leads to major landscape alteration and increased soil erosion rates (Gillmore et al. 2020). This can cause entering of Ni into water bodies such as rivers, later contaminating the sea (Clark 2001). Ni is an essential element that plays an important role in the synthesis of red blood cells. Trace amounts of nickel do not harm cells but at higher concentrations may damage liver and heart and decrease body weight. Nickel at very high concentrations may cause nervous system damage, reduced cell growth, and cancer (WHO 1984). Ni does not have bio-magnification properties and considered to be moderately toxic metal.

8.3.7 Zinc

Zinc (Zn) is found in earth's crust and an essential element for living organisms as it acts as a cofactor in more than 300 enzymes and contributes to the metabolism

(Morel et al. 1994). Excessive introduction of Zn into water bodies from rubber, plastic, cosmetic, and pesticide industries may cause phytoxicity in plants, anemia, lack of muscular coordination, and abdominal pain in humans. Bioaccumulation of Zn is not observed, but higher concentrations in environmental may lead to its bioaccumulation in organisms (De Pooter 2013).

8.3.8 Arsenic

Arsenic (As) has been considered to be one of the most concerning environmental pollutants around the globe (Kaur et al. 2011; Dash et al. 2021). Insecticides, herbicides, fungicides, and other pesticides are the largest source of As pollution in water bodies (El-Sorogy et al. 2016). Marine organisms are mostly contaminated with highly stable pentavalent form of As which is nontoxic and metabolically inert but may get bio-accumulated in some fishes and algae. Water contaminated with As may cause toxicity to blood and central nervous system, breathing problems, nausea, and lung and skin cancer. As is a geogenic issue that affects everyone, but man-made sources such as pesticide manufacturing and metal processing raise the environmental As content.

8.3.9 Silver

Silver (Ag) is a rare but naturally occurring metal that has been used by human civilization for a long time. Ag is extremely toxic in its +1 oxidation state (Ag^{1+}) to plankton and invertebrates (Luoma et al. 1995). Increasing use of Ag as a biocide in recent years raises concerns about its potential as a pollutant (Purcell and Peters 1998). Since Ag^+ ion can be taken up by cell via cell membrane ion transporters, its high concentrations may bio-accumulate in some fish, algae, and shrimps (Clark 2001).

8.3.10 Iron

Iron (Fe) and phosphorus (P) along with some other trace elements act as a limiting nutrient for nitrogen fixation (Mills et al. 2004). As nitrogen-fixing organisms require higher concentrations of Fe, sometimes due to this higher concentration, microbes in remote waters may produce more dimethyl sulfide (DMS) and organic carbon, which in turn may have an impact on the radiative forcing in the atmosphere.

8.3.11 Manganese

Manganese being the 12th most abundant element in earth's crust is a broadly distributed metal in nature (Pinsino et al. 2012). Although low level of manganese intake is necessary for human health, exposure to high Mn concentrations is toxic. Excessive exposure to high levels of Mn for a long period may lead to neurological disorders (Manganism: a disease starting from feeling of weakness and lethargy and leading to speech disturbances, clumsy gait and a masklike face) and respiratory and reproductive problems. The individual metals determine the method by which heavy metals impact human bodies, but eventually, all metals create reactive oxygen radicals, which cause a variety of illnesses in living things. Therefore, setting a threshold at which harmful substances were deemed was crucial (Mahurpawar 2015).

8.4 Permissible Limits of Some Common Heavy Metals

The regulatory limits of heavy metals are defined by several well-known organizations based on their toxicities. Given that humans directly consume water, the residential water supply is regarded as the most significant use of water. The National Water Policy has prioritized drinking as the best use of water resources. Drinking water standards have been developed in India by organizations like the Indian Council of Medical Research (ICMR) and the Bureau of Indian Standards (BIS). Table 8.3 lists the drinking water requirements for hazardous and trace metals according to BIS code 10500-2012 (Hussain and Rao 2018). All water delivery systems must adhere to these restrictions. Some heavy metals are regularly present in naturally occurring water (both surface and groundwater) at concentrations 100 or 1000 times higher than the MCL standards. Additionally, drinking water standards have been established by the USA Environmental Protection Agency (USEPA) and World Health Organization (WHO), which are regarded as global standards. Table 8.4 contains some harmful heavy metal ions that have maximum allowable levels in surface waters in accordance to WHO and USEPA. These heavy metals are important resources for various industrial uses; thus, their removal, recovery, and recycling are more crucial than ever. Regulatory authorities have developed drinking

Table 8.3 Drinking Water Standards for Trace & Toxic metals (BIS-10500-2012; Hussain and Rao 2018)

S. no.	Toxic metal	Requirement (acceptable limit; mg/L)	Permissible limit in the absence of alternative source (mg/L)
1.	Cadmium	0.003	No relaxation
2.	Iron	0.030	No relaxation
3.	Chromium	0.05	1.5
4.	Lead	0.01	No relaxation
5.	Copper	0.05	1.5
6.	Zinc	5	15
7.	Lead	0.01	No relaxation
8.	Nickel	0.0	No relaxation
9.	Arsenic	0.01	0.05

Table 8.4 Based on WHO and US EPA rules, some harmful heavy metal ions have maximum allowable levels in surface waters is tabulated (Hussain and Rao 2018)

Toxicity rank	Heavy metals	USEPA (µg/L)	WHO (µg/L)
1	Arsenic	10	10
2	Lead	15	10
3	Mercury	2	1
8	Cadmium	5	3
17	Chromium	100	50
57	Nickel	100	70
75	Zinc	5000	No guideline
125	Copper	1300	2000

water standards in line with toxicity data gathered from human clinical examinations and numerous other research, such as animal tests. A succinct overview is provided in Table 8.5 in which various international regulatory bodies standards are mentioned.

8.5 Remediation Strategies

Different methods such as membrane filtration, adsorption, ion exchange, chemical precipitation, etc. are some of the most commonly used methods to eliminate heavy metals from wastewaters (Türkmen et al. 2022). Presently, many researchers are working on the remediation of heavy metals from water, soil, and air by natural methods rather than chemical methods. Remediation is the method of removing toxic compounds from environmental media or replacing them with less toxicones. The strategies used to reduce or remove pollutants from soil and other environmental mediums use both in situ and ex situ procedures. The three major categories of remediation techniques are (1) physical approaches, (2) chemical ways, and (3) biological methods (Fig. 8.1).

8.5.1 8.5.1 Physicochemical Methods

These methods involve electrodialysis, chemical precipitation, reverse osmosis, evaporation recovery, physical adsorption, ion exchange, etc. (Qasem et al. 2021). Some of these techniques have been described in Table 8.6. Although these conventional physicochemical methods can remove heavy metal ions, there are certain environmental considerations associated with them. These processes generate some secondary waste products that contaminate the environment (Diep et al. 2018). Besides, certain methods such as chemical precipitation and electrochemical treatment are not effective when concentration of heavy metals is low. Techniques that use membrane, activated carbon, nanoadsorbents, and ion exchange technologies are expensive (Dhankhar and Hooda 2011).

Table 8.5 Standards for drinking water quality for trace elements that might have an impact on public health (Hattingh 1977; Gupta 1999)

Parameter	Lead (Pb)	Arsenic (As)	Cadmium (Cd)	Mercury (Hg)	Zinc (Zn)	Chromium (Cr)	Barium (Ba)	Selenium (Se)	Copper (Cu)
Australia (1973)	50	50	10	–	5000	50	1000	10	10000
FRG (1975)	40	40	6	4	2000	50	–	8	–
Japan (1968)	100	50	–	1	100	50	–	–	10000
NAS (1972)	50	100	10	2	5000	50	1000	10	1000
SABS (1971)	50	50	50	–	5000	50	–	–	1000
US EPA (1975)	50	50	10	2	–	50	1000	10	–
USPHS (1962)	50	10	10	–	5000	50	1000	10	1000
USSR (1970)	100	50	10	5	1000	100	4000	1	100
WHO European (1970)	100	50	10	–	5000	50	1000	10	50
WHO Intern. (1971)	100	50	10	1	5000	–	–	10	50

All values are in µg/L
USPHS US Public Health Service, *SABS* South African Bureau of Standards, *USSR* Russia, *NAS* USA National Academy of Sciences

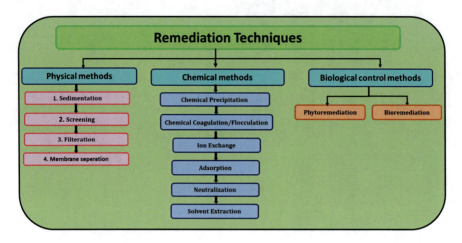

Fig. 8.1 Methods for heavy metal removal

Table 8.6 Physicochemical remediation technologies

S. no.	Physical methods	Description
1.	Screening	This method involves removal of large nonbiodegradable and floating solids
2.	Sedimentation	This technique involves gravity to remove suspended solids from water
3.	Membrane filtration	Membranes are in the form of complex structure which contains dynamic elements on the nanometer scale. In this method different types of membranes are included, namely, reverse osmosis, ultrafiltration, nanofiltration, and electrodialysis
4.	Membrane separation	This process involves membrane to separate the components in a solution by rejecting unwanted substances and allowing others to pass the membrane
5.	Chemical precipitation	This method changes the form of dissolved metal ions into solid particles to facilitate their sedimentation. The coagulant precipitates metal ions by changing pH, electro-oxidizing potential, or co-precipitation
6.	Chemical coagulation/ flocculation	Coagulation is the destabilization of colloids by neutralizing the forces that keep them parted, while flocculation is the agglomeration of destabilized particles.
7.	Electrochemical methods	This method involves the recovery of the heavy metals in the elemental metallic state by using the anodic and cathodic reactions in the electrochemical cell
8.	Ion exchange	This method is a reversible chemical reaction used to replace the undesirable metal ion in an ecofriendly manner
9.	Adsorption	Adsorption method is a solid-liquid mass transfer operation, where the heavy metal (adsorbate) is migrated from the wastewater to the solid surface (called adsorbent) and then bonded due to chemical or physical adsorption over the adsorbent surface

8.5.2 Phytoremediation Technique

Phytoremediation is a technique in which plants act as filters to remove contaminants from soil, water, and air. It is precisely a collection of plant-based technologies that cause removal of contaminants from environment by the use of plants and weeds (Ali et al. 2013). This technique is very cheap and alternative to present treatment methods. In several years, phytoremediation technique is gaining attention due to its cost-effectiveness, eco-friendly nature, and wide acceptance across the globe. Plants accumulate the toxic metals or contaminants from the environment through their body parts (shoots, roots, and stem). Phytoremediation is also identified by different names such as agro-remediation, green remediation, vegetative remediation, green technology, and botano remediation. It is very cheap method that requires technical strategy, knowledge of plants species and cultivars for particular contaminants, and regions. With the help of phytoremediation techniques, a variety of wastewater can be treated as municipal wastewater, industrial water, landfill leachate, paper industry, mine drainage, contaminated lands, and groundwater. This technique reduces the organic and inorganic pollutants from industrial wastewater by using plants. The performance of plants seems to be good in lowering the concentration of pollutants. Phytoremediation techniques follow different mechanisms such as phytoextraction, phytostabilization, phytovolatilization, phytodegradation, and rhizo-filtration during the uptake of heavy metals in the plant (Fig. 8.2). The summary of the phytoremediation techniques and mechanisms is shown in Fig. 8.2 and Table 8.7.

8.5.2.1 Bioremediation

Bioremediation is a technique of converting environmental pollutants into less hazardous forms using living organisms, primarily bacteria. It uses plants, naturally existing bacteria, and fungi to break down or detoxify pollutants that are harmful to the environment or to human health. This technique is advantageous over physicochemical methods due to their eco-friendly nature (Gunatilake 2015). The basis of

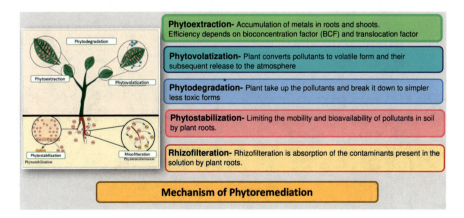

Fig. 8.2 Mechanisms of phytoremediation

Table 8.7 Phytoremediation for different heavy metals using different plant species

Phytotechnology	Pollutants	Plants	Metal-accumulated in plant parts	References
Phytoextraction	Pb, Hg, Cu, Cr, Ni, Zn	Water hyacinth	Roots and shoots	Molisani et al. (2006)
	As	*Hordeum vulgare*	Roots and shoots	Mains et al. (2006)
	Cr and Cd	*Cyperus rotundus*	Roots and shoots	Subhasini and Swamy (2014)
	Pb, Zn, and Cu	*Hordeum hirta*	Roots and shoots	Conesa et al. (2007)
	As	*Eleocharis acicularis*	Shoots	Sakakibara et al. (2011)
	Pahs	*Chrysopogon zizanioides*	–	Un Nisa and Rashid (2015)
	Pb	*Zea mays and Ambrosia artemisiifolia*	–	Shahandeh and Hossner (2000)
	Ni	*Alyssum heldreichii*	Leaves	Rizzi et al. (2004)
	As	*Tagetes minuta*	Shoots	Salazar and Pignata (2014)
	Cu	*Cannabis sativa* L.	Aboveground plant parts	Ahmad et al. (2015)
	Pb and Cd	*Betula occidentalis* and *Thlaspi caerulescens*	Shoots	Koptsik (2014)
	As	*Pteris vittata*	Shoots	Kalve et al. (2011)
	Cd	*Thlaspi caerulescens*	Shoots	Sheoran et al. (2009)
	Zn	*Euphorbia cheiradenia*	Shoots	Chehregani and Malayeri (2007)
Phytovolatilization	Se	*Salicornia bigelovii*	–	Huang et al. (2013)
	Se	*Typha latifolia* L.	–	LeDuc and Terry (2005)
	Se and Hg	*Chara canescens* (musk-grass) and *Brassica juncea*	–	Ghosh and Singh (2005)
	Zn, Mn, Co, Cd, Cr, Ni and As	*Typha latifolia* L.	–	Varun et al. (2011)
	Cd and Zn	*Sorghum bicolor* L.	–	Soudek et al. (2012)
	Cu, Pb, and Zn	*Agrostis castellana*	–	Pastor et al. (2015)

Technique	Metal	Plant species	Mechanism	Reference
Phytodegradation	Trinitrotoluene	*Myriophyllum aquaticum*	–	Rajakaruna et al. (2006)
	Hg	*Azolla caroliniana*	–	Bennicelli et al. (2004)
Phytostabilization	As	*Arundo donax* L.	–	Mirza et al. (2011)
	Pb	*Sorghum halepense* L.	Reduction in rhizosphere	Salazar and Pignata (2014)
	Cu	*Spartina alterniflora*	Accumulation of metal in roots	Chai et al. (2014)
	Cd	*Phragmites australis*	Accumulation of metal in roots	Nunes da Silva et al. (2014)
	Se	*Salicornia bigelovii*	–	Huang et al. (2013)
	Zn	*Halimione portulacoides*	Accumulation of metal in tissues	Andrades-Moreno et al. (2013)
	Pb and Zn	*Suaeda salsa*	Accumulation of metals in roots	Wu et al. (2013)
	Cd	*Salicornia ramosissima*	Accumulation of metal in roots	Pedro et al. (2013)
	Cd	*Arthrocnemum macrostachyum*	Accumulation of metal in roots	Redondo-Gómez et al. (2010)
	Cu	*Commelina communis*	Accumulation of metal in roots	Wang and Zhong (2011)
	Cd, Ni, and As	*Salicornia brachiata*	Accumulation of metal in roots	Xu et al. (2010)
	Cd, Zn, Cu, and Co	*Sarcocornia perennis*	Accumulation of metal in roots	Lefèvre et al. (2010)
	As, Ag, Cr and Sb	*Solanum tuberosum* L.	–	Baghour et al. (2001)
Rhizofilteration	Cd	*Azolla pinnata*	–	Rai (2008)
	Pb	*Noccaea caerulescens*	Aerial part and shoot	Dinh et al. (2018)
	As	*Cynara cardunculus*	–	Llugany et al. (2012)
	Al, Fe, and Mn	*Pistia stratiotes*	–	Veselý et al. (2012)
	Sb	*Cynodon dactylon*	–	Xue et al. (2018)
	Fe	*Typha domingensis*	–	Hegazy et al. (2011)

removal and transportation of wastes for treatment basically is the two methods, namely, biosorption and bioaccumulation (Diep et al. 2018). Biological methods are sustainable and cost-effective in comparison to other methods. Figure 8.2 demonstrates bioremediation strategies commonly used for heavy metal removal.

8.5.2.2 Biosorption

This technique involves a rapid and passive adsorption mechanism. In this method, both living and dead biomass may be present for biosorption process. The biomass is derived from algae, fungi, plants, and bacteria (Ali Redha 2020). This strategy is inexpensive because it is possible to repeatedly renew and repurpose the biomass obtained from industrial waste. Various low-cost adsorbents are those derived from by-products of industry and agriculture such as rice husk, coconut shell, maize cob, etc. Biosorption involves two components: one is solid (biosorbent), and the other is liquid phase that contains an aqueous solution of contaminants such as heavy metals (Abbas Ali et al. 2016). Functional groups present on the solid phase are the major determinants for removal of a particular heavy metal.

8.5.2.3 Bioaccumulation

Bioaccumulation is a natural process by which microorganisms uptake metal ions using specific proteins from their surroundings (Diep et al. 2018). Researchers have also engineered some microorganisms to enhance their bioaccumulation process by introducing metal uptake proteins in their cell envelopes (Ueda 2016). Bioaccumulation utilizes live microorganisms in their metabolically active state, and major concerns associated with this process is the toxicity of wastewater to the organism, expression level of metal sequestering proteins, and exhaustive screening for selection of organism with high bioaccumulative potential. Genetic engineering helps to enhance the bioaccumulative potential of microorganism numerous folds and thus helps in efficient cleanup of water bodies contaminated with heavy metals (Fig. 8.3).

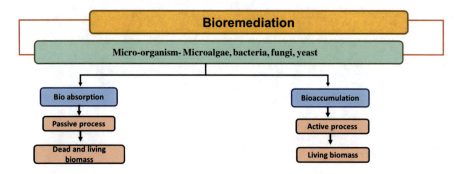

Fig. 8.3 Biosorption and bioaccumulation for heavy metal removal

8.6 Conclusion

Heavy metals become noxious when their concentrations exceed certain level. Increasing concerns regarding their toxicity and hazardous effect on aquatic as well as terrestrial life forms requires extensive study to better understand basic chemistry, toxicity, and selection of appropriate remedial options in order to restore environmental health. To choose an appropriate technique, it is important to consider the type of contaminant, degree of contamination, and cost-effectiveness of the strategy. Besides, efforts to improve their commercialization in developing countries considering environmental sustainability are also important. Implementation of biotechnological aspects and integrated approaches is gaining interest these days. Future research needs to focus on evaluation methods for assessing remediation effectiveness while developing new remediation technologies in future research.

References

Abbas Ali A, Mohamed Sihabudeen M, Zahir Hussain A (2016) Biosorption of heavy metals by Pseudomonas bacteria. Int Res J Eng Technol 3(8):1446–1450

Adler RA, Claassen M, Godfrey L, Turton AR (2007) Water, mining, and waste: an historical and economic perspective on conflict management in South Africa. Econ Peace Secur J 2(2):33–41. https://doi.org/10.15355/epsj.2.2.33

Agency for Toxic Substances and Disease Registry (ATSDR) (2017) Lead toxicity. https://www.atsdr.cdc.gov

Ahmad R, Tehsin Z, Malik ST, Asad SA, Shahzad M, Bilal M et al (2015) Phytoremediation potential of hemp (Cannabis sativa L.): identification and characterization of heavy metals responsive genes. CLEAN Soil Air Water 44:195–201

Ali Redha A (2020) Removal of heavy metals from aqueous media by biosorption. Arab J Basic Appl Sci 27(1):183–193

Ali H, Khan E, Sajad MA (2013) Phytoremediation of heavy metals—concepts and applications. Chemosphere 91(7):869–881

Andrades-Moreno L, Cambrollé J, Figueroa ME, Mateos-Naranjo E (2013) Growth and survival, 739 of Halimione portulacoides stem cuttings in heavy metal contaminated soils. Mar Pollut Bull 75(1–2):28–32

Ansari TM, Marr IL, Tariq N (2003) Heavy metals in marine pollution perspective—a mini review. J Appl Sci 4(1):1–20. https://doi.org/10.3923/jas.2004.1.20

ARL: Cadmium toxicity (2016) Archived from the original on 9 September 2019. http://www.arltma.com

Ayangbenro AS, Babalola OO (2017) A new strategy for heavy metal polluted environments: a review of microbial biosorbents. Int J Environ Res Public Health 14(1):94

Baby J, Raj JS, Biby ET, Sankarganesh P, Jeevitha MV, Ajisha SU, Rajan SS (2010) Toxic effect of heavy metals on aquatic environment. Int J Biol Chem Sci 4(4):939–952

Baghour M, Moreno DA, Hernández J, Castilla N, Romero L (2001) Influence of root temperature on phytoaccumulation of As, Ag, Cr, and Sb in potato plants. J Environ Sci Health Part A 36(7):1389–1401

Bennicelli R, Stepniewska Z, Banach A, Szajnocha K, Ostrowski J (2004) The ability of Azolla caroliniana to remove heavy metals (Hg (II), Cr (III), Cr (VI)) from municipal waste water. Chemosphere 55(1):141–146

Bhatt K, Suyal DC, Kumar S, Singh K, Goswami P (2022) New insights into engineered plant-microbe interactions for pesticide removal. Chemosphere 309(2):136635

Bielicka A, Bojanowska I, Wisniewski A (2005) Two faces of chromium-pollutant and bioelement. Pol J Environ Stud 14(1):5–10

Bielmyer-Fraser GK, Patel P, Capo T, Grosell M (2018) Physiological responses of corals to oceanacidification and copper exposure. Mar Pollut Bull 133:781–790. https://doi.org/10.1016/j.marpolbul.2018.06.048

Brown CJ, Eaton RA (2001) Toxicity of chromated copper arsenate (CCA)-treated wood to non-target marine fouling communities in Langstone Harbour, Portsmouth, UK. Mar Pollut Bull 42(4):310–318. https://doi.org/10.1016/s0025-326x(00)00156-9

Cempel M, Nikel GJPJS (2006) Nickel: a review of its sources and environmental toxicology. Pol J Environ Stud 15(3):375–382

Chai M, Shi F, Li R, Qiu G, Liu F, Liu L (2014) Growth and physiological responses to copper stress in a halophyte Spartina alterniflora (Poaceae). Acta Physiol Plant 36(3):745–754. https://doi.org/10.1007/s11738-013-1452-1

Chehregani A, Malayeri BE (2007) Removal of heavy metals by native accumulator plants. Int J Agric Biol 9:462–465

Chiarelli R, Roccheri MC (2014) Marine invertebrates as bioindicators of heavy metal pollution. Open J Metal 4(4):93–106. https://doi.org/10.4236/ojmetal.2014.44011

Clark RB (2001) Marine pollution. Oxford University Press, Oxford

Conesa HM, Faz A, Arnaldos R (2007) Initial studies for the phytostabilization of a mine tailing from the Cartagena-La Union Mining District (SE Spain). Chemosphere 66(1):38–44. https://doi.org/10.1016/j.chemosphere.2006.05.041

Das S, Patnaik SC, Sahu HK, Chakraborty A, Sudarshan M, Thatoi HN (2013) Heavy metal contamination, physio-chemical and microbial evaluation of water samples collected from chromite mine environment of Sukinda, India. Trans Nonferrous Metals Soc China 23(2): 484–493

Dash B, Sahu N, Singh AK, Gupta SB, Soni R (2021) Arsenic efflux in Enterobacter cloacae RSN3 isolated from arsenic-rich soil. Folia Microbiol 66:189–196

De Pooter D (2013) Heavy metals. http://www.coastalwiki.org/wiki/Heavy_metals

Dhankhar R, Hooda A (2011) Fungal biosorption–an alternative to meet the challenges of heavy metal pollution in aqueous solutions. Environ Technol 32(5):467–491

Diep P, Mahadevan R, Yakunin AF (2018) Heavy metal removal by bioaccumulation using genetically engineered microorganisms. Front Bioeng Biotechnol 6:157

Dinh N, van der Ent A, Mulligan DR, Nguyen AV (2018) Zinc and lead accumulation characteristics and in vivo distribution of Zn^{2+} in the hyperaccumulator Noccaea caerulescens elucidated with fluorescent probes and laser confocal microscopy. Environ Exp Bot 147:1–12. https://doi.org/10.1016/j.envexpbot.2017.10.008

Dixit R, Wasiullah, Malaviya D, Pandiyan K, Singh U et al (2015) Bioremediation of heavy metals from soil and aquatic environment: an overview of principles and criteria of fundamental processes. Sustainability 7(2):2189–2212. https://doi.org/10.3390/su7022189

Dojlido J, Best GA (1993) Chemistry of water and water pollution. Ellis Horwood Limited

Duan L, Song J, Li X, Yuan H, Xu S (2012) Dissolved inorganic tin sources and its coupling with eco-environments in Bohai Bay. Environ Monit Assess 184(3):1335–1349. https://doi.org/10.1007/s10661-011-2044-4

Ducros V (1992) Chromium metabolism, a literature review. Biol Trace Elem Res 32:65–77. https://doi.org/10.1007/BF02784589

Duffus JH (2002) "Heavy metals" a meaningless term? (IUPAC Technical Report). Pure Appl Chem 74(5):793–807

El-Sorogy AS, Youssef M, Al-Kahtany K, Al-Otaiby N (2016) Assessment of arsenic in coastalsediments, seawaters and molluscs in the Tarut Island, Arabian Gulf, Saudi Arabia. J Afr Earth Sci 113:65–72. https://doi.org/10.1016/j.jafrearsci.2015.10.001

Environmental Protection Agency (2017). https://www.epa.gov/clean-air-act-overview/clean-airact-text

8 Heavy Metal Pollution in Water: Cause and Remediation Strategies

Esslemont G (1998) Heavy metals in the tissues and skeleton of scleractiniancorals. PhD thesis, Southern Cross University, Australia, p 257

European Commission (2006) Decision of 12 October 2006 amending, for the purposes of adapting to technical progress, the Annex to Directive 2002/95/EC of the European Parliament and of the Council as regards exemptions for applications of lead and cadmium (notified under document number C (2006) 4790). (October 14, 2006). J Eur Union

Gautam RK, Sharma SK, Mahiya S, Chattopadhyaya MC (2014) Contamination of heavy metals in aquatic media: transport, toxicity and technologies for remediation. In: Heavy metals in water: presence, removal and safety. Royal Society of Chemistry, London

Geisler CD, Schmidt D (1991) An overview of chromium in the marine environment. Deutsche Hydrographische Zeitschrift 44(4):185–196

Ghosh M, Singh SP (2005) A review on phytoremediation of heavy metals and utilization of it's by products. Asian J Energy Environ 6:18

Giga W (2009) The Rhine red, the fish dead-the 1986 Schweizerhalle disaster, a retrospect and long-term impact assessment. Environ Sci Pollut Res Int 16(Suppl 1):S98–S111

Gillmore ML, Gissi F, Golding LA, Stauber JL, Reichelt-Brushett AJ, Severati A et al (2020) Effects of dissolved nickel and nickel-contaminated suspended sediment on the scleractinian coral, Acropora muricata. Mar Pollut Bull 152:110886

Gunatilake SK (2015) Methods of removing heavy metals from industrial wastewater. Methods 1 (1):14

Gupta DC (1999) Environmental aspects of selected trace elements associated with coal and natural waters of Pench Valley coalfield of India and their impact on human health. Int J Coal Geol 40(2–3):133–149

Harada M (1995) Minamata disease: methylmercury poisoning in Japan caused by environmental pollution. Crit Rev Toxicol 25(1):1–24. https://doi.org/10.3109/10408449509089885

Hattingh WHJ (1977) Reclaimed water: a health hazard? Water SA 3(2):104–112

Hegazy AK, Abdel-Ghani NT, El-Chaghaby GA (2011) Phytoremediation of industrial wastewater potentiality by Typha domingensis. Int J Environ Sci Technol 8(3):639–648. https://doi.org/10.1007/BF03326249

Huang JC, Suárez MC, Yang SI, Lin Z-Q, Terry N (2013) Development of a constructed wetland water treatment system for selenium removal: incorporation of an algal treatment component. Environ Sci Technol 47(18):10518–10525

Hussain J, Rao NP (2018) Status of trace and toxic metals in Indian Rivers. Ministry of Water Resources, Government of India, river data Compilation-2, vol 110066. Directorate Planning and Development Organization, New Delhi

ILA: International Lead Association (2018) Lead in aquatic environments: understanding the science. https://www.ila-lead.org/responsibility/lead-in-aquatic-environments–understanding-the-science

Jaishankar M, Tseten T, Anbalagan N, Mathew BB, Beeregowda KN (2014) Toxicity, mechanism and health effects of some heavy metals. Interdiscip Toxicol 7(2):60–72. https://doi.org/10.2478/intox-2014-0009

Kalve S, Sarangi BK, Pandey RA, Chakrabarti T (2011) Arsenic and chromium hyperaccumulation by an ecotype of Pteris vittata–prospective for phytoextraction from contaminated water and soil. Curr Sci:888–894

Kaur S, Kamli MR, Ali A (2011) Role of arsenic and its resistance in nature. Can J Microbiol 57(10):769–774. https://doi.org/10.1139/w11-062

Khan S, Cao Q, Zheng YM, Huang YZ, Zhu YG (2008) Health risks of heavy metals in contaminated soils and food crops irrigated with wastewater in Beijing, China. Environ Pollut 152(3):686–692. https://doi.org/10.1016/j.envpol.2007.06.056

Klaassen CD, Liu J, Diwan BA (2009) Metallothionein protection of cadmium toxicity. Toxicol Appl Pharmacol 238(3):215–220. https://doi.org/10.1016/j.taap.2009.03.026

Koptsik GN (2014) Problems and prospects concerning the phytoremediation of heavy metal polluted soils: a review. Eurasian Soil Sci 47(9):923–939. https://doi.org/10.1134/S1064229314090075

Kumar P, Gupta SB, Anurag, Soni R (2019) Bioremediation of cadmium by mixed indigenous isolates Serratia liquefaciens BSWC3 and Klebsiella pneumoniae RpSWC3 isolated from industrial and mining affected water samples. Pollution 5(2):351–360

LeDuc DL, Terry N (2005) Phytoremediation of toxic trace elements in soil and water. J Ind Microbiol Biotechnol 32(11–12):514–520. https://doi.org/10.1007/s10295-005-0227-0

Lefèvre I, Marchal G, Edmond GM, Correal E, Lutts S (2010) Cadmium has contrasting effects on polyethylene glycol – sensitive and resistant cell lines in the Mediterranean halophyte species. Atriplex halimus L. J Plant Physiol 167(5):365–374

Lenntech (2006) Heavy metals. http://www.lenntech.com/heavy-metals.htm

Llugany M, Miralles R, Corrales I, Barceló J, Poschenrieder C (2012) Cynara cardunculus a potentially useful plant for remediation of soils polluted with cadmium or arsenic. J Geochem Explor 123:122–127

Luoma SN, Ho YB, Bryan GW (1995) Fate, bioavailability and toxicity of silver in estuarine environments. Mar Pollut Bull 31(1–3):44–54. https://doi.org/10.1016/0025-326X(95)00081-W

Mahurpawar M (2015) Effects of heavy metals on human health. Int J Res GRANTHAALAYAH 3(9SE): 1–7. https://doi.org/10.29121/granthaalayah.v3.i9SE.2015.3282

Mains D, Craw D, Rufaut CG, Smith CM (2006) Phytostabilization of gold mine tailings, New Zealand. Part 1: plant establishment in alkaline saline substrate. Int J Phytoremediation 8(2):131–147

Medinews direct (2009) Cadmium exposure can induce early atherosclerotic changes archived 15 March 2012 at the wayback machine

Mills MM, Ridame C, Davey M, La Roche J, Geider RJ (2004) Iron and phosphorous co-limit nitrogen fixation in the eastern topical North Atlantic. Nature 429(6989):292–294. https://doi.org/10.1038/nature02550

Mirza N, Pervez A, Mahmood Q, Shah MM, Shafqat MN (2011) Ecological restoration of arsenic contaminated soil by Arundo donax L. Ecol Eng 37(12):1949–1956. https://doi.org/10.1016/j.ecole

Molisani MM, Rocha R, Machado W, Barreto RC, Lacerda LD (2006) Mercury contents in aquatic macrophytes from two reservoirs in the Paraíba do Sul: Guandú river system, SE Brazil. Braz J Biol 66:101–107

Morel FMM, Reinfelder JR, Roberts SB, Chamberlain CP, Lee JG, Yee D (1994) Zinc and carbon co-limitation of marine phytoplankton. Nature 369(6483):740–742. https://doi.org/10.1038/369740a0

Nunes da Silva M, Mucha AP, Rocha AC, Silva C, Carli C, Gomes CR, Almeida CMR (2014) Evaluation of the ability of two plants for the phytoremediation of Cd in salt marshes. Estuar Coast Shelf Sci 141:78–84

Pastor J, Gutiérrez-Ginés MJ, Hernández AJ (2015) Heavy-metal phytostabilizing potential of Agrostis castellana Boiss and reuter. Int J Phytoremediation 17:988–998. https://doi.org/10.1080/15226514.2014.1003786

Pedro CA, Santos MSS, Ferreira SMF, Gonçalves SC (2013) The influence of cadmium, 1051contamination and salinity on the survival, growth and phytoremediation capacity of the saltmarsh plant Salicornia ramosissima. Mar Environ Res 92:197–205

Pinsino A, Matranga V, Roccheri MC (2012) Manganese: a new emerging contaminant in the environment. In: Srivastava J (ed) Environmental contamination. InTechOpen, London. http://www.intechopen.com/books/environmentalcontamination/manganesea-new-emerging-contaminant-in-the-environment. ISBN: 978-953-51- 0120-8

Pratush A, Kumar A, Hu Z (2018) Adverse effect of heavy metals (As, Pb, Hg, and Cr) on health and their bioremediation strategies: a review. Int Microbiol 21(3):97–106. https://doi.org/10.1007/s10123-018-0012-3

8 Heavy Metal Pollution in Water: Cause and Remediation Strategies

Purcell TW, Peters JJ (1998) Sources of silver in the environment. Environ Toxicol Chem 17(4): 539–546. https://doi.org/10.1002/etc.5620170404

Qasem NA, Mohammed RH, Lawal DU (2021) Removal of heavy metal ions from wastewater: a comprehensive and critical review. Npj Clean Water 4(1):36

Rai PK (2008) Phytoremediation of Hg and Cd from industrial effluents using an aquatic free floating macrophyte Azolla pinnata. Int J Phytoremediation 10(5):430–439

Raiswell R, Canfield DE (2012) The iron biogeochemical cycle past and present. Geochem Perspect 1(1):1–220

Rajakaruna N, Tompkins KM, Pavicevic PG (2006) Phytoremediation: an affordable green technology for the clean-up of metal contaminated sites in Sri Lanka. Cey J Sci 35:25–39

Redondo-Gómez S, Mateos-Naranjo E, Andrades-Moreno L (2010) Accumulation and tolerance characteristics of cadmium in a halophytic Cd-hyperaccumulator, Arthrocnemum macrostachyum. J Hazard Mater 184(1–3):299–307

Reichelt-Brushett A (2012) Risk assessment and ecotoxicology: limitations and recommendations for ocean disposal of mine waste in the coral triangle. Oceanography 25(4):40–51. https://doi.org/10.5670/oceanog.2012.66

Rehman K, Fatima F, Waheed I, Akash MSH (2018) Prevalence of exposure of heavy metals and their impact on health consequences. J Cell Biochem 119(1):157–184

Rieuwerts J (2015) The elements of environmental pollution (1st ed). Routledge, New York

Rizzi L, Petruzzelli G, Poggio G, Guidi GV (2004) Soil physical changes and plant availability of Zn and Pb in a treatability test of phytostabilization. Chemosphere 57(9):1039–1046

Sakakibara M, Ohmori Y, Ha NTH, Sano S, Sera K (2011) Phytoremediation of heavy metal-contaminated water and sediment by Eleocharis acicularis. CLEAN Soil Air Water 39(8): 735–741

Salazar MJ, Pignata ML (2014) Lead accumulation in plants grown in polluted soils. Screening of native species for phytoremediation. J Geochem Explor 137:29–36

Selvi A, Rajasekar A, Theerthagiri J, Ananthaselvam A, Sathishkumar K, Madhavan J, Rahman PK (2019) Integrated remediation processes toward heavy metal removal/recovery from various environments-a review. Front Environ Sci 7:66

Shah SB (2021) Heavy metals in scleractinian corals. Springer International Publishing, Berlin

Shahandeh H, Hossner LR (2000) Plant screening for chromium phytoremediation. Int J Phytoremediation 2(1):31–51

Sheoran V, Sheoran AS, Poonia P (2009) Phytomining: a review. Miner Eng 22(12):1007–1019

Singh A, Sharma A, Verma RK, Chopade RL, Pandit PP, Nagar V et al (2022) Heavy metal contamination of water and their toxic effect on living organisms. In: Dorta D, De Oliveira DP (eds) The toxicity of environmental pollutants. IntechOpen, London

Solomon F (2008) Impacts of heavy metals on aquatic systems and human health. Mining.com. http://www.infomine.com/library/publications/docs/mining.com/Apr2008c.pdf

Soudek P, Petrová Š, Vaněk T (2012) Phytostabilization or accumulation of heavy metals by using of energy crop Sorghum sp. In: 3rd international conference on biology, environment and chemistry IPCBEE. IACSIT Press, Singapore

Subhasini V, Swamy AVVS (2014) Phytoremediation of cadmium and chromium contaminated soils by Cyperus rotundus L. Am Int J Res Sci Technol J Eng Math 6:97–101

Türkmen D, Bakhshpour M, Akgönüllü S, Aşır S, Denizli A (2022) Heavy metal ions removal from wastewater using cryogels: a review. Front Sustain 3:765592

Ueda M (2016) Establishment of cell surface engineering and its development. Biosci Biotechnol Biochem 80(7):1243–1253

Un Nisa W, Rashid A (2015) Potential of vetiver (Vetiveria zizanioides L.) grass in removing selected PAHs from diesel contaminated soil. Pak J Bot 47:291–296

UNEP (United Nations Environment Programme) (2013) Global mercury assessment 2013: sources, emissions, releases and environmental transport. UN Environmental Program Chemicals Branch, Geneva

Varun M, Souza RD, Pratas J et al (2011) Evaluation of phytostabilization, a green technology to remove heavy metals from industrial sludge using Typha latifolia L. experimental design. Biotechnol Bioinformatics Bioeng 1:137–145

Veselý T, Trakal L, Neuberg M, Száková J, Drábek O, Tejnecký V et al (2012) Removal of Al, Fe and Mn by Pistia stratiotes L. and its stress response. Open Life Sci 7(6):1037–1045. https://doi.org/10.2478/s11535-012-0099-z

Wang H, Zhong G (2011) Effect of organic ligands on accumulation of copper in hyperaccumulator and nonaccumulator Commelina communis. Biol Trace Elem Res 143(1):489–499. https://doi.org/10.1007/s12011-010-8850-7

Wang Q, Kim D, Dionysiou DD, Sorial GA, Timberlake D (2004) Sources and remediation for mercury contamination in aquatic systems—a literature review. Environ Pollut 131(2):323–336. https://doi.org/10.1016/j.envpol.2004.01.010

Wang R, Balkanski Y, Boucher O, Bopp L, Chappell A, Ciais P, Tao S (2015) Sources, transport and deposition of iron in the global atmosphere. Atmos Chem Phys 15(11):6247–6270. https://doi.org/10.5194/acp-15-6247-2015

Wase J, Forster C (1997) Biosorbents for metal ions. Taylor & Francis Ltd, London

Wells ML, Price NM, Bruland KW (1995) Iron chemistry in seawater and its relationship to phytoplankton: a workshop report. Mar Chem 48(2):157–182. https://doi.org/10.1016/0304-4203(94)00055-I

World Health Organization (1984) Guidelines for drinking, water quality. WHO, Geneva

World Health Organization (2017) Lead poisoning and health. http://www.who.int/mediacentre/factsheets/fs379/en/

Wu H, Liu X, Zhao J, Yu J (2013) Regulation of metabolites, gene expression, and antioxidant enzymes to environmentally relevant lead and zinc in the halophyte Suaeda salsa. J Plant Growth Regul 32(2):353–361

Wuana RA, Okieimen FE (2011) Heavy metals in contaminated soils: a review of sources, chemistry, risks and best available strategies for remediation. Int Sch Res Netw Ecol 2011: 402647, 20 pp

Xu J, Yin HX, Liu X, Li X (2010) Salt affects plant Cd-stress responses by modulating growth and Cd accumulation. Planta 231(2):449–459

Xue H, Li J, Xie H, Wang Y (2018) Review of drug repositioning approaches and resources. Int J Biol Sci 14(10):1232–1244

Ye BJ, Kim BG, Jeon MJ, Kim SY, Kim HC, Jang TW et al (2016) Evaluation of mercury exposure level, clinical diagnosis and treatment for mercury in toxication. Ann Occup Environ Med 28:5

Zhou Q, Yang N, Li Y, Ren B, Ding X, Bian H, Yao X (2020) Total concentrations and sources of heavy metal pollution in global river and lake water bodies from 1972 to 2017. Global Ecol Conserv 22:e00925

Recent Advances in Biological Wastewater Treatment

9

José Antonio Rodríguez-de la Garza, Pedro Pérez-Rodríguez, Laura María González-Méndez, and Silvia Yudith Martínez-Amador

Abstract

Water is a universal solvent and liquid required in most anthropogenic, natural, and biochemical processes. Once used, water will have chemical or microbiological pollutants (wastewater), and knowing that the amount of water on the planet will always be the same, a treatment process must be chosen that aims to eliminate the greatest amount of pollutants to be able to return it to the receiving bodies of water or to be able to reuse it, reducing the impact on the environment. These wastewater treatment processes can be grouped into chemical, physical, and biological. In the case of biological treatment, it consists of using enzymes, microorganisms, or plants to accumulate, transform, or degrade contaminants through their metabolism. Biological treatment systems can be divided into conventional and advanced. Conventional ones are activated sludge, rotating biological contactors, biofilters, nitrification, and denitrification, among others. As an example of advanced biological treatment systems, we can name all the processes that mineralize nitrogenous compounds to molecular nitrogen (such as ANAMMOX, CANON, OLAND), the bioaugmentation batch-enhanced process (BABE), the use of aerobic granular sludge (AGS), and constructed wetlands (CW), among others. Advanced biological treatment systems can offer some of the following advantages compared to conventional biological systems: decreased hydraulic retention times, fewer space requirements, mineralized contaminants, a better landscape, and nominal investment costs in equipment and infrastructure. The present chapter will focus on discussing the most recent

---

J. A. Rodríguez-de la Garza
Facultad de Ciencias Químicas, Universidad Autónoma de Coahuila, Saltillo, Coahuila, Mexico

P. Pérez-Rodríguez · L. M. González-Méndez · S. Y. Martínez-Amador (✉)
Universidad Autónoma Agraria Antonio Narro, Saltillo, Coahuila, Mexico
e-mail: silvia.martinez@uaaan.edu.mx

© The Author(s), under exclusive license to Springer Nature Singapore Pte Ltd. 2023
R. Soni et al. (eds.), *Current Status of Fresh Water Microbiology*, https://doi.org/10.1007/978-981-99-5018-8_9

205

advances in biological wastewater treatment and how these systems can be applied to deal with existing and emerging pollutants in wastewater.

Keyword

Advanced biological treatment

9.1 Introduction

There are different processes for wastewater treatment; commonly, the two main types are chemical and biological. In biological treatment, the use of microorganisms and plants is considered to transform pollutants either for their assimilation (metabolism) or disassimilation, for which the biological wastewater treatment systems are divided into two large types: microbiological treatment and wetlands (Buha et al. 2015; Pell and Wörman 2009). It is known that the biological treatment of wastewater removes organic matter (carbon), nitrogen, and phosphorus (Christensen et al. 2015) through processes based on the natural cycles of these elements carried out by microorganisms through the conversion of pollutants into gaseous compounds such as carbon dioxide, methane, and molecular nitrogen or to solid compounds such as biomass (Weissbrodt et al. 2020). In the case of wetlands, they are made up of plants, microorganisms, and a substrate; some plants can accumulate or transform heavy metals, degrade organic compounds, volatilize compounds, and remove pathogens, and in a wetland, the removal is done through the plant and microbial activity in conjunction with the substrate that serves as a filter and a support medium for roots and the adhesion of microorganisms (Kataki et al. 2021). As is known, natural cycles have been altered due to anthropogenic activity. This has led humans to novel approaches to understanding and applying these natural occurring processes and, in some cases, improving these processes. The current chapter will describe these processes and how they can be applied to enhance water quality and the environment.

9.2 Biological Nitrogen Removal

Nitrogen pollution of water is mainly due to the excessive use of synthetic nitrogen fertilizers to increase the growth and development of crops in agriculture as well as livestock, the production of fossil fuels, food supplements of animal origin, adhesives, cosmetics, paper, and no less important the increase in human waste as a result of the increase in the population that contributes to the eutrophication of receiving bodies of water due to the excessive growth of algae due to the high nutrient load causing a decrease in oxygen, light, and also of aquatic biodiversity (Bassin 2018; Cárdenas-Calvachi and Sánchez-Ortíz 2013; Huang et al. 2018; Ochoa-Hueso 2017; Tendengren 2021). In this regard, there are various transformation processes for biological nitrogen removal (BNR) through assimilation or

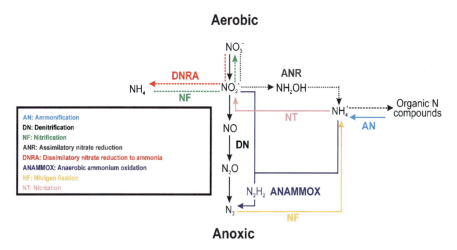

Fig. 9.1 Processes involved in the biological removal of nitrogen

dissimilation, such as biological nitrogen fixation (BNF), nitrification (NF), nitritation (NT), ammonification (AN), denitrification (DN), assimilatory nitrate reduction (ANR), dissimilatory nitrate reduction to ammonium (DNRA), and more advanced processes such as anaerobic ammonia oxidation (ANAMMOX) (Fig. 9.1) from which processes are derived and/or combined such as the single reactor system for high ammonia removal over nitrite (SHARON), completely autotrophic nitrogen removal over nitrite (CANON), oxygen-limited autotrophic nitrification denitrification (OLAND), SHARON/ANAMMOX, partial NT-ANAMMOX (deammonification), denitrifying ammonium oxidation (DEAMOX), simultaneous partial nitrification, anaerobic ammonium oxidation and denitrification (SNAD), single stage removal using anammox and partial nitritation (SNAP), among others (Bassin 2018; Sanabria et al. 2009; Sanchez and Sanabria 2009; Wang et al. 2017). These processes derived or combined from ANAMMOX activity have aroused much interest due to the toxic effect of ammonia on all life forms (Bagchi et al. 2012). Additionally, other technologies can potentially improve BNR processes, such as bioaugmentation batch-enhanced BABE (Gatti et al. 2015) and aerobic granular sludge AGS (Sarma and Tay 2018).

9.3 Advanced Biological Processes for Nitrogen Removal

9.3.1 ANAMMOX

The process based on lithoautotrophic denitrification, known as ANAMMOX, refers to the anaerobic oxidation of ammonium. This process is carried out by bacteria belonging to the phylum Planctomycetes, which are slow-growing and chemolithoautotrophic bacteria that use nitrite as an electron acceptor (although some species of bacteria involved in ANAMMOX have been described that can use organic acids such as propionic as an electron acceptor instead of nitrite) under anoxic conditions producing molecular nitrogen. This process can be implemented in wastewater with high nitrogen content and low carbon content (ANAMMOX process is known to be little tolerant to the presence of organic matter). In other words, it works best in environments with a low carbon/nitrogen ratio, which is one of the factors that govern the nitrogen removal processes and has a major effect on microbial populations. Although the C/N ratio should be low, the combined partial NT-ANAMMOX process has been used to remove nitrogen from carbon-rich wastewater such as leachate from landfills and sludge digesters, coupling processes from DN (where organic matter is used as an electron donor for the reduction of nitrogenous compounds such as nitrate or nitrite to nitrogen) or from anaerobic digestion where most of the organic matter is transformed into methane (Ahmad et al. 2020; Akgul et al. 2013; Bassin 2018; Cervantes-Carrillo et al. 2000; Figueroa et al. 2012; Saha et al. 2022; Sanchez et al. 2014; van Loosdrecht 2008). It should also be considered that ANAMMOX bacteria have good activity in a pH range of 6.7–8.3 (8.0 optimal value) and at a temperature range of 35–40 °C; temperatures over 45 °C may cause cell lysis (Bassin 2018). The ANAMMOX process presents some disadvantages, among which are that the bacteria that oxidize ammonium grow slowly (not only because it is an autotrophic process but also because ANAMMOX bacteria have a low intrinsic conversion rate of ammonium to nitrite or because the enrichment of the culture does not have the ideal conditions for growth). In some cases, the process startup can take over a year, they are sensitive to visible light, and the presence of heavy metals can inhibit the process, high salinity, organic compounds, sulfides, and phosphates; in addition the accumulation of substrates such as nitrite and free ammonia can also be problematic (Bassin 2018; Lacroix et al. 2020; Xie et al. 2017).

9.3.2 SHARON

SHARON comprises a combination of two processes in which aerobic microorganisms participate in the partial NT (ammonium to nitrite) process and facultative anaerobic microorganisms in the DN (nitrite to molecular nitrogen) process. Due to the nature of both processes, they must be carried out in separate reactors (Sanabria et al. 2009; Sánchez and Sanabria 2009). This process is an alternative for wastewater with high ammonium and low organic matter content, which requires low aeration costs and, in some cases, an additional carbon source such as methanol. Among the factors that govern this process are the pH and the redox potential (ORP), which indicate whether the predominant phase is aerobic or anoxic (Claros et al. 2012), and the type of bacteria present in the process, among others. A submerged biofilter (using PVC as carriers) with a partial-SHARON activity for synthetic wastewater treatment (simulating anaerobic digester leachate with a high ammonium concentration and low organic matter concentration) was operated at two hydraulic retention time HRT (0.4 and 0.5 days) in order to know its effect on bacterial biodiversity. At an HRT of 0.5 days, bacterial biodiversity was lower but more specialized since it managed to convert all the ammonium into nitrite, most of the biofilm was formed by *Nitrosomas* sp., while at an HRT of 0.4 days, this biodiversity was higher since the biofilm was formed by *Nitrosomas* sp., *Nitrospira* sp., and *Nitrosovibrio* sp., but the conversion of ammonium to nitrite decreased substantially (González-Martínez et al. 2013). The partial-SHARON activity process has been reported to be used to treat effluents generated during syngas (this type of effluents contain CO, CO_2, H_2S, CH_4, NH_3, organic acids, ammonium formate, phenols, and cyanides synthetic gas) production with a stirred tank reactor (without biomass recirculation and with an induced microbial consortium with synthetic wastewater rich in NH_4^+ at an HRT of 1.25 days), and despite the presence of phenols and cyanides, the NH_4^+ (NT) removal efficiency was maintained, generating an effluent suitable for an ANAMMOX process (Milia et al. 2015a). Also, the partial-SHARON activity process was studied to determine the effect of different Ci (inorganic carbon)/N ratios (1, 1.5, 1.75, and 2) in the sour water treatment (ammonium-rich refinery wastewater) to determine if the effluent from this treatment was suitable for further ANAMMOX (autotrophic) or DN (heterotrophic) process; the study carried out in a water-jacketed continuous stirred tank reactor (CSTR) under the following conditions temperature of 35 °C, pH of 7.0 ± 0.5, HTR of 1.25 days, DO above 2 mg/L and inoculated with activated sludge (AS) from municipal wastewater plant (WWTP). The sour water used in this study contained dissolved organic carbon (DOC), cyanides, sulfides, and phenols (2000 mg NH_4-N was added). The results showed that the Ci/N ratio affects the regulation of SHARON performance; a Ci/N = 1 ratio produces an adequate effluent for an ANAMMOX process, while a Ci/N = 2 ratio produces an adequate effluent for a DN process. Additionally, using adapted biomass showed a reduced inhibition effect of toxic compounds (Milia et al. 2015b).

9.3.3 CANON

CANON is a combination of processes between partial NT (ammonia to nitrite) carried out by aerobic ammonia-oxidizing bacteria (AOB), and the ANAMMOX process (ammonia + nitrite $\rightarrow$ molecular nitrogen) carried out by anaerobic ammonia-oxidizing bacteria (AnAOB), both processes in a single reactor (Sánchez and Sanabria 2009; Wang et al. 2017). The microorganisms involved in these processes are autotrophs that can remove high ammonium concentrations; therefore, adding an organic carbon source is unnecessary. Other advantages of this process are that it requires less aeration and the production of sludge is lower (Bassin 2018; Wang et al. 2017). The factors that must be considered in this process are: aeration control is very important since a DO concentration of 1 mg/L is the ideal value, the thickness and density of the biofilm, the concentration of organic matter, and the temperature in a range between 30 and 35 °C as has been established in different studies, resulting in greater growth of microorganisms (Bassin 2018; Sanabria et al. 2009).

9.3.4 OLAND

OLAND is a similar process to CANON, and even some researchers mention that they are the same process; the reactions are carried out in a single biofilm or biofilter reactor where AOB (ammonium to nitrite) and ANAMMOX anoxic bacteria (Sanabria et al. 2009; Sánchez and Sanabria 2009). The biofilm allows partial nitritation and the ANAMMOX process to occur in a single reactor, reducing costs (capital, operational, and aeration) in the treatment process (Sivalingam et al. 2020). Some important factors of this process are the DO levels and the absence of organic electron donors that ensure that the NT occurs, and later, due to the low concentration of DO, the nitrite is consumed in the oxidation of ammonium to nitrogen; therefore, the coexistence of AOB and ANAMMOX can be carried out in the OLAND reactor (Bassin 2018). OLAND is a process that has been used for the treatment of wastewater with very high or very low ammonium contents, above 1000 or less than 66 mg N L^{-1} depending on the concentration of dissolved oxygen (DO) for which it is recommended that does not exceed 0.2 mg L^{-1} (Sanabria et al. 2009; Sánchez and Sanabria 2009; Hien et al. 2017). This recommendation has been derived from studies where, when using an OLAND process, different DO ranges have been applied (0.4–0.8, 0.2–0.4, and 0.1–0.2 mg L $^{-1}$) for wastewater from a latex processor treated by a Upflow Anaerobic Sludge Blanket (UASB) reactor, which contained 100–200 NH_4^+ mg L^{-1} and 50–80 mg L^{-1} of chemical oxygen demand (COD); the best performance was detected in a DO range of 0.1–0.2 mg L^{-1} with the removal of total nitrogen (TN) and COD of 94% and 61%, respectively (Hien et al. 2017).

9.3.5 DEAMOX

This process combines in a single reactor the ANAMMOX reactions carried out by AnAOB with the autotrophic denitrification carried out by bacteria belonging to the genera Thiobacillus and Thiosphaera, which oxidize reduced forms of sulfur using nitrate as an electron acceptor to produce nitrite in an anoxic/anaerobic environment. and with low organic carbon content (Masłoń and Tomaszek 2009). Among the advantages of this process is that the production of nitrite does not need to be controlled; the formation of granules is stimulated, which is very appropriate for the ANAMMOX process; and high nitrite levels are not detected due to the conversion to molecular nitrogen; however, the process can be limited by a high level of sulfide, which can inhibit the bacterial activity of the DEAMOX process (Kalyuzhnyi et al. 2006; Wang et al. 2018a). Ammonia removal through the DEAMOX process has been applied in industrial wastewater, dewatered sludge, and leachate from sanitary landfills (Masłoń and Tomaszek 2009). The process has been used in continuous flow reactors with AnAOB immobilized in polyethylene glycol in order to reduce the sludge washing in ANAMMOX and with denitrifying bacteria, resulting in a complete nitrate removal and total nitrogen removal of up to 88.5% of synthetic wastewater that contained ammonium, nitrite, nitrate, and sulfate

(Wang et al. 2018a). The DEAMOX process has been compared using planktonic bacteria and bacteria in biofilm (using an SBBR—sequencing biofilm batch reactor—packed with polyurethane foam cubes) with synthetic wastewater (60 mg/L of NH_4^+ and NO_3^-, a C/NO_3^- ratio of 2.5), at a temperature of 30 °C under anoxic condition. The author reported that the ANAMMOX activity (NH_4^+ removal) and TN removal efficiency were significantly higher in the biofilm reactor. This is due to ANAMMOX and denitrifying bacteria occupying different environmental niches, and the increased biofilm thickness creating an environment that favored the ANAMMOX bacteria (due to low O_2 level and COD mass transfer was limitations within the inner layer of the biofilm; these conditions are favorable to ANAMMOX bacteria) (Zhang et al. 2019).

9.3.6 SNAD

This process is the sum of three nitrogen removal processes: NF, ANAMMOX, and DN, in which ammonia and organic carbon contained in wastewater can be removed (Wang et al. 2017). It has been used to remove nitrate formed by AnAOB involved in the CANON process. Through this process, a high nutrient removal has been obtained as a result of heterotrophic denitrification and autotrophic nitrogen removal and has been used to remove nitrogen from wastewater with a low C/N ratio and with a high ammonia content, such as swine wastewater, sludge-digestion liquids, leachate from landfills, and, in a few cases, for domestic wastewater, which contains a low concentration of ammonia (Daverey et al. 2013; Li et al. 2019).

9.3.7 SNAP

Similar to the CANON process, it is carried out in a single reactor, combining the ANAMMOX and partial NT processes, allowing the removal of nitrogen without adding organic carbon and reducing investment and energy costs (Zhang et al. 2014). A membrane-aerated biofilm reactor (MABR) inoculated with AS was tested to remove nitrogen contained in synthetic wastewater through a SNAP process; the results showed a rapid startup of the reactor. In addition, ANAMMOX activity was detected at 48 h, and the removal of NH_4^+ via nitritation was also verified when DO levels were below 0.7 mg/L (it was observed when DO levels increased, there was an imbalance in the microbial community, decreasing the efficiency of MABR), and the total nitrogen removal of 84% was achieved (Augusto et al. 2018). Other reports showed a rapid startup of the SNAP process (46 days), and high removal percentages of 86.9% and 76.8% for NH_4^+ and total nitrogen, respectively, were also obtained in an SBBR inoculated with AS in the synthetic wastewater treatment; the ANAMMOX activity was detected from day 12, and throughout the monitoring, changes in microbial diversity were observed until the SNAP process was established (Cai et al. 2020).

9.3.8 Use of the Combination of Processes to Remove Nitrogen

In the production of carbon shell, large amounts of wastewater are generated, which contain a high concentration of ammonium (600 mg L^{-1}), TN (900 mg L^{-1}), COD (4500 mg L^{-1}), and phenolic compounds (2000 mg L^{-1}), and these waters must be treated in several stages, including the biological removal of nitrogen with advanced processes such as SHARON, ANAMMOX, OLAND, and CANON (Bargieł and Zabochnicka-Świątek 2018). SHARON and ANAMMOX can be potential alternatives to treat the wastewater mentioned above because they are considered economical and sustainable processes for removing ammonia-rich wastewater, such as leachate, from sanitary landfills; they save energy, reduce investment costs, and require less space and operating costs due to lower oxygen consumption and carbon (Shalini and Joseph 2012). Another study focused on the treatment of leachate from landfills, and the treatment was carried out in a multistage system in order to achieve a removal of the high content of organic matter and nitrogen; the multistage process was integrated by a UASB reactor, a membrane bioreactor (MBR), a SHARON reactor, and an ANAMMOX reactor, and it has reported a 99% and 90% removal of BOD_5 and COD, respectively, in the UASB and MBR reactors, while in the SHARON and ANAMMOX reactors, the removal of nitrogen was 92% and 78% for NO_2^- and NH_4^+, respectively (Akgul et al. 2013).

A system consisting of three UASB + SHARON + ANAMMOX reactors was used to know the dynamics of the microbial population during the treatment (314 days) of piggery wastewater with a concentration of 5500–8500 mg/L COD and 500 at 1500 mg/L NH_4^+. The reactors were started up separately to induce the specific activity of each one (gradual increase in the concentration of NH_4^+ and NO_2^- through synthetic residual water), and after 229 days, they were connected and fed with piggery wastewater. In the ANAMMOX reactor, nitrogen removal was detected after 38 days of operation (at this same time, ANAMMOX bacteria were also detected in the biomass samples from the reactor) and increased in time (in the same way the bacteria ANAMMOX and its diversity increased). In the SHARON reactor, bacterial diversity decreased when it was fed with synthetic wastewater (206 days); however, when the influent was changed (change to the effluent from the UASB reactor that treated piggery wastewater), the bacterial diversity increased slightly after 14 days. Finally, it was observed that the microbial composition in the ANAMMOX and SHARON reactors was influenced by the presence and concentration of NH_4^+, NO_2^-, and NO_3^- in addition to other factors such as pH, DO, and COD concentration (Du et al. 2016; Huang et al. 2018).

In the treatment of municipal wastewater, the removal efficiency of organic and nitrogenous compounds, the use of energy, and the production of biosolids of four treatment systems were compared: (a) aerobic organic oxidation AOO + NF followed by the addition of COD and a posterior DN; (b) anaerobic organic removal AOR + AOO followed by an NT process and a subsequent ANAMMOX; (c) DN followed by AOO + NF; C: DN followed by AOO + NF followed by NT and ANAMMOX. The best option was the system described in point "b" due to the lower energy requirement (required oxygen) and production of biosolids in addition to the production of energy in the form of methane (McCarty 2018).

9.3.8.1 BABE

The BABE process is important since nitrifying organisms are very slow growing and sensitive to changes in temperature and pH (Yu et al. 2012). BABE combines the bioaugmentation of nitrifying organisms due to the return of autotrophic nitrifying bacteria from activated sludge that is washed (sludge liquor) from the main treatment process, thereby increasing the NF rate by up to 60%. In addition, the nitrifying organisms grow in flocs of the activated sludge, and without this bioaugmentation, the nitrifiers will tend to grow suspended in the medium; when bioaugmentation is combined with the deammonification process, the negative effects of the nitrogen load of WWTPs are reduced as well as the loss of biomass, so it can be considered that the deammonification process is efficient for the removal of nitrogen and cultivation of nitrifiers for bioaugmentation (Berends et al. 2005; Muszyński-Huhajło et al. 2021; Nsengiyuma et al. 2021; Van Loosdrecht and Salem 2006). A high concentration of nitrite can inhibit denitrifying microorganisms; however, by using the bioaugmentation process in an sequencing batch reactor (SBR) with continuous feeding of nitrite and acetate, the denitrifying bacterial community was enriched, and the nitrite denitrification rate increased from 10 to 275 mg/L h, showing an increase in the microbial biodiversity involved in the denitrification process from 2.16% to 84.26% during the enrichment period (Yao et al. 2019). An SBR operated at 3 different temperatures (20, 15, and 10 °C) was bioaugmented with archaea resulting in up to 20% removal of total nitrogen contained in municipal wastewater compared to other SBR that was not bioaugmented (Szaja and Szulżyk-Cieplak 2020). In addition, SHARON, ANAMMOX, deammonification, and BABE, among other processes, are considered sidestream nutrient removal since they remove a high concentration of nutrients (nitrogen) from the sludge (biosolids) and or the centrate of a treatment plant (Izadi et al. 2021; Nsengiyuma et al. 2021; Van Loosdrecht and Salem 2006).

9.3.8.2 AGS

The AGS is a technology with fewer spatial (lower volume reactor) requirements than activated sludge; due to the compacted biomass, it produces less sludge, requires less energy, and is more tolerant to toxic contaminants and high contaminant loads. In short, it is more effective than AS technology. AGS can simultaneously remove carbon, nitrogen, and phosphorus (Nancharaiah et al. 2019; Sepúlveda-Mardones et al. 2019). The formation of AGS granules is achieved through an adequate establishment of parameters such as the organic load rate (OLR), hydrodynamic shear forces, aeration, temperature, type of microorganisms, and wastewater composition. The granules are formed by different microorganisms distributed in different layers; in the outermost layers is where aerobic microorganisms are located, such as nitrifying bacteria that form nitrate, which will be furthermore denitrified inside the granule where anoxic microorganisms are located. The granules have good sedimentation, so the settling time is concise, which is a great advantage over wastewater treatment systems with activated sludge. Good sedimentation allows the granules to separate between one phase and another (Sepúlveda-Mardones et al. 2019; van der Roest et al. 2011).

9.4 Bioelectrochemical Systems

Bioelectrochemical systems (BES) are a type of technology that allows the conversion of chemical energy in various types of organic substrates to electrical energy through the biological activity as a catalyst. The biological activity is usually provided by microorganisms (bacteria and some yeasts), although, in some cases, the feasibility of some systems to convert solar irradiance to usable energy has been reported (Gul and Ahmad 2019). These types of systems cover different types of cells, and operational configurations, from double-compartment microbial fuel cells (MFC-DC), used almost exclusively for laboratory-scale experimentation, to plant-based microbial fuel cells (PMFC), which allow taking advantage of the photosynthetic processes of plants for the production of electrical energy (Maddalwar et al. 2021). This type of technology presents various advantages over conventional bioenergy processes, among which are its sustainability, zero emissions of greenhouse gases, wide operational range of temperature and pH, and ability to tolerate high organic loads (Cecconet et al. 2020). Currently, interest in this type of technology has increased considerably due to the capacity of this type of system to remediate contaminated water and soil (Varjani 2022), treat persistent pollutants (Bagchi and Behera 2020), and degrade emerging pollutants (Yan et al. 2019).

9.4.1 Types of BES

Bioelectrochemical systems can be classified in different ways, although the most practical is based on the direction of electron transfer and the type of reaction carried out (Zheng et al. 2020).

9.4.1.1 Microbial Fuel Cell (MFC)

Microbial fuel cells are bioelectrochemical systems that allow electrical energy to be generated from the metabolization of organic compounds by electrochemically active microorganisms capable of transferring the generated energy to an external load (Wang et al. 2018b). This process can be described as follows: (1) biocatalyst microorganisms oxidize the organic substrate at the anode, which produces electrons and protons; (2) protons are transported to the cathodic compartment through the cation exchange membrane (CEM); (3) electrons flow out of the system and are redirected toward the cathode compartment; and (4) protons and electrons react with the oxygen present in the cathode compartment to obtain water as the final product (Gajda et al. 2018). The performance of these systems depends on multiple factors; among the most important are the conductivity and microbial biocompatibility of the electrodes (Zhang et al. 2020), nature and concentration of the substrate (Heidrich

9 Recent Advances in Biological Wastewater Treatment

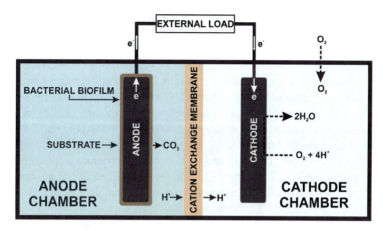

Fig. 9.2 Dual compartment microbial fuel cell design

et al. 2018), internal resistance (Ma et al. 2017), absence or presence of cation exchange membrane (Goel 2018), and overall system design (Bhargavi et al. 2018).

There are different designs of microbial fuel cells; however, the double-compartment design is the most common and recommended for testing on a laboratory scale (Fig. 9.2). Although convenient for testing new configurations and materials, this type of design is counterproductive in the long term by making it difficult to scale to larger dimensions (Flimban et al. 2019).

9.4.1.2 Microbial Electrolysis Cell (MEC)

Microbial electrolysis cells are an anaerobic biological process that converts organic substrates to renewable energy in hydrogen or methane (Lee et al. 2022; Rousseau et al. 2020). In this cell type, organic matter is transformed into CO_2, H^+ and electrons by electrochemically active microorganisms, similar to MFCs. The main difference between the two cells consists in the synthesis of by-products in a controlled manner in the cathode compartment of the cell, seeking, in most cases, to reach sufficient electrochemical potential for the production of hydrogen or methane. For the above to occur properly, it is necessary to overcome the cathodic overpotential present in this type of system, commonly achieving this objective by using an external power source that supplies the necessary electrical current to compensate for what is naturally generated by the microorganisms in the anode compartment during the biodegradation process the organic substrate fed to the cell (Katuri et al. 2019). Among the main applications of MECs is the production of hydrogen, methane, formic acid, and hydrogen peroxide, among others (Aiken et al. 2019; Carrillo-Peña et al. 2022; Hua et al. 2019; Sim et al. 2018). Similarly, due to the ability of this type of system to degrade complex organic matter and change the oxidation state of some inorganic substances, interest in this type of process has increased from an environmental preservation and remediation perspective (Chen et al. 2019).

9.4.1.3 Microbial Desalination Cell (MDC)

Microbial desalination cells are a promising technology for seawater purification and polluted water desalination (Zahid et al. 2022). The functioning and operation of these cells (Fig. 9.3) consist of the following: (1) from an MFC-DC, a third compartment is located that divides the anode and cathode compartments of the cell; (2) said compartment, called desalination chamber, which will be separated from the anodic compartment by an anion exchange membrane (AEM) and of the cathode compartment by a cation exchange membrane; (3) the potential differential that occurs during the oxidation process of organic matter and the consequent flow of electrons from the anodic compartment to the cathodic compartment will cause a migration of the anions through the AEM and toward the anionic compartment and of the cations from the desalination chamber toward the cationic compartment of the cell (Tawalbeh et al. 2020). The final product of this process will be treated water with a significantly lower content of soluble salts compared to the initial solution, and depending on the degree of pollution, a completely treated sample could be obtained (Salehmin et al. 2021). This technology is, therefore, relevant not only for its ability to purify saline water but also for the parallelism it presents when treating wastewater rich in organic matter (anodic compartment) and the consequent generation of electrical energy by oxidizing the substrate (Al-Mamun et al. 2018). Despite this, more in-depth studies are required to define the multiple variables that affect this process.

9.4.1.4 Plant Microbial Fuel Cell (PMFC)

PMFCs are a type of technology that allows a plant system to be coupled to a conventional MFC, seeking to take advantage of the root exudates secreted by plants during the photosynthetic process and convert these substances into electrical energy through microbial action (Kabutey et al. 2019). This technology also turns out to be interesting if it is considered that, indirectly, and using plants as intermediaries; it is

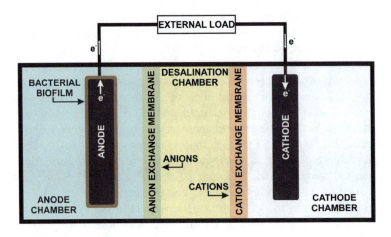

Fig. 9.3 Microbial desalination cell design

possible to convert solar energy (total irradiance) into electrical energy (Chiranjeevi et al. 2019). As this technology depends on multiple elements, there is a large number of factors that can affect the performance of this type of system, among which are the genus and species of the plant used, the source, intensity and spectral balance of the light irradiance applied to the plant, CO_2 concentration in the experimental area, operational parameters of the cell (pH, temperature, electrical conductivity, and electrode material), macro- and micronutrient content in the anodic compartment, type and source of the inoculum, and cell design, among others (Regmi et al. 2018b). One of the main advantages of this process is that it can be carried out indoors or outdoors depending on the site's needs (Osorio-de la Rosa et al. 2019). Similarly, being a technology that can be easily adapted to different environments, the PMFC can be placed on the ground without modifying the area's landscaping, thus reducing the environmental impact (Regmi et al. 2018a).

9.4.2 Recent Advances in BES

Recent research on BES has focused on three main axes: (1) selection and optimization of the materials that make up the electrodes, (2) design and configuration of the devices used, and (3) identification, isolation, and application of new electrochemical active microbial species (Chen et al. 2021).

9.4.2.1 Novel Electrode Materials

Graphene-Based Electrodes
Graphene is a substance with a laminar structure made up of a single layer of two-dimensional oriented carbon atoms. It has a high specific surface area and thermal and electrical conductivity (Zhang et al. 2022). These properties make it an extremely promising material for its application in bioelectrochemical systems by increasing the amount of electrical energy captured and porosity and enhancing substrate degradation (Cui et al. 2018; Tremblay et al. 2020). Anodes modified with graphene deposits have been used to increase the conductivity of these components and maximize the power density achieved in the systems, resulting in significantly higher results when compared to unmodified base materials (Camedda et al. 2019; Lescano et al. 2018; Tian et al. 2019). Graphene coatings have also been applied to obtain biocathodes to increase the porosity and therefore the surface area, of these materials, thus maximizing the mass transfer of the substrate toward the microbial community (Aryal et al. 2019; Fu et al. 2022; Yao et al. 2021). Some other examples of this type of application are shown in Table 9.1.

Conductive Polymer-Based Electrodes
Conducting polymers have generated much interest today due to their chemical stability, biocompatibility, and high conductivity, making these materials suitable for BES as electron acceptors (Narayanasamy and Jayaprakash 2020). Various authors have addressed the advantages of applying this type of materials to increase

Table 9.1 Comparison of application of graphene electrodes in various types of bioelectrochemical systems (BES)

Type of BES	Anode	Cathode	Cationic exchange membrane (CEM)	Inoculum	Substrate	References
MFC	Graphene/carbon brush	Carbon cloth	Nafion 117	Sewage wastewater	Sewage wastewater	Sayed et al. (2021)
MFC	PDDA, graphene, and carbon cloth	Carbon felt	Nafion 117	Anaerobic sludge	Phosphate buffer solution	Chen et al. (2020)
MFC	Graphene	Stainless steel mesh	Nafion 117	Anaerobic sludge	Phosphate buffer solution	Pareek et al. (2019)
MFC	Tungsten carbide/graphene/carbon felt	Carbon paper	Nafion 117	Sugar industrial wastewater	Sugar industrial wastewater	Mohamed et al. (2021)
MEC	Graphite plate	Graphene/nickel oxide nanocomposites	Nafion 117	Sugar industrial wastewater	Sugar industrial wastewater	Jayabalan et al. (2020)
MEC	Graphite fiber brush	Graphene/Pt nanoparticles	–	–	Sodium acetate buffer	Sánchez-Peña et al. (2022)
MEC	Graphite brush	Graphene/graphite plate	Ultrex CMI-7000	Domestic wastewater	Anaerobic sludge	Hu et al. (2019)

Table 9.2 Comparative application of conductive polymer electrodes in various types of bioelectrochemical systems (BES)

Type of BES	Anode	Cathode	Cationic exchange membrane (CEM)	Inoculum	Substrate	References
MFC	CNT/PPy	Carbon cloth	–	Anaerobic slugde	Wastewater and nutrient solution	Zhao et al. (2019)
MFC	nPPy/graphite felt	Carbon cloth	–	*Shewanella putrefaciens*	Glucose-based nutrient solution	Sumisha and Haribabu (2018)
MFC	nNIBP/carbon felt	Carbon felt	–	*Cystobasidium slooffiae*	Xylose-based nutrient solution	Moradian et al. (2022)
MFC	NIBP/CNT sponges	Graphite rod	Nafion 117	Anaerobic slugde	Sodium acetate-based nutrient solution	Xu et al. (2019a)
MFC	nPTh/carbon cloth	Carbon cloth	Nafion 117	*Shewanella putrefaciens*	Glucose-based nutrient solution	Rajendran et al. (2021)

the performance of BES, emphasizing the low production cost, which would allow this type of system to be scaled more easily (Fischer et al. 2018; Sonawane et al. 2018). A bioanode composed of polypyrrole-carboxymethylcellulose-titanium nitride/carbon brush hydrogel (PPy-CMC-TiN/CB) was developed to increase the mechanical stability of the supports and increase the performance of the system, observing highly favorable results (Wang et al. 2020). In chromium (VI) reduction, an abiotic cathode made from carbon/manganese dioxide cloth coated with polypyrrole (PPy) was used to prevent corrosion of the composite, managing to remove 100% of the contaminant in less than 32 h. Therefore, this material is considered suitable for its use on larger scales (Liu et al. 2020a). The viability of PPy/MnO_2 nanocomposites focused on identifying their electrochemical behavior when applied in BES and evaluating their ability to generate electrical power has also been determined (Prakash et al. 2020). More examples of the above are mentioned in Table 9.2.

Electrode-Based Nanomaterials

In recent years, the development of nanomaterials (materials with a size between 1 and 100 nm) has experienced rapid growth due to its great application potential in a wide variety of technological areas such as electronics, data storage, and ceramics, among others (Liu et al. 2020b). These materials facilitate the extracellular transfer

Table 9.3 Application of nanomaterial electrodes in various bioelectrochemical systems (BES)

Type of BES	Anode	Cathode	Cationic exchange membrane (CEM)	Inoculum	Substrate	References
MFC	CNT/carbon cloth	Pt/carbon cloth	Nafion 117	Municipal wastewater	Municipal wastewater	Iftimie and Dumitru (2019)
MFC	Carbon cloth	Co/N-CNT/carbon cloth	–	Anaerobic digester sludge	Sodium acetate nutrient solution	Yang et al. (2019)
MFC	Carbon felt	CNt/MoS$_2$	Nafion 117	Originated on a stable operating MFC	Phosphate and acetate nutrient solution	Xu et al. (2019b)
MEC	Carbon cloth	nFe$_3$O$_4$ / carbon cloth	–	Dairy industry wastewater	Dairy industry wastewater	Rani et al. (2021)
MEC	Carbon felt	Several types of CNTs and metals/carbon felt	–	Originated on a stable operating MFC	Acetate nutrient solution	Choi et al. (2019)

of electrons from microorganisms toward the anode in BES (Kaur et al. 2020). Electrospun carbon nanotube and nanofiber nanocomposites (CNTs/CNFs) were evaluated in an MFC, observing an increase in electrical conductivity, excellent biocompatibility, good hydrophilicity, and high electrocatalytic activity (Cai et al. 2019). A cathode composed of nitrogen-doped carbon nanofibers was developed to maximize the catalytic capacity of the electrode in a single-compartment MFC, increasing cell performance and obtaining similar results to other works where Pt was used as a cathodic catalyst (Massaglia et al. 2019). Carbon cloth anodes coated with gold nanoparticles increased the generation of electrical energy in a double compartment MFC, reaching remarkable values due to the high affinity of these materials with the electrochemically active microorganisms of the cell (Wu et al. 2018). Table 9.3 shows other examples of this type of application in BES.

9.4.2.2 Systems Design and Configuration

Although most of the scientific research related to BES has indeed focused on developing new anodic and cathodic materials to favor the conductivity of this type of materials, in recent years, modifications have been promoted similarly regarding the architecture in this type of systems, seeking to obtain designs and configurations that are easier to scale, lower cost, and generally more efficient (Bakonyi et al. 2018). Different authors have focused on increasing the ionic conductivity of exchange membranes, this being one of the main limitations for obtaining good performance in BES. A cation exchange membrane based on sulfonated polyether ether ketone (SPEEK) and sulfonated titanium nanotubes

9 Recent Advances in Biological Wastewater Treatment

(S-TNT) was synthesized to increase the ionic conductivity of an MFC, thus reducing the resistivity of this component (Kugarajah and Dharmalingam 2020). Ceramic membranes with different pore diameters have been evaluated as separators in MFC, comparing their performance with a membrane more commonly used in this type of study (Nafion 117), verifying the feasibility of applying this type of materials in BES, and managing to increase the total conductivity of the cell (Daud et al. 2018). A sulfonated biocarbon composite membrane was applied as CEM in BES, with good results; additionally, this membrane is low-cost and easy to scale (Chakraborty et al. 2020). Membranes made with natural clays mixed with goethite were used as CEM in a laboratory-scale MFC, achieving efficiencies similar to conventional systems that use commercial CEM (Nafion 117) but at a fraction of the production cost, being this promising for the eventual scaling of this type of systems (Das et al. 2020).

9.4.2.3 Novel Electroactive Microorganisms

A wide variety of microorganisms can produce an electrical current and transfer it to external electron acceptors, such as the anodes in the BES. These exoelectrogens are present in the three taxonomic domains (superkingdoms) and can be used as long as they are provided with the appropriate conditions for their correct development (Logan et al. 2019). Currently, more attention has been paid to the identification, isolation, and use of this type of microorganisms to increase performance in BES. The role of microorganisms of the genus *Clostridium* in generating electrical energy by acting as adjuvants in anaerobic biofilms in BES has been reported (Rivalland et al. 2022). A biofilm of fruit peel leachate was formed onto carbon felt bioanode, probing the versatility and diversity of sources from which microorganisms can be isolated with the capacity of using external materials as final electron acceptors (Kebaili et al. 2020). Up to 17 species of 6 different genera of microorganisms within the rhizosphere and plant tissue of angelica and sweet potato were reported, such as *Bacillus* sp., *Pleomorphomonas* sp., *Rahnella* sp., *Shinella* sp., *Paenibacillus* sp., and *Staphylococcus* sp., (commonly known species), and there may still be other species that could be present that could present superior capacities for the generation and transport of electrons from inside the cell and toward an external acceptor (Ling et al. 2022).

9.5 Membrane Bioreactors

Membrane bioreactors (MBR) combine the activated sludge remediation process with a membrane filtration process. This technology can treat a wide range of organic substrates, generate a lower amount of sludge as a by-product, and have high degradation rates of pollutants (Pervez et al. 2020). These systems allow the elimination of secondary and tertiary treatment processes, reducing costs and increasing the volume of treated water (Al-Asheh et al. 2021). Recent studies have verified the wide versatility of this type of system when applying them for the remediation of water contaminated with microplastics (Bayo et al. 2020), pharmaceutical residues (Femina-Carolin et al. 2021), and hydrocarbons derived from petroleum (Zeirani et al. 2019), among others.

9.5.1 Types of MBR

MBRs are categorized based on their configuration or design, with two main types: (1) submerged membrane bioreactor and (2) lateral flow membrane bioreactor (Goswami et al. 2018).

9.5.1.1 Submerged Membrane Bioreactor (SMBR)

In submerged membrane bioreactors (as their name indicates), the filtration membrane is placed inside the bioreactor and in direct contact with both the substrate and the biomass (Fig. 9.4), thus reducing operating costs. Therefore, the substrate will be forced to interact with both the microorganisms in the system and the membrane, increasing the contaminant removal rate. Despite the above, as the membrane is located within the system and under constant adverse conditions, its care and maintenance will be complex, creating the opportunity for system failures to occur (Miyoshi et al. 2018). One of the main problems when developing and implementing this type of technology is the so-called biofouling, a phenomenon described as the accumulation of microorganisms on unwanted surfaces (Pichardo-Romero et al. 2020). In the case of SMBR, biofouling occurs primarily on the surface of the remediation membrane, as the biomass accumulates and occludes its pores, decreasing the flow of the substrate through the membrane and, therefore, affecting the performance of the system in the long term (Aslam et al. 2018). Therefore, most of the research on this type of system has focused on reducing this phenomenon by modifying the porosity and chemical composition of the filtration membranes (Nazmkhah et al. 2022; Sano et al. 2022; Xiao et al. 2021).

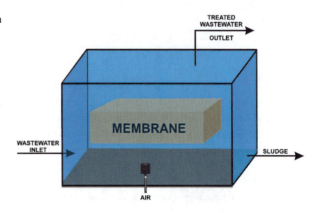

Fig. 9.4 Basic configuration of a submerged membrane bioreactor

9.5.1.2 Side Stream Membrane Bioreactor (SSMBR)

Lateral flow membrane bioreactors remove the filtration membrane from the system and convert it to a secondary treatment. Both the substrate and the biomass are recirculated from the bioreactor to the membrane (and vice versa), increasing the process efficiency and the quality of the treated water (Fig. 9.5). In this typeof design, the maintenance of the filtration membranes is facilitated, but considering that an extra step is added to the treatment process, the operational costs, labor, and complexity of this type of technology will increase (Lim et al. 2021). Another advantage of this type of system, when compared to SMBRs, is that biofouling can be avoided by increasing the feed rate of the substrate from the bioreactor and toward the membrane (inlet flow), using the hydrodynamics of the solution to detach the biomass or other contaminating particles that are occluding the superficial or internal pores of membrane (Martínez et al. 2021).

9.5.2 Recent Advances in MBR

In recent years, research related to MBR has focused on modifying the filtration membranes' manufacturing materials to prevent biofouling, one of the main limitations of scaling and implementing this type of system (Sun et al. 2019). The performance of a filtration membrane coated with a nanofibrous cellulose composite for the treatment of domestic wastewater has been reported, resulting in a great capacity of these materials to prevent biofouling, associating this phenomenon with the high hydrophilicity of these composites, considering these results as promising for the scaling of this type of membranes (Lotfikatouli et al. 2021). Modifications have been made to the surface of a polyvinylidene fluoride (PVDF) ultrafiltration membrane by applying coatings of tannic acid (TA) and copper (Cu^{2+}) in different proportions, evaluating the ability of these coatings to prevent biofouling of the membranes; the modified membranes presented greater permeability and ease of cleaning than their unmodified counterpart, but they highlight the need to continue carrying out complementary studies (Maneewan et al. 2021). Polysulfone

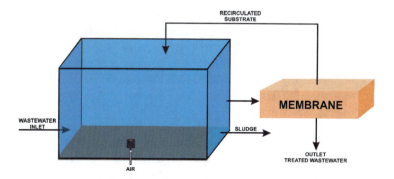

Fig. 9.5 Base configuration of a lateral flow membrane bioreactor

Table 9.4 Comparison of membrane bioreactors and their most notable advances by component

Type of MBR	Filtration membrane	Inoculum	Substrate	References
SMBR	Zinc oxide and sodium alginate on a polyacrylonitrile	Activated sludge	Urban wastewater sludge	Sokhandan et al. (2020)
EMBR	Conductive copper hollow fiber	Anoxic digest sludge	Acetate solution	Liu et al. (2018)
MaMBR	Fe_3O_4-coated ZrO_2/PAN nanocomposite	Activated sludge	Urban wastewater	Noormohamadi et al. (2020)
MBR-like system	Electrospun cellulose acetate PVDF	–	Synthetic wastewater	Bakhsha et al. (2021)
exMBR	AgNPs on a zeolitic imidazolate framework-8	–	Phenol and NaCl solution	Wang et al. (2022)
mcs	Vanillin/CPE	Activated sludge	Activated sludge	Leda et al. (2018)
SSMBR	Lytic phage therapy on a modified SGO-PVDF	–	Municipal wastewater	Ayyaru et al. (2018)
MF-MBR	AgNPs on a sterapore polyethylene hollow fibers	Activated sludge	Glucose, NH_4Cl and KH_2PO_4 solution _	Le et al. (2019)
MBR-like	Quaternary ammonium compounds loaded on silica nanopollens on a PVDF base	*E. coli*	Simulated domestic wastewater	Zhai et al. (2020)
MBR-like	PES modified membrane with semiconductor diode laser	Activated sludge	BSA solution	Polat et al. (2020)
MBR-like	Nanocurcumin/PES	*E. coli*	Nutrient agar medium	Kumar and Arthanareeswaran (2019)

ultrafiltration membranes with metal organic frameworks (MOF) were used to evaluate their ability to treat wastewater from a dairy industry that generated cheese whey, obtaining positive results not only in the reduction of organic load but also in increasing the feeding flow of this substrate, improving the efficiency of the treatment (Bazrafshan et al. 2021). A combination of Ag-MOF was immobilized on the surface of a polydimethylsiloxane (PDMS) nanocomposite to mitigate biofouling in the membrane, resulting in a great anti-biofouling capacity of the synthesized materials, attributing this property to the slow release of Ag^+ ions, counting this type of ions with a high antimicrobial capacity (Yuan et al. 2022). Hydrophilic nylon fiber membranes modified with polyvinyl alcohol (PVA) have been used for the long-term operation of an ANAMMOX membrane bioreactor, verifying the ability of this type of material to reduce the biofouling generated in this type of bioreactor (Ni et al. 2021). More examples of the above are presented in Table 9.4.

9.6 Constructed Wetlands

Constructed wetlands are engineered systems designed and built to use natural processes involving vegetation, soil, and associated microorganisms, allowing wastewater treatment (Vymazal 2014). The UN has reported that wetlands have a high potential to address water problems and offer multiple advantages that lead to sustainable development (Kataki et al. 2021). Constructed wetlands are designed to take advantage of natural wetlands but with a more controlled environment; most research done on constructed wetlands has been designed to treat municipal and domestic wastewater, but lately, constructed wetlands have been employed in the treatment of other types of wastewater from different industries and slaughterhouses (Navarro-Frómeta et al. 2020; Pérez et al. 2022). Under the climate change scenario, constructed wetlands will face challenges, such as increased concentration of pathogens in wastewater due to increased global temperature; increased precipitation, which will cause an increase in water runoffs containing pathogens; and finally the reuse of treated wastewater, related to water scarcity (López et al. 2019; Vymazal 2014). These systems also function as water treatment plants, creation of new habitats, recreational and educational facilities, and ecological art areas (Stefanakis 2020). Constructed wetlands can remove various pollutants (organic, nutrients, microelements, among others) through a series of processes, which improve water quality (Pérez et al. 2022) and have been applied for the secondary and tertiary treatment of urban and industrial wastewater (Marín-Muñiz 2017). The main constructed wetlands used, both at an applied level and in research, are free surface flow, horizontal subsurface flow, and vertical subsurface flow, based on the path of water flow through the system (Asprilla et al. 2020; López et al. 2019; Stefanakis et al. 2014). Free surface flow systems design includes a 10–50 cm water column over a substrate layer (mainly soil); subsurface flow-constructed wetlands are gravel beds (Asprilla et al. 2020; Stefanakis 2020; Vymazal 2014). According to the vegetation type, a classification includes constructed wetlands of emergent macrophytes and submerged macrophytes, respectively, (López et al. 2019; Stefanakis 2020); common systems are those with rooted emergent macrophytes (López et al. 2019). When more than one type of constructed wetland is used at the same facility, it is known as a hybrid wetland system (López et al. 2019; Asprilla et al. 2020).

9.6.1 Free Water Surface-Constructed Wetland

A free water surface constructed wetland (FWSW) consists of a series of flooded channels whose objective is to mimic the natural processes of a natural wetland, marsh, or wetland (Fig. 9.6) (Asprilla et al. 2020). By flowing gently through the wetland, the particles settle, the pathogens are destroyed, and the organisms and plants use the nutrients, allowing the water to flow over the ground and be exposed to the atmosphere and sun radiation (Marín-Muñiz 2017). The channel or dam is lined with an impermeable barrier (clay or geotextile) covered with stones, gravel, and

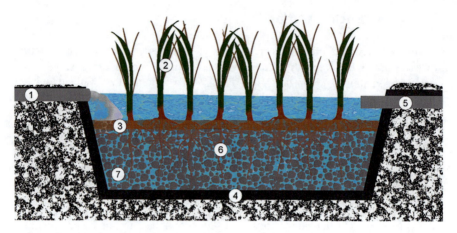

Fig. 9.6 Free water surface constructed wetland. (1) Inlet, (2) plants (macrophytes), (3) sludge, (4) liner, (5) rhizome network, (6) outlet, (7) gravel bed

earth and planted with vegetation from the region. The wetland is flooded with wastewater to a depth of 10–45 cm above ground level; wastewater flows gently through the wetland and goes through simultaneous physical, chemical, and biological processes; solids are filtered, organic matter is degraded, and nutrients are removed (Asprilla et al. 2020). Once in the wetland, the heavier particles settle out, thus removing the nutrients attached to them; plants and the communities of microorganisms (in stems and roots) take in nutrients such as nitrogen and phosphorus; chemical reactions can cause other elements to precipitate; pathogens are removed from water by natural decomposition, predation by higher organisms, sedimentation, and ultraviolet radiation (Asprilla et al. 2020; Marín-Muñiz 2017; Stefanakis 2020). Although the underwater layer of the wetland is anaerobic, the roots of the plants release oxygen in the area surrounding the root hairs, creating an environment conducive to complex chemical and biological activities (Marín-Muñiz 2017). The efficiency of the FWSW also depends on the good distribution of water at the inlet; wastewater can enter the wetland using dams or drilling holes in a distribution pipe to allow it to enter at regular intervals (Asprilla et al. 2020; Marín-Muñiz 2017). The advantages of using this type of wetland are: it is aesthetic; it performs a high reduction of the BOD, a moderate removal of pathogens, the construction, and repair with local materials; it does not require electricity; and it does not generate bad smells. Among the disadvantages are the high reproduction of mosquitoes, the startup of the wetland slowness, the requirement of a large area of land, the monitoring of an expert, and the moderate investment in land and materials (Marín-Muñiz 2017). Plant species commonly used in this type of wetland are *Phragmites australis*, *Typha* spp., *Scirpus* spp., and *Juncus* spp. (Stefanakis et al. 2014; Vymazal 2013). Other macrophytes have also been evaluated in this type of wetland by comparing the behavior of an FWSW and a horizontal subsurface flow (HSFW) wetland for wastewater treatment with other macrophytes *(Eichornia crassipes, Typha domingensis, Paspalum paniculatum, Cyperus articulatus)* and two HRTs to determine the removal of pollutants from the wastewater from the bathrooms and dining hall of a university campus; it was observed that the FWS wetland with *Typha domingensis* had the highest removal efficiency in terms of to turbidity, color, COD, BOD$_5$, TN, TP, and total suspended solids (TSS) (Solís-Silvan et al. 2016). Combinations of different macrophytes have also been made when evaluating the performance of an FWSW to restore the Jialu River in China, using a combination of emergent plants + floating plants + submerged plants for 2 years in 1 area of 7400 m^2 and a hydraulic loading rate (HLR) of 14 cm/day with a high viscosity silt loam substrate, finding that the average removal rate of total nitrogen, total phosphorus, ammoniacal nitrogen (NH$_4^+$-N), COD, and TSS was 75%, 78%, 85%, 40%, and 80%, respectively, in summer and autumn, and in winter, they were 30%, 73%, 45%, 25%, and 78%, respectively. The system demonstrated stability and great potential to improve water quality, in addition to being easy to maintain during its operation, which is why it was considered an effective treatment option to restore a polluted river and to be feasible to apply in a generalized way and any place (Wang et al. 2012).

9.6.2 Horizontal Subsurface Flow Constructed Wetland

The horizontal subsurface flow constructed wetland (HSFW) consists of a large concavity filled with gravel and sand planted with vegetation, where wastewater flows horizontally (Fig. 9.7). The advantages of this system include a high reduction of BOD, suspended solids, and pathogens, no mosquito problems that occur in FWSW, no electricity required, and low operating costs (Cisterna Osorio and Pérez Bustamante 2019; Zurita-Martínez et al. 2011). Among the disadvantages of this type of wetlands are that they require a large area of land, there is little nutrient removal, there are risks of obstructions, the initial adaptation period is long, and

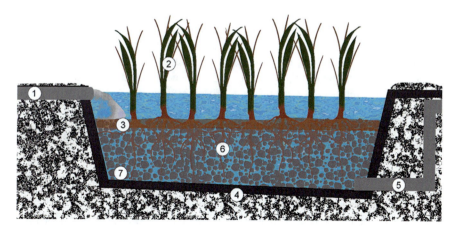

Fig. 9.7 Horizontal subsurface flow constructed wetland. (1) Inlet, (2) plants (macrophytes), (3) sludge, (4) liner, (5) rhizome network, (6) outlet, (7) gravel bed

experience in design and construction is required (Cisterna Osorio and Pérez Bustamante 2019; Vymazal 2014). In the HSFW, the circulation of water is carried out through a granular medium, with a depth that depends on the size and length of the roots of the plants (Asprilla et al. 2020; Vymazal 2014). The vegetation transfers a small amount of oxygen to the root, so aerobic bacteria can colonize the area and degrade organic materials (Cisterna Osorio and Pérez Bustamante 2019). The design of the HSFW depends on the treatment objective and on the quality and quantity of the effluent, including decisions on the number of parallel flow paths and compartmentalization as well as the efficiency of wetland removal, which depends on the surface area (length by width), while the cross-sectional area (width times depth) determines the maximum possible flow. Generally, an area of 5–10 m^2 per person is required; the bed should be lined with an impermeable liner (clay or geotextile) to prevent leaching, and it should be wide and shallow to maximize the flow path of water in contact with the roots of the vegetation, and a wide area should be used for entry, in order to distribute the flow evenly and avoid clogging, and the outlet must be variable so that the water surface can be adjusted and thus optimize the treatment performance (Asprilla et al. 2020). *Phragmites australis* was used in an HSFW fed with different HLR for domestic wastewater treatment; authors determined that the removal rates of the system were closely dependent on the applied HLR levels, and the highest removal rates detected were 64.9%, 62.5%, 86.3%, and 80.34% for BOD_5, COD, TSS, and oils and fats, respectively, with an HLR of 0.050 m^3 day^{-1} m^{-2}, and above this load, the efficiencies decreased (Çakir et al. 2015). Most antibiotics are not efficiently removed in conventional wastewater treatment plants, making implementation efficient and low-cost treatment alternatives necessary. *Phragmites australis* has also been used in HSFW to remove different concentrations of the antibiotic florfenicol (10, 15, 20, and 25 mg/L of florfenicol) at an HTR of 4.2 days, detecting a maximum removal of 77.9% of florfenicol and 85.2% of COD during the first days of exposure, and the results showed that florfenicol was not retained in the granular material of the wetlands, nor the macrophytes and that the removal mechanism was by biological degradation (Rodriguez et al. 2021). Four HSFWs planted with *Cyperus alternifolius* bioactivated with efficient microorganisms for treating slaughterhouse wastewater to assess the effect of HRT, microorganisms activation, and the type of substrate (combinations of brown soil, limestone gravel, zeolite, and river sand) on the efficiency in the removal of contaminants. As a biostimulant element of the autochthonous microbial flora in wetlands, the commercial bioproduct ME-50 (LABIOFAM) was used, consisting of a mixture of autochthonous microorganisms where filamentous fungi and yeasts, *Lactobacillus*, photosynthetic bacteria, and actinomycetes predominate. The commercial bioproduct is described as a brown aqueous suspension with a slight odor of wine or ferment and acidic pH (3.2–3.8). The maximum removal efficiency was achieved in bioactivated systems at 96 h, with a removal of 96% for COD, 96.4% for BOD_5, and 94% for fats and oils, so HSFW can be considered a method efficient for the treatment of wastewater generated in slaughterhouses, achieving high levels of pollutants removal (Esperanza-Pérez et al. 2022). The HSFW has been used as a tertiary treatment of wastewater from a dairy industry on a pilot scale using *Typha domingensis* and river gravel as a substrate, and 32 influent and effluent samples were collected over 7 months, finding significant differences in the removal of suspended solids (SS), NTK, NO_3^-, and TP of 78.4%, 25.7%, 47.8%, and 29.9%, respectively; the concentration of fecal coliform bacteria (FC) decreased in the order of magnitude, and the total biomass increased 4.6 times, recommending the use of this type of wetland for tertiary treatment of wastewater from dairy industries (Schierano et al. 2020).

9.6.3 Vertical Subsurface Flow Constructed Wetland

The vertical subsurface flow constructed wetland (VSFW) is a wastewater and sludge treatment process that is a relatively recent technology (Fig. 9.8) and is still under development. Still, it offers economic, environmental, and social advantages (Stefanakis 2020). The VSFW achieves a high oxygen transfer rate, unlike the horizontal ones, where the nitrification of the wastewater is difficult due to the limitation in the availability of oxygen (Almuktar et al. 2018; Pérez et al. 2022; Stefanakis 2020). This type of wetland allows relatively high volumes of water to be treated per square meter; it is beneficial for the agricultural sector since it is highly efficient in terms of treatment against different pollutants from wastewater, in reducing COD, BOD, and solid particles in wastewater (Almuktar et al. 2018; Pérez et al. 2022). However, they are not very efficient in removing phosphorus due to the insufficient interaction between the wastewater and the system media. Several studies have shown that HSFWs perform well in nitrification, while others indicate their inadequacies in denitrification (Almuktar et al. 2018). This operation can be improved by inserting aeration pipes to favor nitrification and removal of organic matter, compared to HSFW (Pérez et al. 2022; Stefanakis 2020). The VSFW has been used for the removal of ammonium when treating sewage with two

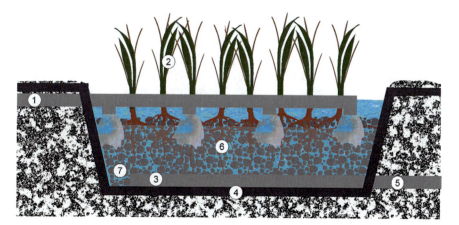

Fig. 9.8 Vertical subsurface flow constructed wetland. (1) Inlet, (2) plants (macrophytes), (3) drenage, (4) liner, (5) rhizome network, (6) outlet, (7) gravel bed

different substrates (gravel and a mixture of gravel and zeolite), finding that the HLR used (3 m day^{-1}) and zeolite produced the removal of a 96.69% of NH_4^+-N through a change in structure and microbial gene expression, promoting nitrification (Zhang et al. 2021). *Typha domingensis* was used to remove ammonium and other contaminants in domestic wastewater in a VSFW to further reuse the treated effluent in a hydroponic culture of *Lactuca sativa*. The wetland was operated for 3 months at a HLR of 83 g COD/m^3 day. The average removal percentages for each parameter were 91%, 64%, 89%, 81%, and 88% for COD, TKN, orthophosphates, NH_4^+, and FC, respectively. The results of the reuse of the treated effluent as a hydroponic culture medium indicate that the effluent from the system allowed the growth of *Lactuca sativa* but in a lower proportion compared to the growth obtained in the conventional hydroponic culture medium (Arias et al. 2021). *Phragmites australis* and *Typha latifolia* were tested in a VSFW as a tertiary treatment for removing emerging contaminants (caffeine, galaxolide, tonalide, parsol MCX, sunscreen UV-15, naproxen, methyl dihydro jasmonate, *alkylphenols, triclosan, alkylphenol mono-*, and diethoxylates); the results showed a high removal of caffeine, which is a compound with high biodegradability, and tonalide, triclosan, sunscreen, and naproxen also presented a high removal, being slightly higher with *Typha latifolia* (Navarro-Frómeta et al. 2020).

9.7 Conclusion

Various biological processes derived from nature and addressed in this work provide an overview of the recent advances in wastewater treatment. Nature can remove pollutants from water, but anthropogenic activities have exceeded their limits by generating enormous amounts of wastewater with high concentrations of biodegradable pollutants, wastewater with recalcitrant, and highly toxic or emerging pollutants. The biological process(es) necessary to remove the pollutants present in the wastewater must be established accordingly to the type of pollutants, the available area for wastewater facilities implementation, the economic resources, the technical support, and the by-products that can be generated (methane, nitrogen, electricity, biomass, etc.), during the process(es). A wide variety of processes, reactors, microorganisms, and plants can be combined to have a good removal percentage. In addition to the factors above, the climate, the landscape, the proportions such as C/N, and the need to meet a certain quality for reuse or discharge of treated water will be very important also.

Advanced processes offer more advantages than conventional processes in terms of saving space, time, and money; this goes hand to hand with the discovery of reactions and metabolic routes of microorganisms and plants since, with the understanding of these, the optimal operating conditions of the reactors will be proposed.

References

Ahmad HA, Ni SQ, Ahmad S, Zhang J, Ali M, Ngo HH, Guo W, Tan Z, Wang Q (2020) Gel immobilization: a strategy to improve the performance of anaerobic ammonium oxidation (anammox) bacteria for nitrogen-rich wastewater treatment. Bioresour Technol 313:123642

Aiken DC, Curtis TP, Heidrich ES (2019) Avenues to the financial viability of microbial electrolysis cells [MEC] for domestic wastewater treatment and hydrogen production. Int J Hydrog Energy 44(5):2426–2434

Akgul D, Aktan CK, Yapsakli K, Mertoglu B (2013) Treatment of landfill leachate using UASB-MBR-SHARON–Anammox configuration. Biodegradation 24(3):399–412

Al-Asheh S, Bagheri M, Aidan A (2021) Membrane bioreactor for wastewater treatment: a review. Case Stud Chem Environ Eng 4:100109

Al-Mamun A, Ahmad W, Baawain MS, Khadem M, Dhar BR (2018) A review of microbial desalination cell technology: configurations, optimization and applications. J Clean Prod 183: 458–480

Almuktar SA, Abed SN, Scholz M (2018) Wetlands for wastewater treatment and subsequent recycling of treated effluent: a review. Environ Sci Pollut Res 25(24):23595–23623

Arias GAT, Duré G, Decoud LV, Arias TRL (2021) Reutilización del efluente de un humedal construido de flujo subsuperficial vertical en un cultivo hidropónico tipo NFT de Lactuca sativa. Rev Soc Cient Parag 26(1):35–48

Aryal N, Wan L, Overgaard MH, Stoot AC, Chen Y, Tremblay PL, Zhang T (2019) Increased carbon dioxide reduction to acetate in a microbial electrosynthesis reactor with a reduced graphene oxide-coated copper foam composite cathode. Bioelectrochemistry 128:83–93

Aslam M, Ahmad R, Kim J (2018) Recent developments in biofouling control in membrane bioreactors for domestic wastewater treatment. Sep Purif Technol 206:297–315

Asprilla WJ, Ramírez JS, Rodriguez DC (2020) Humedales artificiales subsuperficiales: comparación de metodologías de diseño para el cálculo del área superficial basado en la remoción de la materia orgánica. Revista Ingenierías USBMed 11(1):65–73

Augusto MR, Camiloti PR, de Souza TSO (2018) Fast start-up of the single-stage nitrogen removal using anammox and partial nitritation (SNAP) from conventional activated sludge in a membrane-aerated biofilm reactor. Bioresour Technol 266:151–157

Ayyaru S, Choi J, Ahn YH (2018) Biofouling reduction in a MBR by the application of a lytic phage on a modified nanocomposite membrane. Environm Sci Water Res Technol 4(10): 1624–1638

Bagchi S, Behera M (2020) Assessment of heavy metal removal in different bioelectrochemical systems: a review. J Hazard Toxic Radioactive Waste 24(3):04020010

Bagchi S, Biswas R, Nandy T (2012) Autotrophic ammonia removal processes: ecology to technology. Crit Rev Environ Sci Technol 42(13):1353–1418

Bakhsha N, Ahmeda Z, Mahara RB, Khatrib Z (2021) Development and application of electrospun modified polyvinylidene fluoride (PVDF) nanofibers membrane for biofouling control in membrane bioreactor. Desalin Water Treat 217:74–82

Bakonyi P, Koók L, Kumar G, Tóth G, Rózsenberszki T, Nguyen DD, Chang SW, Zhen G, Bélafi-Bakó K, Nemestóthy N (2018) Architectural engineering of bioelectrochemical systems from the perspective of polymeric membrane separators: a comprehensive update on recent progress and future prospects. J Membr Sci 564:508–522

Bargieł P, Zabochnicka-Świątek M (2018) Technologies of coke wastewater treatment in the frame of legislation in force. Environ Protect Nat Resour 29(1):11–15

Bassin JP (2018) New processes for biological nitrogen removal. In: Advanced biological processes for wastewater treatment. Springer, Cham, pp 143–203

Bayo J, López-Castellanos J, Olmos S (2020) Membrane bioreactor and rapid sand filtration for the removal of microplastics in an urban wastewater treatment plant. Mar Pollut Bull 156:111211

Bazrafshan N, Firouzjaei MD, Elliott M, Moradkhani A, Rahimpour A (2021) Preparation and modification of low-fouling ultrafiltration membranes for cheese whey treatment by membrane bioreactor. Case Stud Chem Environ Eng 4:100137

Berends DHJG, Salem S, Van der Roest HF, Van Loosdrecht MCM (2005) Boosting nitrification with the BABE technology. Water Sci Technol 52(4):63–70

Bhargavi G, Venu V, Renganathan S (2018) Microbial fuel cells: recent developments in design and materials. IOP Conf Ser Mater Sci Eng 330(1):012034

Buha DM, Atalia KR, Baagwala WY, Shah NK (2015) Review on wastewater treatment technologies for nitrogen removal. Global J Multidiscip Stud 4(6):299–318

Cai T, Huang M, Huang Y, Zheng W (2019) Enhanced performance of microbial fuel cells by electrospinning carbon nanofibers hybrid carbon nanotubes composite anode. Int J Hydrog Energy 44(5):3088–3098

Cai F, Lei L, Li Y (2020) Rapid start-up of single-stage nitrogen removal using anammox and partial nitritation (SNAP) process in a sequencing batch biofilm reactor (SBBR) inoculated with conventional activated sludge. Int Biodeterior Biodegradation 147:104877

Çakir R, Gidirislioglu A, Çebi U (2015) A study on the effects of different hydraulic loading rates (HLR) on pollutant removal efficiency of subsurface horizontal-flow constructed wetlands used for treatment of domestic wastewaters. J Environ Manag 164:121–128

Camedda C, Hoelzle RD, Carucci A, Milia S, Virdis B (2019) A facile method to enhance the performance of soil bioelectrochemical systems using in situ reduced graphene oxide. Electrochim Acta 324:134881

Cárdenas-Calvachi GL, Sánchez-Ortiz IA (2013) Nitrógeno en aguas residuales: orígenes, efectos y mecanismos de remoción para preservar el ambiente y la salud pública. Universidad y Salud 15(1):72–88

Carrillo-Peña D, Escapa A, Hijosa-Valsero M, Paniagua-García AI, Díez-Antolínez R, Mateos R (2022) Bioelectrochemical enhancement of methane production from exhausted vine shoot fermentation broth by integration of MEC with anaerobic digestion. Biomass Convers Biorefinery: 1–10

Cecconet D, Sabba F, Devecseri M, Callegari A, Capodaglio AG (2020) In situ groundwater remediation with bioelectrochemical systems: a critical review and future perspectives. Environ Int 137:105550

Cervantes-Carrillo F, Pérez J, Gomez J (2000) Avances en la eliminación biológica del nitrógeno de las aguas residuales. Rev Latinoam Microbiol 42(2):73–82

Chakraborty I, Das S, Dubey BK, Ghangrekar MM (2020) Novel low cost proton exchange membrane made from sulphonated biochar for application in microbial fuel cells. Mater Chem Phys 239:122025

Chen J, Xu W, Wu X, Jiaqiang E, Lu N, Wang T, Zuo H (2019) System development and environmental performance analysis of a pilot scale microbial electrolysis cell for hydrogen production using urban wastewater. Energy Convers Manag 193:52–63

Chen X, Li Y, Yuan X, Li N, He W, Liu J (2020) Synergistic effect between poly (diallyldimethylammonium chloride) and reduced graphene oxide for high electrochemically active biofilm in microbial fuel cell. Electrochim Acta 359:136949

Chen P, Guo X, Li S, Li F (2021) A review of the bioelectrochemical system as an emerging versatile technology for reduction of antibiotic resistance genes. Environ Int 156:106689

Chiranjeevi P, Yeruva DK, Kumar AK, Mohan SV, Varjani S (2019) Plant-microbial fuel cell technology. In: Microbial electrochemical technology. Elsevier, Amsterdam, pp 549–564

Choi MJ, Yang E, Yu HW, Kim IS, Oh SE, Chae KJ (2019) Transition metal/carbon nanoparticle composite catalysts as platinum substitutes for bioelectrochemical hydrogen production using microbial electrolysis cells. Int J Hydrog Energy 44(4):2258–2265

Christensen ML, Keiding K, Nielsen PH, Jørgensen MK (2015) Dewatering in biological wastewater treatment: a review. Water Res 82:14–24

Cisterna Osorio PE, Pérez Bustamante L (2019) Propuesta de humedales artificiales, impulsores de biodiversidad, que depuran aguas contaminadas para la recuperación de lagunas urbanas de concepción. Revista Hábitat Sustentable 9(1):20–31

Claros J, Serralta J, Seco A, Ferrer J, Aguado D (2012) Real-time control strategy for nitrogen removal via nitrite in a SHARON reactor using pH and ORP sensors. Process Biochem 47(10): 1510–1515

Cui D, Yang LM, Liu WZ, Cui MH, Cai WW, Wang AJ (2018) Facile fabrication of carbon brush with reduced graphene oxide (rGO) for decreasing resistance and accelerating pollutants removal in bio-electrochemical systems. J Hazard Mater 354:244–249

Das I, Das S, Dixit R, Ghangrekar MM (2020) Goethite supplemented natural clay ceramic as an alternative proton exchange membrane and its application in microbial fuel cell. Ionics 26(6): 3061–3072

Daud SM, Daud WRW, Kim BH, Somalu MR, Bakar MHA, Muchtar A, Jahim JM, Lim SS, Chang IS (2018) Comparison of performance and ionic concentration gradient of two-chamber microbial fuel cell using ceramic membrane (CM) and cation exchange membrane (CEM) as separators. Electrochim Acta 259:365–376

Daverey A, Hung NT, Dutta K, Lin JG (2013) Ambient temperature SNAD process treating anaerobic digester liquor of swine wastewater. Bioresour Technol 141:191–198

Du WL, Huang Q, Miao LL, Liu Y, Liu ZP (2016) Association of running manner with bacterial community dynamics in a partial short-term nitrifying bioreactor for treatment of piggery wastewater with high ammonia content. AMB Express 6(1):1–16

Esperanza-Pérez G, Jiménez-Prieto Y, Manrique-Rasúa MC (2022) Humedales construidos bioactivados con microorganismos eficientes: una vía para la gestión de residuales en el matadero Chichí Padrón. Centro Azúcar 49(3):57–68

Femina-Carolin C, Senthil-Kumar P, Janet-Joshiba G, Vinoth-Kumar V (2021) Analysis and removal of pharmaceutical residues from wastewater using membrane bioreactors: a review. Environ Chem Lett 19(1):329–343

Figueroa M, Vázquez-Padín JR, Mosquera-Corral A, Campos JL, Méndez R (2012) Is the CANON reactor an alternative for nitrogen removal from pre-treated swine slurry? Biochem Eng J 65:23–29

Fischer F, Sugnaux M, Savy C, Hugenin G (2018) Microbial fuel cell stack power to lithium battery stack: pilot concept for scale up. Appl Energy 230:1633–1644

Flimban SG, Ismail IM, Kim T, Oh SE (2019) Overview of recent advancements in the microbial fuel cell from fundamentals to applications: design, major elements, and scalability. Energies 12(17):3390

Fu Q, He Y, Li Z, Li J, Zhang L, Zhu X, Liao Q (2022) Direct CO2 delivery with hollow stainless steel/graphene foam electrode for enhanced methane production in microbial electrosynthesis. Energy Convers Manag 268:116018

Gajda I, Greenman J, Ieropoulos IA (2018) Recent advancements in real-world microbial fuel cell applications. Curr Opin Electrochem 11:78–83

Gatti MN, Giménez JB, Carretero L, Ruano MV, Borrás L, Serralta J, Seco A (2015) Enrichment of AOB and NOB population by applying a BABE reactor in an activated sludge pilot plant. Water Environ Res 87(4):369–377

Goel S (2018) From waste to watts in micro-devices: review on development of membraned and membraneless microfluidic microbial fuel cell. Appl Mater Today 11:270–279

González-Martínez A, Calderón K, Albuquerque A, Hontoria E, González-López J, Guisado IM, Osorio F (2013) Biological and technical study of a partial-SHARON reactor at laboratory scale: effect of hydraulic retention time. Bioprocess Biosyst Eng 36(2):173–184

Goswami L, Kumar RV, Borah SN, Manikandan NA, Pakshirajan K, Pugazhenthi G (2018) Membrane bioreactor and integrated membrane bioreactor systems for micropollutant removal from wastewater: a review. J Water Process Eng 26:314–328

Gul MM, Ahmad KS (2019) Bioelectrochemical systems: sustainable bio-energy powerhouses. Biosens Bioelectron 142:111576

Heidrich ES, Dolfing J, Wade MJ, Sloan WT, Quince C, Curtis TP (2018) Temperature, inocula and substrate: contrasting electroactive consortia, diversity and performance in microbial fuel cells. Bioelectrochemistry 119:43–50

Hien NN, Van Tuan D, Nhat PT, Van TTT, Van Tam N, Que VNX, Dan NP (2017) Application of oxygen limited autotrophic nitritation/denitrification (OLAND) for anaerobic latex processing wastewater treatment. Int Biodeterior Biodegradation 124:45–55

Hu J, Zeng C, Liu G, Lu Y, Zhang R, Luo H (2019) Enhanced sulfate reduction accompanied with electrically-conductive pili production in graphene oxide modified biocathodes. Bioresour Technol 282:425–432

Hua T, Li S, Li F, Zhou Q, Ondon BS (2019) Microbial electrolysis cell as an emerging versatile technology: a review on its potential application, advance and challenge. J Chem Technol Biotechnol 94(6):1697–1711

Huang Q, Du WL, Miao LL, Liu Y, Liu ZP (2018) Microbial community dynamics in an ANAMMOX reactor for piggery wastewater treatment with startup, raising nitrogen load, and stable performance. AMB Express 8(1):1–15

Iftimie S, Dumitru A (2019) Enhancing the performance of microbial fuel cells (MFCs) with nitrophenyl modified carbon nanotubes-based anodes. Appl Surf Sci 492:661–668

Izadi P, Izadi P, Eldyasti A (2021) Towards mainstream deammonification: comprehensive review on potential mainstream applications and developed sidestream technologies. J Environ Manag 279:111615

Jayabalan T, Matheswaran M, Preethi V, Mohamed SN (2020) Enhancing biohydrogen production from sugar industry wastewater using metal oxide/graphene nanocomposite catalysts in microbial electrolysis cell. Int J Hydrog Energy 45(13):7647–7655

Kabutey FT, Zhao Q, Wei L, Ding J, Antwi P, Quashie FK, Wang W (2019) An overview of plant microbial fuel cells (PMFCs): configurations and applications. Renew Sust Energ Rev 110:402–414

Kalyuzhnyi S, Gladchenko M, Mulder A, Versprille B (2006) DEAMOX—new biological nitrogen removal process based on anaerobic ammonia oxidation coupled to sulphide-driven conversion of nitrate into nitrite. Water Res 40(19):3637–3645

Kataki S, Chatterjee S, Vairale MG, Dwivedi SK, Gupta DK (2021) Constructed wetland, an eco-technology for wastewater treatment: a review on types of wastewater treated and components of the technology (macrophyte, biolfilm and substrate). J Environ Manag 283: 111986

Katuri KP, Ali M, Saikaly PE (2019) The role of microbial electrolysis cell in urban wastewater treatment: integration options, challenges, and prospects. Curr Opin Biotechnol 57:101–110

Kaur R, Marwaha A, Chhabra VA, Kim KH, Tripathi SK (2020) Recent developments on functional nanomaterial-based electrodes for microbial fuel cells. Renew Sust Energ Rev 119: 109551

Kebaili H, Kameche M, Innocent C, Benayyad A, Kosimaningrum WE, Sahraoui T (2020) Scratching and transplanting of electro-active biofilm in fruit peeling leachate by ultrasound: re-inoculation in new microbial fuel cell for enhancement of bio-energy production and organic matter detection. Biotechnol Lett 42(6):965–978

Kugarajah V, Dharmalingam S (2020) Investigation of a cation exchange membrane comprising sulphonated poly ether ether ketone and sulphonated titanium nanotubes in microbial fuel cell and preliminary insights on microbial adhesion. Chem Eng J 398:125558

Kumar RS, Arthanareeswaran G (2019) Nano-curcumin incorporated polyethersulfone membranes for enhanced anti-biofouling in treatment of sewage plant effluent. Mater Sci Eng C 94:258–269

Lacroix A, Mentzer C, Pagilla KR (2020) Full-scale N removal from centrate using a sidestream process with a mainstream carbon source. Water Environ Res 92(11):1922–1934

Le HQ, Sowe A, Chen SS, Duong CC, Ray SS, Cao TND, Nguyen NC (2019) Exploring nanosilver-coated hollow fiber microfiltration to mitigate biofouling for high loading membrane bioreactor. Molecules 24(12):2345

Leda KT, Takada K, Jiang SC (2018) Surface modification of a microfiltration membrane for enhanced anti-biofouling capability in wastewater treatment process. J Water Process Eng 26: 55–61

Lee HS, Xin W, Katakojwala R, Mohan SV, Tabish NM (2022) Microbial electrolysis cells for the production of biohydrogen in dark fermentation-a review. Bioresour Technol 363:127934

Lescano MI, Gasnier A, Pedano ML, Sica MP, Pasquevich DM, Prados MB (2018) Development and characterisation of self-assembled graphene hydrogel-based anodes for bioelectrochemical systems. RSC Adv 8(47):26755–26763

Li X, Zhang J, Zhang X, Li J, Liu F, Chen Y (2019) Start-up and nitrogen removal performance of CANON and SNAD processes in a pilot-scale oxidation ditch reactor. Process Biochem 84: 134–142

Lim S, Choi PJ, An AK, Han DS, Phuntsho S, Shon H (2021) Submerged versus side-stream osmotic membrane bioreactors using an outer-selective hollow fiber osmotic membrane for desalination. Desalination 515:115196

Ling L, Yang C, Li Z, Luo H, Feng S, Zhao Y, Lu L (2022) Plant endophytic bacteria: a potential resource pool of electroactive micro-organisms. J Appl Microbiol 132(3):2054–2066

Liu D, Chen X, Bian B, Lai Z, Situ Y (2018) Dual-function conductive copper hollow fibers for microfiltration and anti-biofouling in electrochemical membrane bioreactors. Front Chem 6:445

Liu X, Yin W, Liu X, Zhao X (2020a) Enhanced Cr reduction and bioelectricity production in microbial fuel cells using polypyrrole-coated MnO_2 on carbon cloth. Environ Chem Lett 18(2): 517–525

Liu Y, Zhang X, Zhang Q, Li C (2020b) Microbial fuel cells: nanomaterials based on anode and their application. Energ Technol 8(9):2000206

Logan BE, Rossi R, Ragab A, Saikaly PE (2019) Electroactive microorganisms in bioelectrochemical systems. Nat Rev Microbiol 17(5):307–319

López D, Leiva AM, Arismendi W, Vidal G (2019) Influence of design and operational parameters on the pathogens reduction in constructed wetland under the climate change scenario. Rev Environ Sci Biotechnol 18(1):101–125

Lotfikatouli S, Hadi P, Yang M, Walker HW, Hsiao BS, Gobler C, Reichel M, Mao X (2021) Enhanced anti-fouling performance in membrane bioreactors using a novel cellulose nanofiber-coated membrane. Sep Purif Technol 275:119145

Ma Z, Song H, Wang H, Xu P (2017) Improving the performance of microbial fuel cells by reducing the inherent resistivity of carbon fiber brush anodes. J Power Sources 348:193–200

Maddalwar S, Nayak KK, Kumar M, Singh L (2021) Plant microbial fuel cell: opportunities, challenges, and prospects. Bioresour Technol 341:125772

Maneewan P, Sajomsang W, Singto S, Lohwacharin J, Suwannasilp BB (2021) Fouling mitigation in an anaerobic membrane bioreactor via membrane surface modification with tannic acid and copper. Environ Pollut 291:118205

Marín-Muñiz JL (2017) Humedales construidos en México para el tratamiento de aguas residuales, producción de plantas ornamentales y reúso del agua. Agro Productividad 10(5):90–95

Martínez R, Ruiz MO, Ramos C, Cámara JM, Diez V (2021) Fouling control of submerged and side-stream membrane bioreactors based on the statistical analysis of mid-term assays. J Clean Prod 326:129336

Masłoń A, Tomaszek JA (2009) Anaerobic ammonium nitrogen oxidation in DEAMOX process. Environ Prot Eng 35(2):123–130

Massaglia G, Margaria V, Sacco A, Castellino M, Chiodoni A, Pirri FC, Quaglio M (2019) N-doped carbon nanofibers as catalyst layer at cathode in single chamber microbial fuel cells. Int J Hydrog Energy 44(9):4442–4449

McCarty PL (2018) What is the best biological process for nitrogen removal: when and why? Environ Sci Technol 52:3835–3841

Milia S, Perra M, Cappai G, Carucci A (2015a) SHARON process as preliminary treatment of refinery wastewater with high organic carbon-to-nitrogen ratio. Desalin Water Treat 57(38): 17935–17943

Milia S, Perra M, Muntoni A, Carucci A (2015b) Partial nitritation of nitrogen-rich refinery wastewater (sour water) with different Ci/N molar ratios. Desalin Water Treat 55(3):791–798

Miyoshi T, Nguyen TP, Tsumuraya T, Tanaka H, Morita T, Itokawa H, Hashimoto T (2018) Energy reduction of a submerged membrane bioreactor using a polytetrafluoroethylene (PTFE) hollow-fiber membrane. Front Environ Sci Eng 12(3):1–10

Mohamed HO, Talas SA, Sayed ET, Park SG, Eisa T, Abdelkareem MA, Fadali O, Chae K, Castaño P (2021) Enhancing power generation in microbial fuel cell using tungsten carbide on reduced graphene oxide as an efficient anode catalyst material. Energy 229:120702

Moradian JM, Yang FQ, Xu N, Wang JY, Wang JX, Sha C, Ali A, Yong YC (2022) Enhancement of bioelectricity and hydrogen production from xylose by a nanofiber polyaniline modified anode with yeast microbial fuel cell. Fuel 326:125056

Muszyński-Huhajło M, Zięba B, Janiak K, Miodoński S, Jurga A, Szetela R (2021) Can nitrifiers from the sidestream deammonification process be a remedy for the N-overload of the main-stream reactor? Sci Total Environ 790:148066

Nancharaiah YV, Sarvajith M, Mohan TK (2019) Aerobic granular sludge: the future of wastewater treatment. Curr Sci 117(3):395–404

Narayanasamy S, Jayaprakash J (2020) Application of carbon-polymer based composite electrodes for microbial fuel cells. Rev Environ Sci Biotechnol 19(3):595–620

Navarro-Frómeta AE, Beissos F, Marc-Bec J, Jaumejoan T (2020) Desempeño de humedales construidos de flujo vertical en el tratamiento de aguas residuales municipales. Revista Cubana de Química 32(3):365–377

Nazmkhah A, Oghyanous FA, Etemadi H, Yegani R (2022) Optimizing dose of coagulant and pH values for membrane fouling control in a submerged membrane bioreactor. J Chem Technol Biotechnol 97(10):2794–2804

Ni L, Shi Q, Wu M, Ma J, Wang Y (2021) Fouling behavior and mechanism of hydrophilic modified membrane in anammox membrane bioreactor: role of gel layer. J Membr Sci 620: 118988

Noormohamadi A, Homayoonfal M, Mehrnia MR, Davar F (2020) Employing magnetism of Fe3O4 and hydrophilicity of ZrO2 to mitigate biofouling in magnetic MBR by Fe3O4-coated ZrO2/PAN nanocomposite membrane. Environ Technol 41(20):2683–2704

Nsengiyuma O, Coothen Y, Ikumi D, Naidoo K (2021) The impact of sludge return liquors on South African wastewater treatment plants. Water SA 47(1):106–112

Ochoa-Hueso R (2017) El ciclo del nitrógeno y el hombre: De lo esencial a lo excesivo. Ecosistemas 26(1):1–3

Osorio-de la Rosa E, Vázquez Castillo J, Carmona Campos M, Barbosa Pool GR, Becerra Nuñez G, Castillo Atoche A, Ortegón Aguilar J (2019) Plant microbial fuel cells–based energy harvester system for self-powered IoT applications. Sensors 19(6):1378

Pareek A, Sravan JS, Mohan SV (2019) Fabrication of three-dimensional graphene anode for augmenting performance in microbial fuel cells. Carbon Resour Convers 2(2):134–140

Pell M, Wörman A (2009) Biological wastewater treatment systems. In: Jörgensen SE (ed) Ecosystem ecology. Elsevier, Amsterdam, pp 166–180

Pérez YA, García-Cortés DA, Jauregui-Haza UJ (2022) Humedales construidos como alternativa de tratamiento de aguas residuales en zonas urbanas: una revisión. Ecosistemas 31(1):2279

Pervez M, Balakrishnan M, Hasan SW, Choo KH, Zhao Y, Cai Y, Zarra T, Belgiorno V, Naddeo V (2020) A critical review on nanomaterials membrane bioreactor (NMs-MBR) for wastewater treatment. NPJ Clean Water 3(1):1–21

Pichardo-Romero D, Garcia-Arce ZP, Zavala-Ramírez A, Castro-Muñoz R (2020) Current advances in biofouling mitigation in membranes for water treatment: an overview. Processes 8(2):182

Polat B, Ozay Y, Bilici Z, Kucukkara İ, Dizge N (2020) Membrane modification with semiconductor diode laser to reduce membrane biofouling for external MBR system and modelling study. Sep Purif Technol 241:116747

Prakash O, Mungray A, Chongdar S, Kailasa SK, Mungray AK (2020) Performance of polypyrrole coated metal oxide composite electrodes for benthic microbial fuel cell (BMFC). J Environ Chem Eng 8(2):102757

Rajendran R, Dhakshina Moorthy GP, Krishnan H, Anappara S (2021) A study on polythiophene modified carbon cloth as anode in microbial fuel cell for lead removal. Arab J Sci Eng 46(7): 6695–6701

Rani G, Krishna K, Yogalakshmi KN (2021) Enhancing the electrochemical performance of Fe3O4 nanoparticles layered carbon electrodes in microbial electrolysis cell. J Environ Chem Eng 9(6): 106326

Regmi R, Nitisoravut R, Charoenroongtavee S, Yimkhaophong W, Phanthurat O (2018a) Earthen pot–plant microbial fuel cell powered by vetiver for bioelectricity production and wastewater treatment. CLEAN Soil Air Water 46(3):1700193

Regmi R, Nitisoravut R, Ketchaimongkol J (2018b) A decade of plant-assisted microbial fuel cells: looking back and moving forward. Biofuels 9(5):605–612

Rivalland C, Radouani F, Gonzalez-Rizzo S, Robert F, Salvin P (2022) Enrichment of Clostridia enhances Geobacter population and electron harvesting in a complex electroactive biofilm. Bioelectrochemistry 143:107954

Rodriguez DC, Cardona MA, Peñuela G (2021) Behavior of horizontal sub-surface flow wetlands as an alternative for the wastewater treatment contaminated with florfenicol. Ingeniería y Competitividad 23(1):9703

Rousseau R, Etcheverry L, Roubaud E, Basséguy R, Délia ML, Bergel A (2020) Microbial electrolysis cell (MEC): strengths, weaknesses and research needs from electrochemical engineering standpoint. Appl Energy 257:113938

Saha S, Gupta R, Sethi S, Biswas R (2022) Enhancing the efficiency of nitrogen removing bacterial population to a wide range of C: N ratio (1.5: 1 to 14: 1) for simultaneous C & N removal. Front Environ Sci Eng 16(8):1–15

Salehmin MNI, Lim SS, Satar I, Daud WRW (2021) Pushing microbial desalination cells towards field application: prevailing challenges, potential mitigation strategies, and future prospects. Sci Total Environ 759:143485

Sanabria J, Bedoya L, Sánchez J (2009) Proceso ANAMMOX una aplicación en ingeniería: Revisión general de los aspectos microbianos. Ingeniería de Recursos Naturales y del Ambiente 8:83–92

Sánchez J, Sanabria J (2009) Metabolismos microbianos involucrados en procesos avanzados para la remoción de Nitrógeno, una revisión prospectiva. Rev Colomb Biotecnol 11(1):114–124

Sanchez J, Sanabria J, Jetten M (2014) Faster autotrophic growth of anaerobic ammonium-oxidizing microorganisms in presence of nitrite, using inocula from Colombia. Rev Colomb Biotecnol 16:146–152

Sánchez-Peña P, Rodriguez J, Gabriel D, Baeza JA, Guisasola A, Baeza M (2022) Graphene functionalization with metallic Pt nanoparticles: a path to cost-efficient H2 production in microbial electrolysis cells. Int J Hydrog Energy 47(34):15397–15409

Sano T, Kawagoshi Y, Kokubo I, Ito H, Ishida K, Sato A (2022) Direct and indirect effects of membrane pore size on fouling development in a submerged membrane bioreactor with a symmetric chlorinated poly (vinyl chloride) flat-sheet membrane. J Environ Chem Eng 10(2): 107023

Sarma SJ, Tay JH (2018) Aerobic granulation for future wastewater treatment technology: challenges ahead. Environ Sci Water Res Technol 4(1):9–15

Sayed ET, Alawadhi H, Olabi AG, Jamal A, Almahdi MS, Khalid J, Abdelkareem MA (2021) Electrophoretic deposition of graphene oxide on carbon brush as bioanode for microbial fuel cell operated with real wastewater. Int J Hydrog Energy 46(8):5975–5983

Schierano MC, Panigatti MC, Maine MA, Griffa CA, Boglione R (2020) Horizontal subsurface flow constructed wetland for tertiary treatment of dairy wastewater: removal efficiencies and plant uptake. J Environ Manag 272:111094

Sepúlveda-Mardones M, Campos JL, Magrí A, Vidal G (2019) Moving forward in the use of aerobic granular sludge for municipal wastewater treatment: an overview. Rev Environ Sci Biotechnol 18(4):741–769

Shalini SS, Joseph K (2012) Nitrogen management in landfill leachate: application of SHARON, ANAMMOX and combined SHARON–ANAMMOX process. Waste Manag 32(12): 2385–2400

Sim J, Reid R, Hussain A, An J, Lee HS (2018) Hydrogen peroxide production in a pilot-scale microbial electrolysis cell. Biotechnol Rep 19:e00276

Sivalingam V, Dinamarca C, Janka E, Kukankov S, Wang S, Bakke R (2020) Effect of intermittent aeration in a hybrid vertical anaerobic biofilm reactor (HyVAB) for reject water treatment. Water 12(4):1151

Sokhandan F, Homayoonfal M, Davar F (2020) Application of zinc oxide and sodium alginate for biofouling mitigation in a membrane bioreactor treating urban wastewater. Biofouling 36(6): 660–678

Solís-Silvan R, López-Ocaña G, Bautista-Margulis RG, Hernández-Barajas JR, Romellón-Cerino MJ (2016) Evaluación de humedales artificiales de flujo libre y subsuperficial en la remoción de contaminantes de aguas residuales, utilizando diferentes especies de vegetación macrófita. Interciencia 41(1):40–47

Sonawane JM, Patil SA, Ghosh PC, Adeloju SB (2018) Low-cost stainless-steel wool anodes modified with polyaniline and polypyrrole for high-performance microbial fuel cells. J Power Sources 379:103–114

Stefanakis AI (2020) Constructed wetlands: description and benefits of an eco-tech water treatment system. In: Waste management: concepts, methodologies, tools, and applications. IGI Global, Hershey, pp 503–525

Stefanakis AI, Akratos CS, Tsihrintzis VA (2014) Vertical flow constructed wetlands: ecoengineering systems for wastewater and sludge treatment, 1st edn. Elsevier, Amsterdam

Sumisha A, Haribabu K (2018) Modification of graphite felt using nano polypyrrole and polythiophene for microbial fuel cell applications—a comparative study. Int J Hydrog Energy 43(6):3308–3316

Sun M, Yan L, Zhang L, Song L, Guo J, Zhang H (2019) New insights into the rapid formation of initial membrane fouling after in-situ cleaning in a membrane bioreactor. Process Biochem 78: 108–113

Szaja A, Szulżyk-Cieplak J (2020) Influence of bioaugmentation strategy of activated sludge on the co-treatment of reject water and municipal wastewater at a decreasing temperature. J Ecol Eng 21(5):97–106

Tawalbeh M, Al-Othman A, Singh K, Douba I, Kabakebji D, Alkasrawi M (2020) Microbial desalination cells for water purification and power generation: a critical review. Energy 209: 118493

Tedengren M (2021) Eutrophication and the disrupted nitrogen cycle. Ambio 50(4):733–738

Tian T, Qiao S, Yu C, Yang Y, Zhou J (2019) Low-temperature anaerobic digestion enhanced by bioelectrochemical systems equipped with graphene/PPy-and MnO2 nanoparticles/PPy-modified electrodes. Chemosphere 218:119–127

Tremblay PL, Li Y, Xu M, Yang X, Zhang T (2020) Graphene electrodes in bioelectrochemical systems. In: Microbial electrochemical technologies. CRC Press, Boca Raton, pp 422–443

van der Roest HF, De Bruin LMM, Gademan G, Coelho F (2011) Towards sustainable wastewater treatment with Dutch Nereda® technology. Water Pract Technol 6(3):1–2

van Loosdrecht MC (2008) Innovative nitrogen removal. In: Biological wastewater treatment: principles, modelling and design. IWA Publishing, London, pp 139–155

Van Loosdrecht MCM, Salem S (2006) Biological treatment of sludge digester liquids. Water Sci Technol 53(12):11–20

Varjani S (2022) Prospective review on bioelectrochemical systems for wastewater treatment: achievements, hindrances and role in sustainable environment. Sci Total Environ 841:156691

Vymazal J (2013) Emergent plants used in free water surface constructed wetlands: a review. Ecol Eng 61:582–592

Vymazal J (2014) Constructed wetlands for treatment of industrial wastewaters: a review. Ecol Eng 73:724–751

Wang W, Gao J, Guo X, Li W, Tian X, Zhang R (2012) Long-term effects and performance of two-stage baffled surface flow constructed wetland treating polluted river. Ecol Eng 49:93–103

Wang H, Zhi W, Deng N, Ji G (2017) Review on the fate and mechanism of nitrogen pollutant removal from wastewater using a biological filter. Pol J Environ Stud 26(5):1943–1954

Wang W, Wang X, Wang S, Li J (2018a) Partial denitrification coupled with immobilization of anammox in a continuous upflow reactor. RSC Adv 8(56):32016–32021

Wang R, Yan M, Li H, Zhang L, Peng B, Sun J, Liu D, Liu S (2018b) FeS2 nanoparticles decorated graphene as microbial-fuel-cell anode achieving high power density. Adv Mater 30(22): 1800618

Wang Y, Wen Q, Chen Y, Li W (2020) Conductive polypyrrole-carboxymethyl cellulose-titanium nitride/carbon brush hydrogels as bioanodes for enhanced energy output in microbial fuel cells. Energy 204:117942

Wang B, Tian Y, Yuan G, Liao X, Yang T, You X, Liao Y (2022) Membrane biofouling mitigation via immobilizing Agnps@ Zif-8 nanocapsules on membrane surface for extractive membrane bioreactor. J Environ Chem Eng 11:109899

Weissbrodt DG, Laureni M, van Loosdrecht MCM, Comeau Y (2020) Basic microbiology and metabolism. In: Chen GH, van Loosdrecht MCM, Ekama GA, Brdjanovic D (eds) Biological wastewater treatment: principles, modelling and design, 2nd edn. IWA Publishing, London, pp 11–76

Wu X, Xiong X, Owens G, Brunetti G, Zhou J, Yong X, Xie X, Zhang L, Wei P, Jia H (2018) Anode modification by biogenic gold nanoparticles for the improved performance of microbial fuel cells and microbial community shift. Bioresour Technol 270:11–19

Xiao X, Guo H, Ma F, You S, Geng M, Kong X (2021) Biological mechanism of alleviating membrane biofouling by porous spherical carriers in a submerged membrane bioreactor. Sci Total Environ 792:148448

Xie H, Ji D, Zang L (2017) Effects of inhibition conditions on anammox process. IOP Conf Ser Earth Environ Sci 100(1):012149

Xu H, Wang L, Wen Q, Chen Y, Qi L, Huang J, Tang Z (2019a) A 3D porous NCNT sponge anode modified with chitosan and polyaniline for high-performance microbial fuel cell. Bioelectrochemistry 129:144–153

Xu Y, Zhou S, Li M (2019b) Enhanced bioelectricity generation and cathodic oxygen reduction of air breathing microbial fuel cells based on MoS2 decorated carbon nanotube. Int J Hydrog Energy 44(26):13875–13884

Yan W, Xiao Y, Yan W, Ding R, Wang S, Zhao F (2019) The effect of bioelectrochemical systems on antibiotics removal and antibiotic resistance genes: a review. Chem Eng J 358:1421–1437

Yang W, Lu JE, Zhang Y, Peng Y, Mercado R, Li J, Zhu X, Chen S (2019) Cobalt oxides nanoparticles supported on nitrogen-doped carbon nanotubes as high-efficiency cathode catalysts for microbial fuel cells. Inorg Chem Commun 105:69–75

Yao R, Yuan Q, Wang K (2019) Enrichment of denitrifying bacterial community using nitrite as an electron acceptor for nitrogen removal from wastewater. Water 12(1):48

Yao J, Huang Y, Hou Y, Yang B, Lei L, Tang X, Scheckel K, Li Z, Wu D, Dionysiou DD (2021) Graphene-modified graphite paper cathode for the efficient bioelectrochemical removal of chromium. Chem Eng J 405:126545

Yu L, Peng D, Pan R (2012) Shifts in nitrification kinetics and microbial community during bioaugmentation of activated sludge with nitrifiers enriched on sludge reject water. J Biomed Biotechnol 2012:691894

Yuan G, Tian Y, Wang B, You X, Liao Y (2022) Mitigation of membrane biofouling via immobilizing Ag-MOFs on composite membrane surface for extractive membrane bioreactor. Water Res 209:117940

Zahid M, Savla N, Pandit S, Thakur VK, Jung SP, Gupta PK, Prasad R, Marsili E (2022) Microbial desalination cell: desalination through conserving energy. Desalination 521:115381

Zeirani T, Fallah N, Nasernejad B (2019) Biological treatment of real petrochemical wastewater containing styrene using submerged membrane bioreactor: effects of HRT. Int J Environ Sci Technol 16(7):3793–3800

Zhai Y, Zhang X, Wu Z, Wang Z (2020) Modification of polyvinylidene fluoride membrane by quaternary ammonium compounds loaded on silica nanopollens for mitigating biofouling. J Membr Sci 597:117679

Zhang J, Zhou J, Han Y, Zhang X (2014) Start-up and bacterial communities of single-stage nitrogen removal using anammox and partial nitritation (SNAP) for treatment of high strength ammonia wastewater. Bioresour Technol 169:652–657

Zhang H, Du R, Cao S, Wang S, Peng Y (2019) Mechanisms and characteristics of biofilm formation via novel DEAMOX system based on sequencing biofilm batch reactor. J Biosci Bioeng 127(2):206–212

Zhang K, Ma Z, Song H, Zhang M, Xu H, Zhao N (2020) Macroporous carbon foam with high conductivity as an efficient anode for microbial fuel cells. Int J Hydrog Energy 45(21): 12121–12129

Zhang Q, Yang Y, Chen F, Zhang L, Ruan J, Wu S, Zhu R (2021) Effects of hydraulic loading rate and substrate on ammonium removal in tidal flow constructed wetlands treating black and odorous water bodies. Bioresour Technol 321:124468

Zhang F, Yang K, Liu G, Chen Y, Wang M, Li S, Li R (2022) Recent advances on graphene: synthesis, properties, and applications. Compos Part A Appl Sci Manuf 160:107051

Zhao N, Ma Z, Song H, Xie Y, Zhang M (2019) Enhancement of bioelectricity generation by synergistic modification of vertical carbon nanotubes/polypyrrole for the carbon fibers anode in microbial fuel cell. Electrochim Acta 296:69–74

Zheng T, Li J, Ji Y, Zhang W, Fang Y, Xin F, Dong W, Wei P, Ma J, Jiang M (2020) Progress and prospects of bioelectrochemical systems: electron transfer and its applications in the microbial metabolism. Front Bioeng Biotechnol 8:10

Zurita-Martínez F, Castellanos-Hernández OA, Rodríguez-Sahagún A (2011) El tratamiento de las aguas residuales municipales en las comunidades rurales de México. Revista Mexicana de Ciencias Agrícolas 2(1):139–150

Recent Developments in Wastewater Treatments

10

Marlene Lariza Andrade-Guel, Christian Javier Cabello-Alvarado, Lucía Fabiola Cano-Salazar, Carlos Alberto Ávila-Orta, and Víctor Javier Cruz-Delgado

Abstract

Water is a crucial resource for life and daily life activities. Its demand is rising due to rapid population growth, which, coupled with the misuse and contamination of freshwater, has put this resource under growing stress. Its use for daily human life, agricultural, industrial, and energy sector activities results in wastewater that contains different substances depending on each activity. Such substances include organic (sewage), heavy metals, medicinal drugs, colorants, pesticides, bacteria, viruses, etc. Therefore, wastewater sanitation is a crucial activity for the sustainable use of water. Sanitation treatments depend on the contaminant substance(s) and are divided into three main conventional groups named: physical, chemical, and biological treatments. Besides, in recent years, new developments

M. L. Andrade-Guel
Departamento de Química Macromolecular y Nanomateriales, Centro de Investigación en Química Aplicada, Saltillo, Coahuila, Mexico

C. J. Cabello-Alvarado
Departamento de Materiales Avanzados, Centro de Investigación en Química Aplicada, Saltillo, Coahuila, Mexico

CONAHCYT - Centro de Investigación en Química Aplicada, Saltillo, Mexico

L. F. Cano-Salazar
Facultad de Ciencias Químicas, Universidad Autónoma de Coahuila, Saltillo, Coahuila, Mexico

C. A. Ávila-Orta
Departamento de Materiales Avanzados, Centro de Investigación en Química Aplicada, Saltillo, Coahuila, Mexico

V. J. Cruz-Delgado (✉)
Departamento de Procesos de Transformación, Centro de Investigación en Química Aplicada, Saltillo, Coahuila, Mexico
e-mail: victor.cruz@ciqa.edu.mx

© The Author(s), under exclusive license to Springer Nature Singapore Pte Ltd. 2023
R. Soni et al. (eds.), *Current Status of Fresh Water Microbiology*,
https://doi.org/10.1007/978-981-99-5018-8_10

have been made for water sanitation. This book chapter summarizes conventional and recent developments, aiming to have a broad perspective on wastewater treatments. A particular emphasis is made on recent developments such as filters based on polymeric membranes, plasma treatment, UV radiation, and electrochemical oxidation. Membranes of polymer nanocomposites increase the flow and prevent fouling; some nanoparticles can even have antibacterial properties. Organic compounds such as pharmaceutical and synthetic dyes and pathogenic bacteria can be removed using plasma treatment. UV radiation helps to reduce wastewater treatment times when photocatalytic materials are used. Finally, it has been reported that electrochemical oxidation removes residues from pharmaceutical industries that cannot be removed with conventional treatments.

Keywords

Wastewater · Pollutants · Treatments · Physical process · Chemical process

10.1 Introduction

Water pollution is a problem that continues to grow every day, causing many aquatic ecosystems around the world to be contaminated. This phenomenon has become an environmental problem reported in the literature. The different human activities of land use, mainly deforestation, urbanization, agricultural activities, etc., are the leading causes of contamination of surface water systems.

Why is it essential to treat wastewater, and in what scenario can this process be applied? Currently, drinking water is an essential resource. However, only 30% of existing water is suitable for human consumption. Therefore, the use of this element must be conscious and sustainable. In this sense, once used for the first time, drinking water automatically becomes wastewater. However, this does not mean it should be lost entirely because various treatments allow it to become a potential resource again. Therefore, residual water contains impurities, from human waste and has chemical and biological components. This waste can be categorized as domestic, industrial, and urban.

According to UNESCO, 80% of wastewater returns to the ecosystem without any treatment, increasing the risk of consuming contaminated water.

Why is it important to treat wastewater? Due to the scarcity of this particular element. Wastewater, previously treated, allows its reuse and favors the circulation of flows in various industry and agriculture sectors. This way, a sustainable water resource with a reduction in the water footprint is obtained. Furthermore, this wastewater treatment is an implementation opportunity for different organizations and brings significant benefits to the ecosystem, such as the following:

- The reuse of water for industrial purposes.
- It becomes an excellent resource for agricultural irrigation.
- Solid waste can be used as fertilizer.

10 Recent Developments in Wastewater Treatments

Regarding the industry, they are significant consumers of water. On a global scale, they consume ~22% of the total water produced, while in high-income countries, it can be as much as 60%. It is estimated that by 2050, manufacturing industries could increase their water consumption by 400%. Aqueous discharges in various industrial steps, like cooling towers, boiler heating, purification, etc., can contain numerous suspended or dissolved contaminants, and these effluents are called industrial wastewater. Industries such as chemical and petrochemical, paper and pulp, food processing, tannery, and other manufacturing industries constitute the primary sources of industrial wastewater. These wastewaters usually have a high organic compound concentration, non-neutral pH, different temperatures, salinity, turbidity, and a high content of heavy metals (Dash et al. 2021; Sahu et al. 2022). Wastewater from leather manufacturing, food processing and preservation, textile processing, and petroleum refining can have a high salt concentration. The wastewater composition varies according to the chemical products used in the previous processes and the nature of the treatment they have been subjected to; therefore, classifying industrial wastewater into specific categories is challenging (Ahmed et al. 2022; Prakash et al. 2022; Ranjan et al. 2022).

Conventional management for wastewater treatment includes chemical precipitation, flotation, and ion exchange. However, these processes have several drawbacks, such as low removal efficiency, high energy consumption, and the generation of toxic sludge, which limit their widespread application. Recently, several alternative treatment methods have been investigated to improve the quality of the treated effluent. As a result, different treatment technologies have been developed, such as coagulation-flocculation, adsorption, membrane filtration, reverse osmosis, advanced oxidation, biological processes (activated sludge, anaerobic-aerobic treatment, and membrane bioreactors), electro-technologies, and photocatalytic processes (Singh et al. 2021; Debbarma et al. 2021; Suyal and Soni 2022). Therefore, knowledge of the processes and their effects is fundamental to making a proper plan to address this challenge (Ceretta et al. 2021; Ahmed et al. 2022).

The scientific community has developed various ecological and low-cost nanomaterials for wastewater treatment. These materials are designed with specific unique functional and surface characteristics to purify polluted water. Furthermore, these nanomaterials were also used in the degradation of industrial effluents. In recent decades, nanoparticles have been studied for their ability as adsorbents and tested on a broad spectrum of contaminants (Jayakaran et al. 2019).

10.2 Conventional Methods of Water Treatment

10.2.1 Physical Processes

10.2.1.1 Filtration

One of the most widely used processes for the treatment of contaminated water is filtration, which consists of passing a gas or a liquid through a filter to separate the matter in suspension, because it is a simple, economical, and with high percentages of contaminant removal. Water treatment technology, especially filtration, is one of the main processes of water treatment systems in the world; it has thousands of years of development history (Thuptimdang et al. 2021).

The bank filtration method involves locating pumping wells near a surface spring. The pumping action causes a hydraulic head differential between the surface spring and the groundwater table, further inducing water flow through the porous medium to the producing well. In addition, several physicochemical and biological processes occur due to percolation, which attenuates the pollutants in the body of water. Therefore, the interface between surface and groundwater plays a fundamental role in pollutant attenuation processes, presenting intense activities and functioning as a biogeochemical reactor. Finally, the water derived from bank filtration is a mixture of the infiltrated water from the surface spring and the groundwater in the aquifer (Freitas et al. 2017). As a result, the removal of diclofenac has been successfully achieved. However, complementary treatment techniques are required to achieve 100% removal of this compound (de Carvalho Filho et al. 2022).

Another method that has been used is constructing wetlands, which are resilient, cost-effective, and environmentally friendly eco-technology and are an alternative option for heavy metal removal (Kataki et al. 2021). They are significant for developing countries, especially those seeking green economic development. With extensive research on this technology in the last decade, the specific removal mechanisms, such as adsorption, sedimentation, filtration, precipitation, coprecipitation, plant uptake, and microbial-mediated processes responsible for heavy metal removal, have been gradually revealed. The removal performance is highly dependent on the type of substrate, plant species, and microbial activity, with extensive interactive effects (Yu et al. 2022).

Moreover, compared with traditional mechanical filtration, biological filtration can significantly improve the removal efficiency of pollutants due to biological effects. It is also essential for biological water treatment in many industrialized and developed countries. It is often used to treat municipal wastewater, aquaculture wastewater, leachate from landfills, and dyeing wastewater (Li et al. 2016; Xiao et al. 2019). Biological filtration can be optimized for the removal of various types of contaminants. The main types of biological filtration developed or applied include aerated biological filters, denitrification biofilters, constructed wetlands, etc. Although biofiltration systems are widely used, more than existing studies are needed to explain the intrinsic mechanism of biological filtration operation clearly and on the microorganisms involved (Jin et al. 2023).

Carbon and its derivatives have been successfully used to purify polluted water. Activated carbon (AC) is one of the most common commercial adsorbents, including a wide range of processed amorphous carbon-based materials with high porosity and surface area. The AC preparation process consists of carbonizing the carbonaceous raw material and activating the carbonized product. Activated carbons can be used in a few forms: powdered, granulated, and fibrous. Granular activated carbon (GAC) is an excellent adsorbent for various contaminants and has been used in abundant applications in water and wastewater treatment (Benstoem et al. 2017).

10   Recent Developments in Wastewater Treatments

In addition to adsorption capacity, biological growth on GAC during water and wastewater treatment is also an expected consequence due to the favorable environment provided by this material. The biomass attached to the GAC can carry out the biodegradation of compounds. Biologically activated carbon (BAC) filtration has been used for drinking water; the removal of organic matter, especially the biodegradable fraction responsible for forming by-products of disinfection processes and microbiological regrowth within the systems water distribution; removal of micropollutants such as pesticides, algae toxins, and substances accountable for causing flavor and occurrence of nitrification-denitrification; and essential processes for the removal of ammonia that can cause environmental problems and risks to public health (dos Santos and Daniel 2020).

10.2.1.2 Adsorption

The adsorption process, also known as the surface phenomenon, has some advantages, such as simple design, lower capital cost, and lower operating cost. Recently, there has been a search for low-cost adsorbents that can be produced using agricultural waste, industrial waste, and natural materials. The adsorbent selection depends mainly on low production cost, adsorption capacity, regeneration properties, high surface area, and pore volumes. During the adsorption process, the solute will deposit on the surface of the adsorbent due to attractive forces. Eventually, it can form a monolayer or multilayer, depending on the experimental results. Recent investigations have shown that the adsorption process is spontaneous and requires a low temperature, and the adsorbate concentration will increase (on the adsorbent surface) over more extended periods (Yagub et al. 2014). Therefore, the adsorption process can be classified into two groups: physical and chemical adsorption.

The equilibrium adsorption capacity (q_e, mmol/g) can be determined based on the following equation (Eq. (10.1)).

$$q_e = \frac{V(C_0 - C_e)}{M} \tag{10.1}$$

where V, M, C_o, and C_e represent the volume (L), the mass of adsorbent (g), the initial adsorbate concentration (mmol/L), and the equilibrium adsorbate concentration (mmol/L), respectively.

The Langmuir model and the Freundlich model can be used to describe the adsorption data. The adsorption isotherm explains the amount of solute adsorbed on the surface of the adsorbent under constant temperature, the adsorption capacity of the monolayer, and the heterogeneous surface.

The adsorption process can be represented using the Langmuir isotherm (Eq. (10.2)) and Freundlich model (Eq. (10.3)), respectively.

$$\frac{C_e}{q_e} = \frac{1}{(b)X_m} + \frac{C_e}{X_m} \qquad (10.2)$$

where C_e, X_m, and b represent the equilibrium adsorbate concentration (mmol/L), theoretical maximum adsorption capacity (mmol/g), and constant value (L/mmol), respectively.

$$\text{Log}q_e = \text{Log}k + \frac{1}{n\text{Log}C_e} \qquad (10.3)$$

where C_e, k, and $1/n$ represent equilibrium adsorbate concentration (mmol/L), constant value (mmol/g), and constant value, respectively.

Activated carbon (AC) can be synthesized from any carbonaceous material (agricultural residues, different parts of plants, biomass) that contains high amounts of carbon and less inorganic content. The obtained activated carbon (powdered, granular, fiber, and extruded form) showed high surface area and porosity structure and can be used to remove contaminants from aqueous solutions. Granular activated carbon can be used in the food and beverage industry, wastewater treatment, and air and gas purification. There is great demand for the granular form due to its simple regeneration and reuse. On the other hand, powdered activated carbon showed a higher surface-volume ratio with a particle size of less than 0.177 mm. Therefore, it can eliminate unwanted odors, flavors, and colors. In general, activated carbon can be classified into three groups, namely, microporous (less than 2 nm), mesoporous (2–50 nm), and macroporous (greater than 50 nm), based on the pore size according to the IUPAC (Reza et al. 2020).

In addition to activated carbon and carbon derivatives, other particles have been used as adsorbents, among which are zeolites, nanoclays, and metal oxides, against a wide variety of contaminants, including heavy metals, due to their large surface area and pore volume, as described in Table 10.1.

10.2.2 Chemical Processes

10.2.2.1 Coagulation-Flocculation

Coagulation-flocculation is widely used in the post- and pretreatment of domestic or industrial water. This technique comprises two stages: in the first, the coagulant is added by dispersing it in the residual water with rapid agitation; in the second stage, the flocculant is added, whose function is the agglomeration of small particles that are dissolved in the residual water, and finally, the flocs settle creating a sludge that is removed, and the supernatant is transferred to the subsequent treatment; one of the disadvantages of this process is the production of sludge (Golob et al. 2005; Teh et al. 2016).

This process has been used to treat leachate from sanitary landfills considered highly contaminated water, studying the type and dose of coagulant, the effect of pH, and the optimal treatment conditions. Fresh leachates have shown low pH and high

10 Recent Developments in Wastewater Treatments

Table 10.1 Adsorbents used in the removal of different contaminants

Adsorbent	Pollutant	Test conditions	Removal/ adsorption capacity	Reference
Zeolite	Brilliant blue	pH = 3–5	87%	Pinedo-Hernández et al. (2019)
Activated carbon	Congo red	pH = 2, T = 25 °C	0.47 mg/g	Litefti et al. (2019)
Activated carbon	Reactivate red 198	pH = 2, T = 25 °C	253 mg/g	Mahvi et al. (2014)
Zeolite	Disperse orange 25	–	125 mg/g	Markandeya et al. (2021)
Chitosan/multi-walled carbon nanotube	Direct blue 71	pH 6.25	29 mg/g	Abbasi and Habibi (2016)
Multi-walled carbon nanotube	Maxilon Blue 5G	–	260 mg/g	Alkaim et al. (2015)
Multi-walled carbon nanotube	Malachite green	–	142 mg/g	Shirmardi et al. (2013)
Magnetic zeolite	Crystal purple	pH = 10.3, T = 50 °C, time = 45 min	98%	Shirani et al. (2014)
Zeolite 4A	Methylene blue	Time 180 min	99%	Belachew and Hinsene (2022)
Activated carbon	Rhodamine B	pH = 3	5.6 mg/g	Mousavi et al. (2021)
Zeolite	Rhodamine B	T = 30 °C	7.9 mg/g	Cheng et al. (2016)
NaY zeolite	Phenol	Time 120 min, pH = 4.0	77%	Ba Mohammed et al. (2021)
$FeCl_3$ zeolite	Phenol	Time 100 min, pH = 3.0	65–95%	Ebrahimi et al. (2013)
Coconut activated carbon	Phenol	–	13–24 mg/g	Ma et al. (2013)
Natural montmorillonite	Ametryn	–	188 mg/g	Shattar et al. (2017)
Commercial activated carbon	Carbofuran	–	96 mg/g	Salman and Hameed (2010)
Clay	Carbaryl	Time 120 min, T = 20 °C, pH = 5.1	2.2 mg/g	Ouardi et al. (2013)
N-doped biochar	Pb^{2+}	pH = 2–7	214 mg/g	Jiang et al. (2019)
N-doped biochar	Cr(VI)	pH = 2	100 mg/g	Guo et al. (2019)
N-doped biochar	Ni (II)	–	44 mg/g	Yin et al. (2019)
N-doped biochar	Cd^{2+}	pH = 6	197 mg/g	Yu et al. (2018)
N-doped biochar	Cu^{2+}	pH = 6	104 mg/g	Yu et al. (2018)

(continued)

Table 10.1 (continued)

Adsorbent	Pollutant	Test conditions	Removal/adsorption capacity	Reference
ZnO	Pb^{2+}	pH = 6.5, time = 30 min, T = 25 °C	434 mg/g	Yin et al. (2018)
ZnO	Cd^{2+}	pH = 7, time = 120 min, T = 55 °C	217 mg/g	Khezami et al. (2017)
ZnO	Hg^{2+}	pH = 7, time = 120 min, T = 55 °C	714 mg/g	Sheela et al. (2012)
ZnO	As(III)	pH = 7, time = 105 min, T = 50 °C	52 mg/g	Yuvaraja et al. (2018)
ZnO	Cr(VI)	pH = 2, time = 90 min, T = 50 °C	12.2 mg/g	Pandey and Tripathi (2017)
γ- MnOOH	Cu(II)	pH = 5, T = 40 °C, time = 120 min	11.9 mg/g	Cano-Salazar et al. (2020)
PO_4/CO_3 composite	Tetracycline	pH = 6.5, time = 360 min	118 mg/g	Yukhajon et al. (2023)
Carbon nanotubes	Ciprofloxacin/ Indigo carmine	pH = 4–6, T = 25–50 °C	95 mg/g, 93 mg/g	Elamin et al. (2023)
rGO / nZVI	Doxycycline	pH = 7, T = 25 °C	31.6 mg/g	Abdelfatah et al. (2022)
NH_2-MIL-101 (Al/Chitosan)	Azithromycin	pH = 7.9, time = 64 min	238 mg/g	Azari et al. (2022)

chemical oxygen demand (COD) content, while partially stabilized leachates were characterized by high pH and low COD (Tatsi et al. 2003). Because of this, coagulation-flocculation has been proposed as a pretreatment method for fresh leachates or as a post-treatment process for partially stabilized leachates. Previous studies report using coagulants and flocculants in leachate from landfills to reduce turbidity and membrane fouling (Marañón et al. 2008).

On the other hand, commonly used coagulants and flocculants can cause an increase in the concentration of metals in water, which can have implications for human health, and consequently, the production of large volumes of toxic sludge. As an alternative to this problem, in recent years, biopolymers have been used since they have a less environmental impact and are low cost (Renault et al. 2009). Chitosan is one of the promising biopolymers in the coagulation-flocculation area, and this polymer is used in different effluents from the food industry. Some of the properties that chitosan has that help its application as a coagulant and flocculant are its cationic charge and its ability to bind to certain solids specifically, and it also exhibits good solubility (Roussy et al. 2005a). Roussy et al. studied the effect of the coagulant dose, in this case, chitosan, and the effect of the pH. These authors found that at pH 5, the protonation of the amino groups led to the coagulation effect. For this study, low amounts of biopolymer were needed to achieve the decantation of colloids (Roussy et al. 2005b). Some naturally based flocculants include biopolymers such as starch, chitin, pectin, and cellulose derivatives. The latter has been little studied; one method to produce cellulose flocculants is introducing aldehyde functionalities by aqueous oxidation of periodate. Chemical modification of cellulose with functional groups of carboxylic acids, imines, and aldehydes gives good performance in water-soluble colloids (Suopajärvi et al. 2013).

Coagulation-flocculation is a process widely used in treatment plants as a pretreatment to eliminate contaminants from water. However, the new coagulants and flocculants must be developed with high sludge dehydration characteristics; as mentioned above, it is a by-product of the process and represents a disadvantage to the method; besides, it must be viable at an industrial level and be environmentally friendly (Wei et al. 2018).

10.2.2.2 Disinfection

Water is a primary means of spreading diseases since it contains various pathogenic microorganisms, including bacteria, viruses, intestinal protozoa, and other microorganisms. The disinfection process is the elimination of all microorganisms present in the water through different techniques such as the following:

- Physical treatment (the application of heat or physical agents)
- Radiation (ultraviolet)
- Metal ions (silver or copper)
- Chemicals (surfactants)
- Oxidants (chlorine, chlorine dioxide, ozone, potassium permanganate)

The most widely used technique is with oxidants, since it can be scaled to an industrial level in a treatment plant and eliminates most pathogens. However, the disadvantage of this technique is the production of by-products that can become potentially toxic. In this way, another compound is added to pre-treat the water sample and not generate by-products; Fe(VI) pre-oxidizes the organics in the water to control the formation of by-products when chlorine is added (Liu et al. 2020).

The recent development of new disinfection technologies focused on attacking a wide variety of pathogens is based on advanced oxidation processes characterized by the generation of highly reactive oxidants, particularly hydroxyl (OH) species. Among these advanced oxidation processes are the Fenton reaction, which uses a mixture of iron ions and hydrogen peroxide that generates OH species under mild acid conditions (García-Fernández et al. 2019; García-Espinoza et al. 2021).

10.2.3 Biological Processes

10.2.3.1 Anaerobe

In recent years, municipal wastewater has been considered a source of nutrients, energy, and water. Anaerobic digestion represents an alternative for managing activated sludge contained in sewage. It is characterized by converting organic matter to methane and CO_2 in the absence of oxygen and interacting with different bacterial populations that have to be anaerobic (Vinardell et al. 2020). It has different stages of hydrolysis, acidification, acetogenesis, and methanogenesis. The microorganisms present can convert the sludge into biogas and reduce its volume (Zhang et al. 2020). One of the limitations of this biological process is the low methane production in the methanogenesis stage.

Sun et al. studied the effect of incorporating conductive materials based on carbon and iron to improve the methanogenesis process; the total methane production increased by 9.38%. In addition, conductivity measurements and concentrations of conductive particles indicated that the particles accelerate the metabolism of microorganisms and promote methane production (Sun et al. 2021).

The anaerobic oxic sedimentation process reduces sludge and is dependent on biological enzymatic activity. Xu et al. conducted a study based on an anaerobic oxic sedimentation process for 168 days for the treatment of wastewater from the textile industry, and the results indicated that the process could increase the sludge reduction performance and improve the sludge discoloration rate, with an efficiency of 90% (Xu et al. 2021).

Anaerobic digestion for producing biomethane from anaerobic sludge from wastewater treatment presents some drawbacks, such as the difficulty of cell lysis of the sludge, which entails a long time. Due to this, pretreatments are carried out on the sludge. In recent years, cavitation has been proposed as an emerging technology that helps biological processes. Some of the advantages of this pretreatment are time, greater enzymatic digestibility, and removal of organic contaminants; cavitation can be hydrodynamic or acoustic with the use of ultrasound (Bhat and Gogate 2021).

10.2.3.2 Aerobic

Biological aerobic treatments are applied to convert organic and inorganic compounds in wastewater in the presence of oxygen. Aerobic treatment plants are classified into three generations:

- The first that is mechanical.
- The second consists of large tanks with several thousand cubic meters in which flow conditions are poorly controlled and oxygen transfer is poor.
- Third-generation plants contain bioreactors and sedimentation tanks and have several advantages such as small volumes, closed, emission-proof, high performance, flexibility, no scaling problems, and high utilization of oxygen content in the air (Gavrilescu and Macoveanu 1999).

Zeolite improves the aerobic process and helps nitrogen removal, sludge sedimentation, and biomass retention, capable of removing phosphorus. It is also used to immobilize microorganisms, such as the bacteria *Acinetobacter junii*, which allows a suitable microenvironment for the growth and survival of pure microorganisms. In addition, it improves membrane bioreactors to reduce membrane fouling (Montalvo et al. 2020).

On the other hand, aeration is a critical element in the aerobic treatment process. It provides microorganisms with the dissolved oxygen required, keeps solids in suspension, and controls the fouling of membrane bioreactors. Other advantages of aeration administration through pure oxygen are accelerating enzyme activity, producing less sediment, and minimizing foam formation (Skouteris et al. 2020).

Jiang et al. studied the effect of the concentration of Mg^{2+} in the residual sludge systems for treating residual waters with aniline. The experiment was in two batch reactors of parallel sequencing where Mg^{2+} was placed in only one reactor. They concluded that the high concentration of Mg^{2+} could inhibit the metabolism of aerobic bacteria, and it was also found that some aniline-degrading bacteria decreased, but denitrifying bacteria increased with the presence of Mg^{2+} (Jiang et al. 2020).

10.2.3.3 Enzymatic

Isolated enzymes produce by-products in less quantity and toxicity; the process is environmentally friendly. Enzymes can degrade organic compounds, such as dyes, and offer greater specificity and regulation under certain conditions. They belong to the family of oxidoreductases that catalyze oxidation and reduction reactions, finding applications in the diagnosis and treatment of wastewater (Routoula and Patwardhan 2020).

Enzymes can be administered directly to wastewater, immobilized in a material, or incorporated into a release and dosing device. In recent studies, the use of nanoparticles for enzyme immobilization is considered complex and often requires a surface modification of the enzymes before their integration into nanoparticles. Some disadvantages are the cost of manufacturing, the recovery of the nanoparticles, and the implementation on an industrial scale (Karthik et al. 2021).

The immobilization and the use of enzymatic bioreactors help solve the problems experienced with enzymatic treatment in a water treatment plant, such as enzymatic denaturation and effluent washing. In addition, the enzymatic reactor removes emerging contaminants, such as estrogen, since they facilitate the conversion of estrogens and the purification of the mixture. However, attention should be paid to the porosity of the reactor membrane for effective enzymatic binding and efficient component separation (Zdarta et al. 2022).

10.3 Recent Developments for Water Treatment

10.3.1 Filters Based on Polymeric Membranes

Membranes and filters act as a structure composed of pores that remove contaminants through affinity or size exclusion, allowing the separation of pollutants according to their size and the pore size of the filter or membrane. On the contrary, affinity membranes reject contaminants based on the electrostatic interaction of the functional groups of the pollutants.

Membranes can be classified based on the contaminant size that they can reject. Figure 10.1 shows the classification by size, as well as the contaminants retained in each membrane according to pore size; it is worth mentioning that a smaller pore size

Fig. 10.1 Type of filtration membranes by size and type of rejected contaminants

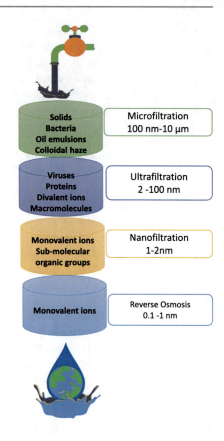

requires a higher pressure needed to force the water through the filter or membrane (Mautner 2020).

Membranes based on natural polymers such as cellulose were among the first used to separate solids. Then in recent years, membranes based on nanocellulose have emerged that help to improve separation due to their size. One of the characteristics of these membranes is that they are made entirely from renewable resources and are biodegradable compared to composite membranes and synthetic membranes. The preparation of nanocellulose membranes has been done by the solvent evaporation method that results in films (Kontturi et al. 2006; Metreveli et al. 2014).

Membranes of synthetic origin have been used due to their excellent thermal and chemical stability, control of the pore formation mechanism, and low cost. Polymers used for the manufacture of membranes for wastewater treatment are polyamide, polyester, poly(acrylonitrile), polyurethane, poly(tetrafluoroethylene), and polysulfone (Hamid et al. 2011; Bhol et al. 2021).

Polysulfone is one of the polymers that stands out due to its mechanical and temperature resistance. Therefore, membranes based on polysulfone and nanoparticles have been developed to cover defects that the membranes may have,

10 Recent Developments in Wastewater Treatments

Table 10.2 Polymer composite membranes and pollutants removed

Nanoparticle	Polymer	Method	Pollutant	Reference
γ-Al_2O_3	Polyamide	Interfacial polymerization	Polyvalent cations such as Mg^{2+}	Hassan et al. (2022)
Titania nanotube	Polyamide	Interfacial polymerization	Endocrine-disrupting compound	Ahmad et al. (2022)
Graphene oxide, Au	Polyamide 6	Nonsolvent-induced phase separation	Desalination and sewage treatment	Xu et al. (2022)
Silica	Polyether–polysulfone	Interfacial polymerization	Heavy metal	Al-Gamal et al. (2022)
Zeolite	Polysulfone Chitosan	Phase inversion	Dyes	Gowriboy et al. (2022)
TiO_2 –SiO_2	Polyether–Polysulfone	Vapor-ventilated in situ chemical deposition	–	Gan et al. (2022)
ZnO	Polysulfone–polyurethane	Phase inversion	Dyes	Kalaivizhi et al. (2022)

such as high hydrophobicity, pore blockages, and low resistance (Esfahani et al. 2015; Yadav et al. 2019).

Membranes composed of polymers and nanoparticles receive more and more attention for wastewater treatment; the inorganic nanostructures used are ZnO, TiO_2, SiO_2, and ZrO_2. These particles mainly increase the flow and prevent fouling; also, some nanoparticles have antibacterial properties for eliminating pathogenic bacteria in wastewater (Sheikh et al. 2020).

The incorporation of ZnO and SiO_2 into a polyethersulfone membrane to separate water and oil has been reported. The SiO_2 membrane had better organic resistance and 25% more permeate flux than the polyethersulfone membrane with ZnO (Ismail et al. 2020). Table 10.2 shows the composite membranes with nanoparticles and the pollutants they remove.

10.3.1.1 Plasma Treatment

The use of plasma has grown for the removal and mineralization of polluting organic compounds (pharmaceuticals and synthetic dyes) and pathogenic bacteria in wastewater in recent years. Plasma technology consists of an electric field source with highly charged particles that generate oxidizing species, reducing species, and electrons (Zeghioud et al. 2020). Plasma in contact with water is part of an advanced oxidation process since it generates OH ions from water molecules through the impact of dissociation electrons and radical reactions. There are different plasma treatments, serpentine-shaped discharge plasma on a solution with argon bubbling, plasma generated within gas bubbles, radiofrequency plasma, and plasma sliding arc, among others (Takeuchi and Yasuoka 2021).

Cold or non-thermal plasma is another type of plasma that is generated where the electrons have a higher temperature than the heavier species. With the passage of energy, the gas dissociates into several reactive species, followed by ionization. This method achieves a more energy-efficient decomposition compared to others. This technique has been used for the decontamination of various harmful elements such as medicinal products, organic dyes, pesticides, herbicides, biomolecules, phenolic compounds, and antibiotic substances. It also decreases chemical oxygen demand and pathogenic bacteria and viruses (Gururani et al. 2021).

Rashid et al. used an air discharge plasma at atmospheric pressure from multiple parallel tubes underwater to decompose residues of dyes from the textile industry, such as remazol blue, red, and yellow. They observed that as the incidence time increased, the percentage of discoloration increased, removing 70% of the dyes. Therefore, the plasma technique is considered an alternative ecological way of degrading dyes from the textile industry (Rashid et al. 2020).

Hafeez et al. fabricated a hybrid plasma and ozone reactor by connecting six hybrid plasma reactors in parallel to create a hybrid plasma microreactor system for ozone generation. The efficiency of this hybrid reactor was evaluated for its application in the degradation or elimination of the methyl orange dye. In addition, the parameters of the effect of pH, concentration, and solution volume were studied, achieving the elimination of 86% of the dye at pH $= 3$ with a concentration of 10 ppm and 900 mL of solution in 8 min in the reactors (Hafeez et al. 2021).

Moreover, non-thermal plasma treatments have been used to eliminate the growth of bacteria, with an antibacterial and sterilizing effect. They could surpass conventional methods because it is low cost, it is not necessary to add chemicals, and it is considered a clean energy (Ekanayake et al. 2021).

10.3.1.2 UV Radiation

Ultraviolet radiation has been used as a water disinfectant to eliminate pathogenic microorganisms from different water effluents. However, the UV irradiation process ranges from a wavelength of 315–400 nm, which causes oxidative damage to cells. This radiation can be direct or indirect using photocatalytic materials that absorb radiation and carry out a process of generation of reactive oxygen species (ROS) (Wang et al. 2020).

Using UV integrated with H_2, O_2, and TiO_2 helps to demobilize pathogenic microorganisms; when UV radiation is contrasted with photocatalytic materials, it helps to reduce wastewater treatment times. On the other hand, TiO_2 promotes the inactivation of the growth of microorganisms via photocatalysis, generating internal cell destruction generated by the actions of ROS species, causing physical destruction followed by an oxidative attack of hydroxyl radicals on cell membranes, inducing oxidative stress and loss of permeability (Ghernaout 2020). A disadvantage of this photocatalyst is that the recycling and recovery of these suspended particles become a costly problem; as an option, supports for these particles have been developed (Garrido-Cardenas et al. 2019).

Salazar et al. functionalized the TiO_2 surface with silver nanoparticles and prepared membranes of electrospun poly(vinylidene-co-hexafluoropropylene fluoride) as support with different concentrations of TiO_2 alone and functionalized with silver. The application of this membrane was in the degradation of norflaxin, an emerging pollutant present in wastewater and rivers polluted with pharmaceutical industries, and the degradation of this pollutant was carried out under UV and visible radiation. The nanocomposites showed a degradation efficiency of 64.2% under UV radiation, contrasting with the visible 80.7% during 90 and 300 min. In addition, the membranes demonstrated recyclability and application to remove an emerging contaminant (Salazar et al. 2020).

Andrade et al. developed a nonwoven fabric membrane with photocatalytic ZnO, nanoparticles manufactured by melt blowing, with different concentrations of nanoparticles. The methylene blue adsorption efficiency was 93% in a time of 60 min. It is worth mentioning that the nonwoven fabric presented antimicrobial activity against two strains, *Staphylococcus aureous* and *Candida albicans* (Andrade-Guel et al. 2021).

The photocatalysts are activated by UV radiation, producing highly reactive photoinduced charge carriers that react with contaminants. This technology would help solve problems of high energy consumption in traditional methods, such as high temperature and pressure required for water decontamination. There is a technique that has attracted attention in recent years. It is based on a photocatalytic process called Fenton and produces reactive oxygen species (ROS). It includes three reaction steps: initiation, propagation, and termination reactions. It effectively eliminates persistent organic compounds, bacteria, and viruses (Liu et al. 2020).

10.3.1.3 Electrochemical Oxidation

Electroxidation is one of the most researched techniques today because it can mineralize different polluting organic compounds. The most widely used anodes are mixed metal oxides based on SnO_2 and PbO_2 coatings, boron-doped diamond anodes, and titanium suboxide ceramic anodes (Radjenovic et al. 2020).

Oxidation processes are based on passing an electric current through wastewater with the help of electrodes, creating oxidizing agents such as hydroxyl radicals and chloride that can oxidize ammonia. The cost associated with this process lies in the electric current or energy consumed compared to other methods. However, the cost of operation is low, and it can destroy persistent organic pollutants (Almomani et al. 2020).

Wang et al. reported the electrochemical oxidation in pulsed current mode to treat wastewater with dyes such as indigo and methyl orange with a PbO_2/Ti anode. In addition, the parameters that influence the degradation of dyes were determined: the current density, pulse duty cycle, and flow rate for dye degradation (Wang et al. 2020).

Electrochemical oxidation has been used to treat wastewater containing residues from the pharmaceutical industries that cannot be removed with conventional methods. For example, the electrochemical oxidation of the antibiotic sulfadiazine has been investigated using a titanium suboxide mesh anode; the results revealed that the optimal time for electrolysis is 60 min with 100% degradation. Furthermore, the anode exhibited oxygen evolution potential for OH radical formation and long-term stability (Teng et al. 2020).

10.4 Conclusions

The growing increase in the world population, as well as the intensive use by the manufacturing industry and various agro-industrial activities, has generated, to a great extent, the contamination of fresh water and the bodies that contain it. The use of water for all the daily activities of human beings is unavoidable, for which the development of new technologies that reduce the water footprint and lead to more efficient treatments for its reuse is vital.

The different primary treatments have shown a trend toward the development of new and improved processes for wastewater treatment, emphasizing physical and biological processes. Furthermore, with the increasing progress in developing low-cost adsorbents from renewable and natural sources, viable alternatives have been proposed to remove pesticides, heavy metals, medicines, and pigments, which can be implemented in sewage treatment plants (Kumar et al. 2021). However, it is known that the persistence of these contaminants in water can cause various adverse health effects. Their bioaccumulation can lead to new and more aggressive forms of cancer, chronic renal failure, metabolic syndrome, and others, for which there is still no effective treatment.

Biological processes have shown significant advances, with increases in biogas production as a by-product; however, sludge production is still a considerable challenge. Possibly, the participation of consortia of microorganisms such as bacteria, fungi, and enzymes, which have shown to be effective for the elimination of different heavy metals, could be of great help for the generation of fertilizers.

For their part, advanced processes for wastewater treatment represent viable alternatives for their implementation in wastewater treatment plants. Moreover, since it has been possible to develop more functional materials, lower cost, and higher profitability, the joint use of some of these techniques would lead to obtaining fresh water suitable for human consumption.

Although the extraordinary efforts of the scientific community to develop new techniques for the treatment of wastewater, one of the main challenges to overcome is the awareness of the population to make a more rational use of the precious liquid. Without a behavioral change, the human being will continue contributing to water pollution, reducing its availability for future generations.

Acknowledgments Se agradece el apoyo brindado por el Consejo Nacional de Humanidades, Ciencia y Tecnología de México (CONHACyT) a través de la beca posdoctoral 387368 y del proyecto No. 320888, "Desarrollo de filtros antimicrobianos a base de nanocompuestos poliméricos económicos, reutilizables con bajo impacto ambiental para la descontaminación de ríos de México como una alternativa para sustituir los filtros comerciales" de la Convocatoria de Ciencia Básica y/o Ciencia de Frontera. Modalidad: Paradigmas y Controversias de la Ciencia 2022.

References

Abbasi M, Habibi MM (2016) Optimization and characterization of Direct Blue 71 removal using nanocomposite of Chitosan-MWCNTs: central composite design modeling. J Taiwan Inst Chem Eng 62:112–121. https://doi.org/10.1016/j.jtice.2016.01.019

Abdelfatah AM, El-Maghrabi N, Mahmoud AED, Fawzy M (2022) Synergetic effect of green synthesized reduced graphene oxide and nano-zero valent iron composite for the removal of doxycycline antibiotic from water. Sci Rep 12:19372. https://doi.org/10.1038/s41598-022-23684-x

Ahmad NA, Goh PS, Azman N, Ismail AF, Hasbullah H, Hashim N, Kerisnan Krishnan ND, NKEM Y, Mohamed A, Mohamed Yusoff MA, Karim J, Abdullah NS (2022) Enhanced removal of endocrine-disrupting compounds from wastewater using reverse osmosis membrane with titania nanotube-constructed nanochannels. Membranes 12:958. https://doi.org/10.3390/membranes12100958

Ahmed M, Mavukkandy MO, Giwa A, Elektorowicz M, Katsou E, Khelifi O, Naddeo V, Hasan SW (2022) Recent developments in hazardous pollutants removal from wastewater and water reuse within a circular economy. NPJ Clean Water 5:12. https://doi.org/10.1038/s41545-022-00154-5

Al-Gamal AQ, Satria M, Alghunaimi FI, Aljuryyed NW, Saleh TA (2022) Synthesis of thin-film nanocomposite membranes using functionalized silica nanoparticles for water desalination with drastically improved properties. React Funct Polym 181:105433. https://doi.org/10.1016/j.reactfunctpolym.2022.105433

Alkaim AF, Sadik Z, Mahdi DK, Alshrefi SM, Al-Sammarraie AM, Alamgir FM, Singh PM, Aljeboree AM (2015) Preparation, structure and adsorption properties of synthesized multiwall carbon nanotubes for highly effective removal of maxilon blue dye. Korean J Chem Eng 32: 2456–2462. https://doi.org/10.1007/s11814-015-0078-y

Almomani F, Bhosale R, Khraisheh M, Kumar A, Tawalbeh M (2020) Electrochemical oxidation of ammonia on nickel oxide nanoparticles. Int J Hydrog Energy 45:10398–10408. https://doi.org/10.1016/j.ijhydene.2019.11.071

Andrade-Guel M, Ávila-Orta CA, Cabello-Alvarado C, Cadenas-Pliego G, Esparza-González SC, Pérez-Alvarez M, Quiñones-Jurado ZV (2021) Non-woven fabrics based on nanocomposite nylon 6/ZnO obtained by ultrasound-assisted extrusion for improved antimicrobial and adsorption methylene blue dye properties. Polymers 13:1888. https://doi.org/10.3390/polym13111888

Azari A, Malakoutian M, Yaghmaeain K, Jaafarzadeh N, Shariatifar N, Mohammadi G, Masoudi MR, Sadeghi R, Hamzeh S, Kamani H (2022) Magnetic NH2-MIL-101(Al)/Chitosan nanocomposite as a novel adsorbent for the removal of azithromycin: modeling and process optimization. Sci Rep 12:18990. https://doi.org/10.1038/s41598-022-21551-3

Ba Mohammed B, Yamni K, Tijani N, Lee H-S, Dehmani Y, El Hamdani H, Alrashdi AA, Ramola S, Belwal T, Lgaz H (2021) Enhanced removal efficiency of NaY zeolite toward phenol from aqueous solution by modification with nickel (Ni-NaY). J Saudi Chem Soc 25:101224. https://doi.org/10.1016/j.jscs.2021.101224

Belachew N, Hinsene H (2022) Preparation of Zeolite 4A for adsorptive removal of methylene blue: optimization, kinetics, isotherm, and mechanism study. SILICON 14:1629–1641. https://doi.org/10.1007/s12633-020-00938-9

Benstoem F, Nahrstedt A, Boehler M, Knopp G, Montag D, Siegrist H, Pinnekamp J (2017) Performance of granular activated carbon to remove micropollutants from municipal wastewater—a meta-analysis of pilot- and large-scale studies. Chemosphere 185:105–118. https://doi.org/10.1016/j.chemosphere.2017.06.118

Bhat AP, Gogate PR (2021) Cavitation-based pre-treatment of wastewater and waste sludge for improvement in the performance of biological processes: a review. J Environ Chem Eng 9: 104743. https://doi.org/10.1016/j.jece.2020.104743

Bhol P, Yadav S, Altaee A, Saxena M, Misra PK, Samal AK (2021) Graphene-based membranes for water and wastewater treatment: a review. ACS Appl Nano Mater 4:3274–3293. https://doi.org/10.1021/acsanm.0c03439

Cano-Salazar LF, Martínez-Luévanos A, Claudio-Rizo JA, Carrillo-Pedroza FR, Montemayor SM, Rangel-Mendez JR (2020) Synthesis, structural characterization and Cu(II) adsorption behavior of manganite (γ-MnOOH) nanorods. RSC Adv 10:179–186. https://doi.org/10.1039/C9RA09652C

Ceretta MB, Nercessian D, Wolski EA (2021) Current trends on role of biological treatment in integrated treatment technologies of textile wastewater. Front Microbiol 12:651025. https://doi.org/10.3389/fmicb.2021.651025

Cheng WY, Li N, Pan YZ, Jin LH (2016) The adsorption of rhodamine b in water by modified zeolites. Mod Appl Sci 10:67. https://doi.org/10.5539/mas.v10n5p67

Dash B, Sahu N, Singh AK, Gupta SB, Soni R (2021) Arsenic efflux in Enterobacter cloacae RSN3 isolated from arsenic-rich soil. Folia Microbiol 66:189–196

de Carvalho Filho JAA, da Cruz HM, Fernandes BS, Motteran F, de Paiva ALR, Pereira Cabral JJ (2022) Efficiency of the bank filtration technique for diclofenac removal: a review. Environ Pollut 300:118916. https://doi.org/10.1016/j.envpol.2022.118916

Debbarma P, Joshi D, Maithani D, Dasila H, Suyal DC, Kumar S, Soni R (2021) Sustainable bioremediation strategies to manage environmental pollutants. In: Shah MP (ed) Removal of refractory pollutants from wastewater treatment plants. CRC Press, Boca Raton. https://doi.org/10.1201/9781003204442-14

dos Santos PR, Daniel LA (2020) A review: organic matter and ammonia removal by biological activated carbon filtration for water and wastewater treatment. Int J Environ Sci Technol 17:591–606. https://doi.org/10.1007/s13762-019-02567-1

Ebrahimi A, Hosseinzadeh E, Asgari G, Mohammadi A (2013) Adsorption of phenol from aqueous solution by modified zeolite with FeCl 3. Int J Environ Health Eng 2:6. https://doi.org/10.4103/2277-9183.107915

Ekanayake UGM, Barclay M, Seo DH, Park MJ, MacLeod J, O'Mullane AP, Motta N, Shon HK, Ostrikov K (2021) Utilization of plasma in water desalination and purification. Desalination 500:114903. https://doi.org/10.1016/j.desal.2020.114903

Elamin MR, Abdulkhair BY, Elzupir AO (2023) Removal of ciprofloxacin and indigo carmine from water by carbon nanotubes fabricated from a low-cost precursor: solution parameters and recyclability. Ain Shams Eng J 14:101844. https://doi.org/10.1016/j.asej.2022.101844

Esfahani MR, Tyler JL, Stretz HA, Wells MJM (2015) Effects of a dual nanofiller, nano-TiO2 and MWCNT, for polysulfone-based nanocomposite membranes for water purification. Desalination 372:47–56. https://doi.org/10.1016/j.desal.2015.06.014

Freitas DA, Cabral JJSP, Rocha FJS, Paiva ALR, Sens ML, Veras TB (2017) Cryptosporidium spp. and Giardia spp. removal by bank filtration at Beberibe River, Brazil. River Res Appl 33:1079–1087. https://doi.org/10.1002/rra.3151

Gan N, Lin Y, Zhang Y, Gitis V, Lin Q, Matsuyama H (2022) Surface mineralization of the TiO2–SiO2/PES composite membrane with outstanding separation property via facile vapor-ventilated in situ chemical deposition. Langmuir 38:12951–12960. https://doi.org/10.1021/acs.langmuir.2c02178

García-Espinoza JD, Robles I, Durán-Moreno A, Godínez LA (2021) Photo-assisted electrochemical advanced oxidation processes for the disinfection of aqueous solutions: a review. Chemosphere 274:129957. https://doi.org/10.1016/j.chemosphere.2021.129957

García-Fernández I, Miralles-Cuevas S, Oller I, Malato S, Fernández-Ibáñez P, Polo-López MI (2019) Inactivation of E. coli and E. faecalis by solar photo-Fenton with EDDS complex at neutral pH in municipal wastewater effluents. J Hazard Mater 372:85–93. https://doi.org/10.1016/j.jhazmat.2018.07.037

Garrido-Cardenas JA, Esteban-García B, Agüera A, Sánchez-Pérez JA, Manzano-Agugliaro F (2019) Wastewater treatment by advanced oxidation process and their worldwide research trends. Int J Environ Res Public Health 17:170. https://doi.org/10.3390/ijerph17010170

Gavrilescu M, Macoveanu M (1999) Process engineering in biological aerobic waste-water treatment. Acta Biotechnol 19:111–145. https://doi.org/10.1002/abio.370190205

Ghernaout D (2020) Advanced oxidation processes for wastewater treatment: facts and future trends. OALib 7:1–15. https://doi.org/10.4236/oalib.1106139

Golob V, Vinder A, Simonic M (2005) Efficiency of the coagulation/flocculation method for the treatment of dyebath effluents. Dyes Pigments 67:93–97. https://doi.org/10.1016/j.dyepig.2004.11.003

Gowriboy N, Kalaivizhi R, Ganesh MR, Aswathy KA (2022) Development of thin film polymer nanocomposite membrane (ZIF-8@PSf/CS) for removal of textile pollutant and evaluating the effect of water samples on human monocytic cell lines (THP-1) using flow cytometer. J Clean Prod 377:134399. https://doi.org/10.1016/j.jclepro.2022.134399

Guo S, Gao Y, Wang Y, Liu Z, Wei X, Peng P, Xiao B, Yang Y (2019) Urea/ZnCl2 in situ hydrothermal carbonization of Camellia sinensis waste to prepare N-doped biochar for heavy metal removal. Environ Sci Pollut Res 26:30365–30373. https://doi.org/10.1007/s11356-019-06194-8

Gururani P, Bhatnagar P, Bisht B, Kumar V, Joshi NC, Tomar MS, Pathak B (2021) Cold plasma technology: advanced and sustainable approach for wastewater treatment. Environ Sci Pollut Res 28:65062–65082. https://doi.org/10.1007/s11356-021-16741-x

Hafeez A, Shezad N, Javed F, Fazal T, Saif Ur Rehman M, Rehman F (2021) Developing multiplexed plasma micro-reactor for ozone intensification and wastewater treatment. Chem Eng Process Process Intensif 162:108337. https://doi.org/10.1016/j.cep.2021.108337

Hamid NAA, Ismail AF, Matsuura T, Zularisam AW, Lau WJ, Yuliwati E, Abdullah MS (2011) Morphological and separation performance study of polysulfone/titanium dioxide (PSF/TiO2) ultrafiltration membranes for humic acid removal. Desalination 273:85–92. https://doi.org/10.1016/j.desal.2010.12.052

Hassan GK, Al-Shemy M, Adel AM, Al-Sayed A (2022) Boosting brackish water treatment via integration of mesoporous γ-Al2O3NPs with thin-film nanofiltration membranes. Sci Rep 12:19666. https://doi.org/10.1038/s41598-022-23914-2

Ismail NH, Salleh WNW, Ismail AF, Hasbullah H, Yusof N, Aziz F, Jaafar J (2020) Hydrophilic polymer-based membrane for oily wastewater treatment: a review. Sep Purif Technol 233:116007. https://doi.org/10.1016/j.seppur.2019.116007

Jayakaran P, Nirmala GS, Govindarajan L (2019) Qualitative and quantitative analysis of graphene-based adsorbents in wastewater treatment. Int J Chem Eng 2019:1–17. https://doi.org/10.1155/2019/9872502

Jiang Q, Xie W, Han S, Wang Y, Zhang Y (2019) Enhanced adsorption of Pb(II) onto modified hydrochar by polyethyleneimine or H3PO4: an analysis of surface property and interface mechanism. Colloids Surfaces A Physicochem Eng Asp 583:123962. https://doi.org/10.1016/j.colsurfa.2019.123962

Jiang Y, Shang Y, Gong T, Hu Z, Yang K, Shao S (2020) High concentration of Mn2+ has multiple influences on aerobic granular sludge for aniline wastewater treatment. Chemosphere 240:124945. https://doi.org/10.1016/j.chemosphere.2019.124945

Jin L, Sun X, Ren H, Huang H (2023) Biological filtration for wastewater treatment in the 21st century: a data-driven analysis of hotspots, challenges and prospects. Sci Total Environ 855:158951. https://doi.org/10.1016/j.scitotenv.2022.158951

Kalaivizhi R, Danagody B, Yokesh A (2022) ACs@ZnO incorporated with a PSF/PU polymer membrane for dye removal. Mater Adv 3:8534–8543. https://doi.org/10.1039/D2MA00794K

Karthik V, Senthil Kumar P, Vo D-VN, Selvakumar P, Gokulakrishnan M, Keerthana P, Audilakshmi V, Jeyanthi J (2021) Enzyme-loaded nanoparticles for the degradation of wastewater contaminants: a review. Environ Chem Lett 19:2331–2350. https://doi.org/10.1007/s10311-020-01158-8

Kataki S, Chatterjee S, Vairale MG, Dwivedi SK, Gupta DK (2021) Constructed wetland, an eco-technology for wastewater treatment: a review on types of wastewater treated and components of the technology (macrophyte, biofilm and substrate). J Environ Manag 283:111986. https://doi.org/10.1016/j.jenvman.2021.111986

Khezami L, Taha KK, Amami E, Ghiloufi M, El Mir L (2017) Removal of cadmium(II) from aqueous solution by zinc oxide nanoparticles: kinetic and thermodynamic studies. Desalin Water Treat 62:346–354. https://doi.org/10.5004/dwt.2017.0196

Kontturi E, Tammelin T, Österberg M (2006) Cellulose—model films and the fundamental approach. Chem Soc Rev 35:1287–1304. https://doi.org/10.1039/B601872F

Kumar P, Dash B, Suyal DC, Gupta SB, Singh AK, Chowdhury T, Soni R (2021) Characterization of arsenic-resistant *Klebsiella pneumoniae* RnASA11 from contaminated soil and water samples and its bioremediation potential. Curr Microbiol 78:3258–3267

Li P, Zuo J, Wang Y, Zhao J, Tang L, Li Z (2016) Tertiary nitrogen removal for municipal wastewater using a solid-phase denitrifying biofilter with polycaprolactone as the carbon source and filtration medium. Water Res 93:74–83. https://doi.org/10.1016/j.watres.2016.02.009

Litefti K, Freire MS, Stitou M, González-Álvarez J (2019) Adsorption of an anionic dye (Congo red) from aqueous solutions by pine bark. Sci Rep 9:16530. https://doi.org/10.1038/s41598-019-53046-z

Liu H, Wang C, Wang G (2020) Photocatalytic advanced oxidation processes for water treatment: recent advances and perspective. Chem Asian J 15:3239–3253. https://doi.org/10.1002/asia.202000895

Ma Y, Gao N, Chu W, Li C (2013) Removal of phenol by powdered activated carbon adsorption. Front Environ Sci Eng 7:158–165. https://doi.org/10.1007/s11783-012-0479-7

Mahvi A, Bazrafshan E, Mostafapour F (2014) Decolorization of reactive red 198 by adsorption onto ZnCl2 activated pistachio hull wastes. Int J Environ Health Eng 3:7. https://doi.org/10.4103/2277-9183.131811

Marañón E, Castrillón L, Fernández-Nava Y, Fernández-Méndez A, Fernández-Sánchez A (2008) Coagulation–flocculation as a pretreatment process at a landfill leachate nitrification–denitrification plant. J Hazard Mater 156:538–544. https://doi.org/10.1016/j.jhazmat.2007.12.084

Markandeya, Shukla SP, Srivastav AL (2021) Removal of disperse orange and disperse blue dyes present in textile mill effluent using zeolite synthesized from cenospheres. Water Sci Technol 84:445–457. https://doi.org/10.2166/wst.2021.216

Mautner A (2020) Nanocellulose water treatment membranes and filters: a review. Polym Int 69:741–751. https://doi.org/10.1002/pi.5993

Metreveli G, Wågberg L, Emmoth E, Belák S, Strømme M, Mihranyan A (2014) A size-exclusion nanocellulose filter paper for virus removal. Adv Healthc Mater 3:1546–1550. https://doi.org/10.1002/adhm.201300641

Montalvo S, Huiliñir C, Borja R, Sánchez E, Herrmann C (2020) Application of zeolites for biological treatment processes of solid wastes and wastewaters – a review. Bioresour Technol 301:122808. https://doi.org/10.1016/j.biortech.2020.122808

Mousavi SA, Kamarehie B, Almasi A, Darvishmotevalli M, Salari M, Moradnia M, Azimi F, Ghaderpoori M, Neyazi Z, Karami MA (2021) Removal of Rhodamine B from aqueous solution by stalk corn activated carbon: adsorption and kinetic study. Biomass Convers Biorefinery. https://doi.org/10.1007/s13399-021-01628-1

Ouardi M, Alahiane S, Qourzal S, Abaamrane A, Assabbane A, Douch J (2013) Removal of carbaryl pesticide from aqueous solution by adsorption on local clay in Agadir. Am J Anal Chem 4:72–79. https://doi.org/10.4236/ajac.2013.47A010

Pandey M, Tripathi BD (2017) Synthesis, characterization and application of zinc oxide nano particles for removal of hexavalent chromium. Res Chem Intermed 43:121–140. https://doi.org/10.1007/s11164-016-2610-z

Pinedo-Hernández S, Sánchez-Mendieta V, Gutiérrez-Segura E, Solache-Ríos M (2019) Efficient removal of brilliant blue by clinoptilolite tuff modified with Fe3+ and Fe-Cu nanoparticles. Desalin Water Treat 144:300–310. https://doi.org/10.5004/dwt.2019.23623

Prakash O, Pathak A, Kuamr A, Juyal VK, Joshi HC, Gangola S, Patni K, Bhandari G, Suyal DC, Nand V (2022) Spectroscopy and its advancements for environmental sustainability. In: Suyal DC, Soni R (eds) Bioremediation of environmental pollutants. Springer, Cham, pp 317–338

Radjenovic J, Duinslaeger N, Avval SS, Chaplin BP (2020) Facing the challenge of poly- and perfluoroalkyl substances in water: is electrochemical oxidation the answer? Environ Sci Technol 54:14815–14829. https://doi.org/10.1021/acs.est.0c06212

Ranjan P, Maithani D, Suyal DC, Singh AK, Giri K, Sharma VK, Soni R (2022) Microbial fuel cells for wastewater treatment. In: Suyal DC, Soni R (eds) Bioremediation of environmental pollutants. Springer, Cham, pp 53–74

Rashid MM, Chowdhury M, Talukder MR (2020) Textile wastewater treatment by underwater parallel-multi-tube air discharge plasma jet. J Environ Chem Eng 8:104504. https://doi.org/10.1016/j.jece.2020.104504

Renault F, Sancey B, Badot P-M, Crini G (2009) Chitosan for coagulation/flocculation processes – an eco-friendly approach. Eur Polym J 45:1337–1348. https://doi.org/10.1016/j.eurpolymj.2008.12.027

Reza MS, Yun CS, Afroze S, Radenahmad N, Bakar MSA, Saidur R, Taweekun J, Azad AK (2020) Preparation of activated carbon from biomass and its' applications in water and gas purification, a review. Arab J Basic Appl Sci 27:208–238. https://doi.org/10.1080/25765299.2020.1766799

Roussy J, Van Vooren M, Dempsey BA, Guibal E (2005a) Influence of chitosan characteristics on the coagulation and the flocculation of bentonite suspensions. Water Res 39:3247–3258. https://doi.org/10.1016/j.watres.2005.05.039

Roussy J, Van Vooren M, Guibal E (2005b) Chitosan for the coagulation and flocculation of mineral colloids. J Dispers Sci Technol 25:663–677. https://doi.org/10.1081/DIS-200027325

Routoula E, Patwardhan SV (2020) Degradation of anthraquinone dyes from effluents: a review focusing on enzymatic dye degradation with industrial potential. Environ Sci Technol 54:647–664. https://doi.org/10.1021/acs.est.9b03737

Sahu B, Srivastava A, Suyal DC, Kumar R, Soni R (2022) Role of biochar in wastewater treatment and sustainability. In: Suyal DC, Soni R (eds) Bioremediation of environmental pollutants. Springer, Cham, pp 339–349

Salazar H, Martins PM, Santos B, Fernandes MM, Reizabal A, Sebastián V, Botelho G, Tavares CJ, Vilas-Vilela JL, Lanceros-Mendez S (2020) Photocatalytic and antimicrobial multifunctional nanocomposite membranes for emerging pollutants water treatment applications. Chemosphere 250:126299. https://doi.org/10.1016/j.chemosphere.2020.126299

Salman JM, Hameed BH (2010) Adsorption of 2,4-dichlorophenoxyacetic acid and carbofuran pesticides onto granular activated carbon. Desalination 256:129–135. https://doi.org/10.1016/j.desal.2010.02.002

Shattar SFA, Zakaria NA, Foo KY (2017) Utilization of montmorillonite as a refining solution for the treatment of ametryn, a second generation of pesticide. J Environ Chem Eng 5:3235–3242. https://doi.org/10.1016/j.jece.2017.06.031

Sheela T, Nayaka YA, Viswanatha R, Basavanna S, Venkatesha TG (2012) Kinetics and thermo-dynamics studies on the adsorption of Zn(II), Cd(II) and Hg(II) from aqueous solution using zinc oxide nanoparticles. Powder Technol 217:163–170. https://doi.org/10.1016/j.powtec.2011.10.023

Sheikh M, Pazirofteh M, Dehghani M, Asghari M, Rezakazemi M, Valderrama C, Cortina J-L (2020) Application of ZnO nanostructures in ceramic and polymeric membranes for water and wastewater technologies: a review. Chem Eng J 391:123475. https://doi.org/10.1016/j.cej.2019.123475

Shirani M, Semnani A, Haddadi H, Habibollahi S (2014) Optimization of simultaneous removal of methylene blue, crystal violet, and fuchsine from aqueous solutions by magnetic nay zeolite composite. Water Air Soil Pollut 225:2054. https://doi.org/10.1007/s11270-014-2054-2

Shirmardi M, Mahvi AH, Hashemzadeh B, Naeimabadi A, Hassani G, Niri MV (2013) The adsorption of malachite green (MG) as a cationic dye onto functionalized multi walled carbon nanotubes. Korean J Chem Eng 30:1603–1608. https://doi.org/10.1007/s11814-013-0080-1

Singh D, Bhasin S, Madan N, Singh M, Suyal N, Singh N, Soni R, Suyal DC (2021) Metagenomics: insights into microbial removal of the contaminants. In: Bhatt P, Gangola S, Udayanga D, Kumar G (eds) Microbial technology for sustainable environment. Springer, Singapore, pp 293–306

Skouteris G, Rodriguez-Garcia G, Reinecke SF, Hampel U (2020) The use of pure oxygen for aeration in aerobic wastewater treatment: a review of its potential and limitations. Bioresour Technol 312:123595. https://doi.org/10.1016/j.biortech.2020.123595

Sun M, Zhang Z, Liu G, Lv M, Feng Y (2021) Enhancing methane production of synthetic brewery water with granular activated carbon modified with nanoscale zero-valent iron (NZVI) in anaerobic system. Sci Total Environ 760:143933. https://doi.org/10.1016/j.scitotenv.2020.143933

Suopajärvi T, Liimatainen H, Hormi O, Niinimäki J (2013) Coagulation–flocculation treatment of municipal wastewater based on anionized nanocelluloses. Chem Eng J 231:59–67. https://doi.org/10.1016/j.cej.2013.07.010

Suyal DC, Soni R (2022) Bioremediation of environmental pollutants. Springer, Cham

Takeuchi N, Yasuoka K (2021) Review of plasma-based water treatment technologies for the decomposition of persistent organic compounds. Jpn J Appl Phys 60:SA0801. https://doi.org/10.35848/1347-4065/abb75d

Tatsi AA, Zouboulis AI, Matis KA, Samaras P (2003) Coagulation–flocculation pretreatment of sanitary landfill leachates. Chemosphere 53:737–744. https://doi.org/10.1016/S0045-6535(03)00513-7

Teh CY, Budiman PM, Shak KPY, Wu TY (2016) Recent advancement of coagulation–flocculation and its application in wastewater treatment. Ind Eng Chem Res 55:4363–4389. https://doi.org/10.1021/acs.iecr.5b04703

Teng J, Liu G, Liang J, You S (2020) Electrochemical oxidation of sulfadiazine with titanium suboxide mesh anode. Electrochim Acta 331:135441. https://doi.org/10.1016/j.electacta.2019.135441

Thuptimdang P, Siripattanakul-Ratpukdi S, Ratpukdi T, Youngwilai A, Khan E (2021) Biofiltration for treatment of recent emerging contaminants in water: current and future perspectives. Water Environ Res 93:972–992. https://doi.org/10.1002/wer.1493

Vinardell S, Astals S, Peces M, Cardete MA, Fernández I, Mata-Alvarez J, Dosta J (2020) Advances in anaerobic membrane bioreactor technology for municipal wastewater treatment: a 2020 updated review. Renew Sust Energ Rev 130:109936. https://doi.org/10.1016/j.rser.2020.109936

Wang J, Yao J, Wang L, Xue Q, Hu Z, Pan B (2020) Multivariate optimization of the pulse electrochemical oxidation for treating recalcitrant dye wastewater. Sep Purif Technol 230:115851. https://doi.org/10.1016/j.seppur.2019.115851

Wei H, Gao B, Ren J, Li A, Yang H (2018) Coagulation/flocculation in dewatering of sludge: a review. Water Res 143:608–631. https://doi.org/10.1016/j.watres.2018.07.029

Xiao R, Wei Y, An D, Li D, Ta X, Wu Y, Ren Q (2019) A review on the research status and development trend of equipment in water treatment processes of recirculating aquaculture systems. Rev Aquac 11:863–895. https://doi.org/10.1111/raq.12270

Xu H, Yang B, Liu Y, Li F, Song X, Cao X, Sand W (2021) Evolution of microbial populations and impacts of microbial activity in the anaerobic-oxic-settling-anaerobic process for simultaneous sludge reduction and dyeing wastewater treatment. J Clean Prod 282:124403. https://doi.org/10.1016/j.jclepro.2020.124403

Xu J, Chen Y, Cao M, Wang C, Guo P (2022) Highly efficient solar steam generation of polyamide 6 membrane modified with graphene oxide and Au nanoparticles. J Mater Res 37:1475–1485. https://doi.org/10.1557/s43578-022-00552-y

Yadav S, Soontarapa K, Padaki M, Balakrishna RG, Lai J-Y (2019) Supplementing multi-functional groups to polysulfone membranes using Azadirachta indica leaves powder for effective and highly selective acid recovery. J Hazard Mater 369:1–8. https://doi.org/10.1016/j.jhazmat.2019.02.010

Yagub MT, Sen TK, Afroze S, Ang HM (2014) Dye and its removal from aqueous solution by adsorption: a review. Adv Colloid Interf Sci 209:172–184. https://doi.org/10.1016/j.cis.2014.04.002

Yin X, Meng X, Zhang Y, Zhang W, Sun H, Lessl JT, Wang N (2018) Removal of V (V) and Pb (II) by nanosized TiO2 and ZnO from aqueous solution. Ecotoxicol Environ Saf 164:510–519. https://doi.org/10.1016/j.ecoenv.2018.08.066

Yin W, Zhang W, Zhao C, Xu J (2019) Evaluation of removal efficiency of Ni(II) and 2,4-DCP using in situ nitrogen-doped biochar modified with aquatic animal waste. ACS Omega 4:19366–19374. https://doi.org/10.1021/acsomega.9b02769

Yu W, Lian F, Cui G, Liu Z (2018) N-doping effectively enhances the adsorption capacity of biochar for heavy metal ions from aqueous solution. Chemosphere 193:8–16. https://doi.org/10.1016/j.chemosphere.2017.10.134

Yu G, Wang G, Chi T, Du C, Wang J, Li P, Zhang Y, Wang S, Yang K, Long Y, Chen H (2022) Enhanced removal of heavy metals and metalloids by constructed wetlands: a review of approaches and mechanisms. Sci Total Environ 821:153516. https://doi.org/10.1016/j.scitotenv.2022.153516

Yukhajon P, Somboon T, Sansuk S (2023) Enhanced adsorption and colorimetric detection of tetracycline antibiotics by using functional phosphate/carbonate composite with nanoporous network coverage. J Environ Sci 126:365–377. https://doi.org/10.1016/j.jes.2022.04.009

Yuvaraja G, Prasad C, Vijaya Y, Subbaiah MV (2018) Application of ZnO nanorods as an adsorbent material for the removal of As(III) from aqueous solution: kinetics, isotherms and thermodynamic studies. Int J Ind Chem 9:17–25. https://doi.org/10.1007/s40090-018-0136-5

Zdarta J, Nguyen LN, Jankowska K, Jesionowski T, Nghiem LD (2022) A contemporary review of enzymatic applications in the remediation of emerging estrogenic compounds. Crit Rev Environ Sci Technol 52:2661–2690. https://doi.org/10.1080/10643389.2021.1889283

Zeghioud H, Nguyen-Tri P, Khezami L, Amrane A, Assadi AA (2020) Review on discharge plasma for water treatment: mechanism, reactor geometries, active species and combined processes. J Water Process Eng 38:101664. https://doi.org/10.1016/j.jwpe.2020.101664

Zhang X, Chen J, Li J (2020) The removal of microplastics in the wastewater treatment process and their potential impact on anaerobic digestion due to pollutants association. Chemosphere 251:126360. https://doi.org/10.1016/j.chemosphere.2020.126360

Wastewater Treatment: Perspective and Advancements

11

Divya Goel, Vineet Kumar Maurya, and Sudhir Kumar

Abstract

Water is a basic necessity on which the human life dwells. Water pollution through human activities is making it unfit for its legitimate use. This makes a huge impact on the environment as well as public health bearing considerable economic and social costs. There are numerous ways of water pollution among which the industrial waste and agricultural pollution, e.g. chemical fertilizers and pesticides, are major contributors. The water pollution negatively affects the declining water table and levels in major water bodies. There is a pressing need for efficient as well as cost-effective wastewater treatment strategies and processes to make the clean water available for domestic and industrial use. In this chapter, we are going to discuss about the developments in wastewater treatment like microalgae, membrane-based nanofiltration, and photocatalysis-based treatment of wastewater. The principle, application, and their effectiveness toward remediation of wastewater to meet the reusable water standards will also be addressed. The processes though conventional are emerging as effective ways to industrialize wastewater treatment with certain modifications and combinations in reactors and the material used for treatment. The design of reactors for these wastewater treatment technologies has also been discussed. Wastewater treatment via microalgae bioremediation has proven to be very cost efficient as it generates biomass, which can be further used in different biological processes. Nanofiltration and photocatalysis have the advantage of minimum contamination,

---

D. Goel · S. Kumar (✉)
Department of Biotechnology, H.N.B. Garhwal University (A Central University), Srinagar Garhwal, Uttarakhand, India

V. K. Maurya
Department of Botany and Microbiology, H.N.B. Garhwal University (A Central University), Srinagar Garhwal, Uttarakhand, India

© The Author(s), under exclusive license to Springer Nature Singapore Pte Ltd. 2023
R. Soni et al. (eds.), *Current Status of Fresh Water Microbiology*,
https://doi.org/10.1007/978-981-99-5018-8_11

and they can be optimized for better removal of pollutants from wastewater. These emerging wastewater treatment technologies can form an outline for development of better processes for wastewater remediation and can play an important role in water conservation through reuse and recycle of treated wastewater.

Keywords

Water pollution · Wastewater treatment · Nanofiltration · Photocatalysis · Microalgae

11.1 Introduction

11.1.1 Water: Basic Source of Life Sustainability

Water is a basic need and resource for human life sustainability and survivability. There is a total of 1400 million km^3 of water on earth. However, only a fraction of this total water is available for human use. The world population doubles in almost every 30 years making the demand for fresh water stratospheric. The availability of water per annual capita is decreasing rapidly. It was about 3300 m^3 in 1960 and then decreased to 1250 m^3 in 1995. It has been estimated that it will further decline by about 50%, leaving only 650 m^3 by 2025. In most of the countries, 70–90% of the available water is used in agriculture (Abdel-Raouf et al. 2012; FAO 2014). It has been documented earlier that though the population of the world is increasing threefold, the water consumption has skyrocketed at sixfold creating an alarming situation. It has been predicted that by 2050 about 40% of the world population will be living in severe water-deficit areas (Leflaive et al. 2012). A normal person requires about 20 l of water per day for drinking, cooking, and washing. However, there are regions where people have less than a liter per person for these purposes. In India, only 20% of the population has access to tap water (Shagun 2019). This precious resource is being converted to waste through human activities including industrial technologies and domestic purposes. Other major factors such as population growth and density, industrialization with nonrenewable sources of water, dwindling economic situation, and institutionalization of majority of human activities also contribute toward water scarcity.

11.1.2 Water Pollution

Water pollution is one of the major reasons behind the declining levels of water resources and scarcity of water that humans are experiencing today. Pollution is one of the worst vices of man-made inventions and industries. All these human activities are releasing a lot of natural and nonnatural, organic, and inorganic compounds into the environment leading to pollution. Water pollution not only affects human health

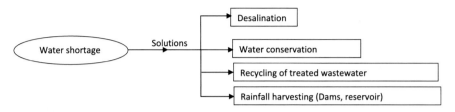

Fig. 11.1 Flowchart depicting solutions to problem of water shortage

and ecosystem negatively but also makes the freshwater resources unfit for use. This decreases the availability of water resources, thus creating a situation of water scarcity (Fig. 11.1). The problem of water shortage has few solutions like desalination, water conservation efforts, recycling and reuse of wastewater, and rainfall conservation projects (Chauhan and Kumar 2020; Vedavyasan 2007; Xie et al. 2009). The current desalination capacity across the globe is 97.4 million cubic meters per day (m^3/day). As of 30 June 2018, more than 20,000 desalination plants had been contracted around the world.

11.1.3 Clean Water and Sanitation: An Important Sustainable Development Goal

In India, lack of proper sanitation results in a lot of human wastewater to be directly released into the water bodies without treatment. In several towns and villages, which are being converted to towns, water supply networks are being setup without proper sanitation facilities like establishment of sewage systems and reforming the existing ones. Though this improves the water supply facility, there is an increase in water pollution and health hazards among the people of these civilizations. The contents that are being released into the water bodies include salts, solid suspension, and pathogens, which together create oxygen deficit in the water.

The wastage of water has a lot of economic as well as health implications. The world is recognizing the efforts that work towards a sustainable and economically reliable treatment of wastewater. Wastewater treatment is a tool, which has the capability to convert this waste to resource with the help of emerging technologies. Proper discharge of wastewater and treatment technologies entails a complicated process with health and economic consequences. Currently, India has the capacity to treat only 37% (approximately 22,693 million l/day) of the sewage waste generated, which is approximately 61,754 million l/day (Sugam et al. 2017). Ineffective wastewater treatment is an example of the inept use of water resources and is contributing negatively to water scarcity our country is facing today. Availability of clean water and proper sanitation is one of the "Sustainable Development Goals" as given by the UN. This goal can be achieved through effective and economic wastewater treatment strategies.

The contents of wastewater are a multifaceted mixture of natural and man-made materials. It consists of both organic and inorganic compounds. Carbohydrates, proteins, fats, and a number of acidic compounds outline the majority of carbon present in the wastewater. The inorganic compounds present in wastewater majorly consist of sodium, calcium, potassium, magnesium, chlorine, sulfur, phosphates, ammonium salts, and heavy metals (Horan 1990; Lim et al. 2010; Tebbutt 1983). Wastewater creates a perfect niche for growth of thousands of microorganisms, which include protozoan, viruses, and bacterial species. Though majority of these microorganisms are harmless, there are pathogenic organisms as well. The pathogenic species in wastewater can cause cholera, typhoid, dysentery, tuberculosis, hepatitis, and many other infections (Henery 1989).

There are numerous types of contaminants in wastewater that need to be removed for making it suitable for reuse. Based on the type of contaminants, the wastewater has been categorized into different classes: domestic wastewater (includes sewage), agricultural wastewater, food-processing wastewater, pharmaceutical wastewater, antibiotic-rich wastewater, academic laboratory wastewater, industrial wastewater (includes textile, packaging, paper, etc.), dairy wastewater, and petroleum sludge. According to the type of contaminants in the wastewater, different remediation and treatment technologies come into play for wastewater treatment.

11.1.4 Conventional Steps of Wastewater Treatment

The treatment of wastewater is a four-step process starting from preliminary treatment followed by primary, secondary, and tertiary treatment processes. Each step is meant to play an important role in removal of pollutants from wastewater. Preliminary treatment involves removal of large solid materials discharged into the water through sewers. These materials can be rags, plastics, wood, heavy stones, fecal discharge, etc., which can block the treatment plant or hamper the functioning of the equipment. These large-sized materials are stopped via bars with a space of about 20–60 mm between them. The stopped material is removed by raking. The grit materials, that is, sand, clay, and pebbles, are removed by decreasing the speed of the wastewater flow. At slow speed, the grit gets collected with large floating material and is removed along with it (Gray 2004; Tebbutt 1983).

The primary treatment of wastewater is meant to remove solids through sedimentation. It includes the passage of wastewater through sedimentation tanks where solids are removed after they settle down via gravity. Almost 70% of solids are removed through this process. However, during primary treatment, the microbiological species removal fluctuates as each microorganism has different sedimentation rate (Sivagurunathan et al. 2015).

The secondary treatment of wastewater includes the removal of biological oxygen demand (BOD). BOD refers to the oxygen needed by microorganisms to convert organic matter into CO_2 and H_2O. The higher BOD suggests that more oxygen is needed by the microorganisms indicating that there is high level of organic matter in the water. BOD is referred to as water quality indicator or water pollution indicator.

BOD is removed by using certain heterotrophic bacteria that metabolize the organic matter for their own growth (Horan 1990). Biological units are used for operation in treatment plants for removal of BOD. Depending upon the mode in which the biological units are utilized, there are majorly two types of reactors:

1. **Fixed film reactors** wherein the microorganisms are used as a biofilm attached to a fixed surface. The organic matter is absorbed onto the surface of the biofilm and gets aerobically metabolized (Sivagurunathan et al. 2015).
2. **Suspension reactors** wherein the microorganisms are suspended or mixed in the wastewater via motorized stirring or air circulation channels. The suspended microorganisms metabolize the organic matter in suspension (Rao and Subrahmanyam 2004).

The tertiary treatment of wastewater helps in the removal of all organic material. It can be achieved by both biological and chemical methods. Tertiary treatment removes excess ammonium, nitrogen, and phosphorous in wastewater (Horan 1990). Though primary and secondary treatment result in a clear effluent, the inorganic and organic content of the effluent can result in eutrophication and degrade the water quality when released into the environment. Nitrogen and phosphorus are generally removed from the wastewater through a number of processes including reverse osmosis, adsorption, ultrafiltration, and biological methods (Oswald 1988).

The primary, secondary, and tertiary treatment of wastewater cannot ensure 100% removal of the contaminant load. There are multiple quaternary treatment processes as well, which are meant to disinfect the water, that is, to eliminate harmful disease-causing microorganisms and remove any heavy metals and other inorganic compounds present in the treated water. Quaternary treatment can also be done via both chemical and biological processes.

Numerous processes are employed in tertiary and quaternary treatment of waste-water. Here are some efficient and economically sustainable wastewater treatment technologies:

1. Reverse osmosis-based membrane filtration (Ben Aim et al. 1993)
2. Nanofiltration (nanoparticles in membrane bioreactor) (Barakat 2011; Choi et al. 2002)
3. Electrodialysis (Caprarescu et al. 2012; Habib et al. 2018; Hell and Lahnsteiner 2002; Lee et al. 2002)
4. Absorption through hydrogels (Bekiari et al. 2008; Ju et al. 2009; Peng et al. 2012)
5. Nanocomposites (Badruddoza et al. 2011; Hasan et al. 2016; Horst et al. 2016; Rajendran et al. 2016; Saravanan et al. 2013)
6. Iron-based nanoparticles (Adegoke and Stenström 2019; Aragaw et al. 2021)
7. Microalgae for bioremediation (Phang 1991; Shi et al. 2007)
8. Ozone treatment (Glaze 1987; Hernández-Leal et al. 2011; Perkowski et al. 1996; Rice 1996)

9. Microbial fuel cells for bioremediation (Logan 2004; Zhang et al. 2016)
10. Photocatalysis (Ren et al. 2021)

In this chapter, we are going to discuss three advanced strategies toward wastewater treatment: (1) microalgae-based bioremediation, (2) nanofiltration, and (3) photocatalysis. We will confer about their principles, reactor design, pollutant removal, and other aspects.

11.2 Microalgae-Based Bioremediation

Microalgae-based treatment is primarily used as a tertiary treatment procedure and has been proposed as a potential secondary treatment process as well. The conventional wastewater treatment methods have certain drawbacks such as large area requirement, huge amount of energy consumption, and heavy monetary burden of operation and maintenance. The microalgae offer an efficient alternative with a lot of advantages.

Microalgae are the micro form of mainly water-based photosynthetic organisms, which cannot be seen with naked eye. Microalgae are found in freshwater resources as well as in saline water sources. They can be categorized as both prokaryotic cyanobacteria (blue green algae) and eukaryotic microorganisms. Microalgae are primarily differentiated through their colors into green algae, red algae, golden algae, and diatoms (Stengel et al. 2011).

Algae are ecologically important organisms as they are major producers of oxygen and food suppliers for the aquatic animals. Biomass obtained through algae production can be used as source of multiple industrial and pharmaceutical products. The principle behind the use of microalgae for wastewater treatment is the use of sunlight and organic matter, inorganic nutrients, and heavy metals from industrial wastewater for algal growth, which in turn produces biomass to be collected and used in further processes (Fig. 11.2) (Oswald et al. 1957).

There are certain factors that affect the nutrient removal from the wastewater and algal growth during the treatment process. These factors include microalgae species and properties of wastewater, that is, N/P ratio, pH, light, and temperature. These factors are optimized for maximum nutrient removal and supreme algal growth via pre-treatment of wastewater, algal acclimatization with wastewater, and blend of

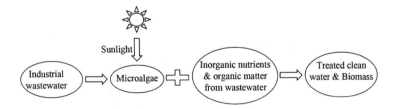

Fig. 11.2 Process of wastewater treatment via microalgae

different wastewater (Cai et al. 2013; Li et al. 2019). There are certain limitations that are posed by the composition of wastewater. High ammonium concentration is one of the major factors that affect the pH of the effluent and the microalgae growth as well. It can have detrimental effects on microalgae diversity in the bioreactor giving rise to resistant species, which have high nitrogen-metabolism-related enzymes. This helps in better detoxification of ammonia in the wastewater (Zhao et al. 2019).

11.2.1 Mechanism of Carbon, Nitrogen, Phosphorus, and Heavy Metal Utilization by Microalgae

Carbon: Microalgae can use both organic carbon, that is, sugars, alcohol, acids, and inorganic carbon, that is, CO_2 and $COOH^-$, which in turn is almost equal to 50% of the biomass that is generated via microalgal growth. Organic carbon is majorly utilized for generation of energy and mass accumulation via glycolysis. Inorganic carbon is mostly consumed through Calvin cycle wherein CO_2 is converted to ATP and NADPH (Singh and Mishra 2019; Uggetti et al. 2018).

Nitrogen: Microalgae requires nitrogen for a number of processes in their growth. It can exploit inorganic nitrogen in the form of ammonia (NH_4), nitrate (NO_3^-), and nitrite (NO_2^-) and organic nitrogen in the form of urea and amino acids. Microalgae produces the basic building blocks of growth, that is, proteins, nucleic acids, and phospholipids via utilization of organic and inorganic nitrogen. Inorganic nitrogen is absorbed through the cell membrane of algae and organic nitrogen is digested through metabolic pathways (Sniffen et al. 2018; Zhu et al. 2013).

Phosphorous: Microalgae consumes both organic and inorganic phosphorus for the production of ATP, nucleic acids, phospholipids, and proteins. It utilizes the inorganic phosphorus in the form of PO_4^{3-}, HPO_3^{2-}, and $H_2PO_4^-$, which is the preferred form of uptake by microalgae. In case of organic phosphorus, it converts it into inorganic form via phosphatase enzyme present on the cell surface and then further utilizes it (Dyhrman et al. 2012; Su 2021).

Heavy metals: Microalgae use certain heavy metals like boron, copper, iron, and zinc for incorporation into enzymatic reactions and cell metabolism processes. Cell wall and membrane of algae have numerous binding sites for heavy metal ions facilitated by large surface area. Primary way of heavy metal removal is biosorption by microalgae wherein the heavy metal ions are transported through cell membrane into the cytoplasm and adhere to the internal binding sites present on the proteins and peptides (Leong and Chang 2020; Monteiro et al. 2012; Singh et al. 2021).

11.2.2 Microalgae Growth and Cultivation System for Wastewater Treatment

There are three modes for microalgae growth and cultivation that can be optimized for maximum biomass production, which in turn gives us treated clean water.

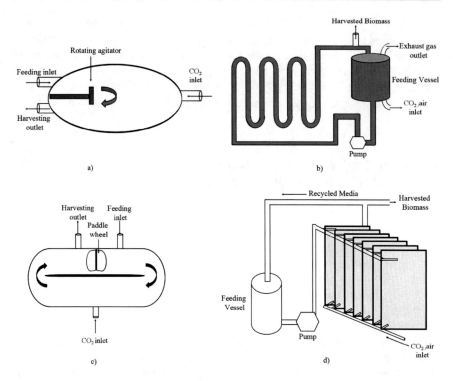

Fig. 11.3 Cultivation systems for microalgae production and wastewater treatment: open cultivation systems, (**a**) circular pond and (**b**) tubular photobioreactor, (**c**) raceway pond; closed cultivation systems, and (**d**) flat plate photobioreactor (Image reference: Faried et al. (2017))

1. Autotrophic: It generally takes place in open ponds where algae utilize natural light sources and inorganic carbon for biomass accumulation.
2. Heterotrophic: Algae cultivation is done in closed chambers where algae utilize organic source of carbon and does not require light for biomass production.
3. Mixotrophic: This mode includes utilization of both organic and inorganic form of carbon by algae in light conditions for maximum biomass production.

For large-scale production of algal biomass and treatment of wastewater, open cultivation system and closed cultivation systems have been designed (Fig. 11.3). Open cultivation system consists of circular ponds and raceway ponds. Both of these systems implement stirring and agitation mechanisms for continuous aeration, suspension of algae, and optimum flow of water. They both are autotrophic cultivation mode with advantage of low-cost construction and operation along with great production area and easy to clean after cultivation. There are certain limitations of open cultivation system, which includes water evaporation, influence of environmental factors, contamination of other microorganisms, and requirement of large area (Lee 2001; Pulz 2001). Requirement of large open area and high evaporation rate through the same is another cause of concern during the microalgae wastewater treatment. An action plan to overcome high evaporation rates includes the usage of deeper ponds and short hydraulic retention time with the limitation of light penetration (Young et al. 2017). Obstruction of light by solids present in the wastewater also impedes the microalgae growth in the reactors. The usage of broad filters as primary treatment to stop the solid material from entering the reactor helps to control the situation and minimize the light obstruction in the reactor. Closed cultivation system consists of vertical tubular photobioreactors and flat plate horizontal photobioreactors. In case of closed bioreactors, the variable conditions can be maintained accordingly, and better results can be obtained with stable parameters. Moreover, there is no contamination through external microorganisms. Though high capital requirement for assembling of a closed photobioreactor remains a concern, the use of wastewater as nutrient and water source for growth of microalgae and recovery of biomass and clean water after the process makes it cost-effective (Gupta et al. 2019; Posten 2009; Ugwu et al. 2008; Wen et al. 2016).

11.3 Nanofiltration

Nanofiltration refers to the process in which membranes, with a pore size of 0.5–2 nm and corresponding molecular weight of 100–1000 Da, are used for wastewater treatment. This process assists in improving the quality of drinking water as it helps in removal of suspended solids, bacteria and viruses, metal ions, and oil particles as well (Guo et al. 2022; Zhao et al. 2021). It is one of the applications of membrane filtration technology with some advanced modifications to achieve better efficiency and lower energy consumption. Certain modes of membrane filtration are size exclusion, ion exchange, and adsorption.

11.3.1 The Design and Specifications of Nanofiltration Unit

In size exclusion mode, the contaminants from wastewater are excluded via the membrane with diameter larger than the pore size of the membrane. The micro-/ultrafiltration techniques are unable to remove heavy metal ions and microorganisms such as viruses. Nanofiltration provided a solution with a smaller pore size, thereby helping in removal of heavy metal ions, viruses, and other pollutants from wastewater. The ion exchange mode of nanofiltration is where nanofiltration is combined with ion-exchange technology. In this process, firstly the water is subjected to ion exchange wherein exchange of equivalent of ions takes place and the osmotic potential of water is governed by the concentration of ions in the effluent. This effluent is then subjected to nanomembranes wherein the maximum of the ions, pollutants, and pathogens are removed via filtration. Nanomembranes display a greater efflux potential even with low osmotic pressure (Sarkar and SenGupta 2009). Adsorption nanofiltration is the process where adsorption is combined with nanofiltration. Adsorption of heavy metals, biological molecules, organic matter, and inorganic ions is followed by nanofiltration via membrane separation

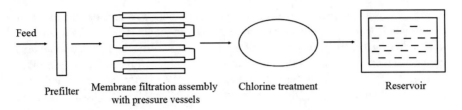

Fig. 11.4 A flow diagram of nanofiltration for wastewater treatment

technology. This increases the efficiency of contaminant removal from wastewater and aids in improving the drinking water quality that can be obtained after this course of action (Hanif et al. 2021). A commonly used nanofiltration unit is composed of a prefilter with a pore size of around 50 μm. This prefilter helps in the removal of the sediments and other pollutants of large size. The effluent is then fed into the membrane filtration unit, which consists of nanomembranes, and the pressure of circulation flow is increased. This results in a clean water stream. This clean water is then subjected to chlorine treatment and the treated water flows to the reservoir (Fig. 11.4) (Aliverti et al. 2011). The membrane filtration unit can be modified according to the quality and properties of the wastewater feed, and it can also be combined with other energy efficient techniques as discussed above.

Recently, Higgins et al. suggested that pH of feedwater affects the adsorption as well as the removal of ibuprofen (IBU) from the wastewater. They illustrated that removal of IBU is directly proportional to the increase in feed pH, whereas adsorption of IBU is indirectly related to the increase in feed pH. It was also conveyed that stainless steel equipment of flat-sheet reactor also affects the adsorption of the IBU up to 27.3% (Higgins and Duranceau 2020). Another development in nanofiltration is the preparation of membrane with ionic liquids (ILs). Ionic liquids are basically molten organic salts with high polarity as well as thermal and chemical stability. ILs are involatile and are known to dissolve both organic and inorganic materials (Zheng et al. 2020).

11.3.2 Thin-Film Composite Structure

Nanomembranes have been optimized for better separation performance via change of material and structure. Recent nanomembranes consist of thin-film composite structure (TFC). Thin-film composite membranes are basically molecular semipermeable membranes, which are custom prepared for water purification systems as a multilayered film and acts as a molecular filter for removal of impurities from water. A composite nanofiltration membrane can be described as a bilayer film, which is constructed via two-step process. Firstly, a thick, porous, and nonselective layer is formed followed by overcoating with an ultrathin barrier layer. The nonselective layer is normally composed of woven fabric, and the ultrathin barrier layer is made up of chemical polymers. Both these layers are always different in their chemical composition (Petersen 1993). Usually, in a thin-film composite membrane, the

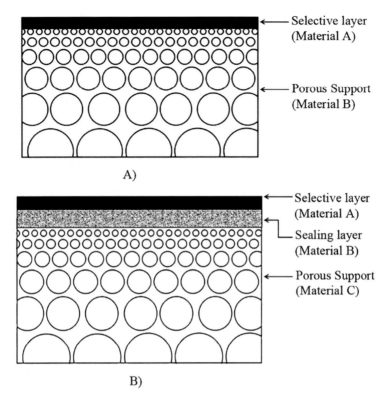

Fig. 11.5 Schematic representation of thin film composite membrane: (**a**) single-layer composite and (**b**) multilayer composite (Pinnau 2000)

selective layer is constructed through a process called as interfacial polymerization (IP). IP takes place at the interface of two immiscible solutions, aqueous and organic, containing different monomers. The most common TFC membrane is made up of polyamide, which is formed by IP between trimesoyl chloride (TMC) and a diamine. Desired membrane structure and separation performance can be obtained by optimization of membrane properties such as pore size, thickness, roughness, and hydrophilicity and understanding its chemical properties (Bui et al. 2011; Hermans et al. 2015; Ismail et al. 2015; Verbeke et al. 2019).

There can be two forms of composite membranes: (1) single-layer and (2) multilayer (Fig. 11.5). The single-layer composite membrane is assembled with two components only. It consists of a porous material over a woven or unwoven fabric support, and on this porous layer, a thin selective layer is formed through solution coating method. During the solution coating method, a dilute polymer solution is poured over the porous layer to form a thin deposition with a thickness of less than 1 µm. However, it is very difficult to form a selective layer of such thickness without loopholes. These defects are formed as the polymer solution is sucked into the porous layer and a uniform coating of selective layer is not formed (Pinnau 2000).

To overcome this difficulty, a multilayer composite membrane is assembled. The suction by porous membrane is avoided by fabricating a sealing layer between porous support and ultrathin selective layer. This sealing layer helps in blocking the suction pores of the porous support and helps in the formation of a uniform and disruption-free selective layer. The sealing layer should be composed of a highly permeable material than the selective layer, so that it does not interfere with the mass transport during the separation process (Pinnau 2000). Another method has also been developed by Malakian et al. wherein the colloidal fouling is minimized by using line and groove patterns on NF membranes by thermal embossing. Colloidal fouling refers to the accumulation of silica, iron, aluminum, and other inorganic compounds, which decrease the permeability of the membrane (Malakian et al. 2020). Ozonation before the nanofiltration has been shown to be beneficial for the wastewater treatment process. Pre-ozonation results in mineralization of organic matter, thus decreasing the membrane fouling and maintaining the membrane permeability. The flux rate can be sustained through pre-ozonation (Amadou-Yacouba et al. 2022).

11.3.3 Water Purification by Nanofiltration Process

Nanofiltration as a technique has been applied on many fronts of water purification process, namely, groundwater treatment, surface water treatment, and brackish water treatment. Nanomembranes are also used at the point of use and water reuse. Nanofiltration excels in removal of most pollutants, pathogens, and natural organic matter from water. The process involves pretreatment, nanofiltration, and post treatment of wastewater. Pretreatment majorly includes coagulation, sedimentation, and micro-/ultrafiltration to largely remove particles, organic matter, and bacteria, thus reducing the membrane damage during nanofiltration. Nanofiltration through tailored separation potential then removes the pollutants up to a significant level. The nanomembranes can be coupled with other techniques for better separation performance and efficient wastewater treatment. Ion exchange, coagulation, electrical filtration, plant-derived nanoparticles, etc. can be combined with nanomembranes for enhancing the efficiency of the treatment (Ang et al. 2016; Kumar and Chauhan 2022; Sarkar and SenGupta 2008). During the post treatment process, disinfecting agents are added to prevent the regrowth of bacteria and other microorganisms. Since nanofiltration is able to remove majority of the infectious pathogens, the concentration of disinfecting agent required is quite low during the post treatment process (Ohkouchi et al. 2013).

11.4 Photocatalytic Wastewater Treatment

Photocatalysis is emerging as a striking solution for degradation of natural organic matter and heavy metal ions in wastewater. Photocatalyst is a combination of word "photo" and "catalyst." Photo means light, and catalyst means accelerator of chemical reaction. A substance that absorbs light and in turn catalyzes a chemical reaction is known as a photocatalyst. The most common photocatalysts that are used for wastewater treatment are metal oxides, metal sulfides, oxysulfides, and oxynitrides. Some of the metal oxides that are used as photocatalysts are iron (III) oxide (Fe_2O_3), zinc oxide (ZnO), niobium pentoxide (Nb_2O_5), titanium oxide (TiO_2), tin (II) oxide (SnO_2), tungsten trioxide (WO_3), bismuth oxide (Bi_2O_3), vanadium oxide (V_2O_5), and zirconia (ZrO_2) (Raizada et al. 2017).

11.4.1 The Concept of Photocatalysis

The principle behind photocatalysis involves the activation of photocatalyst via light source, which instigates the reaction. The photocatalyst produces electron and hole pairs, which take the reaction forward. The electrons (e^-) present in the valence band are excited, and they are transferred to the conduction band leaving holes (h^+) behind in the valence band. The smaller the bandgap between valence band and conduction band, more visible light photons can be captured, and more electrons can be excited. The separation of electron and holes is necessary as electrons and holes present on the surface can directly be involved in the chemical reactions (Fig 11.6). The electrons can reduce the heavy metal ions, while holes can interact with hydroxyl ions OH^- or water molecules H_2O to form ·OH hydroxyl radicals. The excited electrons can also react with dissolved oxygen in water to form $·O_2^-$ superoxide radicals. The pollutants firstly get adsorbed on the surface of the catalyst and increase the redox capability of the catalyst. The radicals generated by the catalyst then participate in a sequence of chemical reactions wherein organic matter, heavy metal ions, and other contaminants present in water are degraded (Ren et al. 2021). However, the catalysis is slowed down by the recombination of electrons and holes. The large number of charge carriers results in recombination, which produces photons or heat, thereby decreasing the rate of photocatalysis. There has been development of different strategies to overcome the problem of recombination. Advances in the assembly of semiconductors with heterojunctions or external circuits with photocatalytic effects have provided an alternative with a solution to

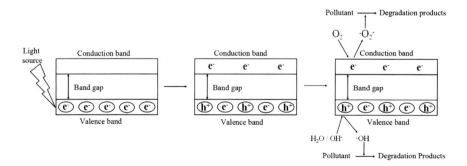

Fig. 11.6 Simplified working principle of photocatalytic unit

recombination and enhance the photocatalytic ability of the semiconductor (Wang et al. 2014; Zhang et al. 2013).

11.4.2 Semiconductor Photocatalysts Used for Degradation of Different Pollutants

Different photocatalytic reactors have been studied to degrade a variety of pollutants in wastewater. The organic matter or organic compounds like dyes, phenolic compounds, and hydrocarbons are decomposed into biodegradable molecules, which are further broken down into harmless CO_2 and H_2O. A cluster of photocatalytic semiconductors such as TiO_2, ZnO, Fe_2O_3, and C_3N_4 have been studied for degrading a variety of organic pollutants (Cao and Zhu 2008; Horikoshi and Serpone 2020; Lee et al. 2016; Moreira et al. 2019). Dyes have been shown to be degraded via TiO_2, ZnO, $BiFeWO_6$, SnO_2, and MoS_2. Similarly, phenolic compounds and petroleum hydrocarbons have been photocatalytically degraded by CuO-TiO_2, Fe_3O_4, or C_3N_4 (Ghasemi et al. 2016; Li and Lin 2020; Yang et al. 2020). The heavy metal ions with high valency are degraded or reduced to low-valence or zero-valence ions via photocatalytic process. The set of photocatalytic compounds that are used for degradation of different heavy metal ions include TiO_2-ZrO_2, Mn_3O_4-ZnO, $NiFeO_4$, Fe_3O_4-TiO_2, Fe_2O_3-C_3N_4, and nanoFe_3O_4 (Bi et al. 2019; Kadi et al. 2020; Li et al. 2017; Liu et al. 2019; Thomas and Alexander 2020; Yan et al. 2020). Pharmaceutical compounds such as antibiotics, anti-inflammatory drugs, lipid regulators, etc. are also degraded into harmless compounds with the help of photocatalysts. The semiconductor photocatalysts used for degradation of pharmaceutical compounds include TiO_2, CuO, Fe_3O_4, ZrO_2, and $NiFe_2O_4$ (Carbuloni et al. 2020; Khanmohammadi et al. 2021; Liu et al. 2020; Wang et al. 2019). The semiconductor photocatalysts that are used for removal or degradation of pesticides are TiO_2, ZnO, and $Ag/LaTiO_3$ (Shawky et al. 2020; Taghizade Firozjaee et al. 2018). The elimination of microorganisms from wastewater is a very crucial step for water to be drinkable. The photocatalysts majorly help in inactivation of microorganisms by production of reactive oxygen species (ROS) such as OH· (hydroxyl free radical) and $\cdot O_2^-$ (superoxide free radical). ROS helps in oxidation of nucleic acids, cell membrane, and other macromolecules leading to inhibition of respiratory cycle of the cell and leakage of important cell ingredients, thereby causing cell death. The photocatalysts that are employed for removal of microorganisms include TiO_2, $BiVO_4$, and Fe_2O_3 (Ng et al. 2016; Regmi et al. 2017; Song et al. 2019).

11.4.3 Structure and Design of Photocatalytic Reactor

Any process for wastewater treatment can be optimized and analyzed for better efficacy and low energy consumption. Similarly, photocatalytic reactors for wastewater treatment can be designed and developed according to the wastewater quality,

Fig. 11.7 Fluidized bed photoreactor (Image reference: Brame (2017))

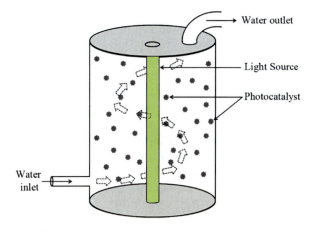

Fig. 11.8 Fixed bed photoreactor (Image reference: Ghuman et al. (2015))

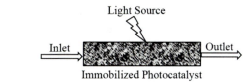

photocatalytic material, and light source. Mainly photoreactors can be classified into (1) fluidized bed reactors and (2) fixed bed reactors. This classification is based on the form of the photocatalyst to be used.

Fluidized bed photoreactors (FBPR) are designed on the principle that the liquid influent is passed over the stationary bed of solid particles (i.e., photocatalyst semiconductor material) with sufficient velocity to suspend the solid particles into the liquid influent. When the velocity of liquid influent is low, it doesn't disturb the stationary bed and simply passes through the holes and gaps in the bed. However, when the velocity of the fluid is increased, there is an expansion of the bed until the solid particles are totally suspended in the liquid and behave like a fluid. In the fluid chamber, there is a light source, which activates the photocatalyst. In FBPR, the catalysts are loaded on the granulated carrier from which they are suspended in the influent wastewater supply. The light source present in the chamber excites the catalyst particles, which then give rise to hydroxyl radicals that degrade the contaminants present in the water. The effluent, that is, treated water, then comes out of the filter (Fig. 11.7). FBPR are advantageous as they have large surface area, rapid reaction rate, and minimum transfer constraint. The only drawback is the recovery of catalyst semiconductor material, which is quite difficult resulting in blockage of the reactor, and reuse of the catalyst is almost impossible, thus making it tricky to scale up in the form of large capacity industry reactors (Dong et al. 2015; Enesca 2021).

Fixed bed photoreactors are based on fixed membrane-type reactors (Fig. 11.8). In this type of reactor, the photocatalytic semiconductor material is immobilized on a surface like glass, silica, clay, polymers, etc. The influent wastewater in low velocity is passed through the fixed bed of catalyst. The light source activates/excites the catalyst, which generates the hydroxyl free radicals in the influent and removes the pollutants from the wastewater. The effluent/clean water then comes out of the filter. Though the efficacy of fixed bed photoreactor is less than the FBPR, it is advantageous in terms of scaling up the reactor system as the photocatalyst is easy to recover and can be recycled and reused for a prolonged period of time (Ren et al. 2021; Vaiano et al. 2015).

Both fluidized bed photoreactor and fixed bed photoreactor are designed in two geometrical shapes: cylindrical and rectangular. In cylindrical photoreactors, the light is source is situated in the center of the cylinder or arranged in a circle inside the cylinder. The photocatalyst can be dispersed in the liquid inside or can be static on a substrate medium. The cylindrical reactors have advantage of circular motion aeration and proper mixing of the mobile catalyst as well as proper circulation of wastewater in case of immobilized catalyst. In rectangular photoreactors, the light sources can be adjusted according to the requirements, for example, on lateral sides of the chamber, evenly placed in the middle of the chamber, corners of the chamber, etc. The immobilized catalyst can also be arranged appropriately so as to obtain maximum reaction rate, for example, on the sides of the chamber, on the cover of the light source, on the roof of the chamber, or floor of the chamber. In case of liquid or fluidized catalyst, the flow rate and direction have to be optimized according to the shape of the chamber so as to obtain a homogeneous distribution of the catalyst (Enesca 2021).

11.5 Conclusions

Water is equivalent to life in all forms. People and governments, all over the world, have been warned about the colossal problem of water scarcity by the scientists and water management since the start of the twenty-first century. These alarm bells have been met with partial solutions and half-cooked attempts. It is now time to put in cohesive efforts and extraordinary steps to save this essential resource. In words of Jamie Linton "Water is what we make of it." Water is an indispensable element, which makes life sustainable on earth. This chapter sheds light on the wastewater treatment processes that can help not only save water but also recycle and reuse water. According to the World Water Council, major investments in wastewater treatment technologies have been able to help many developing countries to impede the water scarcity and improve the surface water quality. The microbial bioremediation, nanofiltration, and photocatalysis have been the most efficacious, cost-effective, low energy consuming, and conventional as well as modern wastewater treatment processes. These methods can be optimized for enhanced efficiency and performance with the proper knowledge of their working principle, which have been discussed in the chapter. Our aim is to develop a superior understanding of pros and cons of different wastewater treatment technologies and how they can be best developed for making water a resource rather than a waste through reuse and recycling of wastewater.

References

Abdel-Raouf N, Al-Homaidan AA, Ibraheem IB (2012) Microalgae and wastewater treatment. Saudi J Biol Sci 19:257–275. https://doi.org/10.1016/j.sjbs.2012.04.005

Adegoke AA, Stenström TA (2019) Chapter 17 - metal oxide nanoparticles in removing residual pharmaceutical products and pathogens from water and wastewater. In: Grumezescu AM (ed) Nanoparticles in pharmacotherapy. William Andrew Publishing, Norwich, pp 561–589. https://doi.org/10.1016/B978-0-12-816504-1.00016-8

Aliverti N, Callegari A, Capodaglio AG, Sauvignet P (2011) Nom removal from freshwater supplies by advanced separation technology. In: Hlavinek P, Winkler I, Marsalek J, Mahrikova I (eds) Advanced water supply and wastewater treatment: a road to safer society and environment. Springer, Dordrecht, pp 49–61

Amadou-Yacouba Z, Mendret J, Lesage G, Zaviska F, Brosillon S (2022) Impact of pre-ozonation during nanofiltration of MBR effluent. Membranes 12:341. https://doi.org/10.3390/membranes12030341

Ang WL, Mohammad AW, Benamor A, Hilal N, Leo CP (2016) Hybrid coagulation–NF membrane process for brackish water treatment: effect of antiscalant on water characteristics and membrane fouling. Desalination 393:144–150. https://doi.org/10.1016/j.desal.2016.01.010

Aragaw TA, Bogale FM, Aragaw BA (2021) Iron-based nanoparticles in wastewater treatment: a review on synthesis methods, applications, and removal mechanisms. Journal of Saudi Chemical Society 25:101280. https://doi.org/10.1016/j.jscs.2021.101280

Badruddoza AZM, Tay ASH, Tan PY, Hidajat K, Uddin MS (2011) Carboxymethyl-β-cyclodextrin conjugated magnetic nanoparticles as nano-adsorbents for removal of copper ions: synthesis and adsorption studies. Journal of Hazardous Materials 185:1177–1186. https://doi.org/10.1016/j.jhazmat.2010.10.029

Barakat MA (2011) New trends in removing heavy metals from industrial wastewater. Arabian Journal of Chemistry 4:361–377. https://doi.org/10.1016/j.arabjc.2010.07.019

Bekiari V, Sotiropoulou M, Bokias G, Lianos P (2008) Use of poly(N,N-dimethylacrylamide-co-sodium acrylate) hydrogel to extract cationic dyes and metals from water. Colloids and Surfaces A: Physicochemical and Engineering Aspects 312:214–218. https://doi.org/10.1016/j.colsurfa.2007.06.053

Ben Aim R, Liu MG, Vigneswaran S (1993) Recent development of membrane processes for water and waste water treatment. Water Science and Technology 27:141–149. https://doi.org/10.2166/wst.1993.0221

Bi J et al (2019) Oil-phase cyclic magnetic adsorption to synthesize Fe3O4@C@TiO2-nanotube composites for simultaneous removal of Pb(II) and Rhodamine B. Chemical Engineering Journal 366:50–61. https://doi.org/10.1016/j.cej.2019.02.017

Brame J (2017) Fluidized bed PhotoReactor (FBPR). US Army corps of Engineers, Engineer research and development centre website

Bui N-N, Lind ML, Hoek EMV, McCutcheon JR (2011) Electrospun nanofiber supported thin film composite membranes for engineered osmosis. Journal of Membrane Science 385-386:10–19. https://doi.org/10.1016/j.memsci.2011.08.002

Cai T, Park SY, Li Y (2013) Nutrient recovery from wastewater streams by microalgae: status and prospects. Renewable and Sustainable Energy Reviews 19:360–369. https://doi.org/10.1016/j.rser.2012.11.030

Cao S-W, Zhu Y-J (2008) Hierarchically nanostructured α-Fe2O3 hollow spheres: preparation, growth mechanism, photocatalytic property, and application in water treatment. Journal of Physical Chemistry C 112:6253–6257. https://doi.org/10.1021/jp8000465

Caprarescu S, Purcar V, Vaireanu D-I (2012) Separation of copper ions from synthetically prepared electroplating wastewater at different operating conditions using electrodialysis separation. Science and Technology 47:2273–2280. https://doi.org/10.1080/01496395.2012.669444

Carbuloni CF, Savoia JE, Santos JSP, Pereira CAA, Marques RG, Ribeiro VAS, Ferrari AM (2020) Degradation of metformin in water by TiO2–ZrO2 photocatalysis. Journal of Environmental Management 262:110347. https://doi.org/10.1016/j.jenvman.2020.110347

Chauhan JS, Kumar S (2020) Wastewater ferti-irrigation: an eco-technology for sustainable agriculture. Sustainable Water Resources Management 6:31. https://doi.org/10.1007/s40899-020-00389-5

Choi J-H, Dockko S, Fukushi K, Yamamoto K (2002) A novel application of a submerged nanofiltration membrane bioreactor (NF MBR) for wastewater treatment. Desalination 146: 413–420. https://doi.org/10.1016/S0011-9164(02)00524-6

Dong S, Zhang X, He F, Dong S, Zhou D, Wang B (2015) Visible-light photocatalytic degradation of methyl orange over spherical activated carbon-supported and Er3+:YAlO3-doped TiO2 in a fluidized bed. Journal of Chemical Technology Biotechnology 90:880–887. https://doi.org/10.1002/jctb.4391

Dyhrman ST et al (2012) The transcriptome and proteome of the diatom Thalassiosira pseudonana reveal a diverse phosphorus stress response. PLOS ONE 7:e33768. https://doi.org/10.1371/journal.pone.0033768

Enesca A (2021) The influence of photocatalytic reactors design and operating parameters on the wastewater organic pollutants removal—a mini-review. Catalysts 11:556. https://doi.org/10.3390/catal11050556

FAO (2014) Water – the most basic resource but also the most essential how much water do we use for...? Food and Agriculture Organization of the United Nations, Rome. https://www.fao.org/zhc/detail-events/en/c/231215/. Accessed 5 March 2022

Faried M, Samer M, Abdelsalam E, Yousef RS, Attia YA, Ali AS (2017) Biodiesel production from microalgae: processes, technologies and recent advancements. Renewable and Sustainable Energy Reviews 79:893–913. https://doi.org/10.1016/j.rser.2017.05.199

Ghasemi Z, Younesi H, Zinatizadeh AA (2016) Preparation, characterization and photocatalytic application of TiO2/Fe-ZSM-5 nanocomposite for the treatment of petroleum refinery wastewater: optimization of process parameters by response surface methodology. Chemosphere 159: 552–564

Ghuman K, Wood T, Hoch L, Mims C, Ozin G, Singh CV (2015) Illuminating CO2 reduction on frustrated Lewis pair surfaces: Investigating the role of surface hydroxides and oxygen vacancies on nanocrystalline In2O3-x(OH)y. Phys Chem Chem Phys 17(22):14623–14635. https://doi.org/10.1039/C5CP02613J

Glaze WH (1987) Drinking-water treatment with ozone. Environmental Science Technology 21: 224–230. https://doi.org/10.1021/es00157a001

Gray NF (2004) Biology of wastewater treatment, vol 4. World Scientific, Singapore

Guo H et al (2022) Nanofiltration for drinking water treatment: a review. Frontiers of Chemical Science and Engineering 16:681–698. https://doi.org/10.1007/s11705-021-2103-5

Gupta S, Pawar SB, Pandey RA (2019) Current practices and challenges in using microalgae for treatment of nutrient rich wastewater from agro-based industries. Science of the Total Environment 687:1107–1126. https://doi.org/10.1016/j.scitotenv.2019.06.115

Habib G, Akhter M, Qamar S (2018) Application of electrodialysis in waste water treatment and impact of fouling on process performance. Journal of Membrane Science 8:1000182. https://doi.org/10.4172/2155-9589.1000182

Hanif A et al (2021) A novel combined treatment process of hybrid biosorbent–nanofiltration for effective Pb(II) removal from wastewater. Water 13:233316. https://doi.org/10.3390/w13233316

Hasan A, Elsaeed S, Kamal R, Abdel-Raouf ME-S (2016) Low cost biosorbents based on modified starch iron oxide nanocomposites for selective removal of some heavy metals from aqueous solutions. Advanced Materials Letters 7:402–409. https://doi.org/10.5185/amlett.2016.6061

Hell F, Lahnsteiner J (2002) The application of electrodialysis for drinking water treatment. In: Rubin H, Shamir U, Nachtnebel P, Fürst J (eds) Water resources quality: preserving the quality of our water resources. Springer, Berlin, pp 315–327. https://doi.org/10.1007/978-3-642-56013-2_18

Henery JG (1989) Water pollution. In: Heinke GW, Henery JG (eds) Environmental science and engineering, 2nd edn. Prentice-Hall, Engelwood Cliffs, pp 297–329

Hermans S, Dom E, Mariën H, Koeckelberghs G, Vankelecom IFJ (2015) Efficient synthesis of interfacially polymerized membranes for solvent resistant nanofiltration. Journal of Membrane Science 476:356–363. https://doi.org/10.1016/j.memsci.2014.11.046

Hernández-Leal L, Temmink H, Zeeman G, Buisman CJN (2011) Removal of micropollutants from aerobically treated grey water via ozone and activated carbon. Water Research 45:2887–2896. https://doi.org/10.1016/j.watres.2011.03.009

Higgins CJ, Duranceau SJ (2020) Removal of Enantiomeric Ibuprofen in a nanofiltration membrane process. Membranes 10:20383. https://doi.org/10.3390/membranes10120383

Horan NJ (1990) Biological wastewater treatment systems: theory and operation. Wiley, Chichester

Horikoshi S, Serpone N (2020) Can the photocatalyst TiO2 be incorporated into a wastewater treatment method? Background and prospects. Catalysis Today 340:334–346. https://doi.org/10.1016/j.cattod.2018.10.020

Horst MF, Alvarez M, Lassalle VL (2016) Removal of heavy metals from wastewater using magnetic nanocomposites: analysis of the experimental conditions. Separation Science and Technology 51:550–563. https://doi.org/10.1080/01496395.2015.1086801

Ismail AF, Padaki M, Hilal N, Matsuura T, Lau WJ (2015) Thin film composite membrane—recent development and future potential. Desalination 356:140–148. https://doi.org/10.1016/j.desal.2014.10.042

Ju X-J, Zhang S-B, Zhou M-Y, Xie R, Yang L, Chu L-Y (2009) Novel heavy-metal adsorption material: ion-recognition P(NIPAM-co-BCAm) hydrogels for removal of lead(II) ions. Journal of Hazardous Materials 167:114–118. https://doi.org/10.1016/j.jhazmat.2008.12.089

Kadi MW, Mohamed RM, Ismail AA, Bahnemann DW (2020) Performance of mesoporous α-Fe2O3/g-C3N4 heterojunction for photoreduction of Hg(II) under visible light illumination. Ceramics International 46:23098–23106. https://doi.org/10.1016/j.ceramint.2020.06.087

Khanmohammadi M, Shahrouzi JR, Rahmani F (2021) Insights into mesoporous MCM-41-supported titania decorated with CuO nanoparticles for enhanced photodegradation of tetracycline antibiotic. Environmental Science and Pollution Research 28:862–879. https://doi.org/10.1007/s11356-020-10546-0

Kumar S, Chauhan JS (2022) Application of plant-based nanoparticles in wastewater decontamination. In: Rai JPN, Saraswat S (eds) Nano-biotechnology for waste water treatment: theory and practices. Springer, Cham, pp 89–113. https://doi.org/10.1007/978-3-031-00812-2_4

Lee Y-K (2001) Microalgal mass culture systems and methods: their limitation and potential. Journal of Applied Phycology 13:307–315. https://doi.org/10.1023/A:1017560006941

Lee H-J, Sarfert F, Strathmann H, Moon S-H (2002) Designing of an electrodialysis desalination plant. Desalination 142:267–286. https://doi.org/10.1016/S0011-9164(02)00208-4

Lee KM, Lai CW, Ngai KS, Juan JC (2016) Recent developments of zinc oxide based photocatalyst in water treatment technology: a review. Water Research 88:428–448. https://doi.org/10.1016/j.watres.2015.09.045

Leflaive X et al (2012) Water. https://doi.org/10.1787/env_outlook-2012-8-en

Leong YK, Chang J-S (2020) Bioremediation of heavy metals using microalgae: recent advances and mechanisms. Bioresource Technology 303:122886. https://doi.org/10.1016/j.biortech.2020.122886

Li F, Lin M (2020) Synthesis of biochar-supported K-doped g-C3N4 photocatalyst for enhancing the polycyclic aromatic hydrocarbon degradation activity. International Journal of Environmental Research and Public Health 17:2065. https://doi.org/10.3390/ijerph17062065

Li N, Tian Y, Zhao J, Zhang J, Zhang J, Zuo W, Ding Y (2017) Efficient removal of chromium from water by Mn3O4@ZnO/Mn3O4 composite under simulated sunlight irradiation: synergy of photocatalytic reduction and adsorption. Applied Catalysis B Environmental 214:126–136. https://doi.org/10.1016/j.apcatb.2017.05.041

Li K et al (2019) Microalgae-based wastewater treatment for nutrients recovery: a review. Bioresource Technology 291:121934. https://doi.org/10.1016/j.biortech.2019.121934

Lim SL, Chu WL, Phang SM (2010) Use of Chlorella vulgaris for bioremediation of textile wastewater. Bioresour Technol 101:7314–7322. https://doi.org/10.1016/j.biortech.2010.04.092

Liu F, Zhang W, Tao L, Hao B, Zhang J (2019) Simultaneous photocatalytic redox removal of chromium(vi) and arsenic(iii) by hydrothermal carbon-sphere@nano-Fe3O4. Environmental Science Nano 6:937–947. https://doi.org/10.1039/C8EN01362D

Liu N et al (2020) Magnetic Fe3O4@MIL-53(Fe) nanocomposites derived from MIL-53(Fe) for the photocatalytic degradation of ibuprofen under visible light irradiation. Materials Research Bulletin 132:111000. https://doi.org/10.1016/j.materresbull.2020.111000

Logan B (2004) Biologically extracting energy from wastewater: biohydrogen production and microbial fuel cells. Environ Sci Technol 38:160

Malakian A, Zhou Z, Messick L, Spitzer TN, Ladner DA, Husson SM (2020) Understanding the role of pattern geometry on nanofiltration threshold flux. Membranes 10:120445. https://doi.org/10.3390/membranes10120445

Monteiro CM, Castro PML, Malcata FX (2012) Metal uptake by microalgae: underlying mechanisms and practical applications. Biotechnology Progress 28:299–311. https://doi.org/10.1002/btpr.1504

Moreira NFF, Sampaio MJ, Ribeiro AR, Silva CG, Faria JL, Silva AMT (2019) Metal-free g-C3N4 photocatalysis of organic micropollutants in urban wastewater under visible light. Applied Catalysis B Environmental 248:184–192. https://doi.org/10.1016/j.apcatb.2019.02.001

Ng TW, Zhang L, Liu J, Huang G, Wang W, Wong PK (2016) Visible-light-driven photocatalytic inactivation of Escherichia coli by magnetic Fe2O3-AgBr. Water Research 90:111–118. https://doi.org/10.1016/j.watres.2015.12.022

Ohkouchi Y, Yata Y, Bun R, Itoh S (2013) Chlorine requirement for biologically stable drinking water after nanofiltration. Water Supply 14:405–413. https://doi.org/10.2166/ws.2013.214

Oswald WJ (1988) Microalgae and wastewater treatment. In: Borowitzka MA, Borowitzka LJ (eds) Microalgal biotechnology. Cambridge University Press, Cambridge, pp 305–328

Oswald WJ, Gotaas HB, Golueke CG, Kellen WR, Gloyna EF, Hermann ER (1957) Algae in waste treatment. Sewage and Industrial Wastes 29:437–457

Peng X-W, Zhong L-X, Ren J-L, Sun R-C (2012) Highly effective adsorption of heavy metal ions from aqueous solutions by macroporous xylan-rich hemicelluloses-based hydrogel. Journal of Agricultural and Food Chemistry 60:3909–3916. https://doi.org/10.1021/jf300387q

Perkowski J, Kos L, Ledakowicz S (1996) Application of ozone in textile wastewater treatment. Ozone 18:73–85. https://doi.org/10.1080/01919519608547342

Petersen RJ (1993) Composite reverse osmosis and nanofiltration membranes. Journal of Membrane Science 83:81–150. https://doi.org/10.1016/0376-7388(93)80014-O

Phang S (1991) The use of microalgae to treat agro-industrial wastewater. In: Proceedings of a Seminar held at Murdoch Univ., Western Australia

Pinnau I (2000) Membrane separations I membrane preparation. In: Wilson ID (ed) Encyclopedia of separation science. Academic, Oxford, pp 1755–1764. https://doi.org/10.1016/B0-12-226770-2/05241-8

Posten C (2009) Design principles of photo-bioreactors for cultivation of microalgae. Engineering in Life Sciences 9:165–177. https://doi.org/10.1002/elsc.200900003

Pulz O (2001) Photobioreactors: production systems for phototrophic microorganisms. Applied Microbiology and Biotechnology 57:287–293. https://doi.org/10.1007/s002530100702

Raizada P, Kumari J, Shandilya P, Dhiman R, Pratap Singh V, Singh P (2017) Magnetically retrievable Bi2WO6/Fe3O4 immobilized on graphene sand composite for investigation of photocatalytic mineralization of oxytetracycline and ampicillin. Process Safety and Environmental Protection 106:104–116. https://doi.org/10.1016/j.psep.2016.12.012

Rajendran S, Khan MM, Gracia F, Qin J, Gupta VK, Arumainathan S (2016) Ce3+-ion-induced visible-light photocatalytic degradation and electrochemical activity of ZnO/CeO2 nanocomposite. Scientific Reports 6:31641. https://doi.org/10.1038/srep31641

Rao KR, Subrahmanyam N (2004) Process variations in activated sludge process – a review. Ind Chem Engr 46:48–55

Regmi C, Kim T-H, Ray SK, Yamaguchi T, Lee SW (2017) Cobalt-doped BiVO4 (Co-BiVO4) as a visible-light-driven photocatalyst for the degradation of malachite green and inactivation of harmful microorganisms in wastewater. Research on Chemical Intermediates 43:5203–5216. https://doi.org/10.1007/s11164-017-3036-y

Ren G, Han H, Wang Y, Liu S, Zhao J, Meng X, Li Z (2021) Recent advances of photocatalytic application in water treatment: a review. Nanomaterials 11:1804. https://doi.org/10.3390/nano11071804

Rice RG (1996) Applications of ozone for industrial wastewater treatment — a review. Ozone Science Engineering 18:477–515. https://doi.org/10.1080/01919512.1997.10382859

Saravanan R, Karthikeyan S, Gupta VK, Sekaran G, Narayanan V, Stephen A (2013) Enhanced photocatalytic activity of ZnO/CuO nanocomposite for the degradation of textile dye on visible light illumination. Materials Science and Engineering C 33:91–98. https://doi.org/10.1016/j.msec.2012.08.011

Sarkar S, SenGupta AK (2008) A new hybrid ion exchange-nanofiltration (HIX-NF) separation process for energy-efficient desalination: process concept and laboratory evaluation. Journal of Membrane Science 324:76–84. https://doi.org/10.1016/j.memsci.2008.06.058

Sarkar S, SenGupta AK (2009) A hybrid ion exchange-nanofiltration (HIX-NF) process for energy efficient desalination of brackish/seawater. Water Supply 9:369–377. https://doi.org/10.2166/ws.2009.634

Shagun (2019) Nearly 80 per cent Indian households without piped water connection. Society for Environmental Communications, New Delhi

Shawky A, Mohamed RM, Mkhalid IA, Youssef MA, Awwad NS (2020) Visible light-responsive Ag/LaTiO3 nanowire photocatalysts for efficient elimination of atrazine herbicide in water. Journal of Molecular Liquids 299:112163. https://doi.org/10.1016/j.molliq.2019.112163

Shi J, Podola B, Melkonian M (2007) Removal of nitrogen and phosphorus from wastewater using microalgae immobilized on twin layers: an experimental study. Journal of Applied Phycology 19:417–423. https://doi.org/10.1007/s10811-006-9148-1

Singh V, Mishra V (2019) Bioremediation of nutrients and heavy metals from wastewater by microalgal cells: mechanism and kinetics. In: Tripathi V, Kumar P, Tripathi P, Kishore A, Kamle M (eds) Microbial genomics in sustainable agroecosystems, vol 2. Springer, Singapore, pp 319–357. https://doi.org/10.1007/978-981-32-9860-6_16

Singh DV, Bhat RA, Upadhyay AK, Singh R, Singh DP (2021) Microalgae in aquatic environs: a sustainable approach for remediation of heavy metals and emerging contaminants. Environmental Technology Innovation 21:101340. https://doi.org/10.1016/j.eti.2020.101340

Sivagurunathan P, Sen B, Lin C-Y (2015) High-rate fermentative hydrogen production from beverage wastewater. Applied Energy 147:1–9. https://doi.org/10.1016/j.apenergy.2015.01.136

Sniffen KD, Sales CM, Olson MS (2018) The fate of nitrogen through algal treatment of landfill leachate. Algal Research 30:50–58. https://doi.org/10.1016/j.algal.2017.12.010

Song J et al (2019) Highly flexible, core-shell heterostructured, and visible-light-driven titania-based nanofibrous membranes for antibiotic removal and E. coil inactivation. Chemical Engineering Journal 379:122269. https://doi.org/10.1016/j.cej.2019.122269

Stengel DB, Connan S, Popper ZA (2011) Algal chemodiversity and bioactivity: sources of natural variability and implications for commercial application. Biotechnology Advances 29:483–501. https://doi.org/10.1016/j.biotechadv.2011.05.016

Su Y (2021) Revisiting carbon, nitrogen, and phosphorus metabolisms in microalgae for wastewater treatment. Science of the Total Environment 762:144590. https://doi.org/10.1016/j.scitotenv.2020.144590

Sugam RK, Jain A, Neog K (2017) Rethinking wastewater management in India. China Dialogue Trust. https://www.thethirdpole.net/en/pollution/waterwaste-management-in-india/. Accessed 6 March 2022

Taghizade Firozjaee T, Mehrdadi N, Baghdadi M, Nabi Bidhendi GR (2018) Application of nanotechnology in pesticides removal from aqueous solutions - a review. International Journal of Nanoscience and Nanotechnology 14:43–56

Tebbutt THY (1983) Principles of water quality control, 3rd edn. Pergamon Press, Oxford

Thomas B, Alexander LK (2020) Removal of Pb2+ and Cd2+ toxic heavy metal ions driven by Fermi level modification in NiFe2O4–Pd nano hybrids. Journal of Solid State Chemistry 288:121417. https://doi.org/10.1016/j.jssc.2020.121417

Uggetti E, Sialve B, Hamelin J, Bonnafous A, Steyer JP (2018) CO2 addition to increase biomass production and control microalgae species in high rate algal ponds treating wastewater. Journal of CO2 Utilization 28:292–298. https://doi.org/10.1016/j.jcou.2018.10.009

Ugwu CU, Aoyagi H, Uchiyama H (2008) Photobioreactors for mass cultivation of algae. Bioresource Technology 99:4021–4028. https://doi.org/10.1016/j.biortech.2007.01.046

Vaiano V, Sacco O, Pisano D, Sannino D, Ciambelli P (2015) From the design to the development of a continuous fixed bed photoreactor for photocatalytic degradation of organic pollutants in wastewater. Chemical Engineering Science 137:152–160. https://doi.org/10.1016/j.ces.2015.06.023

Vedavyasan CV (2007) Pretreatment trends — an overview. Desalination 203:296–299. https://doi.org/10.1016/j.desal.2006.04.012

Verbeke R, Bergmaier A, Eschbaumer S, Gómez V, Dollinger G, Vankelecom I (2019) Elemental depth profiling of chlorinated polyamide-based thin-film composite membranes with elastic recoil detection. Environmental Science Technology 53:8640–8648. https://doi.org/10.1021/acs.est.8b07226

Wang H et al (2014) Semiconductor heterojunction photocatalysts: design, construction, and photocatalytic performances. Chemical Society Reviews 43:5234–5244. https://doi.org/10.1039/C4CS00126E

Wang Z, Zhang X, Zhang H, Zhu G, Gao Y, Cheng Q, Cheng X (2019) Synthesis of magnetic nickel ferrite/carbon sphere composite for levofloxacin elimination by activation of persulfate. Separation and Purification Technology 215:528–539. https://doi.org/10.1016/j.seppur.2019.01.063

Wen X et al (2016) Effective cultivation of microalgae for biofuel production: a pilot-scale evaluation of a novel oleaginous microalga Graesiella sp. WBG-1. Biotechnology for Biofuels 9:123. https://doi.org/10.1186/s13068-016-0541-y

Xie RJ et al (2009) Pre-treatment optimisation of SWRO membrane desalination under tropical conditions. Desalination and Water Treatment 3:183–192. https://doi.org/10.5004/dwt.2009.459

Yan R et al (2020) Simultaneous removal of Cu(II) and Cr(VI) ions from wastewater by photoreduction with TiO2–ZrO2. Journal of Water Process Engineering 33:101052. https://doi.org/10.1016/j.jwpe.2019.101052

Yang X et al (2020) Visible-near-infrared-responsive g-C(3)N(4)H(x)(+) reduced decatungstate with excellent performance for photocatalytic removal of petroleum hydrocarbon. Journal of Hazardous Materials 381:120994

Young P, Taylor M, Fallowfield HJ (2017) Mini-review: high rate algal ponds, flexible systems for sustainable wastewater treatment World. J Microbiol Biotechnol 33:117. https://doi.org/10.1007/s11274-017-2282-x

Zhang X, Chen YL, Liu R-S, Tsai DP (2013) Plasmonic photocatalysis. Reports on Progress in Physics 76:046401. https://doi.org/10.1088/0034-4885/76/4/046401

Zhang G, Lee D-J, Cheng F (2016) Treatment of domestic sewage with anoxic/oxic membrane-less microbial fuel cell with intermittent aeration. Bioresource Technology 218:680–686. https://doi.org/10.1016/j.biortech.2016.07.006

Zhao P et al (2019) The alleviative effect of exogenous phytohormones on the growth, physiology and gene expression of Tetraselmis cordiformis under high ammonia-nitrogen stress. Bioresour Technol 282:339–347. https://doi.org/10.1016/j.biortech.2019.03.031

Zhao Y, Tong X, Chen Y (2021) Fit-for-purpose design of nanofiltration membranes for simultaneous nutrient recovery and micropollutant removal. Environmental Science Technology 55:3352–3361. https://doi.org/10.1021/acs.est.0c08101

Zheng D, Hua D, Hong Y, Ibrahim AR, Yao A, Pan J, Zhan G (2020) Functions of ionic liquids in preparing membranes for liquid separations: a review. Membranes 10:120395. https://doi.org/10.3390/membranes10120395

Zhu LD, Takala J, Hiltunen E, Wang ZM (2013) Recycling harvest water to cultivate Chlorella zofingiensis under nutrient limitation for biodiesel production. Bioresource Technology 144:14–20. https://doi.org/10.1016/j.biortech.2013.06.061

Overview of Methods and Processes Used in Wastewater Treatment

12

Rewa Kulshrestha, Aakriti Sisodiya, and Soumitra Tiwari

Abstract

Water itself is plagued by wastewater, the enormity of which has become a global environmental threat. Continuous and sustained efforts taken for treatment of wastewater have led a paradigm shift from conventional to contemporary methods, which are more convenient to tackle indelible pollutants by faster oxidation and sludge formation. This chapter highlights recent developments in wastewater treatment such as advanced oxidation processes, membrane processes, and nanomaterials.

Keywords

Waste water · Pollution · Adsorption · Electrocoagulation

12.1 Introduction

Water is existential to life, and it plays a pivotal role in sustaining communities, economies, and societies. Water is life, and clean water means health. The quality and quantity of water has now become a matter of our sustainability. Less than 3% of all the water resources available on earth is usable as freshwater in the form of ground or surface water, rainwater, or gray water and is meant for utilization in agricultural, domestic or municipal, and industrial sector.

R. Kulshrestha (✉) · A. Sisodiya · S. Tiwari
Department of Food Processing and Technology, Atal Bihari Vajpayee Vishwavidyalaya, Bilaspur, Chhattisgarh, India
e-mail: rewakulshrestha@bilaspuruniversity.ac.in

© The Author(s), under exclusive license to Springer Nature Singapore Pte Ltd. 2023
R. Soni et al. (eds.), *Current Status of Fresh Water Microbiology*,
https://doi.org/10.1007/978-981-99-5018-8_12

12.2 Water Pollution

Pollution is a threat to this most precious resource. Water pollution occurs when harmful substances, often chemicals or microorganisms, contaminate a stream, river, lake, ocean, aquifer, or other body of water, degrading water quality and rendering it toxic to humans or the environment (https://www.nrdc.org/stories/water-pollution-everything-you-need-know). Water pollution, by definition, is the contamination of available water by pollutants or alien materials that lead to disease and death of livestock, aquatic life, and humans (Abdulrazzak et al. 2020). Collateral damages of wastewater are soil and water pollution and water resources shortage, which ultimately affects the water cycle and food chain. Water pollution has become a life-threatening issue at global level as growth of industries at fast-paced and rapidly increasing manufacturing units is producing billions of tons of toxic and hazardous waste materials that are entering natural water bodies without any treatment. Some 80% of the world's wastewater is dumped, largely untreated, back into the environment, polluting rivers, lakes, and oceans.

12.3 Wastewater

12.3.1 Sources

The impetus of our societal development has become the root cause of water pollution. Among many sources of water pollution, two general categories may be clearly identified as direct or point and indirect or nonpoint contaminant sources. Direct sources include effluent outfalls from industries, refineries, waste treatment plants, etc. Indirect sources include contaminants that enter the water supply from soils/groundwater systems and from the atmosphere via rainwater. Industrial wastes such as dyes, heavy metals, surfactants, mining activities, sewage and wastewater, pesticides and chemical fertilizers, energy use, radioactive waste, urban development, etc. (Crini and Lichtfouse 2019) are different contaminants, which are broadly divided into three main categories on the basis of its nature: contamination by organic compounds, inorganic compounds (e.g., heavy metals), and microorganisms (Yadav et al. 2021). A distinction in municipal or domestic waste water and industrial waste water is presented in Fig. 12.1.

12.3.2 Wastewater Treatment

The scarcity of freshwater, changing climate, rapidly growing population, and environmental hazard associated with water push the horizon of wastewater treatment. To date, several interventions have been developed to minimize the wastewater discharges and mitigate the hazards of pollutants, using conventional and modern methods (Xu et al. 2018). Corominas et al. (2018) mention that wastewater treatment

12 Overview of Methods and Processes Used in Wastewater Treatment

Fig. 12.1 Features of municipal and industrial wastewater

plants are complex systems, which have to maintain high performance at all times, despite suffering from hourly, daily, monthly, and seasonal dynamics. Life cycle assessment (LCA) has been employed to assess the environmental impacts of wastewater treatment, as a tool of sustainability.

12.4 Wastewater Treatment Process

Water treatment technologies are used for three purposes, that is, water source reduction, wastewater treatment, and recycling. Primary treatment includes preliminary purification processes of a physical and chemical nature, while secondary treatment deals with the biological treatment of wastewater. In tertiary treatment processes, wastewater (treated by primary and secondary processes) is converted into good quality water that can be used for different types of purpose, that is, drinking, industrial, or medicinal supplies. An overview of different waste water treatments is given in Table 12.1.

12.4.1 Conventional Methods

Physical Methods Mass transfer strategy is the base of the physical pollutant removal methods (Samsami et al. 2020). Physical methods include sedimentation, adsorption, filtration, and coagulation. The main convenience of physical methods lies in simplicity and flexibility for wide range of contaminants, and the drawback of these methods is generation of heap of sludge. Physical methods such as screening

Table 12.1 Different types of wastewater treatment

S. No.	WWT	Objective	Methods	BOD removal
1.	Preliminary treatment	Removal of coarse solids	Screening Skimming	–
2.	Primary treatment	Removal of suspended solids	Sedimentation flotation	35–40%
3.	Secondary treatment	Additional removal of organics and inorganic	Aerobic biological methods, advanced oxidation process	85–90%
4.	Advanced treatment	Removal of nitrogenous oxygen demand (NOD), nutrient removal, removal of toxic materials	Membrane technology, RO, desalination technologies, electro dialysis, ion exchange, freeze crystallization	95% or more

ensure efficient functioning of advanced methods by prohibiting potential damage, blockage, and process interruption.

Chemical Methods Chemical-based reaction due to specialized chemicals such as chlorine, hydrogen peroxide, sodium chlorite, and sodium hypochlorite assists in removing dissolved metals from wastewater and expedite disinfection. Among chemical methods, chemical precipitation, chemisorption, disinfection, ion exchange, and advanced oxidation process (AOP) are considered noteworthy techniques for wastewater treatment.

Biological Method Biological methods are based on microbial decomposition of organic pollutants via aerobic or anaerobic cycle (Saxena and Bharagava 2017). Aerobic treatment involves algal and bacteria-based oxidation ponds and aeration lagoons. Examples of anaerobic processes are anaerobic filter reactor, anaerobic contact process, fluidized-bed reactor, upflow anaerobic sludge blanket, ADI-BVF process, and expanded granular sludge bed process (Englande Jr et al. 2015). The efficiency of biological processes for degradation depends on the selected microbes' adaptability and enzymes' activity (Rashid et al. 2021). Apart from this, the availability of oxygen, retention time, temperature, and the biological activities of the bacteria judge the extent of biological oxidation (Gupta et al. 2012). Generally, heterotrophic bacteria play an important role in removing organic matters in wastewater treatment system (Dadrasnia et al. 2017). The distinctive features of biological method are relatively cheaper cost, minimal, or no secondary excretion of pollutants and most importantly lower damaging effects on the environment (Mingjun et al. 2009). The drawbacks of biological methods are slow process and low biodegradability and require an optimally favorable environment and proper maintenance of microorganisms.

12.4.2 Recent Advances

12.4.2.1 Adsorption

Adsorption is considered simple to operate and cost-effective wastewater treatment technology (Grassi et al. 2012; Liu et al. 2004) with pollutant removal efficiency of 99%. It is the surface phenomena of adherence of adsorbate on adsorbent via chemical or physical interaction. Chemical interaction involves ionic and covalent bonds, whereas physical interaction is based on van der Waals force between adsorbate and adsorbent. The constituents, porosity, and specific surface area (SSA) of adsorbents are responsible for elimination of pollutants from wastewater (Khattri and Singh 2009). There are a range of natural and synthetic adsorbents for the removal of organic pollutants, heavy metals, and dyes. Among them, the most commonly used adsorbents are activated carbon, silica gel, alumina, clay, metal oxides, polyacrylamide, adsorbent resin, and zeolite (Wadhawan et al. 2020; Prajapati et al. 2020). Adsorption onto activated carbons is often cited as the procedure of choice for wastewater pollutants because it gives the best results in terms of efficiency and excellent regeneration capabilities at industrial level (Crini and Lichtfouse 2019; Rathi and Kumar 2021). However, commercial activated carbon is relatively expensive (Abdullah et al. 2009) and has difficulties in separation from water bodies and generates additional pollution. The promising alternates are biochar; biosorbents; nanosorbents, namely, carbon nanotubes (CNT) and nanochitosan (Tan et al. 2015); and magnetic adsorbents (Moosavi et al. 2020) with adsorption capacity in the range of 800 and 5000 mg/g.

12.4.2.2 Chemical Precipitation

Chemical precipitation is a complex procedure with three distinct stages of nucleation, crystal growth, and flocculation (Zueva 2018). The process targets the removal of ionic constituents from water or wastewater by reducing their solubility with the aid of counter ions. The process relies on the use of an alkaline agent, for example, lime or caustic soda to raise the pH of the water that causes the solubility of metal ions to decrease and thus precipitate out of the solvent. It is usually followed by a solid separation operation that may include coagulation and/or sedimentation, or filtration to remove the precipitates. The process is efficient in water softening and stabilization, metallic cations, heavy metals, phosphate removal, and organic molecules (Wang et al. 2005). Significant reduction in chemical oxygen demand (COD), simplicity, and rapidity feature of this method are advantageous, whereas its operating costs from the chemical expense and the cost of disposing of the precipitated sludge that is produced are major limitations of chemical precipitation (Biver and Degols 1982).

12.4.2.3 Coagulation/Flocculation

Coagulation is defined as the process through which very fine solid suspensions are destabilized so that they can begin to agglomerate if conditions allow, while flocculation is defined as the process by which destabilized particles actually conglomerate into larger aggregates that can be separated from wastewater. The inorganic, organic, and biomaterials that promote aggregation and sedimentation of suspended particles in solution are called coagulants and flocculants (Armenante 2014). The working action of coagulation/flocculation is based on charge neutralization, charge patching, bridging, and sweeping (Yang et al. 2016).

Coagulation and flocculation occur in successive steps and followed by sedimentation. If coagulation is incomplete, flocculation step will be unsuccessful, and if flocculation is incomplete, sedimentation will be unsuccessful. Coagulation and flocculation have been reported to be widely used technologies in textile wastewater treatment (Uzal 2015) and paper and cardboard recycling industry wastewater (Gholami et al. 2020). The crucial factors influencing coagulation–flocculation are temperature, pH, effluent quality, coagulant type and its dosage (Nnaji 2012), mixing speed, and settling time of the floc formed. The most commonly used coagulants in wastewater treatment are aluminum sulfate (alum), ferric chloride, ferric sulfate, ferrous sulfate (copperas), sodium aluminate, polyaluminum chloride, and organic polymers (Shammas 2005).

12.4.2.4 Electrocoagulation (EC)

EC is an electrochemical process that uses a low electrical current in situ to eliminate pollutants, namely, heavy metals, dissolved metals, dyes, and suspended solids from wastewater (Nidheesh et al. 2022). When current passes through a metal (sacrificial) anode (M), metal oxidizes to its cation (M^{n+}), and water is reduced to hydrogen gas and the hydroxyl ion (OH^-) simultaneously. These charged metal cations destabilize any colloidal particles by the formation of polyvalent polyhydroxide complexes. These complexes have high adsorption properties, forming aggregates with pollutants. Evolution of hydrogen gas aids in mixing and hence flocculation. Once the floc is generated, the electrolytic gas creates a flotation effect removing the pollutants to the floc—foam layer at the liquid surface (Holt et al. 1999). Thus, this technique is an amalgam of electrochemistry, coagulation, and flotation (Boinpally et al. 2023). Parameters affecting EC process are operating current density (10–150 A/m^2), electrode material and arrangement, interelectrode distance, pH and conductivity, reaction time and temperature, and design of reactor (Tahreen et al. 2020). The anode materials play a significant role in the oxidation reactions that occur on the sacrificial electrodes (Shahedi et al. 2020).

EC is an effective and safe technology for elimination of contaminants of dairy industry (Bazrafshan et al. 2013), textile industry (Afriani and Tiandho 2020), etc. The cost-effective, easily operable low installation cost, and high treatment efficiency are perquisite benefits of EC technology. While on the contrary, the need to replace the "sacrificial anode" periodically, high cost of electricity, and toxic or harmful sludge produced are the major constraints of EC technology.

12.4.2.5 Ion Exchange Technology

Ion exchange is swapping of ions between two electrolytes or between an electrolyte solution and a complex. Ion exchange systems generally contain either cationic or anionic exchange resins, which are high molecular weight polyacids or polybases and are known for insolubility in aqueous and nonaqueous media. The most commonly used ion exchangers are sodium silicates, zeolites, polystyrene sulfonic acid, and acrylic and metha-acrylic resins (Dorfner 1991). The ion exchange technology has seen evolution from resins to membrane resulting in mass scale utilization for desalination of sea and brackish water and for treating industrial effluents (Xu 2005). The advantage of ion exchange as a water remediation technique is that it is very cost-effective in withdrawal of heavy metals and organic contaminants. Very little amount of energy is required, and regeneration of resins is very economical. Still limitations of fouling, adsorption of organic matter, and bacterial contaminations hamper its prolong usage (Kansara et al. 2016).

12.4.2.6 Membrane Filtration

Membrane technology has been widely used for water treatment. Pressure-driven membrane processes for wastewater are microfiltration (MF), ultrafiltration (UF), nanofiltration (NF), reverse osmosis (RO), and membrane bioreactor (MBR) (Al Mahri et al. 2020). MF is more often chosen as a pretreatment for tighter downstream membrane systems, such as nanofiltration or RO units, to avoid particles that may cause fouling of the tighter membrane units (Paul 2002). UF membranes retain macromolecules and colloids from a solution on the basis of molecular weight cutoff (MWCO) and are often chosen for hybrid applications. RO and NF are highly efficient in separating small particles including bacteria and monovalent ions like sodium ions and chloride ions up to 99.5% (Muro et al. 2012). However, NF is water-softening technology, whereas RO is desalination technology (Fig. 12.2).

Feedwater composition, membrane material and type, operating conditions especially transmembrane pressure (TMP), and membrane fouling influence the success of membrane filtration technique. Fouling declines the membrane's permeability as it causes membrane pore blocking by forming a layer of organic compounds. In spite

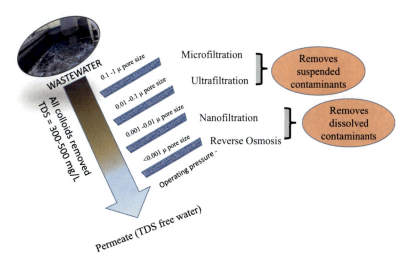

Fig. 12.2 Waste water treatment through different types of membranes

of fouling limitations, membrane technology is beneficial in terms of financial sustainability, low use of chemicals, eco-friendliness, and ease of access (Obotey Ezugbe and Rathilal 2020). Further application in textile, dairy, landfill, and composting plants establishes the coup of membrane technology in wastewater treatment. The waste-intensive nature of membrane fabrication (Razali et al. 2015) and escalating cost of fouling control push the demand for hybrid or integrated membrane technology.

12.4.2.7 Membrane Bioreactor (MBR)

MBR technology was developed in 1969 to overcome the drawbacks of conventional activated sludge process with the aid of membrane. The technology is bestowed with advantages like controlled biomass retention, improved effluent quality, and decreased footprint (Kraume and Drews 2010). WW to be treated by a membrane bioreactor flows into an aeration tank, where biodegradable organic matter and reduced nitrogen compounds are oxidized. Then sludge flow is channeled through a membrane filtration unit where sludge and water are separated. Filtrate is drained of as effluent, and the concentrate is recirculated into the aeration tank. Surplus sludge is discharged via a sludge valve (Van Dijk and Roncken 1997).

12.4.2.8 Advanced Oxidation Processes (AOPs)

Various AOPs are based on the addition or creation of a highly oxidizing species to degrade the organic matter as mentioned in Table 12.2. They have commercial potential for eliminating highly toxic and nonbiodegradable wastes.

12.4.2.9 Microbial Fuel Cell

A microbial fuel cell (MFC) is a galvanic cell in which the biochemical energy contained in the organic matter is directly converted into electricity under anaerobic conditions (Du et al. 2007). According to Munjal et al. (2020), the components of MFC are microbial anodic chamber and a cathodic chamber separated by a proton exchange membrane (PEM). At anode, anaerobic respiring bacteria, particularly *Geobacter* and *Shewanella*, efficiently degrade organic matter of wastewater into carbon dioxide as end product, while generating electrons and protons concomitantly. Electrons pass from the bacteria to the electrode (anode) in the same chamber and then via a circuit to the cathode where they combine with protons and oxygen to form water (Liu et al. 2004). At cathode, oxygen, ferricyanide, and hydrogen peroxide are primarily served as the terminal electron acceptor (TEA). The difference in the potential coupled to electron flow produces electricity in this fuel cell. The long-term stability and performance of MFC is affected by several parameters such as substrate, electrodes' material and distance, effect of cathode material, aeration rate, pH effect, electrolyte, and temperature effect (Jatoi et al. 2021).

Although the current applications of MFCs are still at lab-level, they have been proved to be of great potential industrial applications especially food processing, dairy (Mohan et al. 2010), and brewery industries. MFCs are environmentally friendly technologies (Gude 2016) with energy-saving features resulting from reduced aeration, less sludge production, and ambient temperature requirement,

12 Overview of Methods and Processes Used in Wastewater Treatment

Table 12.2 Description of various AOPs

Type of AOP	Principle	Suitability	Limitation	Reference
Fenton oxidation	Fe^{2+} is used as the catalyst and hydrogen peroxide (H_2O_2) as the oxidant	Organic pollutants	Large volume of iron sludge produced and limited optimum pH range	Palmer and Hatley (2018)
Photo-Fenton	Fenton reaction occurring between H_2O_2 and Fe^{2+}/Fe^{3+} with UV radiations	Emerging contaminants	Requires narrow range of pH	Legrini et al. (1993) Klamerth et al. (2010)
UV-radiation based AOP	Excitation of electrons	Destroys bacteria and degrade aquatic organic substances	High-energy costs or insufficient reductions	Ibrahim et al. (2019) Glaze et al. (1987)
Ozonation	O_3, a potent oxidant and a strong disinfectant	Decolorization of textile wastewaters	high ozone generation cost, short lifetime, pH dependence	
Sonolysis or hydrodynamic cavitation	Uses ultrasonic sound having 20 kHz to 10 MHz frequency range causing cavitation	Degrades volatile organic matter responsible for turbidity	More consumption of energy and less efficiency of mineralization	
Wet air oxidation	Under high temperature (200–325 °C) and pressure (50–150 bar)	Aqueous wastes from the chemical industry and domestic sludge	Corrosion issues	Debellefontaine and Foussard (2000)

which prove MFCs to be exceptional among the existing technologies for wastewater treatment. It promises the removal of heavy metals, ammonia, and COD depletion and recovery of valuable products like silver (Ag) and chromium (Cr) as reviewed by Munoz-Cupa et al. (2021). However, the low power density, concentration polarization, and the high capital cost of MFCs have been debatable for tackling large volumes of wastewater (He et al. 2017). Future scope lies in synergies between MFC and other treatment technologies along with scaling up challenges.

12.5 Nanotechnology

Nano, being the billionth of a meter, has high absorbing, interacting, and reacting capabilities due to its small size and high proportion of atoms at surface (Madhuri et al. 2018). At the nanoscale (1–100 nm), materials demonstrate significantly different physical and chemical properties from their bulk counterparts (Ahmed and Haider 2018).

According to Li et al. (2008), recent nanotechnology-based water treatment targets four areas:

(a) Adsorptive elimination of micro pollutants
(b) Filtration through membranes
(c) Microbial decontamination
(d) Catalytic degradation or photocatalysis

Carbon Nanotube (CNT) Treatment CNT treatment seems possible in wastewater treatment owing to the technology's large adsorption surface area and ability to adsorb a range of challenging materials that are of interest to the wastewater treatment industry (Qu et al. 2013; Lu and Chiu 2006).

Crystallization It is a solid-liquid separation technique, in which the solute crystallizes from the liquid solution and turns into a pure solid crystalline phase (Lu et al. 2017). This technique is effective for desalination and recovery of valuable resources. The panorama of crystallization ranges from ancient evaporation crystallization method to modern hybrid membrane distillation crystallization. Generally, crystallization is used for the wastewater released by cooling towers, coal and gas fired boilers, and the paper and dying industries.

12.6 Conclusion

Wastewater treatment will continue to be a dynamic engineering science necessary for our ecological balance. Wastewater characterization is an important factor in setting up a relevant effective management strategy or treatment process. Various methods highlighted for wastewater treatment have challenges associated with cost and engineering designs along with consideration of BOD. The future prospects lie in energy auditing, computer-aided design of wastewater treatments, and artificial intelligence approach for nearly 99% waste-free discharge.

References

Abdullah AZ, Salamatinia B, Kamaruddin AH (2009) Application of response surface methodology for the optimization of NaOH treatment on oil palm frond towards improvement in the sorption of heavy metals. Desalination 244(1-3):227–238

Abdulrazzak IA, Bierk H, Abdulrazzaq AA (2020) Monitoring and evaluation of the water pollution. IOP Conf Ser Mater Sci Eng 881(1):012101

Afriani F, Tiandho Y (2020) Application of electrocoagulation for textile wastewater treatment: a review. IOP Conf Ser Earth Environ Sci 599(1):012069

Ahmed SN, Haider W (2018) Heterogeneous photocatalysis and its potential applications in water and wastewater treatment: a review. Nanotechnology 29:342001

12 Overview of Methods and Processes Used in Wastewater Treatment

Al Mahri BBA, Balogun HA, Yusuf A, Giwa A (2020) Electro-osmotic thermal process model for performance enhancement of forward osmosis integrated with membrane distillation. Sep Purif Technol. https://doi.org/10.1016/j.seppur.2019.116494

Armenante PM (2014) Coagulation and flocculation. www.cpe.njit.edu/dlnotes/che685/cls07-1.pdf

Bazrafshan E, Moein H, Kord Mostafapour F, Nakhaie S (2013) Application of electrocoagulation process for dairy wastewater treatment. J Chem 2013:640139

Biver D, Degols A (1982) Tech. de l'Eau (Fr.), 428/429, 31, (1982); (abstr) WRC Info. 10, 83-0524 (1983)

Boinpally S, Kolla A, Kainthola J, Kodali R, Vemuri J (2023) A state-of-the-art review of the electrocoagulation technology for wastewater treatment. Water Cycle 4:26–36

Corominas L, Garrido-Baserba M, Villez K, Olsson G, Cortés U, Poch M (2018) Transforming data into knowledge for improved wastewater treatment operation: a critical review of techniques. Environ Model Softw 106:89–103

Crini G, Lichtfouse E (2019) Advantages and disadvantages of techniques used for wastewater treatment. Environ Chem Lett 17(1):145–155

Dadrasnia A, Usman MM, Lim KT, Velappan RD, Shahsavari N, Vejan P, Ismail S (2017) Microbial aspects in wastewater treatment–a technical. Environ Pollut Protect 2(2):75–84

Debellefontaine H, Foussard JN (2000) Wet air oxidation for the treatment of industrial wastes. Chemical aspects, reactor design and industrial applications in Europe. Waste Manag 20(1): 15–25

Dorfner K (1991) Introduction to ion exchange and ion exchangers. In: Dorfner K (ed) Ion exchangers. De Gruyter, Berlin, pp 7–188. https://doi.org/10.1515/9783110862430.7

Du Z, Li H, Gu T (2007) A state of the art review on microbial fuel cells: a promising technology for wastewater treatment and bioenergy. Biotechnol Adv 25(5):464–482

Englande AJ Jr, Krenkel P, Shamas J (2015) Wastewater treatment & water reclamation. In: Reference module in earth systems and environmental sciences. Elsevier, Amsterdam. https://doi.org/10.1016/B978-0-12-409548-9.09508-7

Gholami M, Ghaneian MT, Fatemi SS, Talebi P, Dalvand A (2020) Investigating the efficiency of coagulation and flocculation process in wastewater treatment of paper and cardboard recycling industry. Int J Environ Anal Chem 2020:1–13

Glaze WH, Kang JW, Chapin DH (1987) The chemistry of water treatment processes involving ozone, hydrogen peroxide and ultraviolet radiation. J Ozone Sci Eng 9:335–352

Grassi M, Kaykioglu G, Belgiorno V, Lofrano G (2012) Removal of emerging contaminants from water and wastewater by adsorption process. In: Emerging compounds removal from wastewater: natural and solar based treatments, pp 15–37

Gude VG (2016) Wastewater treatment in microbial fuel cells–an overview. J Clean Prod 122:287–307

Gupta VK, Ali I, Saleh TA, Nayak A, Agarwal S (2012) Chemical treatment technologies for wastewater recycling—an overview. RSC Adv 2(16):6380–6388

He L, Du P, Chen Y, Lu H, Cheng X, Chang B, Wang Z (2017) Advances in microbial fuel cells for wastewater treatment. Renew Sust Energ Rev 71:388–403

Holt P, Barton G, Mitchell C (1999) Electrocoagulation as a wastewater treatment. Austr Environ Eng Res 1000:41–46

Ibrahim N, Zainal SFFS, Aziz HA (2019) Application of UV-based advanced oxidation processes in water and wastewater treatment. In: Advanced oxidation processes (AOPs) in water and wastewater treatment. IGI Global, Hershey, pp 384–414

Jatoi AS, Akhter F, Mazari SA, Sabzoi N, Aziz S, Soomro SA, Ahmed S (2021) Advanced microbial fuel cell for waste water treatment—a review. Environ Sci Pollut Res 28:5005–5019

Kansara N, Bhati L, Narang M, Vaishnavi R (2016) Wastewater treatment by ion exchange method: a review of past and recent researches. ESAIJ 12(4):143–150

Khattri SD, Singh MK (2009) Removal of malachite green from dye wastewater using neem sawdust by adsorption. J Hazard Mater 167(1-3):1089–1094

Klamerth N, Malato S, Maldonado MI, Agüera A, Fernández-Alba AR (2010) Application of photo-fenton as a tertiary treatment of emerging contaminants in municipal wastewater. Environ Sci Technol 44(5):1792–1798. https://doi.org/10.1021/es903455p

Kraume M, Drews A (2010) Membrane bioreactors in waste water treatment–status and trends. Chem Eng Technol 33(8):1251–1259

Legrini O, Oliveros E, Braun AM (1993) Photochemical processes for water treatment. Chem Rev 93(2):671–698

Li Q, Mahendra S, Lyon DY, Brunet L, Liga MV, Li D, Alvarez PJ (2008) Antimicrobial nanomaterials for water disinfection and microbial control: potential applications and implications. Water Res 42(18):4591–4602. https://doi.org/10.1016/j.watres.2008.08.015

Liu H, Ramnarayanan R, Logan BE (2004) Production of electricity during wastewater treatment using a single chamber microbial fuel cell. Environ Sci Technol 38(7):2281–2285

Lu C, Chiu H (2006) Adsorption of zinc (II) from water with purified carbon nanotubes. Chem Eng Sci 61(4):1138–1145. https://doi.org/10.1016/J.CES.2005.08.007

Lu H, Wang J, Wang T, Wang N, Bao Y, Hao H (2017) Crystallization techniques in wastewater treatment: an overview of applications. Chemosphere 173:474–484

Madhuri B, Singh SP, Batra RD (2018) Nanotechnology in wastewater treatment: a review. In: Novel applications in polymers and waste management. Elsevier, Amsterdam, pp 173–182

Mingjun SHAN, Yanqiu WANG, Xue SHEN (2009) Study on bioremediation of eutrophic lake. J Environ Sci 21:S16–S18

Mohan SV, Mohanakrishna G, Velvizhi G, Babu VL, Sarma PN (2010) Bio-catalyzed electrochemical treatment of real field dairy wastewater with simultaneous power generation. Biochem Eng J 51(1-2):32–39

Moosavi S, Lai CW, Gan S, Zamiri G, Pivehzhani AO, Johan MR (2020) Application of efficient magnetic particles and activated carbon for dye removal from wastewater. ACS Omega 5(33): 20684–20697

Munjal M, Tiwari B, Lalwani S, SharmaM SG, Sharma RK (2020) An insight of bioelectricity production in mediator less microbial fuel cell using mesoporous Cobalt Ferrite anode. Int J Hydrog Energy 45:12525–12534

Munoz-Cupa C, Hu Y, Xu C, Bassi A (2021) An overview of microbial fuel cell usage in wastewater treatment, resource recovery and energy production. Sci Total Environ 754:142429

Muro C, Riera F, del Carmen Díaz M (2012) Membrane separation process in wastewater treatment of food industry. In: Food industrial processes–methods and equipment. IntechOpen, Rijeka, pp 253–280

Nidheesh PV, Khan FM, Kadier A, Akansha J, Bote ME, Mousazadeh M (2022) Removal of nutrients and other emerging inorganic contaminants from water and wastewater by electrocoagulation process. Chemosphere 307:135756

Nnaji PC (2012) An investigation of the performance of various coagulants/flocculants in removing the turbidity of coal washery effluents. M. Eng. Thesis, Federal University of Technology, Owerri

Obotey Ezugbe E, Rathilal S (2020) Membrane technologies in wastewater treatment: a review. Membranes 10(5):89

Palmer M, Hatley H (2018) The role of surfactants in wastewater treatment: Impact, removal and future techniques: a critical review. Water Res 147:60–72

Paul DH (2002) MF, UF, NF, RO defining the four basic membrane processes. Opflow 28(5):10–15

Prajapati AK, Das S, Mondal MK (2020) Exhaustive studies on toxic Cr (VI) removal mechanism from aqueous solution using activated carbon of Aloe vera waste leaves. J Mol Liq 307:112956

Qu X, Alvarez PJJ, Li Q (2013) Applications of nanotechnology in water and wastewater treatment. Water Res 47:3931–3946. https://doi.org/10.1016/j.watres.2012.09.058

Rashid R, Shafiq I, Akhter P, Iqbal MJ, Hussain M (2021) A state-of-the-art review on wastewater treatment techniques: the effectiveness of adsorption method. Environ Sci Pollut Res 28:9050–9066

Rathi BS, Kumar PS (2021) Application of adsorption process for effective removal of emerging contaminants from water and wastewater. Environ Pollut 280:116995

Razali M, Kim JF, Attfield M, Budd PM, Drioli E, Lee YM, Szekely G (2015) Sustainable wastewater treatment and recycling in membrane manufacturing. Green Chem 17(12): 5196–5205

Samsami S, Mohamadi M, Sarrafzadeh M-H, Rene ER, Firoozbahr M (2020) Recent advances in the treatment of dye-containing wastewater from textile industries: overview and perspectives. Process Saf Environ Prot 143:138–163

Saxena G, Bharagava RN (2017) Organic and inorganic pollutants in industrial wastes: ecotoxicological effects, health hazards, and bioremediation approaches. In: Environmental pollutants and their bioremediation approaches. CRC Press, pp 23–56

Shahedi A, Darban AK, Taghipour F, Jamshidi-Zanjani A (2020) A review on industrial wastewater treatment via electrocoagulation processes. Curr Opin Electrochem 22:154–169

Shammas NK (2005) Coagulation and flocculation. In: Physicochemical treatment processes. Springer, Cham, pp 103–139. https://doi.org/10.1385/1-59259-820-x:103

Tahreen A, Jami MS, Ali F (2020) Role of electrocoagulation in wastewater treatment: a developmental review. J Water Process Eng 37:101440

Tan KB, Vakili M, Horri BA, Poh PE, Abdullah AZ, Salamatinia B (2015) Adsorption of dyes by nanomaterials: recent developments and adsorption mechanisms. Sep Purif Technol 150:229–242

Uzal N (2015) Effluent treatment in denim and jeans manufacture. Denim 2015:541–561. https://doi.org/10.1016/b978-0-85709-843-6.00019-6

Van Dijk L, Roncken GCG (1997) Membrane bioreactors for wastewater treatment: the state of the art and new developments. Water Sci Technol 35(10):35–41

Wadhawan S, Jain A, Nayyar J, Mehta SK (2020) Role of nanomaterials as adsorbents in heavy metal ion removal from waste water: a review. J Water Process Eng 33:101038

Wang LK, Vaccari DA, Li Y, Shammas NK (2005) Chemical precipitation. In: Physicochemical treatment processes. Springer, Cham, pp 141–197. https://doi.org/10.1385/1-59259-820-x:141

Xu T (2005) Ion exchange membranes: state of their development and perspective. J Membr Sci 263(1-2):1–29

Xu Z, Wei C, Jin J, Xu W, Wu Q, Gu J, Ou M, Xu X (2018) Development of a novel mixed titanium, silver oxide polyacrylonitrile nanofiber as a superior adsorbent and its application for MB removal in wastewater treatment. J Braz Chem Soc 29:560–571

Yadav D, Singh S, Sinha R (2021) Microbial degradation of organic contaminants in water bodies: technological advancements. In: Pollutants and water management: resources, strategies and scarcity. Springer, Cham, pp 172–209

Yang R, Li H, Huang M, Yang H, Li A (2016) A review on chitosan-based flocculants and their applications in water treatment. Water Res 95:59–89

Zueva SB (2018) Current legislation and methods of treatment of wastewater coming from waste electrical and electronic equipment processing. In: Waste electrical and electronic equipment recycling. Woodhead Publishing, Sawston, pp 213–240

Role of Microorganisms in Polluted Water Treatment

13

Inoka C. Perera, K. A. G. de Alwis, and P. I. T. Liyanage

Abstract

Access to clean water is a problem for over 3 billion people in the world. Annually, around 300,000 children lose their lives due to ailments related to the consumption of polluted water. Therefore, neglecting this crisis may lead to severe consequences for the use of clean water to meet basic human needs. Dissolved, suspended, and deposited organic and inorganic substances and microorganisms can pollute water. This hinders the services provided by a pristine aquatic ecosystem. As the neglect of water security is expected to cost five times more than it is being addressed, the number of studies carried out in this field has been enhanced over time. In this context, the concept of bioremediation to address water pollution has become the cynosure, as it is a sustainable and less costly method in comparison to many other alternatives. Bioremediation is a treatment measure for contaminated water. Here, favorable environmental factors are facilitated to stimulate the growth of selected microorganisms that are capable of metabolizing intended pollutants. This way the quality of water can be improved by either removing or reducing the number of harmful substances in water. There are both in situ and ex situ bioremediation techniques, where the former is less costly while the latter is costly. A variety of microorganisms such as bacteria, fungi, and algae are used in this process. It is a vast subject area that is difficult to summarize to a single chapter. However, we have summarized existing knowledge on various aspects of bioremediation to treat polluted water, including the fundamentals, recent advances, and applications, and critically discuss their pros and cons emphasizing future perspectives.

I. C. Perera (✉) · K. A. G. de Alwis · P. I. T. Liyanage
Synthetic Biology Laboratory, Department of Zoology and Environment Sciences, Faculty of Science, University of Colombo, Colombo, Sri Lanka
e-mail: icperera@sci.cmb.ac.lk

© The Author(s), under exclusive license to Springer Nature Singapore Pte Ltd. 2023
R. Soni et al. (eds.), *Current Status of Fresh Water Microbiology*,
https://doi.org/10.1007/978-981-99-5018-8_13

Keywords

Bioremediation · Ecosystem · BOD · COD

13.1 Introduction

Removal of contaminants from the environment has become a global issue. The excessive cost associated with most of the conventional cleanup processes may cause neglection of this process entirely or partially paving the way for many adversities. However, the existence of microorganisms has shed some bright light on addressing this matter. This way environmental pollutants are successfully eliminated by incorporating the diverse metabolic reactions of the microorganisms in a technique called bioremediation. In general, bioremediation implies the eradication of pollutants by biological processes, with the use of microorganisms such as fungi and bacteria.

Here, they use enzymes to catalyze the degradation of environmental pollutants. Some of these digest organic matter and give out substances such as carbon dioxide and water. This way microbes consume these pollutants either to gain energy or to enhance their growth and survival denoting mutual benefits. Biodegradation may depend on either indigenous or exogeneous microbes where the former refers to the microbes at the local site while the latter refers to the foreign organisms introduced externally to the site. This could be conducted either at the contaminated site itself or by transporting them to a separate location away from the initial contaminated site, where the former is called in situ bioremediation and the latter is called ex situ bioremediation. Either way, bioremediation facilitates the growth of suitable microorganisms to eliminate a massive amount of pollutants in contaminated water.

13.2 Global Impact of Water Pollution

Water pollution can be considered a critical environmental issue faced worldwide, and people all over the world face circumstances caused by this issue. This statement can be further explained by analyzing the impacts of water pollution on several sectors.

Both surface water and underground water sources can get contaminated by pollutants (Hasan et al. 2019). Water pollution can be divided into three main categories: contamination by organic compounds, inorganic compounds, and microorganisms (Coelho et al. 2015). Several fields are directly connected to water pollution. Agriculture, industries, and oil spills are some. Agriculture is a field where many pollutants can be emitted to water sources if not handled correctly according to the guidelines. When discussing this matter, chemical fertilizers, insecticides, and other organic and inorganic pollutants would come into the topic. Various types of harmful dyes and other chemicals used in manufacturing clothes might be emitted into water resources in textile industries. Oil spills, automobile, and stationery

13 Role of Microorganisms in Polluted Water Treatment

industries are some other sources that can emit water pollutants. Heavy metals such as Cu (II) and Cr (VI) (Breida et al. 2019), dyes, and bleaching chemicals are some examples of the water pollutants that can be released to water bodies by industries.

13.2.1 Effect on the Human Health

Human health is seriously affected by water pollution. According to the WHO, around 829,000 people are estimated to die each year from diarrhea. This condition results from unsafe drinking water, sanitation, and hand hygiene. Considering "unsafe drinking water," it is directly linked to water pollution.

According to a study conducted in the Turag river in Dhaka, Bangladesh, the maximum concentration of turbidity, hardness, BOD, TDS, and COD are much higher than the standard limits. Moreover, it has been revealed that the local communities suffer from various health conditions, including diarrhea and yellow fever (Hasan et al. 2019). The following are some of the diseases that the consumption of polluted water can cause.

Cholera, which is caused by the infection of toxicogenic *Vibrio cholerae*, can be considered a severe diarrheal disease that can rise to epidemic and pandemic levels (Mandal et al. 2011). Cholera is also considered a reflection of social inequalities since it is primarily found in underprivileged populations worldwide (Somboonwit et al. 2017). Therefore, to a considerable extent, cholera is an indicator that shows the socioeconomic and health facilities of a particular country. Diarrhea is a term that is commonly associated with water pollution. Diarrhea can be considered more a symptom than a disease and can be described as loose and watery bowel movements. Diarrhea usually lasts for a few days. However, if it lasts longer, it could be a symptom of more severe conditions such as inflammatory bowel disease (IBD) and irritable bowel syndrome (IBS). Consumption of polluted water can be a significant cause of diarrhea. In a study done in Shenzen, China, stool samples were collected from 412 patients having diarrhea. Nineteen pathogens had been detected in those samples. Some detected bacterial pathogens are *Salmonella, Campylobacter jejuni, Shigella, Listeria monocytogenes, Vibrio parahaemolyticus*, and *Vibrio cholera*. According to the study, bacterial infections are the dominant cause of diarrhea (Shen et al. 2016). Typhoid fever is another condition that can be caused by water pollution. Considering its epidemiology, *Salmonella enterica* subspecies *enterica serovar Typhi (Salmonella typhi)* is the cause of typhoid fever (Crump 2019). Feces is considered the major source of exit of *Salmonella typhi*, and scientific studies show the availability of this microorganism in urine as well (Sears et al. 1924). People can get infected with *S. typhi* due to the consumption of unsafe food and water contaminated with *S. typhi* through fecal matter.

Other than the diarrheal conditions described above, the conditions such as respiratory diseases, cancers, cardiovascular diseases, and neurological disorders are also associated with polluted water (Ullah et al. 2014). Reproductive failures can also be caused by the consumption of polluted water (Currie et al. 2013). It is said that there are many water contaminants (e.g., disinfected byproducts, fluorinated compounds), which are also endocrine-disrupting chemicals (EDCs). These EDCs can potentially cause detrimental effects on the endocrine system, which would finally lead to the impairment of the fertility and growth of humans and other animals (Gonsioroski et al. 2020).

Another important fact that should be considered is that, for a human to get affected by polluted water, he/she does not need to be in direct contact with polluted water. Pollutants get accumulated in human bodies via food chains as well. When the contaminants are passed through each trophic level, the accumulating concentration of the contaminant in the body of the organism is increased. The scientific term that is used to describe this term is biomagnification. Biomagnification can be defined as the condition when the chemical concentration in an organism exceeds the concentration of its food when the main route of exposure takes place via the diet of the organism (Svanback and Bolnick 2019). Since human is at the last trophic level of many urban food chains, we are in grave danger.

The above facts show that humans are severely affected by the infections associated with water pollution while shedding light on the necessity to implement urgent measures.

13.2.2 Mortality Caused All Over the World Due to Polluted Water and Lack of Hygiene

One of the most tragic aspects of this man-made water pollution is when it costs the lives of the people. So far, infections associated with polluted water and lack of hygiene have cost thousands of lives worldwide. According to the World Bank, the below map indicates the mortality rates caused due to unsafe water, unsafe drinking, and lack of hygiene (per 100,000 population), considering 2016 as the most recent year. By going through this map, it can be realized that many countries in the African region have mortality rates of more than 51.20 (per 100,000 population), which is a concerning issue. According to WHO, around 829,000 people die yearly from diarrhea caused by polluted and unsafe drinking water, lack of sanitation, and inadequate hand hygiene. WHO mentions that since diarrhea is highly preventable, higher mortality rates of children can be avoided if the necessary measures are implemented (Fig. 13.1).

13.2.3 Effect on Ecosystems

The polluted water would also deteriorate the life of aquatic ecosystems, which would ultimately result in the disruption of food chains and food webs. This condition would finally create an imbalance in environmental equilibrium and ecological balance. When considering the impacts of water pollutants on fish species, it is reported that high amounts of suspended water pollutants can interrupt the usual behaviors of fish species. Fish species that rely mainly on sight to catch

13 Role of Microorganisms in Polluted Water Treatment

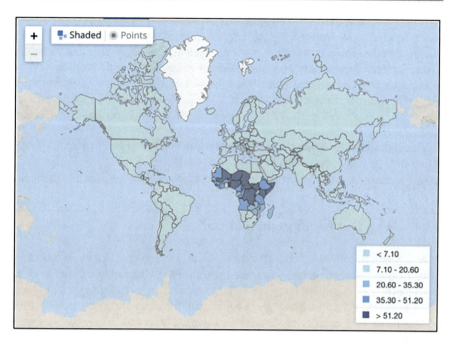

Fig. 13.1 Mortality rate attributed to unsafe water, unsafe drinking, and lack of hygiene (per 100,000 population). Source: https://data.worldbank.org/indicator/SH.STA.WASH.P5

their prey tend to show prominent avoidance behavior since they are highly susceptible to the high amounts of suspended solids in the water bodies (Malik et al. 2020).

On the other hand, aquatic ecosystems can be considered the ultimate sinks of pollutants. As a result of industrialization and urbanization, many environmental pollutants are emitted into water resources. It is known that water has a natural capacity to neutralize the effect of contaminants. However, once this natural capacity is exceeded, the self-generating capacity of the water will be lost. That can be mentioned as one of the main reasons to turn toward other methods to remediate the polluted water (Bashir et al. 2020).

13.3 What Is Water Pollution?

Water pollution can be simply defined as the alterations of biological, chemical, and physical properties of water in a way it would hazardously affect public health and the environment. Water pollution is one of the most critical issues in the twenty-first century that must be addressed immediately. Urbanization and industrialization caused by the increasing population are the main reasons for the emission of various pollutants to water resources. When considering the causes and means of water pollution, oil spills, medical waste, less-organized sewage disposal systems, and fertilizer runoff are noteworthy.

It is known that nearly 9 million barrels of oil are released into the ocean every year (Cohen 2013). DeBofsky et al. (2020) reported that in July 2016, a Husky Energy pipeline spilled, causing a portion of oil to enter the North Saskatchewan River, Canada. Then muscle, bile, and intestinal samples were collected from several fish species from the affected area. It had been revealed that the concentrations of polycyclic aromatic hydrocarbons (PAHs, a crude oil constituent) in dorsal muscles significantly correlated with the gut microbial composition of walleyes (*Sander vitreus*).

The authorities have implemented general rules and regulations for the proper disposal of garbage and other waste materials. Utilizing an efficient method to reduce water pollution is a timely need. Hence, bioremediation comes into the topic.

13.4 Why Opt for Bioremediation?

Several methods (physicochemical and biological) are available to remove the heavy metals from the water before releasing them into the environment (Coelho et al. 2015). Isolation and minimizing the mobility of metal ions using physical barriers made of materials such as steel and cement, solidification and stabilization of the contaminants, size selection processes to remove the large and cleaner particles from the small and more polluted particles, and electrokinetic processes are some of the examples for the physiochemical methods that are being used to treat contaminated water (Mulligan et al. 2001). However, these methods might not be cost-effective and environmentally friendly. Therefore, bioremediation can be suggested as a better method for the same purpose.

13.5 Bioremediation as an Answer to Mitigate Water Pollution

Bioremediation is the process of utilizing specific biomolecules or various biomasses to get bound and concentrate ions or molecules present in an aqueous solution (Coelho et al. 2015). In simpler words, it provides an option to minimize the detrimental impacts of contaminants using biological methods.

Considering the basic principles behind bioremediation, living organisms (primarily microorganisms) are utilized to treat waste materials to convert them into less toxic forms (Vidali 2001). As described under "1. Global Impact of Water Pollution," it is evident that adequate measures must be taken to minimize water pollution. Otherwise, it would result in irreversible damage to the world in many sectors such as health and the environment. In that light, bioremediation comes into the topic mainly due to its advantages over many other methods.

13.6 Balance of Pollutants in a Natural Ecosystem

There are threshold levels for the nutrients in natural aquatic ecosystems. Once the threshold level is exceeded, a nutrient may be considered a pollutant. When the nutrient ratios are altered, the aquatic ecosystems might respond to them in various ways. Phosphorus is considered the limiting nutrient factor for phytoplankton growth in freshwater ecosystems. Atmospheric nitrogen also affects the acidification of freshwater ecosystems, resulting in detrimental effects. It could be challenging to isolate the effects of a particular nutrient (e.g., nitrogen) from the effects of the other nutrients. It is said that higher amounts of nutrients cannot be isolated and examined in the presence of other nutrients that are essential for the growth of plants, such as phosphorus and silicon. Altered stoichiometric values in nutrients would cause changes in the aquatic ecosystem (Rabalais 2002).

Nutrient cycling is important when considering the balance of nutrients in aquatic ecosystems. Water plays a prominent role in nutrient cycling processes (Malik et al. 2020). Animals are also involved in the nutrient cycling in freshwater ecosystems. Animals can supply nutrients (e.g., nitrogen and phosphorus) via excretory processes. It is known that the nutrient cycling by animals fulfills a substantial amount of the nutrient demand of the primary producers in ecosystems (Vanni 2003). According to a study by Pennifold and Davis (2001), in the presence of macrofauna, the release of phosphate and ammonium into the water column has been increased, while the release of nitrate has been decreased, which means nitrate has been uptaken by the sediments. Climatic conditions might also affect the nutrient cycling in freshwater ecosystems to a considerable level. According to Pennifold and Davis (2001), the season and the dissolved oxygen levels influence the nutrient concentrations. According to their study, ammonium has been released to the overlying water column at considerably higher rates in winter, and nitrogen has been released at higher rates in summer than in the winter.

Anthropogenic activities hugely affect the nutrient/pollutant balance in freshwater ecosystems. The disposal of domestic garbage in freshwater ecosystems can be commonly observed in many developing countries. Other than that, the emission of wastewater and chemicals from industries has also become a severe issue that directly affects the nutrient balance in freshwater ecosystems. Eutrophication is an important mechanism when it comes to nutrient imbalance in freshwater ecosystems. Eutrophication can be defined as the occurrence of an excessive accumulation of nutrients in a water body, which can finally result in higher growth of plants. This can be considered a growing issue in many freshwater bodies, which causes adverse effects on their ecosystem (Rathore et al. 2016). While causing detrimental effects on the water quality, it also hugely affects the biotic components. In eutrophication, extensive algal blooms are caused due to the increment of suspended particles in the water bodies, and the water clarity is also decreased. The rapidly increasing rate of precipitation leads to the destruction of benthic habitats (Dorgham 2014).

According to the above facts, it is clear that the balance of nutrients/pollutants plays a vital role in natural ecosystems.

13.7 Examples of the Microorganisms Used in Bioremediation

Microorganisms play a vital role in the biological degradation of materials. Hence, they are important in treating polluted water. Bacteria, protozoa, metazoa, filamentous bacteria, algae, and fungi are more prominent groups of microorganisms that are identified in bioremediation (Rani et al. 2019) (Table 13.1).

13.8 Mechanism in the Use of Microorganisms in Bioremediation of Polluted Water

Microbes have an innate capability to metabolize and remove toxic xenobiotics from their surrounding. The mechanism involved in this process breakdown substances with higher complexity into simpler forms such as CO_2 and H_2O. This way harmful substances in polluted water are converted into eco-friendly, safer forms by bioremediation. Apart from irradicating the toxins, this also aids to monitor the optimal levels of inorganic substances in water at equilibrium (Misal et al. 2011). The process of bioremediation relies on the temperature, nutrition, pH, concentration of various substances, time, and the types of microbes used (Rani et al. 2019).

The treatment process incorporates three key stages, that is, primary treatment, secondary treatment, and tertiary treatment. Initially, the undissolved substances that are suspended in polluted water are eliminated in the primary treatment. Wastewater has a comparatively higher biological oxygen demand compared to pure water. Thus, in the secondary stage of the treatment, process microbes are used to mitigate this issue. Here, the dissolved water content in water may be increased by degrading organic matter using both aerobic and anaerobic microbes. Also at this level, the turbidity of water is increased due to the purification process. In the final stage, a diverse set of approaches such as ion exchange and reverse osmosis are used to purify the effluent (Rani et al. 2019).

Problems such as toxicity to humans and scaling in pipes/containers arise due to the presence of some ions in water. Here, the former is caused by heavy metal pollution, whereas the latter is due to the high levels of minerals in the water. Therefore, resins made out of artificial or organic substances with ionic functional groups are used to eliminate these ions from water. Above ionic groups hold ions that facilitate ion exchange in contaminated water. Followed by several rounds of ion exchange, the resin can be treated and reused (Crist et al. 2010).

Reverse osmosis is a technique that uses a membrane to separate contaminants from wastewater. This is used to remove both organic and inorganic pollutants from water (Jamil et al. 2019; Madsen and Søgaard 2014; Nikbakht Fini et al. 2019; Urtiaga et al. 2013). The incorporation of sand filtration techniques together with membrane separation methods has reduced the amount of water that needs to undergo bioremediation, as the longer retention time in the sand filter is permitting the degradation of the contaminants in water (Schostag et al. 2022).

13 Role of Microorganisms in Polluted Water Treatment

Table 13.1 Selected microorganims used in bioremediation

Microorganism group	Species	Property investigated	Reference
Bacteria	*Cronobacter sakazakii, Enterobacter asburiae, Leclercia adecarboxylata, Klebsiella oxytoca, K. pneumoniae, Bacillus* sp., *Enterococcus thailandicus, Chromobacterium vaccinii, Serratia* sp., *Kosakonia oryzae,* and *Escherichia coli*	• Removal of organic materials • Heterotrophs consume and use the energy of organic matter for cellular regeneration. This accelerates their growth in the medium • Concentration used: 10^8 to 10^9/ml	Lefebvre et al. (2006), Silva-Bedoya et al. (2016)
Protozoa	*Trachelophyllum pusillum, Vorticella microstoma, Carchesium polypinum, Chilodonella uncinata, Opercularia coarctata,* and *Aspidisca costata*	• Feed on pathogenic bacterial strains and suspended matter • Some depend on oxygen for survival, whereas others need very little to no oxygen • Flagellates and ciliates are prominent groups of protozoa that are employed in the bioremediation process where the former feeds on organic material and the latter feeds on the free-floating bacteria • Concentration used : 5×10^3 to 2×10^4	Amaral et al. (2004), Cacciò et al. (2003)
Metazoa		• Act as indicators • Rotifers and nematodes are metazoan groups that play dynamic roles in activated sludge • They can feed on other microbes • Higher concentrations of toxic materials are known to be detrimental to their survival	Rani et al. (2019)
Filamentous bacteria	*Microthrix parvicella, Thiothrix* sp., *Alcanivorax borkumensis, Sphaerotilus natans,* and *Beggiatoa* sp.	• Important in forming floc, which represents a loose clump of granules • The abundance of filamentous bacteria depends on factors such as the dissolved oxygen content, temperature, nutrition, etc.	Rani et al. (2019)
Algae	*Oscillatoria, Chlamydomonas, Phormidium autumnale,*	• Extract harmful substances (heavy metals and pesticides) from polluted water	Hoeger et al. (2005), Martins et al. (2010)

(continued)

Table 13.1 (continued)

Microorganism group	Species	Property investigated	Reference
	Limnothrix, Synechocystis, Microcystis, and *Lyngbya*	• Reduce the nutrient overload in wastewater by using it to produce biomass	
Fungi	*Sphaerotilus natans, Aspergillus, Penicillium, Fusarium*, and *Absidia*	• Break down organic compounds at lower pH values • Convert ammonia into nitrite via an oxidation reaction that hinders the development of bacteria • Hyphae, which is a characteristic feature of fungi adsorb suspended matter in the wastewater • Secrete certain types of enzymes that aid in the degradation process	Hossain Molla et al. (2004), Rani et al. (2019)

13.9 Microbial Water Treatment Systems

The use of microorganisms is a promising approach for the purification of contaminated water. These microbial water treatment systems can be classified into two main types based on the oxygen demand. They are aerobic, anaerobic, and facultative methods (Shah and Roguez-Couto n.d.). Microorganisms break down substances to harvest energy via respiration. Microbes that require and do not require oxygen for this process can be incorporated into aerobic and anaerobic bioremediation, respectively. Furthermore, in the presence of oxygen, the former will take place, while the latter will not as oxygen may halt the process by interfering with the metabolic reactions of the anaerobes.

13.9.1 Aerobic Systems

These systems have a higher capability of limiting both pathogens and the biological oxygen demand. Aerobic systems can eliminate a diversity of organic materials. Also, this is a convenient method that is easy to establish and maintain. The sludge production in aerobic systems is minimal. The "membrane bioreactor system" is an example of an aerobic system in which the treated water can straightly be used in the recycling process.

Oxygen is considered to be one of the key growth retardation factors for the bacteria that degrade hydrocarbons. The excessive release of hydrocarbons to waterways depletes the dissolved oxygen levels in wastewater attenuating the

13 Role of Microorganisms in Polluted Water Treatment

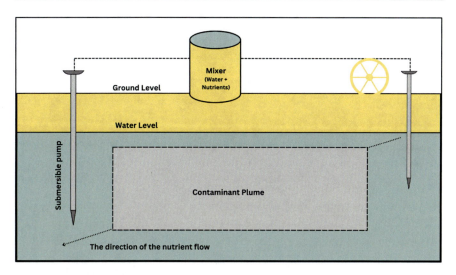

Fig. 13.2 Conceptual diagram: aerobic bioremediation

aerobic biodegradation process. Thus, measures are taken to aerate the wastewater to accelerate biodegradation (Fig. 13.2). This principle is applied in the technique called "enhanced aerobic biodegradation," which is typically used for water contaminated with high levels of petroleum hydrocarbons (Epa and of Underground Storage Tanks n.d.; Mikkonen et al. 2018).

However, the contamination of the membranes and the hindrance to aeration are plausible drawbacks of membrane bioreactors. Furthermore, another approach called "the conventional Activated Sludge Plants" demands higher levels of energy, and the resulting sludge is needed to follow complex disposal protocols. Also, some aerobic systems need a huge space, which is a limiting factor in many scenarios. "The rotating biological contactor" is an aerobic system that retains high levels of opaqueness in the treated water (Rani et al. 2019).

13.9.2 Anaerobic Systems

This is normally used to treat wastewater enriched with organic matter as signified with high biological oxygen demand (BOD), chemical oxygen demand (COD), and total suspended solids (TSS). In anaerobic systems, microbes degrade organic substances without using oxygen. This is a cyclic process that starts with the acceptance of wastewater into a bioreactor receptacle. This bioreactor has a gelatinous layer of sludge, which is the home of a variety of microbes. The anaerobes in sludge cause the biological degradation of substances in wastewater. This results in the formation of effluent with a low biological and chemical oxygen demand. Also, this accounts for low levels of the total suspended solids in the treated water. Hence, this approach is used in the treatment of wastewater generated from various

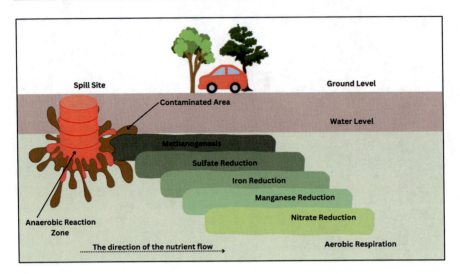

Fig. 13.3 Conceptual diagram: anaerobic bioremediation

industries such as agriculture, food, milk, paper, and clothing. Furthermore, an anaerobic system can also be used in the treatment of sewage (Cyprowski et al. 2018; Qadri et al. 2021; Scott 1995; Samcotech n.d.).

In anaerobic systems, the amount of biological waste formed in the process is less. The demand for nutrition is also minimum in this approach. Thus, anaerobic systems are user-friendly methods that require minimum management (Fig. 13.3). Furthermore, the production of methane as a terminal product is an added advantage. The growth retardation of microorganisms and the pungent smell are drawbacks of anaerobic systems (Rani et al. 2019).

13.9.3 Facultative Systems

Some microorganisms can survive both in oxygen-rich and poor environments. These are called facultative microorganisms. Thus, in the absence of dissolved oxygen, they may rely on other substances such as sulfates, nitrates, or other mechanisms to conduct biological degradation.

Hydrocarbons can be degraded in both aerobic and anaerobic environments. However, the rate of degradation is greater in the former compared to that of the latter (Coates et al. 1997; Holliger and Zehnder 1996). Facultative anaerobes, those that reduce nitrates, iron, and manganese, can degrade hydrocarbons (Coates et al. 1997; Fries et al. 1994). Interestingly, the ability of facultative anaerobes to survive and thrive under aerobic as well as anaerobic conditions has paved the way to maintain a balanced population status aiding the catabolism of pollutants in water. Therefore, studies are conducted to study the features of such bacterial strains that are used in facultative bioremediation (Grishchenkov et al. 2000). Facultative systems need less energy and are convenient to manage. On the contrary, the sludge deposits should be cleaned from time to time (Rani et al. 2019).

13.10 Concerns Regarding Bioremediation

Despite being a versatile technique, bioremediation is facing some limitations. For instance, substances saturated with chorine and polycyclic aromatic hydrocarbons are not successfully eliminated via microorganisms. Furthermore, it is plausible for microbial degradation to form more harmful byproducts upon metabolizing the xenobiotics in the medium. The formation of vinyl chloride after reacting with trichloroethylene exemplifies this. Here, the resultant product, vinyl chloride, is a harmful carcinogen (Megharaj et al. 2014). Thus, in some scenarios, the products generated following biodegradation could be more harmful than the initial contaminant.

Furthermore, the applicability of bioremediation to eliminate only biodegradable compounds limits the use of this approach. Therefore, some contaminants will not undergo partial degradation or will not be degradable at all. Compared to other conventional approaches of treatment, bioremediation is a time-consuming method. In the case of ex situ bioremediation, actions such as excavating and pumping are required, making the process lengthier than usual. Also, the control of the release of organic matter that is volatile is not feasible in an ex situ setting (Hlihor et al. 2017; Sharma 2021). Being a biological process, biodegradation demands specific requirements to yield optimal results. Therefore, the success of the process may depend on factors such as the availability of microorganisms with the optimum metabolic capacity, appropriate growth conditions with nutrients for the enrichment of their population, and so on. Also, contradictions may arise when extrapolating the outcomes of both laboratory and pilot studies into absolute sites. Thus, further studies are needed to formulate novel or modified bioremediation techniques to approach sites with a diversity of pollutants in the form of solids, liquids, and gases. Also, the reliability of the process has to be confirmed by establishing performance specifications including endpoints for bioremediation treatment. This signifies bioremediation as a field of study that urges intense research on the outcomes of the processes governed by microorganisms (Megharaj et al. 2014).

13.11 Costs Associated with Bioremidiation

Many factors such as the type of contaminant, location where the purification is carried out (ex situ/in situ), degree of contamination, and the type of the bioremediation technique used may influence the cost associated with bioremediation.

Since bioremediation depends on the ability of microorganisms to digest waste, the digestion of some contaminants might be difficult for them over others. To address this matter, research should be carried out to select a better species to be used in the process. This significantly increases the cost associated with the bioremediation project. In the case of ex situ bioremediation, wastewater should be transported from the site of origin to the treatment plant, making the process more complex and costly. Furthermore, depending on the level of contamination, the intensity of the purification will have to change. Thus, the approach for bioremediation may have to be advanced or intensified in the case of treating highly contaminated waster. In addition, it might be necessary to repeat some steps or the entire purification process to ensure that the desired purity is acquired. Hence, when using heavily contaminated water, the ingredient cost, labor cost, energy consumption, and many more will account for the expenses. Furthermore, the type of bioremediation technique used will also have an impact on the cost. Therefore, appropriacy, viability, and sustainability should be under scrutiny before implementing a bioremediation project.

13.12 Future Advances in the Use of Bioremediation on Controlling Water Pollution

Bioremediation is used as a treatment measure for industrial effluents, marine oil spills, etc. This is an economic method to mitigate many environmental issues. One measure that can be used to increase the efficiency of this method is by enhancing the interaction between the microbes and the pollutants. This is important, especially when dealing with hydrophobic contaminants. Among the many toxins present in wastewater, heavy metals cause various repercussions. They are environmental pollutants with detrimental impacts on human health. Heavy metals account for cancer, neurotoxicities, and much more, leading to the malfunction of the organ systems. Thus, advances in co-remediation techniques such as microbial electrochemical systems could be used to enhance the efficiency of the removal of heavy metals. Not only for heavy metals but also for many contaminants the combination of diverse technologies alongside the conventional bioremediation process has shed promising insights. However, the identification of suitable technologies to be coupled is a critical process. Also, there is an urge to search for novel microbes with higher capacity and efficiency in the removal of noxious heavy metals and other pollutants that are not currently eradicated efficiently (Rosanti et al. 2020). Hence, studies have been conducted to identify different approaches in which enzymes can be engineered to address a particular need. Methene monooxygenase and hydroxylating dioxygenase are some instances of this where the former can be used to clean up oil and the latter for polycyclic aromatic hydrocarbons in water. Such enzyme production can be efficiently regulated with bioengineered microorganisms with the capability of identifying and responding to the pollutant levels in the environment (Hemamali et al. 2022). In the future with the advanced application of in silico methods and synthetic biology, enzyme-mediated bioremediation could be novel approachs to mitigate water pollution (Dutta and Shityakov 2021).

13.13 Conclusion

In conclusion, many anthropogenic activities such as industrial effluents, urbanization, and many more have risked the balance of the ecosystems by causing environmental pollution. To date, many waterways are polluted with various chemicals such as hydrocarbons, pesticides, dioxanes, and many other xenobiotics that are toxic to flora and fauna. Hence, proper remediation methods are crucial to re-establish and maintain the sustainability of the environment. In that light bioremediation, the use of a variety of microorganisms serves as a solution in treating polluted water over other conventional remediation methods. Currently, most bioremediation approaches are centered on the innate capabilities of the existing microorganisms to metabolize organic matter. Therefore, in the future, a synthetic biology approach to redesigning enzymes will outsmart the capabilities of the existing microbes in the remediation of polluted water. Thus, there is a vacuum of knowledge that needs to be filled by novel studies on catalysts that can be used in bioremediation to ameliorate the ecosystem.

References

Amaral AL, Da Motta M, Pons MN, Vivier H, Roche N, Mota M, Ferreira EC (2004) Survey of protozoa and metazoa populations in wastewater treatment plants by image analysis and discriminant analysis. Environmetrics 15:381–390. https://doi.org/10.1002/env.652

Bashir I, Lone FA, Bhat RA, Mir SA, Dar ZA, Dar SA (2020) Concerns and threats of contamination on aquatic ecosystems. Bioremed Biotechnol 2020:1–26. https://doi.org/10.1007/978-3-030-35691-0_1

Breida M, Younssi SA, Ouammou M, Bouhria M, Hafsi M (2019) Pollution of water sources from agricultural and industrial effluents: special attention to NO_3^-, Cr(VI), and Cu(II). Water Chem. https://doi.org/10.5772/INTECHOPEN.86921

Cacciò SM, De Giacomo M, Aulicino FA, Pozio E (2003) Giardia cysts in wastewater treatment plants in Italy. Appl Environ Microbiol 69:3393–3398. https://doi.org/10.1128/AEM.69.6.3393-3398.2003

Coates JD, Woodward J, Allen J, Philp P, Lovley DR (1997) Anaerobic degradation of polycyclic aromatic hydrocarbons and alkanes in petroleum-contaminated marine harbor sediments. Appl Environ Microbiol 63:3589. https://doi.org/10.1128/AEM.63.9.3589-3593.1997

Coelho LM, Rezende HC, Coelho LM, de Sousa PAR, Melo DFO, Coelho NMM (2015) Bioremediation of polluted waters using microorganisms. In: Advances in bioremediation of wastewater and polluted soil. IntechOpen, London. https://doi.org/10.5772/60770

Cohen MA (2013) Water pollution from oil spills. Encycl Energy Nat Resour Environ Econ 3–3:121–126. https://doi.org/10.1016/B978-0-12-375067-9.00094-2

Crist DR, Crist RH, Oberholser K, Erikson J, Bennett J, Noto L (2010) Ion exchange basis for biosorption and bioremediation of heavy metals. Miner Process Extr Metall Rev 19:253–263. https://doi.org/10.1080/08827509608962444

Crump JA (2019) Progress in typhoid fever epidemiology. Clin Infect Dis 68(1):S4. https://doi.org/10.1093/CID/CIY846

Currie J, Zivin JG, Meckel K, Neidell M, Schlenker W (2013) Something in the water: contaminated drinking water and infant health. Can J Econ 46(3):791–810. https://doi.org/10.1111/CAJE.12039

Cyprowski M, Stobnicka-Kupiec A, Ławniczek-Wałczyk A, Bakal-Kijek A, Gołofit-Szymczak M, Górny RL (2018) Anaerobic bacteria in wastewater treatment plant. Int Arch Occup Environ Health 91:571. https://doi.org/10.1007/S00420-018-1307-6

DeBofsky A, Xie Y, Jardine TD, Hill JE, Jones PD, Giesy JP (2020) Effects of the husky oil spill on gut microbiota of native fishes in the North Saskatchewan River, Canada. Aquat Toxicol 229: 105658. https://doi.org/10.1016/J.AQUATOX.2020.105658

Dorgham MM (2014) Effects of eutrophication. Eutrophication 2:29–44. https://doi.org/10.1007/978-94-007-7814-6_3/COVER/

Dutta K, Shityakov S (2021) New trends in bioremediation technologies toward environment-friendly society: a mini-review. Front Bioeng Biotechnol 9:520. https://doi.org/10.3389/FBIOE.2021.666858/BIBTEX

Epa and of Underground Storage Tanks (n.d.) How to evaluate alternative cleanup technologies for underground storage tank sites - a guide for corrective action plan reviewers, Chapter 12, Enhanced Aerobic Bioremediation

Fries MR, Zhou J, Chee-Sanford J, Tiedje JM (1994) Isolation, characterization, and distribution of denitrifying toluene degraders from a variety of habitats. Appl Environ Microbiol 60:2802–2810. https://doi.org/10.1128/AEM.60.8.2802-2810.1994

Gonsioroski A, Mourikes VE, Flaws JA (2020) Endocrine disruptors in water and their effects on the reproductive system. Int J Mol Sci 21(6):1929. https://doi.org/10.3390/IJMS21061929

Grishchenkov VG, Townsend RT, McDonald TJ, Autenrieth RL, Bonner JS, Boronin AM (2000) Degradation of petroleum hydrocarbons by facultative anaerobic bacteria under aerobic and anaerobic conditions. Process Biochem 35:889–896. https://doi.org/10.1016/S0032-9592(99)00145-4

Hasan K, Shahriar A, Jim KU (2019) Water pollution in bangladesh and its impact on public health. Heliyon 5(8):e02145. https://doi.org/10.1016/J.HELIYON.2019.E02145

Hemamali EH, Weerasinghe LP, Tanaka H, Kurisu G, Perera IC (2022) LcaR: a regulatory switch from Pseudomonas aeruginosa for bioengineering alkane degrading bacteria. In: Biodegradation. Springer. Jan-17 2022

Hlihor RM, Gavrilescu M, Tavares T, Favier L, Olivieri G (2017) Bioremediation: an overview on current practices, advances, and new perspectives in environmental pollution treatment. Biomed Res Int. https://doi.org/10.1155/2017/6327610

Hoeger SJ, Hitzfeld BC, Dietrich DR (2005) Occurrence and elimination of cyanobacterial toxins in drinking water treatment plants. Toxicol Appl Pharmacol 203:231–242. https://doi.org/10.1016/j.taap.2004.04.015

Holliger C, Zehnder AJB (1996) Anaerobic biodegradation of hydrocarbons. Curr Opin Biotechnol 7:326–330. https://doi.org/10.1016/S0958-1669(96)80039-5

Hossain Molla A, Fakhru'L-Razi A, Zahangir Alam M (2004) Evaluation of solid-state bioconversion of domestic wastewater sludge as a promising environmental-friendly disposal technique. Water Res 38:4143–4152. https://doi.org/10.1016/J.WATRES.2004.08.002

Jamil S, Loganathan P, Listowski A, Kandasamy J, Khourshed C, Vigneswaran S (2019) Simultaneous removal of natural organic matter and micro-organic pollutants from reverse osmosis concentrate using granular activated carbon. Water Res 155:106–114. https://doi.org/10.1016/J.WATRES.2019.02.016

Lefebvre O, Vasudevan N, Thanasekaran K, Moletta R, Godon JJ (2006) Microbial diversity in hypersaline wastewater: the example of tanneries. Extremophiles 10:505–513. https://doi.org/10.1007/s00792-006-0524-1

Madsen HT, Søgaard EG (2014) Applicability and modelling of nanofiltration and reverse osmosis for remediation of groundwater polluted with pesticides and pesticide transformation products. Sep Purif Technol 125:111–119. https://doi.org/10.1016/J.SEPPUR.2014.01.038

Malik DS, Sharma AK, Sharma AK, Thakur R, Sharma M, Sharma AK (2020) A review on impact of water pollution on freshwater fish species and their aquatic environment. In: Advances in environmental pollution management: wastewater impacts and treatment technologies.

Agriculture and Environmental Science Academy, Haridwar. https://doi.org/10.26832/aesa-2020-aepm-02

Mandal S, Mandal MD, Pal NK (2011) Cholera: a great global concern. Asian Pac J Trop Med 4(7): 573–580

Martins J, Peixe L, Vasconcelos V (2010) Cyanobacteria and bacteria co-occurrence in a wastewater treatment plant: absence of allelopathic effects. Water Sci Technol 62:1954–1962. https://doi.org/10.2166/WST.2010.551

Megharaj M, Venkateswarlu K, Naidu R (2014) Bioremediation. Encycl Toxicol 2014:485–489. https://doi.org/10.1016/B978-0-12-386454-3.01001-0

Mikkonen A, Yläranta K, Tiirola M, Dutra LAL, Salmi P, Romantschuk M, Copley S, Ikäheimo J, Sinkkonen A (2018) Successful aerobic bioremediation of groundwater contaminated with higher chlorinated phenols by indigenous degrader bacteria. Water Res 138:118–128. https://doi.org/10.1016/J.WATRES.2018.03.033

Misal SA, Lingojwar DP, Shinde RM, Gawai KR (2011) Purification and characterization of azoreductase from alkaliphilic strain Bacillus badius. Process Biochem 46:1264–1269. https://doi.org/10.1016/J.PROCBIO.2011.02.013

Mulligan CN, Yong RN, Gibbs BF (2001) Remediation technologies for metal-contaminated soils and groundwater: an evaluation. Eng Geol 60(1–4):193–207. https://doi.org/10.1016/S0013-7952(00)00101-0

Nikbakht Fini M, Madsen HT, Muff J (2019) The effect of water matrix, feed concentration and recovery on the rejection of pesticides using NF/RO membranes in water treatment. Sep Purif Technol 215:521–527. https://doi.org/10.1016/J.SEPPUR.2019.01.047

Pennifold M, Davis J (2001) Macrofauna and nutrient cycling in the swan river estuary, Western Australia: experimental results. Hydrol Process 15(13):2537–2553. https://doi.org/10.1002/HYP.294

Qadri RZ, Ali M, Rajpal A, Kazmi AA, Tawfik A, Tyagi VK (2021) Anaerobic wastewater treatment for energy recovery and water reclamation. Clean Energy Resour Recover Wastewater Treat Plants Biorefin 2:95–104. https://doi.org/10.1016/B978-0-323-90178-9.00030-5

Rabalais NN (2002) Nitrogen in aquatic ecosystems. Ambio 31(2):102–112. https://doi.org/10.1579/0044-7447-31.2.102

Rani N, Sangwan P, Joshi M, Sagar A, Bala K (2019) Microbes: a key player in industrial wastewater treatment. Microb Wastewater Treat 2019:83–102. https://doi.org/10.1016/B978-0-12-816809-7.00005-1

Rathore SS, Chandravanshi P, Chandravanshi A, Jaiswal K (2016) Eutrophication: impacts of excess nutrient inputs on aquatic ecosystem. IOSR J Agric Vet Sci 9(10):89–96. https://doi.org/10.9790/2380-0910018996

Rosanti D, Wibowo YG, Safri M, Maryani AT, Ramadhan BS (2020) Bioremediations technologies on wastewater treatment: opportunities, challenges and economic perspective. Sainmatika J Ilm Mat Ilmu Pengetah Alam 17:142–156. https://doi.org/10.31851/SAINMATIKA.V17I2.5085

Samcotech (n.d.) What is anaerobic wastewater treatment and how does it work? https://www.samcotech.com/anaerobic-wastewater-treatment-how-it-works/. Accessed 15 July 2022

Schostag MD, Gobbi A, Fini MN, Ellegaard-Jensen L, Aamand J, Hansen LH, Muff J, Albers CN (2022) Combining reverse osmosis and microbial degradation for remediation of drinking water contaminated with recalcitrant pesticide residue. Water Res 216:118352. https://doi.org/10.1016/J.WATRES.2022.118352

Scott K (1995) Industrial waste water and effluent treatment. Handb Ind Membr 1995:575–629. https://doi.org/10.1016/B978-185617233-2/50014-3

Sears HJ, Garhart RW, Mack DW (1924) A milk borne epidemic of typhoid fever traced to a urinary carrier. Am J Public Health 14(10):848. https://doi.org/10.2105/AJPH.14.10.848

Shah MP, Roguez-Couto S (n.d.) Microbial wastewater treatment. Elsevier, Amsterdam

Sharma I (2021) Bioremediation techniques for polluted environment: concept, advantages, limitations, and prospects. Trace Met Environ. https://doi.org/10.5772/INTECHOPEN.90453

Shen H, Zhang J, Li Y, Xie S, Jiang Y, Yanjie W, Ye Y, Yang H, Mo H, Situ C, Qinghua H (2016) The 12 gastrointestinal pathogens spectrum of acute infectious diarrhea in a Sentinel Hospital, Shenzhen, China. Front Microbiol 7:1926. https://doi.org/10.3389/FMICB.2016.01926/BIBTEX

Silva-Bedoya LM, Sánchez-Pinzón MS, Cadavid-Restrepo GE, Moreno-Herrera CX (2016) Bacterial community analysis of an industrial wastewater treatment plant in Colombia with screening for lipid-degrading microorganisms. Microbiol Res 192:313–325. https://doi.org/10.1016/J.MICRES.2016.08.006

Somboonwit C, Menezes LJ, Holt DA, Sinnott JT, Shapshak P (2017) Current views and challenges on clinical cholera. Bioinformation 13(12):405. https://doi.org/10.6026/97320630013405

Svanback R, Bolnick DI (2019) Food speciation. In: Fath B (ed) Encyclopedia of ecology. Elservier, Amsterdam, pp 1636–1642

Ullah S, Muhammad Shafique M, Javed W, Khan SF (2014) An integrated approach for quality assessment of drinking water using GIS: a case study of lower Dir. J Himalayan Earth Sci 47: 163–174

Urtiaga AM, Pérez G, Ibáñez R, Ortiz I (2013) Removal of pharmaceuticals from a WWTP secondary effluent by ultrafiltration/reverse osmosis followed by electrochemical oxidation of the RO concentrate. Desalination 331:26–34. https://doi.org/10.1016/J.DESAL.2013.10.010

Vanni MJ (2003) Nutrient cycling by animals in freshwater ecosystems. Annu Rev 33:341–370. https://doi.org/10.1146/ANNUREV.ECOLSYS.33.010802.150519

Vidali M (2001) Bioremediation. An overview. Pure Appl Chem 73(7):1163–1172. https://doi.org/10.1351/PAC200173071163/MACHINEREADABLECITATION/RIS

Bioremediation of Polluted Water

14

U. M. Aruna Kumara, N. V. T. Jayaprada, and N. Thiruchchelvan

Abstract

Most human involvements such as domestic, agricultural, and industrial have an impact toward water and the ecosystem. Current situation water pollution has become a global crisis and affect serious threat to sustainable development for the whole world. Agriculture is one of the backbones of the economy for many developed countries and developing countries. Crucial agricultural activities are causing surface and groundwater pollution. Therefore, they are rendered unsafe, causing issues for human consumption, soil erosion including irrigation, industrial needs, change in nitrogen cycle, carbon sequestration, and relevant ecological patterns. Normally, water pollution happens due to excessive use of pesticides and chemical fertilizers, causing changes in the physiochemical activities of the water. Eutrophication is the main drawback that happens in the aquatic bodies near the agricultural yards due to the runoff of nutrients. As a result, algal blooms can grow on the water body's surface, causing very low dissolved oxygen levels , which can be problematic for aquatic animals. The current approach of this topic is to reduce the environment and ecological impact of using chemical and pesticides in agriculture and promote the necessity of organic farming. Bioremediation is a biological process that helps to recycle waste into a form of a valuable product or non-harm product. Microorganisms directly participate in converting waste into another form that can be recycled or reused. Due to the ability of microorganisms to survive in all places in the world, they are the most suitable

---

U. M. A. Kumara (✉) · N. V. T. Jayaprada
Department of Agricultural Technology, Faculty of Technology, University of Colombo, Homagama, Sri Lanka
e-mail: umarunakumara@at.cmb.ac.lk

N. Thiruchchelvan
Department of Agricultural Biology, Faculty of Agriculture, University of Jaffna, Jaffna, Sri Lanka

© The Author(s), under exclusive license to Springer Nature Singapore Pte Ltd. 2023
R. Soni et al. (eds.), *Current Status of Fresh Water Microbiology*,
https://doi.org/10.1007/978-981-99-5018-8_14

ones for use as a biotic agent for bioremediation. Temperature and pH like abiotic factors also influence bioremediation. Degradation, eradication, immobilization, and detoxification are the main processes in bioremediation. With the improvement of biotechnological aspects in the world, different methods and strategies like biostimulation, bioaugmentation, bioventing, bio-piles, and bio-attenuation are used. Microorganisms play a major role in bioremediation.

Keywords

Pollution · Waste · Bioremediation · Phytoplankton

14.1 Introduction

Water bodies in the world are facing severe pollution due to rapid population growth and accelerated industrialization. It is reported that agricultural, industrial, and urban emissions are the main sources of pollutants released into the coastal marine and freshwater environment (Chen et al. 2022; Faroque and South 2022). Since the effective controls of industrial and urban domestic sewage are available, agricultural nonpoint source pollution from agricultural production has become the main source of water pollution (Chervier et al. 2022; Xiao et al. 2021). Aquatic ecosystem pollution causes severe repercussions in all living organisms along with human health (Zamora-Ledezma et al. 2021; Datta et al. 2022). This pollution can be due to organic compounds (e.g., hydrocarbons) and inorganic compounds (e.g., heavy metals). Sustainable removal of these pollutants is a greater concern. Organic contaminants like hydrocarbons and textile dyes are a substantial threat to ecosystems with detrimental effects on the biosphere. Water contamination by heavy metals especially through industrial activities is also one of the major causes of pollution (Zamora-Ledezma et al. 2021).

The conventional pollutant removal methods are considered to be expensive and also reported to cause secondary contamination. There is a need to shift from the conventional ways to new approaches requiring biological resources, which are eco-friendly (Noubactep 2019; Ahmed et al. 2022). Bioremediation is such a phenomena, which involves the use of living microorganisms to degrade environmental pollutants into harmless end products (Khalid et al. 2021). A diverse range of microorganisms, including algae, fungi, yeasts, and bacteria, can function as biologically active methylators, capable of modifying toxic species. Technical aspects of bioremediation involve various mechanisms such as biosequestration, biodegradation, phytohydraulics, biological extraction, and volatilization by which microbes immobilize or transform the complex groups of the pollutants remediating the water (More et al. 2022). Microorganisms play a crucial role in heavy metal bioremediation. The bioremediation strategy is based on the high metal binding ability of biological agents, which helps extract heavy metals highly and efficiently from polluted areas (Husain et al. 2022). Although microorganisms cannot destroy metals, they can alter their chemical properties through a surprising array of mechanisms.

14 Bioremediation of Polluted Water 323

Several works have been carried out in this direction using microorganisms for the uptake of heavy metal from solutions. In addition, large-scale treatment of petroleum hydrocarbon contamination in oceans has been reported to be successfully done through microbial-based bioremediation (Jimoh et al. 2022; Dell'Anno et al. 2021).

Apart from using natural microorganisms for bioremediation, microorganisms can be created specifically for the bioremediation process using genetic engineering techniques (Tran et al. 2021; Janssen and Stucki 2020). The studies so far have been reported of inserting two types of genes into the microorganism in this aspect. One is the degradative gene, which will be encoding proteins required for degradation of the pollutant, and the second is the reporter genes that will help in monitoring pollution levels. Another aspect of integrating novel technology into bioremediation is the use of nanotechnology (Thangavelu and Veeraragavan 2022; Tripathi et al. 2022). It reported the possibility of producing environmentally friendly nanomaterials to alleviate these contaminants. The use of microorganisms in nanoparticle synthesis gives green biotechnology a positive impetus to cost reduction and sustainable production as a developing nanotechnology sector. Furthermore, there are reports of using other advanced bioremediation techniques like microencapsulation technology for in situ bioremediation of polluted groundwater applications, particularly with respect to bio-augmentation and bio-stimulation (Ethica et al. 2021).

There are certain limitations associated with bioremediation. Especially with respect to genetically engineered microorganisms, microbial population should have metabolic capacity to degrade contaminants (Liu et al. 2019). In addition to that generally, environmental conditions should be proper for their growth and activity such as temperature, pH, etc. Also, there should be the right amount of contaminants and nutrients. If the process is not controlled, then there are chances that those organic contaminants may not be degraded completely and may result in toxic by-products (Sharma 2020). Bioremediation has a great potential with notable achievements, but still, this eco-friendly low-input biotechnology has been underutilized. The global market scenario of bioremediation technology and services is showing an elevation with compound annual growth rate of 8.3% from 2017 to 2025 (Basak et al. 2020). The main purpose of this chapter is to provide an update concerning the current status of bioremediation of polluted fresh and marine water bodies.

14.2 Importance of Water for Life

The embryonic development up to the last minutes of the floral and faunal organisms needs water as an essential element. Most of the biological processors are mediated by water as a universal solvent. Discarded chemical waste from industries and agricultural setup, surface water runoff, poorly maintained septic systems, maritime logistics, and many land disturbing activates create qualitative and quantitative damages to water and water-based systems such as streams, rivers, lakes, and ocean. The hydrology has given immense support to understand complex nature of water-based systems of the earth. We must understand all of the physical, chemical, and biological processes involving water as it travels through the water cycle if we are to learn how to protect it (USGS 2006; Winter et al. 1998).

Water withdrawal also impacts how a watershed functions and interacts with the water cycle. Use might range from a few homeowners or businesses pumping small amounts of water to irrigate lawns. It could also include large municipalities, industries, mining operations, and agricultural producers pumping large amounts of water to support water demands in the region (USGS 2005). Either way, withdrawing water will affect the rate of evaporation, transpiration, and infiltration in a watershed.

14.3 Potential Means of Water Pollution

Water pollution is considered to be a greater concern all around the world, which requires significant attention from international level to the individual. Agricultural runoffs, chemical spills from industries, and sewage leaks are the main sources for the pollution of surface and groundwater bodies. The agricultural runoffs consist of pesticides, herbicides, and fertilizers (Yang et al. 2022). Irrigation in agriculture is considered to be a carrier of excessive salt from fertilizers to water bodies (Cui et al. 2020). When animal husbandry farms are considered chemical oxygen demand (COD) and biological oxygen demand (BOD), ammonia nitrogen (NH3-N), total phosphorus (TP), total nitrogen (TN), and metals are considered to be the main water-polluting concerns (Muratoglu 2020). Following proper management practices in agriculture plays a significant role in controlling this ways of water pollution.

Industries are a major source of water contamination. There are evidences of releasing heavy metals and other toxic chemicals by industries due to the activities like mining and refining metals and other manufacturing processes (Arif et al. 2020; Yuan et al. 2019). When chemical industries are considered, most of the chemicals that are being released are considered to be toxic, carcinogenic, and mutagenic (Issakhov et al. 2021; Aldalbahi et al. 2021). When industrial and municipal wastes are not properly treated and finally are being released into the environment, it becomes a huge concern for the survival of clean water bodies (Liu et al. 2016). In addition, sewage leakage from domestic waste is also another source of contamination because most of the houses are not connected to the main sewage system in cities (Ringo 2016; Naveen et al. 2018). Adding to that, urbanization also becomes intense day by day.

Other than the abovementioned sources of contamination, deforestation, marine dumping, radioactive waste, and atmospheric deposition are considered to be other means of water body pollution. When water bodies become contaminated, they are not further safe for human consumption because of the presence of certain toxic chemicals, heavy metals, and some pathogenic bacterial spp.

14.4 Agricultural Wastes and Its Contribution for Water Pollution

Most human activities have an impact on water and the ecosystem, including domestic, agricultural, and industrial activities. Water pollution has now become a global crisis, posing a serious threat to the entire world's sustainable development (Mali et al. 2016).

Agriculture is one of the most important economic sectors in both developed and developing countries. Surface and groundwater are polluted by significant agricultural activities (Naveen et al. 2021). As a result of that, they are unsafe for human consumption; soil erosion, including irrigation; industrial needs; nitrogen cycle changes; carbon sequestration; and relevant ecological patterns. Water pollution is typically caused by the excessive use of pesticides and chemical fertilizers, which cause changes in the physiochemical activities of the water (Naveen et al. 2021). This is the primary disadvantage that occurs in aquatic bodies nearby (Mali et al. 2016).

The current population's depletion of freshwater resources, as well as the degradation of water quality, is becoming a major issue. Biological, organic, toxic, and inorganic pollutants commonly contaminate surface water resources and a growing percentage of groundwater reserves as a result of improper disposal of industrial effluents, domestic wastes, and agricultural pollutants (Mali et al. 2016). According to previous research, when harmful substances are released in large quantities into water bodies, they cause damage to people, wildlife, or habitats and naturally result in phenomena such as volcanoes, algal blooms, storms, and earthquakes (Mali et al. 2016). These changes have a significant impact on both the water quality and the water's ecological status. As a result, water is life and plays an important role in all types of life (Naveen et al. 2021).

14.4.1 Water Pollution

Water pollution, in essence, highlights contamination in bodies of water (lakes, seas, groundwater, surface water) (Naveen et al. 2021). Water contamination has a large impact on plants and organic habitats, and it is influenced by the harming of specific species or populations in natural networks. Water contamination can occur when toxins are released directly into bodies of water without adequate treatment to remove harmful mixtures. Nitrates, phosphorus, pesticides, soil sediments, salt, pathogens, toxins, agrochemicals, and some organic substances are examples of water pollutants (Norman 2016). In other words, they can also be referred to as agricultural wastes that pollute the water (Naveen et al. 2021).

14.4.1.1 Wellspring of Water Contamination

Water contamination is visible as surface water or groundwater contamination. It is further subdivided into marine contamination and supplemental contamination. The contamination of wellspring water occurs as follows (Naveen et al. 2021).

14.4.1.2 Point Source

Release into surface water from a single point using a pipeline, outfall, or discard. Surface water discharges from feedlots, food processing plants, and agrochemical handling plants, and chemical spills pollute groundwater (Naveen et al. 2021).

14.4.1.3 Nonpoint Source

Pollutants enter receiving water resources through a variety of human activities that do not have a clear entry point. Large agricultural practices and land use typically pollute water (Naveen et al. 2021).

14.4.2 The Relation Between Agricultural Wastes and Water Pollution

Several agricultural activities, including increased use of fertilizers and pesticides, as well as allied livestock activities, have had a negative impact on water quality (Mali et al. 2016). Nitrates, phosphorus, and some pesticides are the most significant water pollutants (Naveen et al. 2021). Rising nitrate concentrations have a significant impact on drinking water quality, while pesticide use contributes to indirect emissions of toxic substances into the water. Increased nitrate and phosphorus levels in surface water bodies reduce their ability to support plant and animal life (Wubetu 2016).

As a result, agricultural pollution is currently difficult to control because it occurs over a large area and has diffuse and difficult-to-identify sources. It is affected by the following environmental factors: patterns of rainfall, land slope, and soil characteristics (Mali et al. 2016).

14.4.3 Public Health Effect

A variety of human illnesses are caused by contaminated water. According to WHO reports, 4 million people die each year as a result of diarrhea, which is common in waterborne saline (Naveen et al. 2021). Furthermore, minor rural contamination can have a virtuous circle effect on human health. According to WHO, as cultivation practices have expanded, nitrogen in groundwater has increased and has been impacted globally (World Health Organization 1993). Human tumors and childhood disorders such as methemoglobinemia can develop as a result of this nitrogen contamination. And newborn children are even more vulnerable to nitrate contamination (World Health Organization 1993). The following are some of the health consequences of developing countries: many vectors, such as mosquitos, thrive in environments that are hostile to their reproduction.

14.4.4 Technologies to Control Agricultural Water Pollution

14.4.4.1 Reducing Leaching and Erosion of Fertilizer

Several technologies have been developed to reduce fertilizer losses and increase nutrient use efficiency, such as optimizing time, methods, and required doses in fertilizer application (NUE) (Mali et al. 2016). Fertilizer application practices that are well designed to match crop needs and soil fertility significantly reduce fertilizer pollution and leach into groundwater (Jeong et al. 2016).

14.4.4.2 Optimal Pesticide Management Practices

Pesticides enter bodies of water via sediments caused by soil erosion. To control pests, a system should be developed to collect surface runoff and leach to reduce the unmethodical use of pesticides. These activities can help to reduce pesticide levels in bodies of water. Adoption of appropriate soil and water conservation measures, as well as the prohibition of certain activities, is timely important (Mali et al. 2016).

There is an urgent need to prevent water resource pollution and promote environmentally friendly reuse of a large amount of wastewater in the agricultural sector. It is possible to reduce agricultural waste contamination of water through appropriate practices such as improved nutrients, pesticides, crops, soil, and other wastewater management (Mali et al. 2016). The primary focus of this review is the impact of agricultural action on water pollution (Naveen et al. 2021). These numerous exercises can alter the water quality boundaries, affecting physicochemical properties. This demonstrates yet another benefit of farmers adopting integrated aquaculture-agriculture, forestry in fish crops, and livestock, which can collectively grow in a cultivating procedure (Naveen et al. 2021). This is because it can reduce the amount of pollution produced. Farmers should also avoid using synthetic fertilizers (India Economic Survey 2018).

14.5 The Toxic Effect of Pollutants on Marine Phytoplankton

14.5.1 Marine Phytoplankton

Marine phytoplankton, or the autotroph portion of the plankton, exists in very well topmost layers of the ocean and obtains its energy from photosynthesis in the clean oceans down to a level of 200 m. The majority of phytoplankton types are minuscule, unicellular creatures that range in size from 0.4 to 200 μm. Less than 1% of the planet's photosynthetic biomass is constituted of marine phytoplankton. Furthermore, this compartment is where more than 45% of the planet's yearly net primary production is formed. The biomass of this extremely productive compartment is maintained low in comparison to the biomass of terrestrial photosynthetic creatures by constant grazing and recycling (Simon et al. 2009). Initiating the emergence of photosynthesis in the Archaean epoch, marine photoautotrophs began to evolve (Katz et al. 2007). The diversified photosynthetic biota, which is the foundation of all complex life, originated with these primordial organisms. Along with significantly altering the biological chemistry of the oceans and atmosphere, they are also accountable for the oxygenation of the atmosphere. With the combination of information from the domains of biology, ocean bio-geochemistry, and atmosphere interactions, comprehension of the origin of phytoplankton has substantially evolved over the last several decades, and awareness of the diversity of this category has been strengthened (Simon et al. 2009).

14.5.2 Marine Pollutants

The maritime ecosystem may be exposed to chemical pollution from both land- and ocean-based human activities. This can come from both discrete sources, such as discharges, dumping, or unintentional spills, and diffuse sources, like runoff from the atmosphere or rivers, which can travel great distances. Pollutants contain nutrients, which may promote undesired algae production, as well as a variety of potentially dangerous compounds, some of which are tenacious, toxic to living creatures, or bio-accumulative. International and national priorities are used to identify the most hazardous materials, as well as quantities in water, biota, and sediment, which are evaluated to make sure they are within justifiable standards. Trace metals, flame retardants, hydrocarbons, pesticides, dioxins and PCBs, and compounds that could interfere with the endocrine system are some of the pollutants the Marine Institute has identified (Marine Pollutants 2022).

14.5.3 Pollutants' Toxic Effects on Marine Phytoplankton

Throughout the previous two centuries of social and economic development, various synthesized organic compounds have been discharged into the environment as a consequence of agricultural, commercial, and household processes. Persistent organic pollutants (POPs), which are resistant to decomposition and semi-volatile, can withstand long-distance atmospheric transmission before being dumped in remote places, like the global oceans. Certain POPs have an impact on marine phytoplankton by altering the quantity of chlorophyll, the acquisition of carbon and silicate, or by impeding photo-system I and the downward transport of electrons from photo-system II. The toxicity of organic impurities on marine phytoplankton has been studied in the literature through laboratory and field testing processes, as well as investigations of particular contaminants and basic compositions of pollutants (Echeveste et al. 2016).

Specifically in the situation of rising sea surface temperature and stratification, atmospheric deposition plays a major influence in supplying nourishment and toxins to the ocean ecosystem (Yang et al. 2019). Research on the impact of natural aerosols, such as volcanic ash and dust, on marine phytoplankton, has received the most attention (Chien et al. 2016). As anthropological activity increases, more chemical constituents are released and transferred to the oceans, altering the chemistry of the water and impacting phytoplankton production (Mahowald et al. 2018).

Aerosols comprising metals have significant influences on climate and ocean bio-geochemistry. Mainly due to their iron content, dust transferred to oceans with high nutrients, low chlorophyll could encourage phytoplankton development (Yoon et al. 2018). Cu, which is essential for living things in contrast to Fe, has favorable and harmful effects on marine phytoplankton depending on the concentration. Cu-rich aerosols may restrict the growth of phytoplankton, and the effects varied among the various phytoplankton taxa.

The sustainability of an entire ecosystem can be damaged by chemical effects on these primary producers. By being exposed to the pollutants of newer generations and chemicals, phytoplankton populations have less ecological efficiency (Romero et al. 2020). Antibiotics, personal care aromas, and plastic nanoparticles are a few of the new generation substances that have an effect on aquatic ecosystems all over the world. They are a pervasive danger to ecosystem conservation due to the outflow they cause from urban wastewater treatment plants (Broccoli et al. 2021). Ecosystems' functioning may be significantly impacted by changes in the diversity and function of phytoplankton and bacterioplankton caused by chemical contamination. Research revealed that sediment resuspension and chemical contamination had a major impact on phytoplankton-bacterioplankton interactions, which can change how anthropogenic coastal ecosystems function (Pringault et al. 2021).

14.6 Technologies for Refine Water Bodies

14.6.1 Chemical Precipitation

Heavy metals are considered to be a serious concern in the context of water pollution due to the development of many industries and due to their nondegradable quality. They need to be removed as much as possible from water bodies to prevent the harm on ecosystems. Chemical precipitation is such one method that is being used for heavy metal removal (Zhang and Duan 2020). In chemical precipitation, changes in the form of heavy metals into solid particles have been made. In chemical precipitation of heavy metals, the ionic forms have been reduced by the addition of counterions (Chen et al. 2018). By doing so, solubility becomes reduced, and removal becomes easy. In conventional heavy metal removal in wastewater treatment, reagents such as sodium sulfide (Na_2S), lime ($Ca(OH)_2$), and soda ash (Na_2CO_3) are being used. Other novel methods are also in the experimented stage in the purpose of value adding this technique (Fu et al. 2012).

14.6.2 Reverse Osmosis

Reverse osmosis has been considered as an efficient technology for separation of pollutants from industrial and municipal wastewater (Li et al. 2020; Lan et al. 2019; Porré et al. 2022, Mejía et al. 2022). It is being highly employed in wastewater treatment plants due to its fine pores in filtration system along with low energy consumption. Several conventional reverse osmosis setups are being in use like tubular, spiral wound, plate, and frame and hollow fiber techniques. Other novel methods, such as developing high-throughput membrane systems, are also in the experimental stage, with the purpose of adding value to this technique (Jiang et al. 2021). But pressure-driven nature of reverse osmosis is considered to be a limitation of this technique.

14.6.3 Electro Fenton's Dialysis

Hydrogen peroxide and ferrous salt are considered to be Fenton's reagents and are being used for a long time for the purpose of oxidizing organics. This concept has been advanced by integrating electrogeneration of H_2O_2 and Fenton's catalyst to produce OH radicals (Hu et al. 2021). This process facilitates degradation of any organic pollutant (Mousset and Hatton 2022). Other sophisticated techniques have also been integrated to increase the efficiency of electro-Fenton dialysis technique (Khalifa et al. 2021).

14.6.4 Adsorption Electrocoagulation

The adsorption and coagulation processes are used for the persistent organic pollutants, which are difficult to remove from wastewater (Titchou et al. 2021). In the process of coagulation flocculation, the solids that are being in suspended state are transformed into more observable and removable flocs by the addition of chemical flocculants. In electrocoagulation, the coagulation flocculation and adsorption techniques are being integrated together. Here, colloidal particles have been destabilized and have been converted into aggregates by electrical double-layer compression. This novel technique facilitates removing a wide range of pollutants from different types of wastewater sources (GilPavas et al. 2019; Nigri et al. 2020).

14.6.5 Chemical Oxidation

In chemical oxidation, hydrogen peroxide (H_2O_2) is used to oxidize pollutants in wastewater. Usually, H_2O_2 is used independently or in some instances is used with available other physical and biological treatment processes. Usually, in chemical oxidation of pollutants in wastewater, catalysts are added. Examples for such catalysts are ion (Fe^{2+} or Fe^{3+}), UV light, and ozone (O_3) (Han et al. 2022). In H_2O_2-based chemical oxidation, in order to accelerate the OH radical formation process, several techniques are used (Dombrowski et al. 2018). Addition of peroxydisulfate and peroxymonosulfate is such highly employed technique. Apart from this, several other techniques like electrochemical oxidation are integrated into chemical oxidation to increase its process efficiency (Umadevi et al. 2022).

In addition to the above described technologies, chemical oxidation, microfiltration (MF), ultrafiltration (UF), and nanofiltration (NF) are several techniques that are being used for refining wastewater in different wastewater plants.

14.7   The Role of Microorganisms in Bioremediation

Bioremediation is a biological process that helps recycle waste into a form of a valuable product or non-harm product. Microorganisms directly participate in converting waste into another form that can be recycled or reused. Due to the ability of microorganisms to survive in all places in the world, they are the most suitable ones for use as a biotic agent for bioremediation. Temperature and pH like abiotic factors also influence bioremediation. Degradation, eradication, immobilization, and detoxification are the main processes in bioremediation. With the improvement of biotechnological aspects in the world, different methods and strategies like biostimulation, bioaugmentation, bioventing, bio-piles, and bio-attenuation are used. Microorganisms play a major role in bioremediation.

Microorganisms can be used as bioremediation agents to remove and reduce the harm of pollutants in soil, water, and sediments. Microorganisms live in a wide range of biosphere using their impressive metabolic activities, the capability of using a wide range of nutrients, and the ability to establish or grow in different environmental conditions quickly (Abatenh et al. 2017). Using bioremediation, pollutants like heavy metals, metalloids, hydrocarbons, phenolic compounds, and municipal waste like inorganic pollutants can be remediated (Buckner et al. 2018; Al-Mailem et al. 2017). Bioremediation techniques can be classified based on the applied strategies. These are ex situ bioremediation and in situ bioremediation. In the ex situ bioremediation, polluted or contaminated material is removed from the site and transported into the offsite treating place and treated them with particular microorganisms for remediation. In the in situ bioremediation, treating process is done at the contamination site. This method is more preferred than ex situ bioremediation due to less risk of spill out of contaminant while transportation and no need to excavate and pump out contaminant (Buckner et al. 2018). Bioremediation properties having microorganisms can be found in their native environment like oil spilled lake and discharged industrial effluent. But those native microbial can't be used in a foreign environment. Due to that, to select an effective agent microorganisms should be genetically engineered. This is a good approach for biodegradation of pollutants everywhere like surface water, groundwater, marine environments, and soil and sediments (Brar et al. 2017). For the bioremediation of complex pollutants, mixed microbial populations are needed and effective rather than a single population due to the broad enzymatic capabilities of mixed populations. The novel approach is using a consortium of genetically engineered microorganisms rather than using single species. Bacteria, fungi, algae, and yeast are the microorganisms used in bioremediation. Microorganisms react on inorganic pollutants, heavy metal contaminations, antibiotic contamination, organic pollutants, aquaculture waste, and petroleum contamination by acting as bio-catalysts using their enzymatic pathways and enhancing the biochemical reactions of degrading the pollutants (Pal et al. 2020; Abatenh et al. 2017).

14.7.1 Factors Affecting Microbial Remediation

All the factors that affect microbial growth are directly controlling the efficiency of microbial bioremediation. Those are the chemical nature of the pollutant, available nutrients, environmental temperature, pH, salinity, available oxygen amount, and moisture. Microorganisms should have access to the ample amount of compounds that are required to generate energy and nutrients to grow. If the requirements are fulfilled, the growth of the microorganism population happens optimally. It will lead to efficient biochemical reactions of degrading pollutants (Abatenh et al. 2017).

14.7.2 The Role of Microorganisms in Different Bioremediation Methods

When using fungus as a bioremediator, it is called a mycoremediation. Fungi release enzymes in the polluted sites, degrade pollutants, and remove contaminants (Buckner et al. 2018). Also, fungi act as a biocatalyst for heavy metal due to their ability on absorbing heavy metals into the mycelium and spores. After that, heavy metals accumulate intracellularly and precipitate extracellular. Fungi have the ability to decompose lignin, de-colorize dyes, detoxify contaminated water, and degrade hydrocarbons and polychlorinated biphenyls like complex chemicals into nontoxic chemicals (Pal et al. 2020).

Pollutant	Bioremediator
Heavy metals	
Cr	• *Aspergillus* sp.
	• *Rhizopus* sp.
Heavy metals	
Zn	• *Rhizopus arrhizus*
	• *Penicillium spinulosum*
Pentachlorophenol	• *Rhizopus oryzae* CDBB-H-1877
Polychlorinated biphenyls (PCBs)	• *Penicillium chrysogenum*
	• *Scedosporium apiospermum*
Oil	• *Fusarium* sp.

León-Santiesteban et al. (2016), Pal et al. (2020), and Abatenh et al. (2017)

Bacteria can be used as bioremediator to degrade hydrocarbons, pesticides in the soil, and different textile dyes; utilize heavy metals; and purify crude oil-contaminated water. The abilities of bacteria like sorption, hydrolysis, photolysis, oxidation, and reduction reactions, are directly involving the degradation of pollutants by bacteria. Bacteria act as an absorbent by supplying the proteins and

14 Bioremediation of Polluted Water

lipids contained in the peptidoglycan layers as active binding sites for heavy metals (Pal et al. 2020). Some bacteria like *Nitrosomonas* and *Nitrococcus* involve nitrification process of sediments in ponds. It helps to clean the toxic compound like ammonia and nitrate, which reduce the quality of the water (Pal et al. 2020). The bioremediation process done by aerobic bacteria can be enhanced by supplying oxygen.

Pollutant	Bioremediator
Oil	• *Pseudomonas aeruginosa*
	• *Pseudomonas putida*
	• Arthobacter sp.
	• *Bacillus* sp.
Hydrocarbons (benzene, xylene)	• *Pseudomonas putida*
Textile dyes	• *Bacillus* spp. ETL-2012
	• *Pseudomonas aeruginosa*
Heavy metals like Cu, Ni, Cr	• *Pseudomonas aeruginosa*
Endosulfan contained in pesticides	• *Bacillus*
	• *Staphylococcus*

Abatenh et al. (2017) and Pal et al. (2020)

Algae and cyanobacteria are also effective and efficient bioremediators. The method that uses algae and cyanobacteria for bioremediation is called phytoremediation. Their ability on growing at a faster rate and can be cultivated on nonarable land caused the popularity of phytoremediation. Algae uptake nutrients from pollutants and use atmospheric CO_2. Due to that, pollutants are degrading. After the process, algal biomass can be used to produce biofuel, can be used as fertilizer, or can be used as feed (Brar et al. 2017).

Pollutant	Bioremediator
Dairy manure effluent	• Nostoc sp.
Textile effluent	• *Chlorella pyrenoidosa*
Agro-industrial waste water	• *Chlorella sorokiniana*
Effluent of leather processing	• *Chlorella vulgaris*
Benzene, bisphenol, toluene	• *Selenastrum capricornutum*
	• *Monoraphidium braunii*

Pal et al. (2020) and Brar et al. (2017)

Yeast species like *Saccharomyces cerevisiae, Candida tropicalis,* and *Candida utilis* have the ability to bio-absorb heavy metals like Cu, Ni, and Cd and to remove the pollutant from the contaminant sites (Pal et al. 2020).

Microorganisms play a major role in bioremediation. Using microorganisms in bioremediation is very cost-effective and efficient method rather than doing it with different mechanical and chemical methods. The efficiency of microorganisms can be improved using genetic engineering methods.

14.7.3 Microbial Contribution for Eutrophication and Their Metabolic Activities

The term eutrophication refers to the increase of the production of organic materials leading to a progressive rise in the biomass of primary producers in aquatic ecosystems such as lakes, ponds, rivers, oceans and streams, wetlands, swamp, estuaries, etc. Eutrophication is a natural phenomenon, which accompanies the evolution of an aquatic ecosystem over geologic time; however, it has been induced by anthropogenic activities in recent past (Le Moal et al. 2019). The main drivers of anthropogenic eutrophication are diffuse nitrogen and phosphorus pollution due to application of agricultural inputs (Beusen et al. 2016). This article reviews different ways of how microorganisms contribute and response to eutrophication.

14.7.3.1 The Contribution of Microorganisms for Eutrophication of Freshwater Ecosystem and Their Metabolic Activities

The increase of phytoplankton biomass including frequently dominant cyanobacteria is the main effects of lake eutrophication. Therefore, it creates two water layers such as upper layer, the warmer and lighted epilimnion, where primary production occurs, and a colder deeper layer, the hypolimnion. The epilimnion was rich in oxygen, and hypolimnion was deoxygenated. The hypoxia condition leads to increase the internal phosphorus load released by sediment. Further, increase of cyanobacteria blooms is directly linked to ongoing climate change, and diastrophic species of cyanobacteria has the ability to fix atmospheric nitrogen. Moreover, cyanobacteria are altering the food web in lake due to low nutrient levels for zooplanktons (Vinçon-Leite and Casenave 2019).

The characterization of bacterial and microbial eukaryotic communities associated with an ephemeral hypoxia event in Taihu Lake, a shallow eutrophic Chinese lake, was studied by Cai et al. (2018). The hypoxia is defined as a dissolved oxygen level below than 1.42 ml/L and hypoxia zone inhospitable to macrobiota; however, these zones allow to thrive unique bacterial communities, which they can cycling of bulk nutrients and trace elements. The results revealed that metabolic function of bacterial communities were phosphotransferase system, oxygen respiration, ABC transporters and fermentation, phototrophy, nitritation (HAO), and methane oxidation. The hypoxic site commonly enriched anaerobic functional categories was fermentation, fumarate respiration, denitrification (norBC and norZ), dissimilatory sulfur metabolism, dissimilatory nitrite reduction to ammonium, hydrogen oxidoreduction, methanogenesis, nitrogen fixation, aromatic compound degradation, nitrite respiration (nirKS), cellulolysis, and chitinolysis. The vadinBC27 subgroup of Bacteroidetes bacteria has capability to digest algal detritus. The nitrogen cycling is a one of the major geochemical transformations at hypoxic zones. However, authors have reported that nitrifies (*Nitrosomonas* and *Nitrosospira*) activity is lower in hypoxic sites due presence of low oxygen. In contrast to nitrifies, the activity of denitrifies were dominated in hypoxic sites. For instance, *Pseudomonas* species have the ability to do both heterotrophic nitrification and denitrification in simultaneously. Another pathway of removal of nitrogen is ANNOMAX process during algae blooms. Further, sulfur-reducing bacteria (Desulfovibrionales or other common sulfate-reducing bacteria (SRB)) activity is more common in hypoxic sites, and they are responsible to produce H2S gas. Although many eukaryotes such as algae, protozoa, copepods, and fungi prefer to live in oxygen-rich environment, many of them have an antihypoxia capacity. The accumulation of algal scums provides suitable habitat for aquatic fungi and feeds by attachment to organic aggregates and phytoplankton-derived biopolymers. The decomposition of algal biomasses (largely comprised by *Microcystis* spp.) has released the high concentration of dissolved microcystins. Mucor is a more common fungal in hypoxic sites. They have ability to hydrolyze xylose, straw materials, and polysaccharides and directly consume acetic acid under anaerobic condition. Lei et al. (2021) stated that algal bloom containing cyanobacteria releases high amount of algal organic matter (AOM), which will enhance the growth of archaea (methanogens). These methanogens significantly contributed to increase the methylmercury (MeHg) production in eutrophic shallow lakes. Yang et al. (2020) studied the influence of eutrophication on methanogenesis pathways. Authors reveled that eutrophic lake could present a high potential of methanogenesis and more common methanogens in lake were Methanoregula, Methanolinea, Methanobacterium, and Methanosaeta. Further, long-term eutrophication causes to release high ammonia level in lake.

The phytoplankton consisting of diatoms, cyanobacteria, dinoflagellates, and green algae are responsible for eutrophication, and cyanobacteria lead a major role. Eutrophication in freshwater lake leads to hypoxia condition that is a contribution by bacterial and eukaryotic communities. Algal bloom contains cyanobacteria release high amount of AOM that will enhance the growth of methanogens, which contribute to MeHg production in eutrophic shallow lakes.

14.8 Marine Bacteria: Potential Candidates for Enhanced Bio-Remediation

Bacteria are abundant in nature owing to their capacity to adapt to extreme circumstances and perform a variety of physiological activities. Among all the ubiquitous bacteria, marine bacteria have been identified as playing a major role in a variety of biogeochemical processes. Because environmental conditions are constantly changing, the microorganisms in that environment have complicated characteristics of adaptation because they are better suited to the challenging conditions. Therefore, by forming biofilms and producing extracellular polymeric molecules, the bacteria isolated from marine habitats are expected to be more effectively used in the bioremediation of heavy metals, hydrocarbons, and many other refractory chemicals and xenobiotics. The structure, variety, and functional potential of marine microbial communities found in saltwater and sediment have recently been revealed at a global level thanks to novel developments in molecular ecology techniques. This review places particular emphasis on the use of marine bacteria in the field of bioremediation and identifies the process by which adaptive responses obtain their unique features.

Microorganisms are crucial to the maintenance and sustainability of any ecosystem because they can adapt more quickly to environmental changes and degradation. Since they are found everywhere, including in volcanic eruptions, Antarctic glaciers, and conditions on Mars, marine ecosystems are not unlike several other types of habitats. Oceans are considered a source of untold riches, which have been covered by 71% of the globe's surface by water and contain salt water (97%). Also, this water contributes 32% of global net primary production. Additionally, the seas serve as the primary global trade route and the primary climate stabilizer. The oceans are rich in a high diversity of different species, and 178,000 species of marine organisms belong to 34 phyla (UNEP report on the global biodiversity assessment). The oceans have been contaminated globally by substances like petroleum hydrocarbons (PHCs), polycyclic aromatic hydrocarbons (PAHs), antibiotics, heavy metals, and excessive nitrogen and phosphorus; also as a range of pollutants like radioactive wastes, plastics are contaminating the marine environment (Gosai et al. 2017; Dudhagara et al. 2016) as a result of urbanization and the landscape's subsequent industrialization over the past few decades. These contaminants may build up in seawater and sediments, especially in coastal locations, and eventually endanger marine ecosystems and human health. Simply put, the oceans have been indirectly impacted by overpopulation, urbanization, and increased industrialization because waste from all sources eventually makes its way to the sea or ocean. In addition, pollution is caused by offshore drilling and related activities (Gosai et al. 2017, 2018). Therefore, there is a huge requirement to find methods to remove pollutants that have become a serious issue around the globe. Previously, a variety of physicochemical techniques have been used to remove contaminants. These techniques, however, cannot be a long-term solution due to one or both of their shortcomings. To solve this issue, microorganism engagement has been given top priority, and microbial bioremediation has recently received attention (Vala and Dave 2017; Vala et al. 2018). Due to their wide range of catalytic activities and capacity to survive in challenging environments (such as hypersaline, low temperature, acidic or alkaline pH, and high pressure), marine microorganisms are the best candidates for the bioremediation of ocean toxins (Homaei 2017). In order to prove that marine bacteria can be used in enhanced bioremediation, this review summarizes the most recent findings on the unique properties of marine bacteria, their physiologic and genetic adaptation to the dynamic environmental conditions, biogeography and diversity, and the role of marine bacteria in various remediation aspects with the metagenomic approaches.

14.8.1 What Is Bioremediation?

The bioremediation process can be defined as a biological mechanism for recycling waste into another form that can be used and reused by other organisms. Currently, most parts of the world face this problem with various kinds of environmental pollution accumulation on water and land surfaces. It is primarily based on nutrient versatility, but it also relies on the ability of certain microorganisms to convert, modify, and utilize toxic pollutants in order to produce energy and biomass.

Although it can be considered as a well-organized procedural activity, which is applied to break down or transform contaminates into less toxic or nontoxic elemental or compound forms, through the use of microorganisms, bioremediation involves the removal, modification, immobilization, or detoxification of different chemicals and physical pollutants from the environment. Mainly, there are two types of contaminant that can be identified within the polluted surfaces as heavy metal inorganic compounds and organic aromatic carbon compounds. By acting as biocatalysts and accelerating the biochemical processes that breakdown the intended contaminant, microorganisms play a role in the process. Microorganisms can only effectively combat contaminants provided they have access to a variety of materials and chemicals that will enable them to produce energy and nutrients for the growth of new cells. The chemical makeup and concentration of pollutants, the environment's physicochemical properties, and the availability of the pollutants to microorganisms are only a few of the variables that affect how effective the process of bioremediation is (Abatenh et al. 2017).

14.8.2 Potential of Marine Bacteria to Adapt for Environmental Changes

They are perfect for prospective bioremediation and bio-indicator uses because they react fast to changing environmental patterns such as the sea surface temperature, the pH of the immediate surroundings, the shifting UV and light patterns, the sea level rise, tropical storms, and terrestrial inputs, which are only a few of the periodic variations that affect the marine ecosystem. Alphaproteobacteria make up the majority of the microorganisms found in the marine environment. This could be because they are better able to adapt to the changing conditions there. Numerous investigations have been made to determine the molecular basis of adaptability in this group of bacteria with the novel techniques (Gutierrez et al. 2018).

The pattern of growth rates, gene expression, physiological or enzymatic activity, and changes in close or symbiotic relationships with other species are all ways that bacteria adapt to a variety of environmental conditions. When exposed to extreme conditions like pressure, temperature, salinity, and the depletion of micronutrients, some groups of marine bacteria have also been reported to develop many novel mechanisms, such as the synthesis of bioactive compounds, while some other types of bacteria displayed mechanisms to overcome the situations of adaptation toward elevated temperature in seawater including chemotaxis and adhesion to α-galactoside receptor in the coral mucus penetration into epidermal cell differentiation into a viable-but-not-culturable state, intracellular multiplication, developmental defects, and symbiosis with other organisms, which is mostly found in pathogenic microorganisms (Beygmoradi and Homaei 2017).

14.8.3 Marine Bacteria Being a Potential Candidate in Bioremediation Process

By utilizing the ambient genome pool of microbes with incredibly precise catalytic capabilities, metagenomic approaches increase the likelihood of discovering new genes and enzymatic pathways that are supportive of bioremediation processes. On the polymer surfaces (plastic substances), dense microbial biofilms were found using fluorescence in situ hybridization and high-resolution microscopy imaging. With the help of amplicon sequencing of the 16S rRNA gene, it was revealed that the orders Flavobacteriales, Rhodobacterales, Cytophagales, Rickettsiales, Alteromonadales, Chitinophagales, and Oceanospirillales made up the majority of the bacterial communities on all types of plastic in the Mediterranean Sea (Annika et al. 2021). Proteogenomic and metabolomic methods have been used to narrow down the pathways and enzymes that the marine bacteria Mycobacterium sp. DBP42 and Halomonas sp. ATBC28 use to break down plasticizers (Wright et al. 2020). The use of bacterial species, genes, and enzymes derived from microbial communities has been found to be quite effective in the remediation of sites contaminated with diesel and petroleum pollution (Garrido-Sanz et al. 2019).

Some research has discussed the significance of biofilm-forming bacteria in the bioremediation of heavy metals, particularly mercury, in a case study. Notably, growing mercuric pollution in the ocean not only contaminates food chains but also accelerates ocean acidification (Wang et al. 2018). According to the authors, extracellular polymerase-producing bacteria like *Bacillus cereus* BW-03 are promising candidates for inorganic mercury remediation because they can produce extracellular enzymes, biosurfactants, polysaccharides, and amyloids.

Marine bacteria *Pseudomonas fluorescens* BA3SM1 was shown to be highly resistant to cadmium, copper, and zinc with the help of genomics approaches. This bacterium is considered a promising agent for the remediation of polluted waters and sediments.

In a study, it was shown that the marine actinobacterium *Brevibacterium linens* BS258 that produces urease can precipitate and dissolve calcite under various Ca^{2+} concentrations. Calcite precipitation and dissolution were connected to carbonic anhydrase (CA) activities, quorum sensing (QS), and other crucial energy metabolic activity, according to genomic sequencing, transcriptome profiling, and other assays of *B. linens* BS258. Further analysis focused on the heavy metal resistance and removal capacities of *B. linens* BS258. The final outcomes of this study provided fresh insight into the mechanisms at play the potential of bacterial carbonate bio mineralization in marine environment's metal bioremediation.

The results of an analysis of a research study of native microbial communities in seawater samples of the Northwest coast of the Iberian Peninsula showed that the oil-enriched microbial communities were dominated by hydrocarbon-degrading bacteria with no significant differences in geographical locations, such as *Alcanivorax*, *Pseudomonas*, *Acinetobacter*, *Rhodococcus*, *Flavobacterium*, *Oleibacter*, *Marinobacter*, and *Thalassospira*, which could represent prototype consortia for mitigating oil spills. Through production of H2S, sulfate-reducing microorganisms may cause oil souring in oil reservoirs, while nitrate-reducing microorganisms can inhibit sulfate reduction that leads to biosouring mitigation. It has been suggested that mitigation strategies for biosouring could be improved by monitoring volatile fatty acid concentration and microbial diversity in oil reservoirs.

The ability of marine microorganisms to adapt to rapidly changing environmental conditions is well understood, but little is known about how they can resist a toxic environment with culturable techniques. Therefore, applying genomic analysis (culture-independent) research in this area will aid in our understanding of the genetic basis of prodigies of nature. They may have a high potential to be more effective bacterial entities for enhanced bioremediation if they make some useful genetic system modifications (Beygmoradi and Homaei 2017).

14.9 System Biology Approaches in Microbial Reconstruction for Bioremediation

Environmental pollution is a serious problem today, and bioremediation can play an important role in cleaning up polluted sites. Remediation strategies such as chemical and physical approaches are not sufficient to mitigate pollution problems due to the continuous generation of new recycling pollutants due to human activities. Bioremediation using microorganisms is an environmentally friendly and socially acceptable alternative to conventional remediation approaches. Although many microorganisms with bioremediation potential have been isolated and characterized, in most cases, the target pollutants are either completely degraded or ineffective in mixed waste situations. This review envisions advances in systems biology (SB) that enable the analysis of microbial behavior at the community level under various environmental stresses.

Agricultural practices and industrial development are significant features of human civilization. Over the last few decades, excessive use of pesticides and chemical fertilizers in agricultural practices has led to environmental (water, land, and air) pollution. The industrialization has the unfortunate side effect of intentionally or accidentally releasing toxic organic and inorganic chemicals and heavy metals that adversely affect the environment. The leading pollutants from these sources are chemical solvents, paints, industrial by-products, petroleum hydrocarbons, polycyclic aromatic hydrocarbons (PAHs), nitroaromatic compounds, industrial solvents, polychlorinated biphenyls (PCBs), trichlorethylene (TCE), phthalates, benzene, ethylbenzene, toluene and xylene (BTEX), heavy metals, and pesticides. To maintain a healthy environment and to treat existing highly polluted sites, it is necessary to remove these pollutants from waste to prevent their dispersion in the environment. Contaminant remediation was in some cases achieved using physicochemical methods such as solidification, filtration, combustion, evaporation, oxidation and reduction, reverse osmosis, chemical precipitation, electrochemical treatment, and ion exchange approaches. On the other hand, bioremediation, is a more reliable and eco-friendly approach that utilizes the natural ability of microbes and plants to remove or neutralize pollutants in the environment.

Plants and many microorganisms, including bacteria, fungi, and algae, have been reported to have high neutralizing capacity (Avlopoulos et al. 2016)

The successful use of systems biology (SB) and metabolic engineering (ME) approaches in various fields of biology makes it attractive for ecologists to use these approaches for bioremediation. These approaches to pollutant neutralization. The systems biology approach consequently provides valuable detailed information about biological processes. Then, ME can be used to exploit this information to modify microbial metabolic pathways to neutralize specific single or multiple pollutants simultaneously. This review provides insight into SB and facilitates the design and optimization of microbial cell factories for the bioremediation of pollutants at significant levels (The Financial Express 2018).

14.9.1 Systems Biology in Bioremediation

Bioremediation optimization through microbial SB approaches is novel and offers great potential. In general, SB is used to study complex biological systems to explore complex networks and their interconnections in various biological processes at the molecular, cellular, population, community, and ecosystem levels. It provides significant insight into gene expression, enzymes, biosynthetic pathways, and secondary metabolism of microbes and can reveal the alteration of pathways already present under conditions of stress caused by various pollutants. "Omics" techniques (genomics, proteomics, transcriptomics, and metabolomics) are widely used in systems biology for studying the pollutant neutralizing capacity of microorganisms. A comprehensive list of bioremediation examples with toxic chemicals using omics (Avlopoulos et al. 2016).

14.9.2 Genomics

Genomics involves gene sequencing and bioinformatics analysis using a variety of tools and algorithms. The genomes of several microorganisms involved in bioremediation have been sequenced. Genome sequence (6.2 MB) analysis of *Pseudomonas* sp. KT2440 showed the presence of genes encoding many enzymes or proteins such as dehydrogenases, oxidoreductases, oxygenases, ferredoxins, cytochromes, glutathione-S-transferases, sulfur-metabolizing proteins, and efflux pumps. These biochemicals play a significant role in the degradation of several chemicals released from industrial wastes (Zhu et al. 2017).

14.9.3 Transcriptomics

Transcriptomics is mainly used to investigate the differential expression of upregulated and downregulated genes in response to environmental pollutants. It also helps to infer the function of previously unannotated genes. Techniques such as microarrays and RNA sequencing can be applied to quantify a predetermined set of sequences. Analysis of microbial community transcriptomes requires (1) isolation and enrichment of total cellular mRNA, (2) synthesis of cDNA from total mRNA, and (3) sequencing of total cDNA or the use of microarrays for cDNA hybridization (Zhu et al. 2017).

14.9.4 Proteomics

Analysis of the set of proteins produced by bacterial cultures (proteomics) and environmental samples (metaproteomics) can be used to detect changes in the composition and production of proteins and to identify many proteins important to the physiological response of microorganisms in the presence of contaminants.

14.9.5 Metabolomics and Metabolic Flux Analysis

Metabolomics is another rapidly developing field of SB at the interface of biological sciences and chemistry. In this approach, whole set of cellular metabolites produced by microorganisms is analyzed. Microbial metabolism is tightly regulated at various genomic, transcriptional, and posttranslational levels. Metabolic flux analysis, sometimes called "fluxomics," measures the rate of metabolic reactions on a real-time basis to study cellular metabolites produced under specific environmental conditions (Klassen et al. 2017).

This article highlights the various techniques used in SB that help the bioremediation of environmental pollutants through microorganisms. Various omics-based tools in SB, such as genomics, transcriptomics, proteomics, and metabolomics, provide significant information for understanding the complex behavior of microorganisms that play an important role in bioremediation. But in natural conditions, the process is very slow and not very efficient, so SB provides a platform for the reproduction of microorganisms, and metabolic pathways are reconstructed to improve this process. Thus, the integration of SB has opened the way to realize the true potential of bioremediation (Bonifay et al. 2017).

References

Abatenh E et al (2017) The role of microorganisms in bioremediation- a review. Open J Environ Biol 2(1):38–46. https://doi.org/10.17352/ojeb.000007

Ahmed SF, Kumar PS, Rozbu MR, Chowdhury AT, Nuzhat S, Rafa N, Mahlia TMI, Ong HC, Mofijur M (2022) Heavy metal toxicity, sources, and remediation techniques for contaminated water and soil. Environ Technol Innov 25:102114

Aldalbahi A, El-Naggar ME, El-Newehy MH, Rahaman M, Hatshan MR, Khattab TA (2021) Effects of technical textiles and synthetic nanofibers on environmental pollution. Polymers 13(1):155

Al-Mailem DM et al (2017) Biostimulation of indigenous microorganisms for bioremediation of oily hypersaline microcosms from the Arabian Gulf Kuwaiti coasts. J Environ Manag 193:576–583. https://doi.org/10.1016/j.jenvman.2017.02.054

Annika KK, Alejandro AA, Maaike G, Andreas E, Witte Harry J, Ina V, Florian M, Christian L, Miriam W, Engelmann JC, Helge N (2021) Microbial communities on plastic polymers in the Mediterranean Sea. Front Microbiol 12:673553

Arif A, Malik MF, Liaqat S, Aslam A, Mumtaz K, Afzal A, Nisa K, Khurshid F, Arif F, Khalid MSZ (2020) Water pollution and industries. Pure Appl Biol 9(4):2214–2224

Avlopoulos GA, Malliarakis D, Papanikolaou N et al (2016) Visualizing genome and systems biology: technologies, tools, implementation techniques and trends, past, present and future. Giga Sci 4:38

Basak G, Hazra C, Sen R (2020) Biofunctionalized nanomaterials for in situ clean-up of hydrocarbon contamination: a quantum jump in global bioremediation research. J Environ Manag 256:109913

Beusen AH, Bouwman AF, Van Beek LP, Mogollón JM, Middelburg JJ (2016) Global riverine N and P transport to ocean increased during the 20th century despite increased retention along the aquatic continuum. Biogeosciences 13(8):2441

Beygmoradi A, Homaei A (2017) Marine microbes as a valuable resource for brand new industrial biocatalysts. Biocatal Agric Biotechnol 11:131–152. https://doi.org/10.1016/j.bcab.2017.06.013

Bonifay V, Aydin E, Aktas DF et al (2017) Metabolic profiling and metabolomic procedures for investigating the biodegradation of hydrocarbons. In: McGenity TJ, Timmis KN, Nogales B (eds) Hydrocarbon and lipid microbiology protocols: genetic, genomic and system analyses of communities. Springer, Berlin, pp 111–161

Brar A et al (2017) Photoautotrophic microorganisms and bioremediation of industrial effluents: current status and future prospects. 3 Biotech 7(1):1–8. https://doi.org/10.1007/s13205-017-0600-5

Broccoli A, Anselmi S, Cavallo A, Ferrari V, Prevedelli D, Pastorino P, Renzi M (2021) Ecotoxicological effects of new generation pollutants (nanoparticles, amoxicillin and white musk) on freshwater and marine phytoplankton species. Chemosphere 279:130623

Buckner CA et al (2018) Bioremediation techniques for soil pollution. Curr Oncol 25(4):e275–e281. https://doi.org/10.3747/co.25.3884

Cai J, Bai C, Tang X, Dai J, Gong Y, Hu Y, Shao K, Zhou L, Gao G (2018) Characterization of bacterial and microbial eukaryotic communities associated with an ephemeral hypoxia event in Taihu Lake, a shallow eutrophic Chinese lake. Environ Sci Pollut Res 25(31):31543–31557

Chen Q, Yao Y, Li X, Lu J, Zhou J, Huang Z (2018) Comparison of heavy metal removals from aqueous solutions by chemical precipitation and characteristics of precipitates. J Water Process Eng 26:289–300

Chen SS, Kimirei IA, Yu C, Shen Q, Gao Q (2022) Assessment of urban river water pollution with urbanization in East Africa. Environ Sci Pollut Res 29:40812–40825

Chervier C, Amblard L, Déprés C (2022) The conditions of emergence of cooperation to prevent the risk of diffuse pollution from agriculture: a case study comparison from France. J Environ Plan Manag 65(1):62–83

Chien C, Mackey K, Dutkiewicz S, Mahowald N, Prospero J, Paytan A (2016) Effects of African dust deposition on phytoplankton in the western tropical Atlantic Ocean off Barbados. Glob Biogeochem Cycles 30(5):716–734

Cui N, Cai M, Zhang X, Abdelhafez AA, Zhou L, Sun H, Chen G, Zou G, Zhou S (2020) Runoff loss of nitrogen and phosphorus from a rice paddy field in the east of China: effects of long-term chemical N fertilizer and organic manure applications. Glob Ecol Conserv 22:e01011

Datta S, Sinha D, Chaudhary V, Kar S, Singh A (2022) Water pollution of wetlands: a global threat to inland, wetland, and aquatic phytodiversity. In: IGI Global (ed) Handbook of research on monitoring and evaluating the ecological health of wetlands. Hershey, pp 27–50

Dell'Anno F, Rastelli E, Sansone C, Brunet C, Ianora A, Dellanno A (2021) Bacteria, fungi and microalgae for the bioremediation of marine sediments contaminated by petroleum hydrocarbons in the omics era. Microorganisms 9(8):1695

Dombrowski PM, Kakarla P, Caldicott W, Chin Y, Sadeghi V, Bogdan D, Barajas-Rodriguez F, Chiang SY (2018) Technology review and evaluation of different chemical oxidation conditions on treatability of PFAS. Remediat J 28(2):135–150

Dudhagara UR, Rajpara RK, Bhatt JK, Gosai HB, Sachaniya BK, Dave BP (2016) Distribution, sources and ecological risk assessment of PAHs in historically contaminated surface sediments at Bhavnagar coast, Gujarat, India. Environ Pollut 213:338–346. https://doi.org/10.1016/j.envpol.2016.02.030

Echeveste P, Galbán-Malagón C, Dachs J, Berrojalbiz N, Agustí S (2016) Toxicity of natural mixtures of organic pollutants in temperate and polar marine phytoplankton. Sci Total Environ 571:34–41

Ethica SN, Firmansyah A, Muchlissin SI, Sulistyaningtyas AR, Ernanto AR, Saptaningtyas R, Widyawardhana RBI, Darmawati S (2021) Current application of microencapsulation technology in bioremediation of polluted groundwater. World J Agric Soil Sci 4:2020

Faroque S, South N (2022) Water pollution and environmental injustices in Bangladesh. Int J Crime Justice Soc Democr 11(1):1–13

Fu F, Xie L, Tang B, Wang Q, Jiang S (2012) Application of a novel strategy—advanced Fenton-chemical precipitation to the treatment of strong stability chelated heavy metal containing wastewater. Chem Eng J 189:283–287

Garrido-Sanz D, Redondo-Nieto M, Guirado M, Pindado Jiménez O, Millán R, Martin M et al (2019) Metagenomic insights into the bacterial functions of a diesel-degrading consortium for the rhizoremediation of diesel-polluted soil. Gene 10:456

GilPavas E, Dobrosz-Gómez I, Gómez-García MÁ (2019) Optimization and toxicity assessment of a combined electrocoagulation, H2O2/Fe2+/UV and activated carbon adsorption for textile wastewater treatment. Sci Total Environ 651:551–560

Gosai HB, Sachaniya BK, Dudhagara DR, Rajpara RK, Dave BP (2017) Concentrations, input prediction and probabilistic biological risk assessment of polycyclic aromatic hydrocarbons (PAHs) along Gujarat coastline. Environ Geochem Health 2017:1–13

Gosai, HB, Sachaniya, BK, Panseriya, HZ, Dave BP (2018) Functional and phylogenetic diversity assessment of microbial communities at Gulf of Kachchh, India: an ecological footprint. (In communication)

Gutierrez T, Teske A, Ziervogel K, Passow U, Quigg A (2018) Microbial exopolymers: sources, chemico-physiological properties, and ecosystem effects in the marine environment. Front Microbiol 9:1822

Han M, Wang H, Jin W, Chu W, Xu Z (2022) The performance and mechanism of iron-mediated chemical oxidation: advances in hydrogen peroxide, persulfate and percarbonate oxidation. J Environ Sci 128:181–202

Hu J, Wang S, Yu J, Nie W, Sun J, Wang S (2021) Duet Fe3C and FeN x Sites for H2O2 generation and activation toward enhanced electro-Fenton performance in wastewater treatment. Environ Sci Technol 55(2):1260–1269

Husain R, Vikram N, Yadav G, Kumar D, Pandey S, Patel M, Khan NA, Hussain T (2022) Microbial bioremediation of heavy metals by Marine bacteria. In: Development in wastewater treatment research and processes. Elsevier, Amsterdam, pp 177–203

India Economic Survey (2018) Farmers gain as agriculture mechanization speeds up, but more R&D needed. The Financial Express. www.financialexpress.com/budget/india-economic-survey-2018-for-farmers

Issakhov A, Alimbek A, Zhandaulet Y (2021) The assessment of water pollution by chemical reaction products from the activities of industrial facilities: numerical study. J Clean Prod 282:125239

Janssen DB, Stucki G (2020) Perspectives of genetically engineered microbes for groundwater bioremediation. Environ Sci Process Impacts 22(3):87–499

Jeong H et al (2016) Irrigation water quality standards for indirect wastewater reuse in agriculture: a contribution toward sustainable wastewater reuse in South Korea. Water 8(4):169. https://doi.org/10.3390/w8040169

Jiang G, Li H, Xu M, Ruan H (2021) Sustainable reverse osmosis, electrodialysis and bipolar membrane electrodialysis application for cold-rolling wastewater treatment in the steel industry. J Water Process Eng 40:101968

Jimoh AA, Ikhimiukor OO, Adeleke R (2022) Prospects in the bioremediation of petroleum hydrocarbon contaminants from hypersaline environments: a review. Environ Sci Pollut Res 2022:1–28

Katz ME, Fennel K, Falkowski PG (2007) Geochemical and biological consequences of phytoplankton evolution. In: Evolution of primary producers in the sea. Academic Press, New York, pp 405–430

Khalid FE, Lim ZS, Sabri S, Gomez-Fuentes C, Zulkharnain A, Ahmad SA (2021) Bioremediation of diesel contaminated marine water by bacteria: a review and bibliometric analysis. J Mar Sci Eng 9(2):155

Khalifa O, Banat F, Srinivasakannan C, AlMarzooqi F, Hasan SW (2021) Ozonation-assisted electro-membrane hybrid reactor for oily wastewater treatment: a methodological approach and synergy effects. J Clean Prod 289:125764

Klassen A, Faccio AT, Canuto GAB et al (2017) Metabolomics: definitions and significance in systems biology. In: Sussulini A (ed) BT-metabolomics: from fundamentals to clinical applications. Springer, Cham, pp 3–17

Lan L, Kong X, Sun H, Li C, Liu D (2019) High removal efficiency of antibiotic resistance genes in swine wastewater via nanofiltration and reverse osmosis processes. J Environ Manag 231:439–445

Le Moal M, Gascuel-Odoux C, Ménesguen A, Souchon Y, Étrillard C, Levain A, Pinay G (2019) Eutrophication: a new wine in an old bottle? Sci Total Environ 651:1–11

Lei P, Zhang J, Zhu J, Tan Q, Kwong RW, Pan K, Jiang T, Naderi M, Zhong H (2021) Algal organic matter drives methanogen-mediated methylmercury production in water from eutrophic shallow lakes. Environ Sci Technol 55(15):10811–10820

León-Santiesteban HH et al (2016) Pentachlorophenol removal by Rhizopus oryzae CDBB-H-1877 using sorption and degradation mechanisms. J Chem Technol Biotechnol 91(1):65–71. https://doi.org/10.1002/jctb.4566

Li Y, Li M, Xiao K, Huang X (2020) Reverse osmosis membrane autopsy in coal chemical wastewater treatment: evidences of spatially heterogeneous fouling and organic-inorganic synergistic effect. J Clean Prod 246:118964

Liu Y, Guo D, Dong L, Xu Y, Liu J (2016) Pollution status and environmental sound management (ESM) trends on typical general industrial solid waste. Procedia Environ Sci 31:615–620

Liu L, Bilal M, Duan X, Iqbal HM (2019) Mitigation of environmental pollution by genetically engineered bacteria—current challenges and future perspectives. Sci Total Environ 667:444–454

Mahowald N, Hamilton D, Mackey K, Moore J, Baker A, Scanza R, Zhang Y (2018) Aerosol trace metal leaching and impacts on marine microorganisms. Nat Commun 9(1):2614

Mali S, Sanyal S, Bhatt B, Pathak H (2016) Water pollution and agriculture. https://www.researchgate.net/publication/305617702_Water_Pollution_and_Agriculture

Marine Pollutants (2022). Available at https://www.marine.ie/site-area/areas-activity/marine-environment/marine-pollutants. Accessed 16 September 2022

Mejía HFG, Toledo-Alarcón J, Rodriguez B, Cifuentes JR, Porré FO, Haeger MPL, Ovalle NV, Astudillo CL, García A (2022) Direct recycling of discarded reverse osmosis membranes for domestic wastewater treatment with a focus on water reuse. Chem Eng Res Des 184:473–487

More VS, Jacob Samuel Sehar AE, Sheshadri AP, Rajanna S, Kurupalya Shivram A, Fasim A, Rao A, Acharya P, Mulla S, More SS (2022) Bioremediation of pesticides containing soil and water. In: Biotechnology for zero waste: emerging waste management techniques. Wiley, Hoboken, pp 83–94

Mousset E, Hatton AT (2022) Advanced hybrid electro-separation/electro-conversion systems for wastewater treatment, reuse and recovery: compromise between symmetric and asymmetric constraints. Curr Opin Electrochem 35:101105

Muratoglu A (2020) Grey water footprint of agricultural production: An assessment based on nitrogen surplus and high-resolution leaching runoff fractions in Turkey. Sci Total Environ 742:140553

Naveen BP, Sumalatha J, Malik RK (2018) A study on contamination of ground and surface water bodies by leachate leakage from a landfill in Bangalore, India. Int J Geo-Eng 9(1):1–20

Naveen K, Kumar D, Kumar A, Jayabaian I (2021) Agricultural activities a causing water pollution and its mitigation. Int J Mod Agric 10:590–609

Nigri EM, Santos AL, Rocha SD (2020) Removal of organic compounds, calcium and strontium from petroleum industry effluent by simultaneous electrocoagulation and adsorption. J Water Process Eng 37:101442

Norman (2016) List of emerging substances. Network of Reference Laboratories, Research Centres and related Organisations for Monitoring of Emerging Environmental Substances. Available www.norman-network.net/qnode/19

Noubactep C (2019) Metallic iron for environmental remediation: prospects and limitations. In: A handbook of environmental toxicology: human disorders and ecotoxicology. CABI, Wallingford, pp 531–544

Pal AK et al (2020) The role of microorganism in bioremediation for sustainable environment management. In: Bioremediation of pollutants. Elsevier, Amsterdam. https://doi.org/10.1016/b978-0-12-819025-8.00010-7

Porré M, Pisano G, Nahra F, Cazin C (2022) Synthetic access to aromatic alpha-haloketones. Molecules 27(11). https://doi.org/10.3390/molecules27113583

Pringault O, Bouvy M, Carre C, Mejri K, Bancon-Montigny C, Gonzalez C, Leboulanger C, Hlaili A, Goni-Urriza M (2021) Chemical contamination alters the interactions between bacteria and phytoplankton. Chemosphere 278:130457

Ringo J (2016) Status of sewage disposal in Dodoma municipality, Tanzania. Int J Mar Atmos Earth Sci 4(1):24–34

Romero N, Visentini F, Márquez V, Santiago L, Castro G, Gagneten A (2020) Physiological and morphological responses of green microalgae Chlorella vulgaris to silver nanoparticles. Environ Res 189:109857

Sharma I (2020) Bioremediation techniques for polluted environment: concept, advantages, limitations, and prospects. In: Trace metals in the environment-new approaches and recent advances. IntechOpen, London

Simon N, Cras A, Foulon E, Lemée R (2009) Diversity and evolution of marine phytoplankton. C R Biol 332(2-3):159–170

Thangavelu L, Veeraragavan GR (2022) A survey on nanotechnology-based bioremediation of wastewater. Bioinorg Chem Appl 2022:5063177

The Financial Express (2018). https://www.financialexpress.com/budget/india-economic-survey-2018-for-farmers-agriculture-gdp-msp/1034266/. Accessed 8 January 2019

Titchou FE, Zazou H, Afanga H, El Gaayda J, Akbour RA, Hamdani M (2021) Removal of persistent organic pollutants (POPs) from water and wastewater by adsorption and electrocoagulation process. Groundw Sustain Dev 13:100575

Tran KM, Lee HM, Thai TD, Shen J, Eyun SI, Na D (2021) Synthetically engineered microbial scavengers for enhanced bioremediation. J Hazard Mater 419:126516

Tripathi S, Sanjeevi R, Anuradha J, Chauhan DS, Rathoure AK (2022) Nano-bioremediation: nanotechnology and bioremediation. In: Research anthology on emerging techniques in environmental remediation. IGI Global, Hershey, pp 135–149

Umadevi M, Rathinam R, Brindha T, Dheenadhayalan S, Pattabhi S (2022) Application of electrochemical oxidation for the treatment of reactive red 195 using graphite electrode. Asian J Biol Life Sci 10(3):620–625

USGS. August 2005. Watersheds. http://ga.water.usgs.gov/edu/watershed.html

United States Geological Survey (USGS). September (2006) The water cycle: summary from water science basics. http://ga.water.usgs.gov/edu/watercyclesummary.html

Vala, AK & Dave, BP (2017) Marine-Derived Fungi: Prospective Candidates for Bioremediation. Mycoremediation& Environmental Sustainability Springer, Cham 17-37. https://doi.org/10.1007/978-3-319-68957-9_2

Vala AK, Sachaniya B, Dudhagara D, Panseriya HZ, Gosai H, Rawal R, Dave BP (2018) Characterization of L-asparaginase from marine-derived Aspergillus niger AKV-MKBU, its antiproliferative activity and bench scale production using industrial waste. Int J Biol Macromol 108:41–46. https://doi.org/10.1016/j.ijbiomac.2017.11.114. Epub 2017 Nov 23. PMID: 29175524

Vinçon-Leite B, Casenave C (2019) Modelling eutrophication in lake ecosystems: a review. Sci Total Environ 651:2985–3001

Wang Q, Zhang WJ, He LY, Sheng XF (2018) Increased biomass and quality and reduced heavy metal accumulation of edible tissues of vegetables in the presence of Cd-tolerant and immobilizing Bacillus megaterium H3. Ecotoxicol Environ Saf 148:269–274. https://doi.org/10.1016/j.ecoenv.2017.10.036

Winter TC, Harvey JW, Franke OL, Alley WM (1998) Ground water and surface water: a single resource. U.S. Geological Survey (USGS) Circular 1139. USGS, Denver, CO. http://pubs.usgs.gov/circ/circ1139

World Health Organization (1993) Guidelines for drinking-water quality, volume 1: recommendations, 2nd edn. WHO, Geneva. https://apps.who.int/iris/handle/10665/259956

Wright RJ, Bosch R, Gibson MI, Christie-Oleza JA (2020) Plasticizer degradation by marine bacterial isolates: a proteogenomic and metabolomic characterization. Environ Sci Technol 54:2244–2256

Wubetu A (2016) Review paper on irrigation water pollution and its minimization measures. J Biol Agric Healthc 6(21):72–81

Xiao L, Liu J, Ge J (2021) Dynamic game in agriculture and industry cross-sectoral water pollution governance in developing countries. Agric Water Manag 243:106417

Yang T, Chen Y, Zhou S, Li H (2019) Impacts of aerosol copper on marine phytoplankton: a review. Atmos 10(7):414

Yang Y, Chen J, Tong T, Xie S, Liu Y (2020) Influences of eutrophication on methanogenesis pathways and methanogenic microbial community structures in freshwater lakes. Environ Pollut 260:114106

Yang YY, Tfaily MM, Wilmoth JL, Toor GS (2022) Molecular characterization of dissolved organic nitrogen and phosphorus in agricultural runoff and surface waters. Water Res 219:118533

Yoon J, Yoo K, Macdonald A, Yoon H, Park K, Yang E, Kim H, Lee J, Lee M, Jung J, Park J, Lee J, Kim S, Kim S, Kim K, Kim I (2018) Reviews and syntheses: ocean iron fertilization experiments – past, present, and future looking to a future Korean Iron Fertilization Experiment in the Southern Ocean (KIFES) project. Biogeosciences 15(19):5847–5889

Yuan F, Wei YD, Gao J, Chen W (2019) Water crisis, environmental regulations and location dynamics of pollution-intensive industries in China: a study of the Taihu Lake watershed. J Clean Prod 216:311–322

Zamora-Ledezma C, Negrete-Bolagay D, Figueroa F, Zamora-Ledezma E, Ni M, Alexis F, Guerrero VH (2021) Heavy metal water pollution: a fresh look about hazards, novel and conventional remediation methods. Environ Technol Innov 22:101504

Zhang Y, Duan X (2020) Chemical precipitation of heavy metals from wastewater by using the synthetical magnesium hydroxy carbonate. Water Sci Technol 81(6):1130–1136

Zhu Y, Ma N, Jin W et al (2017) Genomic and transcriptomic insights into calcium carbonate biomineralization by marine Actinobacterium Brevibacterium linens BS258. Front Microbiol 8:602

Pollution in Freshwater: Impact and Prevention

15

Nandan Singh and Maitreyie Narayan

Abstract

Water is the life blood of all living things and the most important natural resource on the earth planet. The water is available in liquid, solid and the gaseous forms at the surface of the earth. In the earth, only 2.5% water is present as freshwater. Of this 2.5% freshwater, 68.7% freshwater is locked in ice caps and glaciers, 30.1% is present as groundwater, and only 1.2% is present as other surface freshwater; of this 1.2% water, approximately only 0.5% freshwater is present in the rivers. It is difficult to increase the water supply to the earth, but it is possible to reduce the impact and extend of water resource pollution by managing the supply better. Water pollution is a phenomenon of contamination that means drop of water quality of land waters such as rivers, lakes, seas, oceans and groundwater. It is a major environmental problem across the world. Pollutant transport is the most important linkage between freshwater due to its severe impact on the health and integrity of ecosystems. People are not much aware of the remedies, causes and control of water pollution. The water pollution management and control measures could be through awareness, practice efforts, prevention and minimizing waste, with the help of programmes and projects, monitoring, regulations and legal framework and by avoiding the use of pesticides, fertilizers and non-toxic cleaning material.

Keywords

Pollution · Water · Glaciers · Wetland

N. Singh (✉) · M. Narayan
Department of Forestry and Environmental Science, Kumaun University, Nainital, Uttarakhand, India

© The Author(s), under exclusive license to Springer Nature Singapore Pte Ltd. 2023
R. Soni et al. (eds.), *Current Status of Fresh Water Microbiology*,
https://doi.org/10.1007/978-981-99-5018-8_15

347

15.1 Introduction

The future of the earth depends on environmental components such as water, and the sustainability of these components is very important for all living organisms in the earth. All the organisms in the ecosystem are interrelated and connected to each other for sustainability. Therefore, any changes or deterioration in a part of the ecosystem can affect the whole ecosystem over time. Water is far distinct from other environmental components as it is the main source of life and cannot be substituted (Kılıc 2021). A large portion of the earth is covered with water, but the amount of freshwater available is very low. Besides all the natural causes, the damages caused by anthropogenic activities have created pressure on limited freshwater resource availability and global water problems. All domains of water quality are impacted by the natural interactions between rock and soil heterogeneities and water, including anthropogenic activities that alter natural water cycles (Trabelsi and Zouari 2019; Akhtar et al. 2021). The operation of every living thing and human health may be severely impacted by these changes in water quality. The physicochemical and biological features and the quality, quantity and availability of water resources constantly change as a result of the influence of natural and anthropogenic activities (Khatri and Tyagi 2015; Akhtar et al. 2021).

15.2 Distribution of Water

Between two-thirds and three-fourths of the earth's surface is covered with water that is present in the liquid, solid and the gaseous forms. Of the 100% water available on the earth, 96.5% is present in the oceans, 1% as other saline water, and only 2.5% is present as freshwater. Of this 2.5% freshwater, 68.7% freshwater is locked in polar ice caps and glaciers, 30.1% is available as groundwater, and only 1.2% is present as other surface freshwater; of this 1.2%, approximately only 0.5% freshwater is present in the rivers (Fig. 15.1).

15.3 Types of Water

15.3.1 Atmospheric Water

The atmospheric water mainly comes from evaporation in the atmosphere from the surface of both the ocean and the land such as transpiration and direct evaporation from the surfaces of green plants (evapotranspiration, the sum of all processes by which water moves from the land surface to the atmosphere via both evaporation and transpiration) and evaporation from ice (sublimation), which leads to an increase in the water content of the atmosphere (Table 1.3).

15 Pollution in Freshwater: Impact and Prevention

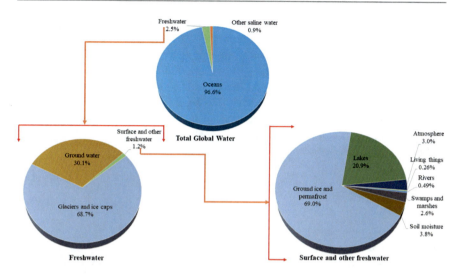

Fig. 15.1 Proportionate distribution of the earth's water (total global water, freshwater, surface water and other freshwater)

15.3.2 Oceans, Inland Seas, Costal Zones and Estuaries

An ocean is a continuous body of salt water that is contained in an enormous basin on the earth's surface. The important oceans and their marginal seas cover almost 71% of the earth's surface, with a mean intensity of 12,100 feet. Inland seas are landlocked seas that are only connected to the ocean through narrow channels. They generally contain many channels, islands, sounds and straits (Barrie et al. 2012). Due to the waves and ensuing currents, the water in the coastal area is quite erratic, the marine border of this zone is defined and is constrained by depths equivalent to ten times the wave height (Khublaryan 2009). An estuary is a partially enclosed coastal water area freely connected with the ocean, within which the seawater is considerably diluted by freshwater flowing from a river catchment area; these are transient zones between the seas and land (Table 1.3).

15.3.3 Rivers, Reservoirs, Lakes and Wetlands

A river, lake, or wetland is a body of water that flows through a self-developed bed augmented by groundwater and surface. The character and development of the water body are influenced by various factors such as the climate, its geological structure and the dimensions of its basin. The water body can be divided into mountain and plain rivers. Plain rivers flow more slowly in wide-relying valleys. A water reservoir is an artificial body of water that can be formed in a river valley to regulate its use for the benefit of the natural economy. It can be divided into three types: mixed, lacustrine and fluvial. Water in permanent or temporary reservoirs can be used for days, weeks, or even a year (Table 1.3).

A lake is a naturally occurring water storage area with a lake basin that is not immediately connected to the ocean. Basins are classified into tectonic, glacial, fluvial, coastal, sinkhole (in karst and thermokarst), volcanic and dammed (made reservoirs and ponds) categories based on their origin (Khublaryan 2009). A wetland is a region of land that experiences a constant or an excessive moistening, which encourages the growth of hydrophilic plants and particular soil processes (Table 1.3).

15.3.4 Groundwater

Water present in the earth's crust of any physical state, including massively crystalline rock cracks and sedimentary rock layers, is referred to as groundwater. There are numerous groundwater classifications based on the geological age, lithological composition, hydrodynamics, temperature, chemical composition and groundwater infiltration and dispersion (Table 1.3).

15.3.5 Soil Water

The term "soil water" typically refers to water that is both closely associated with soil skeletal particles and freely permeable through the soil profile. It may also refer to the surface area of the land as water that is localized in soil pore space, in the liquid moisture form, in the ice form as a solid component and in the soil air form, as a gaseous water form or both (Table 1.3).

15.3.6 Glaciers, Icebergs and Ground Ice

A glacier is a moving, naturally occurring build-up of ice on the land, when the solid phase of water is outweighed by the liquid phase. Only places on the earth's surface where solid precipitation exceeds evaporation are home to glaciers. When a glacier or a glacial barrier breaks, massive pieces of continental ice float in the polar and adjacent oceans as icebergs. Ground ice refers to all types of ice that form in frozen and frozen–cold ground. Ground ice, which forms in soil or rock pores, holes, cavities, or other openings, includes massive ice as one of its constituents. Buried glacier, lake, river and snow bank ice are all categorized as a single form of ground ice (Khublaryan 2009) (Table 15.1).

Table 15.1 Distribution of water over the world (Source: Chapter on "World Fresh Water Resources" by Igor Shiklomanov in Peter H. Gleick (editor), 1993's Water in Crisis: A Guide to the World's Fresh Water Resources (Oxford University Press, New York)

Water source	Water volume (miles^3)	Water volume (km^3)	Percent of freshwater	Percent of total water
Oceans, seas and bays	321,000,000	1,338,000,000	–	96.54
Ice caps, glaciers and permanent snow	5,773,000	24,064,000	68.7	1.74
Groundwater	5,614,000	23,400,000	–	1.69
Freshwater	2,526,000	10,530,000	30.1	0.76
Saline water	3,088,000	12,870,000	–	0.93
Soil moisture	3959	16,500	0.05	0.001
Ground ice and permafrost	71,970	300,000	0.86	0.022
Lakes	42,320	176,400	–	0.013
Freshwater	21,830	91,000	0.26	0.007
Saline water	20,490	85,400	–	0.006
Atmosphere	3095	12,900	0.04	0.001
Swamp water	2752	11,470	0.03	0.0008
Rivers	509	2120	0.006	0.0002
Biological water	269	1120	0.003	0.0001

15.4 Water Pollution

Water flows all over the world, regardless of borders, crossing state lines and country, which means that pollution from one part of the world can affect the other part (Kılıç 2021). This makes it more difficult to set any standard in own ways of using and conserving the earth's water. The word "pollution" can be defined as contamination, dirtying, desecration, spoiling, soiling and destruction (Karatas and Karatas 2016). Many of the sorts of pollutions caused by water pollutants are primarily widespread and affect the health of all residing organisms, mainly mankind. In the recent years, one of the biggest problems that humans are facing is related to water quantity and water quality (UNESCO 2009). Water needs to be kept safe and shielded from pollution of all kinds. Water is among the indispensable ingredients in the centre of the life. Without water, life is not possible, but because water resources are polluted by a variety of hazardous industrial pollutants, it is unhealthy for humans to drink; irrigation activities make water scarce for both people and the environment. Water sources are often unknowingly utilized and poisoned by people, endangering their offspring (Kılıç 2021). Contaminated water is unhealthy for humans to consume since it includes harmful or toxic compounds and bacteria and organisms that cause disease. The quality of water is mainly threatened by agricultural activities, cities, industries waste, mining areas and other so many causes. This pollution is then transferred to surface and groundwater (Dwivedi 2017).

15.4.1 Freshwater

Freshwater environments are a few of the most beneficial biological systems in the world (Ghermandi et al. 2008) and account for many significant services to human culture. In addition, they are extremely environmentally sensitive and flexible ecosystems (Turner et al. 2000). According to their origin, geographical location, water management strategies, water chemistry, dominant species and soil and sediment physiognomies, freshwater ecosystems exhibit a wide range of diversity (Kumaraswamy et al. 2020). In the light of their hydrological, natural and land characteristics, the freshwater ecosystems were divided into three groups: lacustrine (lakes), riverine (along waterways and streams) and palustrine (marshes and lowlands).

The world is concerned about the unfavourable economic, social and environmental consequences of declining water quality in freshwater ecosystems. In the case of small water bodies, such as tanks, lakes and lakes in the past, the problem of declining water quality is notably more alarming, affecting several social (water supply), economic (livestock, fisheries and forestry) and ecological functions (nutrient recycling and groundwater recharge).

15.4.2 Freshwater Pollution

In the recent years, there has been an obvious increase in the demand for freshwater due to the rapid growth of population and fast industrialization (Ramakrishnaiah et al. 2009). Most agricultural development activities endanger living organisms, ecosystems and human health, particularly in terms of excessive fertilizer application and unsanitary conditions (Okeke and Igboanua 2003). Human-based activities have deteriorated water quality in most part of the earth. Freshwater pollution is among the most crucial issue all over the world. Therefore, protecting freshwater sources and freshwater quality is the most urgent need, because of serious negative water pollution issues and global scarcity of freshwater resources. The freshwater pollution grade depends on the pollutant's ecological impact, its abundance and water use. Rainfall can run across the land, collecting contaminants such as pesticides and fertilizers from agricultural lands, forestry and lawns; animal waste, oil and urban road salt (Toth 2009) and toxic elements from abandoned mines are the various causes that lead to freshwater pollution.

15.4.3 Sources of Freshwater Pollution

Among the natural resources, water is the most vulnerable to pollution. Contaminants enter surface water from a specific, identifiable source or from a dense, poorly defined region. Contaminants that originate at a distinct location, such as a pipe, ditch, tank, or sewer, are examples of point source pollution. Point sources are easy to identify and hence relatively easy to stop. Direct pollution from a

15 Pollution in Freshwater: Impact and Prevention

Table 15.2 Properties of the point source water pollution and non-point source water pollution (Source: Kumaraswamy et al. 2020)

Point source water pollution	Non-point source water pollution
Industrial effluent and municipal pollution	Runoff from forest areas, agriculture fields and lawns
Leachate and runoff from dump sites for garbage	Runoff from the range and meadow
Runoff from stables and livestock	Runoff from locations both with and without sewers areas
Runoff from industrial locations, oil refineries and mining sites	Putrefying tank leachate
Storm water sewer overflow	Runoff from abandoned mines and building establishment
Sewage overflows caused by drainage and human waste	Deposition of atmospheric materials on a water surface
Runoff from the construction areas	Impurity-producing activities on the land including logging, wetland modification, building and the development of land or waterways

point source may enter the water. Large lagoons may have vessels that leak lubricant or deposit waste (Bhat et al. 2018). Invasive plants and animals can be transported through routine vessel operations, such as the discharge of ballast water (water needed to steady a ship), which will inevitably lead to the extinction of native species. Contaminants that come from agricultural field, livestock pens and atmosphere are examples of non-point source pollution (Kumaraswamy et al. 2020). Non-point source pollution is more difficult to control than point source pollution since it is caused by a variety of contaminants (Table 15.2).

Population growth, industrialization, urbanization, plastics materials and polythene bags, domestic sewage, pesticides and fertilizers, eutrophication, mining, agrochemical wastes, nutrient enrichment, thermal pollution, oil spills, sediment disruption, acid rain pollution, radioactive waste and climate change are the primary causes of freshwater pollution. Other causes include acid rain pollution, thermal pollution, nutrient enrichment and radioactive waste. It is recorded that 75–80% water pollution is caused by the domestic sewage. Wastes from the electroplating industries, including pesticides, sugar, textiles, paper and pulp, are causes of water pollution (Kamble 2014). Polluted water and water sources smell terrible and have fewer plant and animal species. Nearly about 80% of the world's population is facing threats to water security. Most domestic sewages are discharged into rivers, and the majority of these are untreated. Solid trash, toxins, plastic garbage and bacterial pollutants are all found in domestic sewage, and these hazardous substances pollute the water. One of the main causes to the contamination of freshwater is the discharge of various industrial wastes into rivers without any treatment (Desai 2014; Kılıç 2021).

Increasing population has a detrimental impact on freshwater pollution since it increases the production of solid waste generation (Jabeen et al. 2011). Rivers get both liquid and solid waste discharges, and human excrement contaminates the water. In contaminated water, a large number of bacteria are also found, which are detrimental for human health (Haseena et al. 2017). As more people move into towns and cities, these factors cause freshwater pollution, and there is more water consumption in these areas. Deforestation and urban growth often lead to water pollution. Wetlands are nature's means of filtering and damming water; their destruction eliminates the natural habitats and filters that may store and degrade numerous pollutants; therefore, destroying wetlands is also the cause of water pollution (Kılıc 2021).

15.5 Impact of Freshwater Pollution

Pollution destroys water's natural structure and causes a change in their properties that will harm human and living being's health (Kılıc 2021), and water pollution also has many negative effects on the entire ecosystem. An increase in water pollution has a negative impact on plant development; forests cannot grow, and the minerals required for photosynthesis cannot be obtained from groundwater sources. Freshwater pollution causes many bacterial, viral and parasitic diseases in human body. The growth of algae prevents fish and other freshwater organisms from taking oxygen and negatively affects the ecosystem. Thick, green muck generated by harmful algal blooms has an adverse effect on clear water, leisure activities, companies and property values (US Environmental Protection Agency 2022). It is a natural and an essential process for clay, silt and sand from the ground to wash off into streams and lakes. Floodplains and wetlands are built up by sediments, which also carry nutrients. However, logging, ploughing and other disturbances to the land caused by the construction of roads or buildings can cause excess sediment runoff.

Environmental risk factors that cause diseases with a high mortality in developing countries include unsafe water, lack of wastewater system and inadequate hygiene practices. Intestinal diseases are among the evidentiary results of these factors. Diseases such as diarrhoea, hepatitis and cholera stand out as the main diseases caused by freshwater pollution (World Health Organization 2022). The free circulation of toxic waters in nature poisons the soil and especially pollutes the groundwater (Kılıc 2021). Damage to the soil threatens the lives of living things in the region. With the discharge of sewage water into the seas, the lives of the creatures in the sea are adversely affected. The root causes of water pollution are improper housing, overfishing, and improper use of technology. Exposure to various chemicals causes illness and death as a result (Wear et al. 2021). The natural balance of living things that take these chemical substances into their bodies is disturbed. There is a big association between water pollution and health problems. Many waterborne illnesses and infections are transmitted from one person to another (Halder and Islam 2015).

Lakes and streams with a low pH support a lower quantity and variety of life; if water is polluted, its pH level changes. Most marine plants grow best in water with a pH of 7.0–9.2. As the pH of the water decreases, plant population also declines, which decreases nourishment for some aquatic birds (Gentili et al. 2018). If the pH continues to lower, the numbers of freshwater shrimp, crayfish, clams and some fish start to decline continuously (Food and Agriculture Organization of the United Nations 2020). The bacteria that decompose leaf litter and other waste begin to die at a pH of 5.5, cutting off the nutrient source for plankton. Waterways that already have a low pH are at the risk of serious damage from these temporary increases in acidity. Temporary acidification can completely devastate an ecosystem and result in significant fish extinction (Goolsby and Battaglin 2001). To cool a thermal plant, water is drained from the watercourse, channelled through the area that essentially to be cooled and then reverted to the river. This leads to a rise in the temperature of approximately 10 °C of the water, which is closer to the thermal plant in comparison to the other part of the water bodies. Heated water has a variety of effects on the surrounding ecosystem. Warmer temperatures upsurge the capability of plants to photosynthesize, which may outgrowth an algal bloom (Kumaraswamy et al. 2020). Warmer temperatures also put more strain on aquatic plants and animals. Cool water holds more dissolved oxygen as compared to warm water, making it more difficult for aquatic animals to breathe (Qadri et al. 2020). In water that is warmer than 35 °C, certain organisms perish; lower oxygen levels and higher temperatures could make an animal more susceptible to consequences from infections and harmful substances (Kumaraswamy et al. 2020); as a result, the biodiversity, or variety of species, of the environment may decline. Instead of producing electricity, the plant wastes two-thirds of its energy as heat. As the cooling water passes through the plant, it loses plants, eggs, larvae and young fish, including around 35% of the young striped bass in that section of the river. Some larger fish are killed or injured when they are trapped on the screens that prevent them from being sucked into the cooling system (Qadri et al. 2020).

Nutrients accumulate into lakes and streams through runoff from the land, atmospheric fallout and the recycling of plant and animal tissues within the aquatic environment. Freshwater ecosystems are directly being impacted by recent increases in fertilizer intake. Methemoglobinemia or "blue baby" syndrome, generally known as the illness in infants, is caused by high nitrate concentrations in drinking water. Due to this illness, the baby's digestive system acclimates the nitrate to nitrite, a process that interferes with the blood's ability to carry oxygen (Qadri et al. 2020). Smaller amounts of nutrients in a lake do not immediately kill fishes but speed up the eutrophication process.

Pesticides, flame retardants, industrial solvents and cleaning fluids contain man-made organic compounds that are commonly found in aquatic environments (Folke et al. 2002). Although these persistent organic pollutants (POPs) have many applications, some are lethal even in trace amounts. They are extremely toxic to fish and invertebrates, even in small concentrations, and are strong endocrine disruptors. They prevent both avian and mammal reproduction and development, which lowers the number and survival rates of the offspring (Qadri et al. 2020). In mammals, polychlorinated biphenyls interfere with the metabolism of thyroid hormones, which regulate a diversity of physiological processes, including brain development and metabolism. They also reduce immune system function; for example, polar bears are losing their ability to fight against common infections and are also beginning to show some endocrine effects (Okeke and Igboanua 2003) such as masculinization in some females. The exposure of polychlorinated biphenyls in humans has been linked to neurodevelopmental problems in children (Goolsby and Battaglin 2001), liver dysfunction, skin and respiratory problems, dizziness and possibly cancer. A recent study discovered that the levels of three chemicals (polychlorinated biphenyls, hexachlorobenzene and chlordane) were higher in the mothers of men with a testicular cancer than in a control group, implying that the cancer began in uterus (Carpente et al. 1998).

15.6 Prevention from Pollution in Freshwater

To prevent the negative effects of freshwater pollution, it is primarily the most important step to educate each and every individual around the world. In addition, the sensitivity and importance of the subject can be emphasized with such studies. Water is one of the most important needs of all the living organisms, and for some, it is life itself (Kılıc 2021). It keeps them alive, but polluted water is extremely dangerous. When humans drink polluted water, it has serious effects on their health. The air, the water and the soil are the main elements of our environment (Obafemi et al. 2012).

Environmental education is a key factor in an attempt to address environmental pollution and its harmful effects. The main objective of environmental and water pollution education is to equip learners with knowledge, values and skills that promote the protection and conservation of the environment (Karatas and Karatas 2016). The best way to prevent water pollution is to educate yourself and your surrounding environment. Environmental education is a process that aims to create a population that is aware of the environment as a whole and the problems that it is associated with and that has the knowledge, attitudes, motivation and skills to work both individually and collectively towards solving existing issues and preventing the emergence of new ones (Skanavis 2004). To effectively safeguard the water resources, one of the main goals of water education is to instil the knowledge of water contamination. Water pollution education is a very important and an effective tool to promote public awareness. Water pollution prevention training should be done as follows: producers should be informed by professionals about proper utilization; people should be informed about recycling process and waste materials and should be informed about the harms of excessive consumption; media should be used for promoting water pollution awareness; people should be informed about the importance of all natural resources for life; activities and presentations about the protection of water and water resources should be organized regularly. The ideal procedure to reduce diffuse pollution of waterways is to minimize or avoid the use of chemicals for agricultural, industrial and domestic purposes (Scheierling 1995).

The whole world community is currently experiencing the severe effects of water contamination, which is a global problem. It is advised to have a suitable system for disposing of trash and to treat waste before it enters rivers. To reduce pollution, seminars, workshops and educational programmes should be organized across the global level. Some measures have to be taken before such a major problem occurs.

15 Pollution in Freshwater: Impact and Prevention

Water pollution causes many negative effects such as diseases, death of aquatic animals, economic costs of cleaning processes, the destruction of ecosystems and the disruption of food chains.

Here, education on the dangers of freshwater pollution is extremely important, as it helps people to apply the right attitudes when dealing with the environment (Karatas and Karatas 2016). It is important to support and fund educational initiatives that inform and organize people to help safeguard water resources. Environmental education teaches individuals to weigh different sides of an environmental issue to make informed and responsible decisions (US Environmental Protection Agency 2022). Awareness-raising activities are of great importance for people before they contaminate water sources. Water contamination is the result of our impulsive behaviour and carelessness. We all know that water is so important and we should not pollute water. Without water, neither we nor any living thing can survive. Therefore, people should protect, keep saving and help prevent water pollution. Nature provides us many tangible and intangible resources in our daily lives, so let us be responsible and disciplined enough to save, protect, preserve and maintain not only water sources but also other natural resources. There are a number of things we can do in our daily lives to lessen water pollution, even though there is not a single quick action that can be made to halt it. The best way to clean polluted water is not to clean it, but to cease polluting it.

References

Akhtar N, Syakir Ishak MI, Bhawani SA, Umar K (2021) Various natural and anthropogenic factors responsible for water quality degradation: a review. Watermark 13(19):2660. https://doi.org/10.3390/w13192660

Barrie JV, Greene HG, Conway KW, Picard K (2012) Inland tidal sea of the Northeastern Pacific. In: Harris PT, Baker EK (eds) Seafloor geomorphology as benthic habitat. Elsevier, Amsterdam, pp 623–634. https://doi.org/10.1016/B978-0-12-385140-6.00044-X

Bhat RA, Dervash MA, Mehmood MA, Hakeem KR (2018) Municipal solid waste generation and its management, a growing threat to fragile ecosystem in Kashmir Himalaya. Am J Environ Sci 2(11):145–167

Carpente SR, Caraco NF, Correll DL, Howarth RW, Sharpley AN, Smith VH (1998) Non point pollution of surface waters with phosphorus and nitrogen. Ecol Appl 8:559–568

Desai VN (2014) A Study on the water pollution based on the Environmental problem. PARIPEX Indian J Res 3(12):95–96

Dwivedi AK (2017) Researches in water pollution: a review. Int Res J Nat Appl Sci 4:118–142

Folke C, Carpenter S, Elmqvist T, Gunderson L, Holling CS, Walker B (2002) Resilience and sustainable development: Building adaptive capacity in a world of transformations. Ambio 31(5):437–440

Food and Agriculture Organization of the United Nations (2020) The state of world fisheries and aquaculture 2020. Sustainability in action. FAO, Rome. https://doi.org/10.4060/ca9229en

Gentili R, Ambrosini R, Montagnani C, Caronni S, Citterio S (2018) Effect of soil pH on the growth, reproductive investment and pollen allergenicity of Ambrosia artemisiifolia L. Front Plant Sci 9:1335

Ghermandi A, van den Bergh JC, Brander LM, de Groot HL, Nunes PA (2008) The economic value of wetland conservation and creation: a meta-analysis, pp 1–20

Goolsby DA, Battaglin WA (2001) Long-term changes in concentrations and flux of nitrogen in the Mississippi River Basin, USA. Hydrol Process 15(7):1209–1226

Halder JN, Islam MN (2015) Water pollution and its impact on the human health. J Environ Hum 2(1):36–46

Haseena M, Malik MF, Javed A, Arshad S, Asif N, Zulfiqar S, Hanif J (2017) Water pollution and human health. Environ Risk Assess Remed 1(3):16–19

Jabeen S, Mahmood Q, Tariq S, Nawab B, Elahi N (2011) Health impact caused by poor water and sanitation in district Abbottabad. J Ayub Med Coll Abbottabad 23(1):47–50

Kamble SM (2014) Water pollution and public health issues in Kolhapur city in Maharashtra. Int J Sci Res Publ 4(1):1–6

Karatas A, Karatas E (2016) Environmental education as a solution tool for the prevention of water pollution. Survey Fish Sci 3(1):61–70

Khatri N, Tyagi S (2015) Influences of natural and anthropogenic factors on surface and ground-water quality in rural and urban areas. Front Life Sci 8(1):23–39

Khublaryan MG (2009) Types and properties of water. EOLSS 1:468

Kılıc Z (2021) Water pollution: causes, negative effects and prevention methods. Istanbul Sabahattin Zaim Univ J Inst Sci Technol 3(1):129–132

Kumaraswamy TR, Javeed S, Javaid M, Naika K (2020) Impact of pollution on quality of freshwater ecosystems. In: Fresh water pollution dynamics and remediation. Springer, Singapore, pp 69–81. https://doi.org/10.1007/978-981-13-8277-2_5

Obafemi AA, Eludoyin OS, Akinbosola BM (2012) Public perception of environmental pollution in Warri, Nigeria. J Appl Sci Environ Manag 16(3):233–240

Okeke CO, Igboanua AH (2003) Characteristics and quality assessment of surface water and groundwater recourses of Akwa Town, Southeast, Nigeria. J Niger Assoc Hydrogeol Geol 14: 71–77

Qadri H, Bhat RA, Mehmood MA, Dar GH (2020) Fresh water pollution dynamics and remediation. Springer, Singapore, p 76

Ramakrishnaiah CR, Sadashivaiah C, Ranganna G (2009) Assessment of water quality index for the groundwater in Tumkur Taluk, Karnataka State, India. J Chem 6(2):523–530

Scheierling SM (1995) Overcoming agricultural pollution of water: the challenge of integrating agricultural and environmental policies of the European Union. Technical Paper 269. World Bank, Washington

Skanavis C (2004) Environmental education applied to agricultural education. ecological agriculture and rural development in Central and Eastern European Countries pp 59–73

Toth J (2009) Origin, distribution, formation, and effects. Groundwater 20:27

Trabelsi R, Zouari K (2019) Coupled geochemical modeling and multivariate statistical analysis approach for the assessment of groundwater quality in irrigated areas: a study from North Eastern of Tunisia. Groundw Sustain Dev 8:413–427

Turner RK, Van Den Bergh JC, Söderqvist T, Barendregt A, Van Der Straaten J, Maltby E, Van Ierland EC (2000) Ecological-economic analysis of wetlands: scientific integration for management and policy. Ecol Econ 35(1):7–23

UNESCO (2009) The United Nations world water development report 3: water in a changing world. UNESCO, Paris

US Environmental Protection Agency (2022) Driving toward a cleaner future. https://www.epa.gov/

Wear SL, Acuna V, McDonald R, Font C (2021) Sewage pollution, declining ecosystem health, and cross-sector collaboration. Biol Conserv 255:109010

World Health Organization (2022) Drinking-water report 2020 Geneva Switzerland. https://www.who.int/news-room/fact-sheets/detail/drinking-water

Freshwater Pollution: Overview, Prevention, and Control

16

Pragati Srivastava and Manvika Sahgal

Abstract

All living creatures depend on water as an essential natural resource. However, increasing human population, development, and technology severely strained freshwater ecosystems, changing the freshwater's quality status by introducing massive amounts of contaminants. Increasing population tends a pressure of higher requirement for high-quality water for household use and commercial growth, which is affecting this valuable resource in becoming more and more endangered. Consequently, it is crucial to monitor the quality of this valuable resource. Freshwater pollution can be particularly harmful to the health of people and aquatic life since it is utilized as the primary source of portable water by a large portion of the global population. Chemical contaminants have long-term effects on aquatic and related biota in freshwater ecosystems. Therefore, it is vital to regularly check on the health of the freshwater ecosystem, and an attention should be paid to the treatment of effluents before they are released into the freshwater ecosystem.

Keywords

Pollution · Bioremediation · Freshwater

P. Srivastava (✉) · M. Sahgal
Department of Microbiology, G. B. Pant University of Agriculture & Technology, Pantnagar, Uttarakhand, India

© The Author(s), under exclusive license to Springer Nature Singapore Pte Ltd. 2023
R. Soni et al. (eds.), *Current Status of Fresh Water Microbiology*,
https://doi.org/10.1007/978-981-99-5018-8_16

359

16.1 Introduction

The oceans, having an average depth of 3.8 km with water which is extremely saline, encompass the majority of the earth's surface (71%) and are extremely large. The remaining 2.5% of the world's water, which is freshwater, is contained in the oceans and seas, and 68.9% of this freshwater is trapped in glaciers and ice sheets, with the majority of the remaining freshwater (29.9%) being groundwater. Rivers and lakes hold the majority of the 0.3% of freshwater in the world (Okafor 2011). However, freshwater, which makes up a very small percentage by the volume of the earth's water, is required by human population for their domestic and industrial needs. Although it is possible to evaporate the plentiful ocean and seawater to produce freshwater, the current techniques are not cost-effective. As a result, freshwater is a rather rare resource. Because of this, it is cleansed and reused in several nations. Water bodies are a crucial component in the production of electricity and are used for the disposal of sewage and industrial effluents (Eunice et al. 2017), but the regular disposal of these effluents has raised environmental concerns due to its pollution and causes enormous threats to aquatic life and decrepit for general human use.

Biotic, chemical, and physical elements from various sources are a major cause of water pollution (Richardson et al. 2007). There are innumerable distinct reasons for water contamination, including the discharge of field and industrial waste into bodies of water, oil tanker leaks, the overuse of fertilizers and pesticides for crop protection, sewage sludge, and many more (Aboyeji 2013; Bhat et al. 2017). According to Bhat et al. (2014, 2018), the excessive use of chemicals, filth, dust, and debris and overcrowding in metropolitan areas are the main causes of water body pollution (Master et al. 1998; Carpenter et al. 1998; Kalff 2002; Moss 2008). Assimilative capacity (the ability of water bodies to assimilate contaminants) to a degree may be harmful to the life of biota in some form, variety, or structure (Adekunle 2009; Adekunle and Eniola 2008). There is a direct corelation between contaminants and organism health (Otokunefor and Obiukwu 2005). Various authorities employ a wide range of factors as biomarkers of water pollution. Biochemical oxygen demand (BOD), dissolved oxygen (DO), pH, specific conductance, water temperature, and chemical components such as nitrates and phosphates are some of the influencing factors. Across the globe, an estimated value of municipal and industrial discharge in freshwater bodies is calculated to be 80% without any antecedent ministration (Lin et al. 2022). The shortfall of adequate sanitization of freshwater bodies led to increased commencement of threatful waterborne diseases such as cholera, trachoma, helminthiasis, and gastrointestinal illness enabling the absorption of nutrients in children causing malnutrition (Wright et al. 2004).

16.2 Origin of Pollution

16.2.1 Agricultural Wastes

To feed the increasing population, an immense pressure is provoked on enhancing the productivity of agricultural crops via the usage of chemical pesticides and fertilizers, which has dramatically led to deterioration of water quality, making it unfit for drinking and farm sluicing (Mali et al. 2015). Although the use of pesticides has a number of advantages, such as improving food quality and quantity and reducing damage attributed by insects, but it has also aroused concerns about possible negative impacts on the environment, notably those on water supplies. The corresponding environmental effects are primarily caused by the various herbicides' pervasive endurance in the environment, leading to the disturbance of biodiversity. The pesticide chemical composition and climate components decide its dissolution in the soil, and if the process of dissolution is not readily done it will move from one environment entity to other, leading to its pollution "eutrophication" with unknown health risks in the food chain (Sharma et al. 2019). Notably, long-term absorption of pesticides through water can imitate human hormones, impair hormone balance, causes problems with reproduction, have cancerous effects, and impair intelligence, especially in youngsters still in the development stage of body (Yadav and Samadder 2018; Syafrudin et al. 2021).

Coagulation–flocculation, adsorption, filtering, and sedimentation are common pesticide treatment procedures that depend on the phase transfer of contaminants. These techniques frequently have a considerable production expenses and may result in secondary contamination, such as the production of sludge (López-Loveira et al. 2017). For the treatment of water containing tenacious and biorefractory contaminants such as pesticides, advanced oxidation processes (AOPs) are acknowledged as clean technology. Because of its wide range of applications and thermodynamic viability, it has become a new popular water purification method (Galeano et al. 2019). The fundamental hypothesis that governs the water purification via AOPs is based on the production of highly reactive hydroxyl radicals that randomly oxidize a variety of resistant organic pollutants in order to completely break down the organic contaminants into carbon dioxide, water, and mineral salts and efficient conversion of pesticides into more biodegradable species (Zapata et al. 2010).

Establishing regulatory bodies to regularly check on pesticide release in freshwater bodies, this development could be a significant step toward addressing health concerns related to pesticide pollution of freshwater bodies. The management of pesticides and insecticides in rural areas can benefit from an effective integrated pest management approach that ensures proper application for a long-term reduction of pesticides naturally. Water bodies where pesticide compounds have been detected should be regularly monitored for national water security, and drinking water should, as needed, undergo advanced water treatment methods.

16.2.2 Sewage

Sewage discharge in freshwater bodies is the major setback acquired in changes of species composition and deterioration of biodiversity and a drift in aquatic habitats, which ultimately affect the country's economy, environment sustainability, and downfall in the quality of life. Various components in sewage potentially disturb the microbiota through various means (Bhatt et al. 2021). Many physical changes are observed in the freshwater bodies due to sewage discharge. There is a temperature shift above the optimum temperature, which is undesirable for the sustenance of aquatic life. An increase in temperature also affects the solubility of gases such as oxygen (O_2), which is a major requirement for the breakdown of waste by bacteria (Lecerf and Chauvet 2008).

Sewage consists of domestic waste and effluent discharge from commercializing industrial setup. It attains 99.9% water and 0.1% suspended solids and dissolved solids. Microbial population is diverse in sewage water. It consists of many pathogenic population such as *Shigella*, *Escherichia coli*, *Salmonella*, *Vibrio cholerae*, cysts, fungi, viruses, and parasitic eggs (Grandclément et al. 2017; Yu et al. 2019). Freshwater bodies encountering more of untreated and halfway treated sewage have a higher incidence of biodiversity loss and species physiological alteration and fish mortality. Nutrient-rich sewage has a higher ratio of nitrogen (N) and phosphorous (P), which are major governing factors responsible for eutrophication in freshwater bodies. Untreated sewage has immensely affected the trophic status of freshwater habitats (Kirchmann et al. 2017).

16.3 Urbanization and Hiking Population

Most of the growing population is accommodating themselves in urban cities for the scope of development and exposure to new technologies (Strokal et al. 2021). A 2:3 urbanization ratio contributes to global domestic products while altering the composition of freshwater bodies and increasing their use in industries and agriculture (Best 2019; Keeler et al. 2019). A multipollutant approach should be taken into consideration rather than a single-pollutant approach. Advanced research is required to gain a thorough understanding of the interactions that occur between pollutants, their driving sources, and pressures (e.g., nutrients N and P, pathogens, and plastics) (Strokal et al. 2021).

16.4 Metal Mining Impact on Global Freshwater Resources

Although freshwater usage in the mining sector is very low both worldwide and at the national level, when the usage exceeds the carrying capacity set by the standard quantity, it has an impact on the scarce supply of freshwater resources (Meißner 2021). The process of mining extraction and purification requires a large amount of water for the processing (extraction and purification) of the mining plant, which is an energy-intensive process (Gunson et al. 2012).

16.5   Nutrient Imbalance (N and P)

The science behind the nutrient pollution is rather simple. Phosphorus and nitrogen are both found in soil and water, but nitrogen is also found in the air we breathe. Through intensive farming practices and chemical fertilizers, these nutrients are deposited in the ecosystem (Jwaideh et al. 2022). Despite the fact that its use increases the productivity of various agricultural crops, its ineffective soil absorption causes it to run off through rain into freshwater bodies, where it promotes algal bloom and fish mortality, depletes oxygen concentration, and reduces water quality, disrupting aquatic metabolism (Liu et al. 2022; Mishra et al. 2022).

16.6   Heavy Metal Contamination

The environment contains trace amounts of heavy metals in major forms. Lead and mercury, for instance, are fundamental elements of all rocks created in the crust of the earth, whether intrusive, igneous, or volcanic in origin. Plants only need heavy metals such as vanadium, copper, cobalt, zinc, and iron (Fe) to perform biological reactions (Mehmood et al. 2019). For instance, Fe in hemoglobin is substantially necessary to transmit O_2 in the blood. However, after or over a certain threshold value, it is biologically harmful to humans, animals, and plants (Singh et al. 2018).

Mercury, the most dangerous heavy metal, is released into the environment when coal is burned. Hazardous forms of mercury are released into the atmosphere when household, municipal, and medical trash is burned. These toxic forms of mercury bioaccumulate in the form of aerosol droplets. These aerosol droplets have a propensity to travel thousands of miles before depositing in bodies of water and land. Sediment deposition of this mercuric aerosol is acquired by the bacteria that convert mercury into its more toxic form, which is eventually consumed via fish, leading to its mortality. It is most menacing for humans' brain, kidney, and liver (Rashid et al. 2019; Kumaraswamy et al. 2020).

16.7   Conclusion

The physiological and biological characteristics of the freshwater bodies may be hampered by untreated sewage discharge. Untreated sewage, oil spills, industrial and mining effluent, and domestic waste (fecal matter and urine) discharges all contain persistent contaminants that degrade the quality of quiescent freshwater bodies and cause eutrophication. As a result, lakes and wetlands have less capacity to hold water, and dissolved oxygen levels are brought down to a level where its bioavailability to aquatic life is decreased. Due to all of these limitations, freshwater resources are not utilized to serve mankind and other forms of life. Given the value of the scarce, nonrenewable resource to the biodiversity of the ecosystem and to humanity, multidisciplinary methodologies should be used to target many pollutants simultaneously before they are dumped into freshwater bodies, both spatially and temporally.

References

Aboyeji OO (2013) Freshwater pollution in some Nigerian local communities, causes, consequences and probable solutions. AJIS 2(13):111

Adekunle AS (2009) Effects of industrial effluent on quality of well water within Asa Dam industrial estate, Ilorin, Nigeria. Nat Sci 7(1):39–43

Adekunle AS, Eniola ITK (2008) Impact of industrial effluents on quality of segment of Asa river within an industrial estate in Ilorin, Nigeria. N Y Sci J 1(1):17–21

Best J (2019) Anthropogenic stresses on the world's big rivers. Nat Geosci 12(1):7–21

Bhat RA, Nazir R, Ashraf S, Ali M, Bandh SA, Kamili AN (2014) Municipal solid waste generation rates and its management at Yusmarg forest ecosystem, a tourist resort in Kashmir. Waste Manag Res 32(2):165–169

Bhat RA, Mehmood MA, Dervash MA, Mushtaq N, Bhat JIA, Dar GH (2017) Current status of nutrient load in Dal Lake of Kashmir Himalaya. J Pharmacogn Phytochem 6(6):165–169

Bhat RA, Dar SA, Dar DA, Dar GH (2018) Municipal solid waste generation and current scenario of its management in India. Int J Adv Res Sci Eng 7(2):419–431

Bhatt P, Pathak VM, Bagheri AR, Bilal M (2021) Microplastic contaminants in the aqueous environment, fate, toxicity consequences, and remediation strategies. Environ Res 200:111762

Carpenter SR, Caraco NF, Correll DL, Howarth RW, Sharpley AN, Smith VH (1998) Nonpoint pollution of surface waters with phosphorus and nitrogen. Ecol Appl 8(3):559–568

Eunice OE, Frank O, Voke U, Godwin A (2017) Assessment of the impacts of refinery effluent on the physico-chemical properties of Ubeji creek, Delta State, Nigeria. J Environ Anal Toxicol 7(1):1000428

Galeano LA, Guerrero-Flórez M, Sánchez CA, Gil A, Vicente MÁ (2019) Disinfection by chemical oxidation methods. In: Applications of advanced oxidation processes (AOPs) in drinking water treatment. Springer, Berlin, pp 257–295

Grandclément C, Seyssiecq I, Piram A, Wong-Wah-Chung P, Vanot G, Tiliacos N et al (2017) From the conventional biological wastewater treatment to hybrid processes, the evaluation of organic micropollutant removal: a review. Water Res 111:297–317

Gunson AJ, Klein B, Veiga M, Dunbar S (2012) Reducing mine water requirements. J Clean Prod 21(1):71–82

Jwaideh MA, Sutanudjaja EH, Dalin C (2022) Global impacts of nitrogen and phosphorus fertiliser use for major crops on aquatic biodiversity. Int J Life Cycle Assess 27:1058–1080. https://doi.org/10.1007/s11367-022-02078-1

Kalff J (2002) Limnology: inland water ecosystems, vol 592. Prentice Hall, Upper Saddle River

Keeler BL, Hamel P, McPhearson T, Hamann MH, Donahue ML, Meza Prado KA et al (2019) Social-ecological and technological factors moderate the value of urban nature. Nat Sustain 2(1):29–38

Kirchmann H, Börjesson G, Kätterer T, Cohen Y (2017) From agricultural use of sewage sludge to nutrient extraction: a soil science outlook. Ambio 46:143–154

Kumaraswamy TR, Javeed S, Javaid M, Naika K (2020) Impact of pollution on quality of freshwater ecosystems. In: Qadri H, Bhat R, Mehmood M, Dar G (eds) Fresh water pollution dynamics and remediation. Springer, Singapore. https://doi.org/10.1007/978-981-13-8277-2_5

Lecerf A, Chauvet E (2008) Diversity and functions of leaf-decaying fungi in human-altered streams. Freshw Biol 53(8):1658–1672

Lin L, Yang H, Xu X (2022) Effects of water pollution on human health and disease heterogeneity: a review. Front Environ Sci 10:880246

Liu M, Lei X, Zhou Y et al (2022) Save reservoirs of humid subtropical cities from eutrophication threat. Environ Sci Pollut Res 29:949–962. https://doi.org/10.1007/s11356-021-15560-4

López-Loveira E, Ariganello F, Medina MS, Centrón D, Candal R, Curutchet G (2017) Degradation alternatives for a commercial fungicide in water: biological, photo-Fenton, and coupled biological photo-Fenton processes. Environ Sci Pollut Res 24:25634–25644

Mali SS, Sanyal SK, Bhatt BP, Pathak H (2015) Water pollution and agriculture. In: Pathak H, Bhatt BP, Gupta SK (eds) State of Indian agriculture-water. National Academy of Agricultural Sciences, New Delhi, pp 39–47

Master LL, Flack SR, Stein BA (eds) (1998) Rivers of life: critical watersheds for protecting freshwater biodiversity. Nature Conservancy, Arlington, p 71

Mehmood MA, Qadri H, Bhat RA, Rashid A, Ganie SA, Dar GH (2019) Heavy metal contamination in two commercial fish species of a trans-Himalayan freshwater ecosystem. Environ Monit Assess 191(2):104

Meißner S (2021) The impact of metal mining on global water stress and regional carrying capacities—a GIS-based water impact assessment. Resources 10(12):120

Mishra A, Kumari M, Kumar R, Iqbal K, Thakur IS (2022) Persistent organic pollutants in the environment: risk assessment, hazards, and mitigation strategies. Bioresour Technol Rep 19: 101143

Moss B (2008) Water pollution by agriculture. Philos Trans R Soc B Biol Sci 363(1491):659–666

Okafor N (2011) Environmental microbiology of aquatic and waste systems. Springer Science & Business Media, Berlin

Otokunefor TV, Obiukwu C (2005) Impact of refinery effluent on the physicochemical properties of a water body in The Niger delta. Appl Ecol Environ Res 3(1):61–72

Rashid A, Bhat RA, Qadri H, Mehmood MA (2019) Environmental and socioeconomic factors induced blood lead in children: an investigation from Kashmir, India. Environ Monit Assess 191:1–10

Richardson SD, Plewa MJ, Wagner ED, Schoeny R, DeMarini DM (2007) Occurrence, genotoxicity, and carcinogenicity of regulated and emerging disinfection by-products in drinking water: a review and roadmap for research. Mutat Res Rev Mutat Res 636(1–3):178–242

Sharma B, Vaish B, Singh UK, Singh P, Singh RP (2019) Recycling of organic wastes in agriculture: an environmental perspective. Int J Environ Res 13:409–429

Singh DV, Bhat JIA, Bhat RA, Dervash MA, Ganei SA (2018) Vehicular stress a cause for heavy metal accumulation and change in physico-chemical characteristics of road side soils in Pahalgam. Environ Monit Assess 190:1–10

Strokal M, Bai Z, Franssen W et al (2021) Urbanization: an increasing source of multiple pollutants to rivers in the 21st century. npj Urban Sustain 1:24. https://doi.org/10.1038/s42949-021-00026-w

Syafrudin M, Kristanti RA, Yuniarto A, Hadibarata T, Rhee J, Al-Onazi WA, Algarni TS, Almarri AH, Al-Mohaimeed AM (2021) Pesticides in drinking water-a review. Int J Environ Res Public Health 18(2):468. https://doi.org/10.3390/ijerph18020468. PMID: 33430077; PMCID: PMC7826868

Wright J, Gundry S, Conroy R (2004) Household drinking water in developing countries: a systematic review of microbiological contamination between source and point-of-use. Tropical Med Int Health 9(1):106–117

Yadav P, Samadder SR (2018) Environmental impact assessment of municipal solid waste management options using life cycle assessment: a case study. Environ Sci Pollut Res 25:838–854

Yu C, Huang X, Chen H, Godfray HCJ, Wright JS, Hall JW et al (2019) Managing nitrogen to restore water quality in China. Nature 567(7749):516–520

Zapata A, Malato S, Sánchez-Pérez JA, Oller I, Maldonado MI (2010) Scale-up strategy for a combined solar photo-Fenton/biological system for remediation of pesticide-contaminated water. Catal Today 151(1–2):100–106

Iron-Degrading Bacteria in the Aquatic Environment: Current Trends and Future Directions

17

David Waikhom, Soibam Ngasotter, Laishram Soniya Devi, Soibam Khogen Singh, and Sukham Munilkumar

Abstract

Iron (Fe) is an abundant element found in the earth's crust and is required by organisms to perform their metabolic activity. However, its toxicity in excess amounts in the aquatic environment can negatively impact the organism's habitat. Therefore, the extra amount of iron in the aquatic environment needs to be checked or converted into a nontoxic form to sustain the well-being of the organisms. Biologically, this can be achieved by targeting iron-degrading microbes that can degrade iron in the environment. In this chapter, we present the role of various iron-degrading bacteria and their importance in the environment. Moreover, we highlight some essential mechanisms involved in the degradation of iron through microbes. *Gallionellaceae* sp. in the KS culture, for example, could oxidize iron in such a way that an electron is gained through iron oxidation and transmitted through the electron transport chain (ETC) where nitrate (NO_2) is reduced step by step to nitric oxide (NO). Thus, this chapter explores the possibility of key microbial groups in remediating iron-toxic environments and the underlying mechanisms involved in the process to benefit the degrading environment.

Keywords

Anaerobic photosynthetic · Acidophilic · Aquatic environment · Iron-degrading microbes · Neutrophilic · *Gallionellaceae*

D. Waikhom · S. Ngasotter · S. Munilkumar
ICAR-Central Institute of Fisheries Education, Mumbai, India

L. S. Devi · S. K. Singh (✉)
College of Fisheries, Central Agricultural University (Imphal), Lembucherra, Tripura, India

© The Author(s), under exclusive license to Springer Nature Singapore Pte Ltd. 2023
R. Soni et al. (eds.), *Current Status of Fresh Water Microbiology*,
https://doi.org/10.1007/978-981-99-5018-8_17

367

17.1 Introduction

Iron is one of the fourth most abundant key elements and also the redox-active metal in the earth's crust. The iron requirement in the organisms is also remarkable; without it, the microorganisms cannot conduct their metabolic activity significantly. For instance, it is involved in cellular respiration and oxygen transport (Bury et al. 2001). Although it is imperative for organisms, excess of it affects the organisms. Most countries face the problem of excess iron in their environment, such as water bodies and sediment, which subsequently lead to iron toxicity to the organism. The ability of microbes to oxidize iron may convert the negative impact of iron to the positive impact in the environment that ultimately utilized by the microbes. Therefore, in this chapter, we discuss the key role of iron-degrading bacteria in nature and their importance in the aquatic environment. In addition, we also highlight some important mechanisms concerned in the degradation of iron by using microbes.

17.2 Toxicity of Iron

The toxicity of iron to organisms, particularly fish, can be broadly classified as acute or chronic depending on the time and strength of the toxic substance that the organisms are exposed to. While the first addresses a brief duration with a greater concentration of the toxic substance that causes adverse effects and leads to mortality in organisms, and the second addresses a prolonged exposure with a lower concentration of the agent that causes gradual adverse effects in the organism. Many researchers have studied iron toxicity to organisms, including fishes. For instance, according to Bury et al. (2001), the ferrous form (Fe^{2+}) is more harmful to European plaice (*Platichthys flesus*) than the ferric form (Fe^{3+}), and iron intake occurs largely in the latter section of the gut. The iron toxicity in the fish mainly affects the fish's gills, subsequently causing respiratory problems (Zahedi et al. 2014). In addition, animals on iron-rich diets may experience stunted development, low feed conversion, food refusal, death, and histopathological injury in the liver cells, which retain excessive iron (Bury and Grosell 2003). Fish can be employed as ecological indicators, displaying physiological changes including opercular beating, abnormal swimming, gill modifications, and feeding changes or problems responding to changes in the aquatic ecosystem (Hundley et al. 2018). According to Gemaque et al. (2019), both the forms of iron ions (Fe^{2+} and Fe^{3+}) are detrimental to the "piau" (*Leporinus friderici*), with Fe^{2+} being the most toxic.

17.3 Classification

Iron-degrading bacteria may be classified broadly into two groups: iron-reducing bacteria (IRB) and iron-oxidizing bacteria (IOB), based on their potential to reduce and oxidize iron, respectively, in the nature. While the first one applies to the ability of the bacteria to convert ferric (Fe^{3+}) ions to ferrous (Fe^{2+}) ions under both aerobic

17 Iron-Degrading Bacteria in the Aquatic Environment: Current Trends... 369

and anaerobic environments, the latter one is the ability of the bacteria to transform ferrous ions to ferric (Fe^{3+}) ions under anaerobic and/or microaerobic conditions (Ebrahiminezhad et al. 2017). The majority of IOB are found in a variety of phyla within the bacteria domain, along with Firmicutes and Nitrospirae, the lion's share being Proteobacteria (Hedrich et al. 2011). Moreover, IOB may be classified into four classes based on their physiological diversity: (a) acidophilic, aerobic iron oxidizers; (b) neutrophilic, aerobic iron oxidizers; (c) aerobic iron oxidizers; and (d) anaerobic photosynthetic iron oxidizers. With the exception of nitrate-dependent iron oxidizers, the majority of the species in this phylum belong to one of the Proteobacteria classes. Furthermore, the Zetaproteobacteria are iron-oxidizing neu-trophilic chemolithoautotrophs found in estuarine and marine settings all over the world.

17.3.1 Acidophilic, Aerobic Iron-Oxidizing Proteobacteria

These groups play an essential role in creating acidic and metal-fortified mine drainage waters. Acidophilic prokaryotes utilize the pH gradient across their cell membranes to drive ATP synthesis through F_1F_0 ATP synthase, but the accompanying proton influx is balanced by gaining electrons from the oxidation of ferrous iron (Hedrich et al. 2011). For example, *Acidithiobacillus ferrooxidans*, *Acidithiobacillus ferrivorans*, *Acidiferrobacter thiooxydans*, *Thiobacillus prosperus*, *Mariprofundus ferrooxydans*, *Ferrovum myxofaciens*, and *Thiomonas* spp are included in this group of proteobacteria (Colmer et al. 1950; Hallberg et al. 2010; Hedrich et al. 2011).

17.3.2 Neutrophilic, Aerobic Iron-Oxidizing Proteobacteria

These are the stalk-forming organisms where they divide at a limited partial pressure of oxygen in anti-gradients of oxygen and Fe^{2+}, viz. *Crenothrix*, Ferritrophicales, *Ferritrophicum radicicola*, *Gallionella*, *Leptothrix cholodnii*, *Leptothrix discophora*, *Leptothrix mobilis*, *Metallogenium*, *Siderocapsa*, *Sideroxydans* spp., *Sideroxydans*, *Sideroxydans* sp. ES-1, *Sideroxydans paludicola*, and *Sphaerotilus natans* (Emerson and Moyer 1997). Transmission electron microscopy of *Gallionella ferruginea* or *Mariprofundus ferrooxydans* revealed that stalks are made up of certain fibrils, which accommodate few millimicrometer-sized iron oxyhydroxide crystals (Chan et al. 2011). In 1838, *Gallionella ferruginea* was first proclaimed by Ehrenberg as a leading neutrophilic iron oxidizer, where it can grow mixotrophically or autotrophically utilizing an electron donor (ferrous iron) (Hallbeck and Pedersen 1991). *Mariprofundus ferrooxydans* is an autotrophic, mesophilic, marine iron oxidizer, which has an immediate morphological resem-blance to *G. ferruginea*, although phylogenetically it is amazingly isolated from the freshwater bacterium and is related to Zetaproteobacteria (Emerson et al. 2010). In addition, it was announced that such heterotrophic *Pseudomonas* or

Pseudoalteromonas-like gammaproteobacteria separated from a seamount of volcanic sands under microaerobic conditions could catalyze ferrous iron oxidation and consequently lead to the development of iron mats in the deep seas (Sudek et al. 2009).

17.3.3 Neutrophilic, Anaerobic Iron Oxidizers

These groups have been created in freshwater, brackish, marine, and anaerobic sediments. Microbes that have a characteristic to oxidize iron and reduce nitrate under anaerobic conditions are classified as autotrophic and heterotrophic. These unique bacteria may thrive in organic acids when nitrate or oxygen served as an electron acceptor (Straub et al. 1996) such as *Acidovorax* sp., *Aquabacterium*, *Dechlorosoma suillum* strain PS, *Dechloromonas* spp., and *Geobacter* (Chaudhuri et al. 2001). In the presence of nitrate, *Thiobacillus denitrificans* also oxidized iron sulfide (FeS) (Straub et al. 1996). Comprehensive research is still being done to convert nitrate to nitrogen gas under anoxic conditions, in a process known as "nitrate-dependent anaerobic ferrous oxidization (NAFO)" with the help of ferrous iron as the electron donor (Su et al. 2017; Zhang et al. 2015). This is because iron is the most redox-active metal in organisms, sediment, and soil (Li et al. 2014).

Pseudogulbenkiania strain 2002, an isolate from a sediment of a freshwater lake, could reduce nitrate and oxidize ferrous iron during the time of growing as an autotroph (Weber et al. 2006), but it can also grow under heterotrophic conditions on a variety of organic substances. 16S rRNA gene sequence analysis disclosed that this isolate was nearly linked to the beta proteobacterium, *Pseudogulbenkiania subflava* (99.3% sequence similarity) (Weber et al. 2009). Surprisingly, the discovery of iron-oxidizing proteobacteria explained nitrate-dependent iron (Fe^{2+}, ferrous) oxidation by the stringent anaerobe *Geobacter metallireducens* (Finneran et al. 2002). Moreover, Raiswell and Canfield (2012) proposed that, anaerobically, ferrous form Fe^{2+} is oxidized with nitrate in the nitrate-containing oxygen minimum zones (OMZs) of present sea.

Anaerobic iron oxidation with nitrate (NO_3) produces nitrite (NO_2^-), nitrogenous gases (nitrogen [N_2] or nitrous oxide [N_2O]), or ammonium (NH_4^+) (Carlson et al. 2013), and a range of authigenic Fe minerals, based on surrounding water chemistry and pH (Carlson et al. 2013), for example,

$$2Fe^{2+} + NO^{-3} + 5H_2O \rightarrow 2Fe(OH)_3 + NO^{-2} + 4H^+ \tag{17.1}$$

$$6Fe^{2+} + 2NO^{-2} + 14H_2O \rightarrow 6Fe(OH)_3 + N_2 + 10H^+ \tag{17.2}$$

In the presence of nitrite, such as via heterotrophic nitrate reduction, iron oxidation accompanied by nitrite in Eq. (17.2) has been demonstrated to occur abiotically at around neutral pH, peculiarly when reactive Fe oxyhydroxide mineral surfaces are accessible to catalyze the process (Picardal 2012). The phrase "nitrate-dependent iron oxidation" refers to the microbial and partially abiotic mechanisms with the aid of ferrous iron oxidized using nitrate as the terminal electron acceptor (Picardal 2012; Klueglein et al. 2014).

Nitrate-dependent iron oxidation has been occurred in a diverse freshwater and marine sediments and also shown in laboratory cultures (Straub et al. 1996; Scholz et al. 2016). This might be a crucial mechanism in biogeochemical cycling, especially in OMZs. Despite the fact that sediments under the Peru OMZ discharge large quantities of iron into the anoxic column of water (Noffke et al. 2012), primary production in the eastern equatorial Pacific high-nitrate-low-chlorophyll (HNLC) zone off Peru has been demonstrated to be iron-restricted (Hutchins et al. 2002). Furthermore, ample of iron liberated from the sediments of Peru margin is restored and buried adjacent to its origin rather than being moved offshore within the OMZ (Scholz et al. 2014). Both of these surprising data might be explained by nitrate-dependent iron oxidation inside the anoxic column of water (Scholz et al. 2016).

17.3.4 Anaerobic Photosynthetic Iron (Fe) Oxidizers

The first proof that microbes might oxidize ferrous iron (Fe^{2+}) in anaerobic settings came from phototrophic purple proteobacteria. The class Alphaproteobacteria is placed in the majority of iron-oxidizing phototrophs that have been discovered. *Thiodictyon* strain L7, a gammaproteobacterium, is the striking exception. This type of bacterium uses ferrous iron as a cause of carbon dioxide reductant, as revealed in the following equation, where CH_2O represents fixed biomass carbon:

$$4Fe^{2+} + CO_2 + 11H_2O + hv \rightarrow CH_2O + 4Fe(OH)_3 + 8H^+ \qquad (17.3)$$

At the same time, the majority of photosynthetic microbes that oxidize ferrous iron consume such a reaction for the assimilation of carbon; it can further be utilized as a detoxification mode of action. Moreover, phototrophic iron oxidizers can employ soluble ferrous iron and other minerals, namely FeS or ferrous carbonate ($FeCO_3$), as origins of reductants although they are not allowed to access ferrous iron in more crystalline minerals like magnetite (Fe_3O_4) or pyrite (FeS_2) (Kappler and Newman 2004). So far, most of the identified and validated phototrophic iron oxidizers have been placed under the thoroughly varied family of Rhodobacteraceae within the class of Alphaproteobacteria including aerobic and anaerobic facultative heterotrophs, facultative methylotrophs, fermentative bacteria, and photoheterotrophs (Imhoff 2005).

Rhodobacter sp. strain SW2, the foremost iron-oxidizing phototroph identified, oxidizes ferrous iron exclusively when supplied with an organic carbon origin and also uses hydrogen (H_2) and organic molecules (Ehrenreich and Widdel 1994). Additional iron-oxidizing photosynthetic purple bacteria in the Rhodobacteraceae family comprise *Rhodovulum iodosum* and *Rhodovulum robiginosum*; both of them are found in marine environments and oxidize sulfide and ferrous iron when fed an organic cosubstrate like acetic acid (Hedrich et al. 2011). Heising and Schink (1998) revealed that a phototrophic isolate of *Rhodomicrobium vannielii*, a heterotrophic nonsulfur purple bacterium from the Hyphomicrobiaceae family, can oxidize ferrous iron, a trait that was later validated in the species' type strain. Tentatively, Widdel et al. (1993) had already identified *Rm. vannielii* as the sole of the iron-oxidizing phototrophic isolates from freshwaters they had collected. Acetate or succinate was introduced to strain BS-1 as cosubstrates to promote growth when ferrous iron was present. According to Heising and Schink (1998), ferrous Fe oxidation is just a small act for *Rm. Vannielii* strain BS-1. Alternatively, the Hyphomicrobiaceae species, *Rhodopseudomonas palustris* strain TIE-1, segregated from an iron-rich mat was utilized as a model species for a genetic research by Jiao et al. (2005). The two photosynthetic iron-oxidizing gammaproteobacteria of *Thiodictyon* strains have been reported. One of the two, strain L7, was desegregated from the identical origin as *Rhodobacter* sp. SW2 (Ehrenreich and Widdel 1994), while the other *Thiodictyon* sp. strain f4 was from a marsh (Croal et al. 2004). In addition, among all the isolated phototrophic iron-oxidizing bacteria, *Thiodictyon* sp. strain f4 shows iron oxidation at elevated rates (Hegler et al. 2008). However, Widdel et al. (1993) observed no one at all of the confirmed *Thiodictyon* spp. they experimented with which could oxidize ferrous iron, implying that such a feature is unusual in this genus.

17.4 Iron-Reducing Bacteria's Role

In the sediment column, the presence of anaerobic dissimilatory Fe(III) iron-reducing bacteria makes distinguishing the type of reactive iron challenging (Lovley and Phillips 1986). These organisms seem to be present in both bacteria and archaea and have a broad phylogenetic range (Kashefi and Lovley 2003). The Geobacteraceae, a wide phylogenetically coherent group that oxidizes acetate to carbon dioxide (CO_2) using Fe(III) as the only electron acceptor (Lonergan et al. 1996), and also H_2-oxidizing Fe(III) reducers including *Shewanella putrefaciens* and *S. alga*, are among them (Lonergan et al. 1996). It was proved that Fe(III) reducers may decrease Fe(III) sheet silicates and magnetite (Kostka et al. 1999). Because Fe sheet silicates and magnetite are resistant to inorganic acid leaching, they are normally allocated as unreactive or weakly reactive iron phases in sequential leaching systems. Numerous sulfate-reducing bacteria (SRB) are linked to Fe(III)-reducing bacteria (FeRB) on a phylogenetic level, and many Geobacteraceae species reduce S(0). The sulfate and iron reducers may be an element of a compact ecosystem as the FeRB uses acetate, which is a typical by-product of SRB metabolism. As a result, these organisms are expected to have a big influence on the iron in the sediment's oxic and transition zones. The proportion of Fe(III) that is reduced by iron-reducing bacteria and reacting with the Fe(II) produced will be determined by the relative rates of the bacterial reduction reactions and sulfidation and the time spent by Fe(III) before being exposed to S(-II) in the bacterial iron-reducing subenvironment. Sequentially, this will change depending on sedimentation characteristics such as the degree of advection in the column of water, sediment grain size, the concentration of organic matter, and the velocity of sedimentation (Rickard 2012).

Moreover, the Fe-reducing bacteria may play a vital role in oil degradation. For instance, ferric iron is used by the hyperthermophile *Ferroglobus placidus* to consume benzene at 85 °C (Holmes et al. 2011; Waikhom et al. 2020). In oil field brine enrichment cultures, iron-reducing bacteria have been detected (Magot 2005); nevertheless, the availability of iron oxide surfaces may restrict the degree of oil biodegradation in the deep below. In sandstones, Fe(III) is found in the range of 0.4–4.0 oxide weight percent on pore filling or grain coating material (Pettijohn 1963). Although the energy gained by Fe(III) reduction is equivalent to aerobic respiration or NO_3 reduction, the oxidant amount required is significantly more. According to the volume, around 10 L of hematite would be required to biodegrade 1 L of hydrocarbon approximately. Due to the inadequate biological availability of ferric iron in hematite, such a procedure would be extremely slow, made even more difficult by the fact that hematite is frequently destroyed in course of diagenesis, and the majority of the iron in sedimentary reservoir rocks is constrained in surprisingly less reactive silicates and sulfides (Prince and Walters 2016).

17.5 Mechanism of Iron Oxidation

17.5.1 Mechanisms

Mechanisms utilized by anaerobically nitrate-reducing iron oxidation (NRFeOx) bacteria to oxidize Fe(II) are yet to be more explored, even though it appears to have different mechanisms, even if the bacteria are autotrophs, chemodenitrifiers, or mixotrophs.

For autotrophic NRFeOx bacteria, three mechanisms have been put forward for Fe(II) oxidation (Fig. 17.1) such as (a) a devoted Fe(II) oxidoreductase, (b) a nonspecific performance of the nitrate reductase, or (c) the *bc1* complex that receives electrons from Fe(II) and passages down the quinone pool (Bryce et al. 2018). The first scenario has dominated research in the recent years, intending to discover a particular outermost membrane Fe(II) oxidoreductase seen in these bacteria (Beller et al. 2013). Metagenomics investigation of the NRFeOx in the KS culture revealed homologs of the cytochrome c putative Fe(II) oxidase *Cyc2*, which have been detected in other identified and validated Fe(II) oxidizers in the draft genomes of *Gallionellaceae sp.* and *Rhodanobacter* sp. observed in KS culture (He et al. 2016).

Furthermore, the *Gallionellaceae* sp. and *Dryobalanops aromatica* RCB in the KS culture, which was suspected to be autotrophic but was not confirmed, both include homologs of the porin cytochrome c porin complex MtoAB (He et al. 2017). Figure 17.1a depicts a possible process for the autotrophic Fe(II) oxidizer to oxidize Fe(II) in the KS culture, in which an electron is gathered through Fe(II) oxidation and transmitted through the ETC, where nitrate is reduced step by step to nitric oxide (NO). NO has the ability to be absorbed by the surrounding population or to react with aqueous Fe(II) outside the cell.

Some NRFeOx bacteria may undergo an enzymatic process to oxidize Fe(II), but whether or not these organisms are real mixotrophs is still up for debate. Many

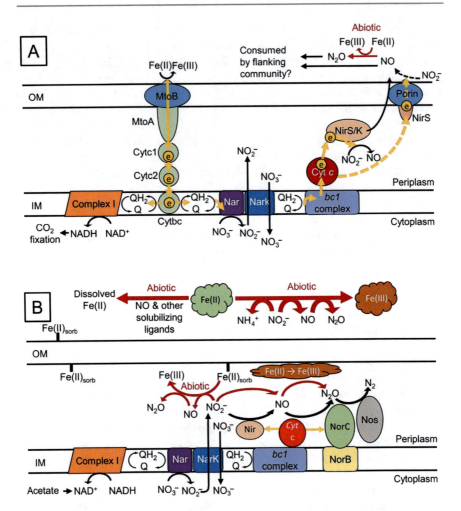

Fig 17.1 The existing hypotheses on the mechanism of iron oxidation in (**a**) *Gallionellaceae* species, considered autotrophic nitrate reducers and Fe(II) oxidizers in the KS culture and (**b**) the reduction of nitrate might be mediated by *Nap* in place of *Nar*, and the reduction of nitric oxide could be aided by *NorZ* instead of *NorC*. Notably, *NorZ* utilizes electrons from quinols as opposed to cytochrome c (Reprinted from Bryce et al. (2018), copyright 2022 Society for Applied Microbiology, with permission from John Wiley and Sons). Abbreviations: OM, outer membrane; IM, inner membrane; Cyt c, cytochrome c; NO, nitric oxide; N_2O, nitrous oxide; NAD, Nicotinamide adenine dinucleotide

people assume that Fe(II) oxidation is powered by denitrification (Brons et al. 1991). It has been hypothesized by Carlson et al. (2013) that all heterotrophic denitrifiers can increase nitrate-dependent Fe(II) oxidation. Several nitrate reducers, including *Escherichia coli*, can oxidize an organic molecule when it is combined with Fe (II) (Brons et al. 1991).

Although the rate of dissolved Fe(II)-mediated abiotic nitrate reduction by nitrite to N_2O is a slow process, the rate of Fe(II)-mediated nitrite reduction to N_2O is kinetically advantageous under environmental conditions if reactive chemical substrates act as catalysts (Colman et al. 2008). Heterogeneous surface catalysis, for instance, can decrease nitrogen species by Fe(II) on viable surfaces such as cell surfaces, green rust, crystalline Fe(III) oxyhydroxides, and pyrite (Bryce et al. 2018). Many researchers have shown that microbially driven NRFeOx, heterotrophic nitrate reduction, and Fe(III)-coupled ammonium oxidation, that is, iron-ammox can result in the development of reactive nitrogen species (nitrite, NO_2^-, or nitric oxide [NO]) as a metabolic intermediate (Picardal 2012), providing a sufficient stock of compounds that could react rapidly with Fe(II) (Bryce et al. 2018).

17.5.2 Mechanisms of Ferrous Iron Oxidation by Photoferrotrophs

The insight mechanisms engaged in Fe(II) oxidation by phototrophic anoxygenic Fe(II)-oxidizing bacteria are yet a mystery. However, *Rp. palustris* TIE-1 has been the subject of the most scientific research for the process of phototrophic bacterial Fe(II) oxidation. Besides, the *pioABC* operon, which stands for "photosynthetic Fe(II) oxidation," is assumed to be essential for electron transport by Fe(II) oxidation in this species. This three-gene operon encodes three proteins: PioA, PioB, and PioC (Jiao and Newman 2007), where PioA is a periplasmic decaheme c-type cytochrome, PioB is an outer membrane beta-barrel protein, and PioC is a periplasmic, strong potential iron–sulfur cluster protein. MtrA and MtrB, which are expressed by the Fe(III) reducer *Shewanella oneidensis* MR-1, are the same as PioA and PioB. (Jiao and Newman 2007). Iro, the putative Fe(II) oxidoreductase of *A. ferrooxidans*, is linked to PioC. *Rp. palustris* TIE-1 carries electrons from *PioA* to *PioC*, which gives them to the *bc1* complex. The electrons may be shifted to the phototrophic reaction center in the inner membrane, according to some studies (Bird et al. 2014). The pioABC operon in *Rhodomicrobium vannielii* acts similarly to the operon in *Rp. palustris* TIE-1 (Bryce et al. 2018).

In *Rp. palustris* TIE-1, the deletion of *pioA* consequences in nearly full loss of Fe(II)-oxidizing capacity, although the deletion of *PioB* and *PioC* consequences in only partial loss when correlated to the wild type (Bose and Newman 2011). When Fe(II) is utilized as an electron donor, the pio genes are expressed at their highest levels, but they are transcribed and translated under all anoxic growth conditions. The global regulator *FixK* controls the expression of the pio operon (Bose and Newman 2011). Supplementary clue in what way Fe(II) influences cellular activities in *Rp. palustris* TIE-1 has come from transcriptome and proteome analyses. These findings reveal that high levels of Fe(II) cause stress reactions unexpectedly in anoxic settings, despite the fact that Fe(II) toxicity caused by oxidative stress created by the Fenton's reaction is not be a concern. The induction of many metal efflux pathways characterizes the cellular response, which was observed throughout both short- and long-term periods of time (Bird et al. 2013). The evidence of PioA's location in the cell is currently contradictory. Based on the sequencing results, Jiao and Newman (2007) suggested that PioA appears to be a periplasmic protein.

Nevertheless, Bose et al. (2014) revealed that *Rp. palustris* TIE-1 can oxidize the crossed-valent Fe(II)–Fe(III) mineral magnetite (Fe_3O_4) in the solid phase and utilize electrons straight from positioned electrodes. PioA can oxidize Fe(II) with the association of its electron transport mechanism, which must thus be existing on the cell's outer membrane, according to this discovery. Furthermore, *Rp. palustris* TIE-1 can only approach surface-bound Fe(II) in magnetite, emphasizing the necessity for a straight surface–mineral interaction mechanism (Byrne et al. 2015, 2016). *Rhodobacter ferrooxidans* SW2 is another well-studied "anoxygenic phototrophic Fe(II) oxidation" pathway. This organism's foxEYZ operon permits it to oxidize Fe (II) (Croal et al. 2007).

The genomic sequencing of *Chlorobium phaeoferrooxidans* reveals that these green sulfur bacteria have yet another mechanism. In *Mariprofundus ferrooxidans* PV-1, which is a microaerophilic Fe(II) oxidizer, the outer membrane cytochrome (*Cyc2PV-1*) encoded by this genome is thought to be engaged in Fe(II) oxidation (Crowe et al. 2008). *Cyc2PV-1* is a far relative of *Cyc2*, a lithotrophic Fe(II)-oxidizing bacteria found in a broad range of obligatory lithotrophic bacteria (He et al. 2017). Besides, *Cyc2* is encoded in the genome of *C. ferrooxidans* DSM13031 (He et al. 2017). The electron transport pathways and proteins involved in Fe(II) oxidation are likely varied, and no one mechanism exists across every single one of the physiological groups of Fe(II) oxidizers, neither within a single group of Fe(II) oxidizers, like phototrophs (Bryce et al. 2018).

17.6 Environmental Significance

The NRFeOx mechanism aids biochemically and microbially driven redox shifts of iron and nitrate species (Liu et al. 2019). The potency and remodeling of pollutants like chlorinated organic compounds and heavy metals (arsenic [As], antimony [Sb], and uranium [U]) are created by the "Fe III–Fe II redox wheel."

Nitrate reduction is intimately linked to N pollution in numerous natural and anthropogenic systems of waters, and the generation of N_2O (greenhouse gas) has the ability to affect global climate change.

As a result, an in-depth investigation of the redox reactions that may occur in the middle of reactive Fe and N species will assist in an improved insight into the Fe–N cycle in nature. It has long been understood that immobilizing these pollutants requires the absorption of heavy metals and radioactive nuclides onto Fe and manganese (Mn) oxides (Means et al. 1978a, b). A practical bioremediating technique for heavy metals including iron and radioactive nuclide fixation in reducing settings might be devised depending on the reported anaerobic NRFeOx employing *Dechlorosoma* species. The findings show that the oxidation of Fe(II) by *D. suillum* is a unique strategy for heavy metal and radionuclide stabilization and immobilization in the ecosystem. Senn and Hemond (2002) revealed that nitrate altered the arsenic (As) cycling in anoxic circumstances in an urban Upper Mystic Lake located in Massachusetts, the United States, by oxidizing Fe(II) to generate arsenic-adsorbing particulate hydrous iron oxides and creating the oxidation of As(III) to As(V).

After large-scale Fe(II) oxidation started, arsenate was discarded from the water adjacent to the injection at the U.S. Geological Survey (USGS) test site on Cape Cod, Massachusetts, the United States, by integration into hydrous Fe oxide, which was precipitated in the iron-reduction zone in a sandy aquifer under an anoxic condition (Senn and Hemond 2002; Höhn et al. 2006), revealing the significance of NRFeOx to arsenic immobilization. The inclusion of nitrate in paddy field soils decreased the Fe(II) concentration in the soil solution while increasing NRFeOx bacterial density, resulting in considerably reduced dissolved Fe(II) concentrations in the rhizosphere soil solution (Liu et al. 2019). It was proposed that NRFeOx bacteria activation might result in arsenic coprecipitation with or adsorption to Fe (III) ions in the soil, followed by reduced arsenic absorption utilizing rice plants (Chen et al. 2008). The findings of continuous-flow sand-filled columns revealed that microbial oxidation of arsenate and Fe(II) associated with denitrification boosted aqueous As immobilization in anaerobic conditions by creating hydrous iron oxides–coated sands with adsorbed pentavalent arsenic (Sun et al. 2009). Despite the fact that it does not affect the anaerobic NRFeOx bacteria metabolism, arsenate efficiently immobilized arsenic during Fe(II) oxidation (efficient for more than 96%), lowering the residual dissolved arsenic concentrations to levels approaching or even below the acceptable limit of drinking water of 10 μg/L (Hohmann et al. 2010). The ability of a single anaerobic NRFeOx *Citrobacter freundii* strain PXL1 to reduce and remove As(III) and nitrate from water synchronously was investigated, and it was disclosed that these bacteria are potential microorganisms for in situ remediation of arsenite- and nitrate-contaminated groundwater in China (Li et al. 2015). The concurrent usage of Fe(II) and nitrate successfully reduced the inflation of arsenic (As) in rice paddy plants by intensifying arsenic oxidation or immobilization intervened by biotic or abiotic iron redox transformation and mineralization (Wang et al. 2018), and the conclusions delivered an understanding about the As/Fe/N biogeochemical cycles, which are essential from the perspective of agricultural crop management activity of arsenic toxicity and its mitigation in arsenic-affected area.

Even though nitrate respiration has long been established as a general process in microbes (Ducluzeau et al. 2009), little has been written about the effects of iron oxidation on nitrate reduction and releasing of a greenhouse gas (N_2O) due to a lack of understanding of nitrogen cycle through abiotic and biotic procedures. However, these abiotic N_2O generation mechanisms have been recognized since the nineteenth century; they are typically disregarded in contemporary environmental studies, making it critical to analyze their importance (Zhu-Barker et al. 2015). Following an early idea (the ferrous wheel theory) for abiotic nitrate immobilization in forest soils, reactive Fe(II) species might convert nitrate to nitrite. Nitrite was then combined alongside dissolved organic materials to form dissolved organic nitrogen (Davidson et al. 2003). Despite the fact that the abiotic and biotic interactions were unclear, the data highlighted the realness activity and significance of NRFeOx in the renowned N cycle scenario. The potential of iron-dependent nitrogen cycling in riparian forest sediments also discussed the Fe–N interaction (Clement et al. 2005). Under a variety of environmental circumstances, dual (N and O) isotope structures were used to investigate the reduction of abiotic nitrite by Fe(II). The findings showed that Fe(II)-mediated nitrite reduction could be a significant origin of abiotic environmental N_2O, particularly in an iron-rich environmental condition with potent redox fluctuations (Buchwald et al. 2016).

17.7   Bioremediation

Various biological approaches, like bioremediation including phytoremediation, have been used to remediate environmental toxins, and they are not only cost-effective but also environmentally beneficial. Bioremediation is a broad word that refers to the transformation of harmful or complicated pollutants into simpler, less toxic forms, or even full degradation, using biological mechanisms including bacteria or microbial products (Datta et al. 2020; Singh et al. 2021). The process of eliminating pollutants from the environment in the absence of the use of chemicals is known as intrinsic bioremediation (Lovley 2003). By altering the environment in which the microorganisms are grown at the polluted location, microbial activity can be increased to achieve higher clean-up rates. The bioremediation technique is regarded as designed in these circumstances (Mishra et al. 2021).

Iron oxidation occurs in most aquatic habitats that include elemental or reduced forms of iron, that is, ferrous iron (Eggerichs et al. 2020). Microorganisms that accelerate iron oxidation gain the energy they need to multiply when they use either nitrate or oxygen as an electron acceptor (Emerson et al. 2010). Iron oxidation of microorganisms can be divided into four categories that have been already mentioned above somewhere in the content. Zagury et al. (1994) demonstrated heavy metal removal by native IOB from polluted soil. Bacteria such as *Acidithiobacillus ferrooxidans* can be used to reduce the soil solution pH lower than 2.5 and to increase the oxidation–reduction potential at room temperature when adjusted to a pH of 4. Iron oxidation through such bacteria was accelerated by the addition of ammonium sulfate $[(NH_4)_2SO_4]$ and potassium phosphate (K_2HPO_4). Zinc and manganese had higher percent removal rates than copper. Metal speciation, together with the soil types studied, affected metal removal by IOB. A potential technique for the efficient removal of arsenic from groundwater was discovered to be the biotic oxidation of iron by the bacteria *Gallionella ferruginea* and *Leptothrix ochracea*, which offered an economical and environmentally beneficial alternative. The microorganisms and iron oxides that were formed in the upflow filtration columns, during this process, provide an ideal environment for the absorption and arsenic removal from the aqueous streams (Katsoyiannis and Zouboulis 2004). Blais et al. (1993) described their research on the function of IOB in the sequestration of toxic metals from sewage sludge. Iron-oxidizing microorganisms lower the pH of sewage sludge and increase its oxidation–reduction potential. These were found to be the metals removed from sludge by *A. ferrooxidans* (previously known as *Thiobacillus ferrooxidans*) in the following ascending order: manganese, zinc, nickel, cadmium, copper, lead, and chromium, with the sludge form used during the experiment having a little effect. Xiang et al. (2000) conducted another batch of investigation aimed at the heavy metal removal from sewage sludge utilizing indigenous IOB. When iron was supplemented to inoculated sets as the ferrous sulfate form, the sludge pH fell from 2.0 to 2.5. Following 16 days of treatment with iron oxidizers, there was a significant decline in the metal contents (chromium, copper, lead, nickel, and zinc) available in sewage sludge, attaining agriculturally acceptable levels, indicating that biological agents can be used to remediate metal contaminants. Recently, the contribution of microaerophilic IOB in arsenic (As) removal was reported by Tong et al. (2019) through the adsorbed/immobilized of arsenite and arsenate that resulted in ferric oxyhydroxide formation following the prominent activity of iron oxidizers with a significant arsenic adsorption affinity. The As content was reduced from 600 to 4.8 g/L following iron oxidizer treatment. The presence of an arsenite oxidase gene, which is accountable for the formation of arsenite from arsenate in microbes, was also discovered in the study. The generation of biogenic iron oxides as a result of the iron oxidation by the bacteria *Gallionella* and *Leptothrix* has a tremendous potential for phosphate elimination from the solution phase (Buliauskaite et al. 2020). Moreover, *Gallionella* sp. has been shown to have an overall phosphate removal capacity than *Leptothrix* sp. When compared to chemically manufactured iron oxides, biologically derived iron oxides had a greater phosphate sequestration performance, implying that they could be used in phosphate recovery treatment systems. Adsorption and precipitation play a substantial role in phosphate elimination by biogenic iron oxides, according to their findings. Another recent study showed that IOB can fortunately bioremediate groundwater contaminated with prominent levels of iron and manganese (Aziz et al. 2020). Not long ago, the ability of bacteria *A. ferrooxidans* to extract heavy metals (chromium, copper, nickel, and zinc) from electroplating sludge under low-voltage stimulation paves the door for industrial-scale metal contamination treatment (Wu et al. 2020). Under laboratory conditions with a pH of 2.0–3.0 and Fe concentrations near 11 g/L, *A. ferrooxidans* catalyzes the synthesis of crystalline biogenic tooeleite, which is an iron-arsenic-based mineral and is thought to remove roughly 95% of As(III) (Li et al. 2020a, b). The research revealed key principles driving biogenic mineral formation, which could lead to the advancement of a treatment system that uses rapidly proliferating IOB to decontaminate As-damaged environments (Li et al. 2020a, b; Singh et al. 2021).

17.8 Future Direction and Conclusion

In the future, nitrate-reducing bacteria may be a valuable microbe for the Biofloc Technology of fish culture to reduce nitrate, which is a severe concern in the system. However, more research into this topic is required. Besides, the pathogenic nature of these bacteria needs to be thoroughly studied to avoid any disease outbreak in the future. Moreover, bioremediation with iron-degrading bacteria can minimize iron toxicity, which is both economical and ecologically sound. Furthermore, this technology will be a boon. However, it is critical to investigate all of the mechanisms involved in the process. Furthermore, the microorganisms must be completely defined before they can be deployed in the bioremediation of iron in an environment where in situ bioremediation has limited. In addition, the remediation of environments contaminated by metalloids, metals, and hazardous organics may benefit from the use of IOB that are resilient to greater metal concentrations and can swiftly adapt to changing natural environmental circumstances.

References

Aziz HA, Tajarudin HA, Wei THL, Alazaiza MYD (2020) Iron and manganese removal from groundwater using limestone filter with iron-oxidized bacteria. Int J Environ Sci Technol 17(5): 2667–2680. https://doi.org/10.1007/s13762-020-02681-5

Beller HR, Zhou P, Legler TC, Chakicherla A, Kane S, Letain TE, O'Day PA (2013) Genome-enabled studies of anaerobic, nitrate-dependent iron oxidation in the chemolithoautotrophic bacterium *Thiobacillus denitrificans*. Front Microbiol 4:249. https://doi.org/10.3389/fmicb.2013.00249

Bird LJ, Coleman ML, Newman DK (2013) Iron and copper act synergistically to delay anaerobic growth of bacteria. Appl Environ Microbiol 79(12):3619–3627. https://doi.org/10.1128/AEM.03944-12

Bird LJ, Saraiva IH, Park S, Calçada EO, Salgueiro CA, Nitschke W, Louro RO, Newman DK (2014) Nonredundant roles for cytochrome c 2 and two high-potential iron-sulfur proteins in the photoferrotroph *Rhodopseudomonas palustris* TIE-1. J Bacteriol 196(4):850–858. https://doi.org/10.1128/JB.00843-13

Blais JF, Tyagi RD, Auclair JC (1993) Metals removal from sewage sludge by indigenous iron-oxidizing bacteria. J Environ Sci Health A 28(2):443–467. https://doi.org/10.1080/10934529309375888

Bose A, Newman DK (2011) Regulation of the phototrophic iron oxidation (pio) genes in *Rhodopseudomonas palustris* TIE-1 is mediated by the global regulator, FixK. Mol Microbiol 79(1):63–75. https://doi.org/10.1111/j.1365-2958.2010.07430.x

Bose A, Gardel EJ, Vidoudez C, Parra EA, Girguis PR (2014) Electron uptake by iron-oxidizing phototrophic bacteria. Nat Commun 5(1):1–7. https://doi.org/10.1038/ncomms4391

Brons HJ, Hagen WR, Zehnder AJ (1991) Ferrous iron dependent nitric oxide production in nitrate reducing cultures of *Escherichia coli*. Arch Microbiol 155(4):341–347. https://doi.org/10.1007/BF00243453

Bryce C, Blackwell N, Schmidt C, Otte J, Huang YM, Kleindienst S, Tomaszewski E, Schad M, Warter V, Peng C, Byrne JM (2018) Microbial anaerobic Fe (II) oxidation–ecology, mechanisms and environmental implications. Environ Microbiol 20(10):3462–3483. https://doi.org/10.1111/1462-2920.14328

Buchwald C, Grabb K, Hansel CM, Wankel SD (2016) Constraining the role of iron in environmental nitrogen transformations: dual stable isotope systematics of abiotic NO2− reduction by Fe (II) and its production of N2O. Geochim Cosmochim Acta 186:1–12. https://doi.org/10.1016/j.gca.2016.04.041

Buliauskaite R, Wilfert P, Suresh Kumar P, de Vet WW, Witkamp GJ, Korving L, van Loosdrecht MC (2020) Biogenic iron oxides for phosphate removal. Environ Technol 41(2):260–266. https://doi.org/10.1080/09593330.2018.1496147

Bury N, Grosell M (2003) Iron acquisition by teleost fish. Comp Biochem Physiol C Toxicol Pharmacol 135(2):97–105. https://doi.org/10.1016/S1532-0456(03)00021-8

Bury NR, Grosell M, Wood CM, Hogstrand C, Wilson RW, Rankin JC, Busk M, Lecklin T, Jensen FB (2001) Intestinal iron uptake in the European flounder (*Platichthys flesus*). J Exp Biol 204(21):3779–3787. https://doi.org/10.1242/jeb.204.21.3779

Byrne JM, Klueglein N, Pearce C, Rosso KM, Appel E, Kappler A (2015) Redox cycling of Fe (II) and Fe (III) in magnetite by Fe-metabolizing bacteria. Science 347(6229):1473–1476. https://doi.org/10.1126/science.aaa4834

Byrne JM, Van Der Laan G, Figueroa AI, Qafoku O, Wang C, Pearce CI, Jackson M, Feinberg J, Rosso KM, Kappler A (2016) Size dependent microbial oxidation and reduction of magnetite nano-and micro-particles. Sci Rep 6(1):1–13. https://doi.org/10.1038/srep30969

Carlson HK, Clark IC, Blazewicz SJ, Iavarone AT, Coates JD (2013) Fe (II) oxidation is an innate capability of nitrate-reducing bacteria that involves abiotic and biotic reactions. J Bacteriol 195(14):3260–3268. https://doi.org/10.1128/JB.00058-13

Chan CS, Fakra SC, Emerson D, Fleming EJ, Edwards KJ (2011) Lithotrophic iron-oxidizing bacteria produce organic stalks to control mineral growth: implications for biosignature formation. ISME J 5(4):717–727. https://doi.org/10.1038/ismej.2010.173

Chaudhuri SK, Lack JG, Coates JD (2001) Biogenic magnetite formation through anaerobic biooxidation of Fe (II). Appl Environ Microbiol 67(6):2844–2848. https://doi.org/10.1128/AEM.67.6.2844-2848.2001

Chen XP, Zhu YG, Hong MN, Kappler A, Xu YX (2008) Effects of different forms of nitrogen fertilizers on arsenic uptake by rice plants. Environ Toxicol Chem 27(4):881–887. https://doi.org/10.1897/07-368.1

Clement JC, Shrestha J, Ehrenfeld JG, Jaffé PR (2005) Ammonium oxidation coupled to dissimilatory reduction of iron under anaerobic conditions in wetland soils. Soil Biol Biochem 37(12):2323–2328. https://doi.org/10.1016/j.soilbio.2005.03.027

Colman BP, Fierer N, Schimel JP (2008) Abiotic nitrate incorporation, anaerobic microsites, and the ferrous wheel. Biogeochemistry 91(2):223–227. https://doi.org/10.1007/s10533-008-9281-9

Colmer AR, Temple KL, Hinkle ME (1950) An iron-oxidizing bacterium from the acid drainage of some bituminous coal mines. J Bacteriol 59(3):317–328. https://journals.asm.org/doi/pdf/10.1128/jb.59.3.317-328.1950

Croal LR, Johnson CM, Beard BL, Newman DK (2004) Iron isotope fractionation by Fe (II)-oxidizing photoautotrophic bacteria. Geochim Cosmochim Acta 68(6):1227–1242. https://doi.org/10.1016/j.gca.2003.09.011

Croal LR, Jiao Y, Newman DK (2007) The fox operon from *Rhodobacter* strain SW2 promotes phototrophic Fe (II) oxidation in *Rhodobacter capsulatus* SB1003. J Bacteriol 189(5):1774–1782. https://doi.org/10.1128/JB.01395-06

Crowe SA, Jones C, Katsev S, Magen C, O'Neill AH, Sturm A, Canfield DE, Haffner GD, Mucci A, Sundby B, Fowle DA (2008) Photoferrotrophs thrive in an Archean Ocean analogue. Proc Natl Acad Sci 105(41):15938–15943. https://doi.org/10.1073/pnas.0805313105

Datta S, Singh S, Kumar V, Dhanjal DS, Sidhu GK, Amin DS, Kumar S, Singh J, Singh J (2020) Endophytic bacteria in xenobiotic degradation. In: Microbial endophytes. Woodhead Publishing, Sawston, pp 125–156. https://doi.org/10.1016/B978-0-12-818734-0.00006-1

Davidson EA, Chorover J, Dail DB (2003) A mechanism of abiotic immobilization of nitrate in forest ecosystems: the ferrous wheel hypothesis. Glob Chang Biol 9(2):228–236. https://doi.org/10.1046/j.1365-2486.2003.00592.x

Ducluzeau AL, Van Lis R, Duval S, Schoepp-Cothenet B, Russell MJ, Nitschke W (2009) Was nitric oxide the first deep electron sink? Trends Biochem Sci 34(1):9–15. https://doi.org/10.1016/j.tibs.2008.10.005

Ebrahiminezhad A, Manafi Z, Berenjian A, Kianpour S, Ghasemi Y (2017) Iron-reducing bacteria and iron nanostructures. J Adv Med Sci Appl Technol 3(1):9–16. https://doi.org/10.18869/nrip.jamsat.3.1.9

Eggerichs T, Wiegand M, Neumann K, Opel O, Thronicker O, Szewzyk U (2020) Growth of iron-oxidizing bacteria *Gallionella ferruginea* and *Leptothrix cholodnii* in oligotrophic environments: Ca, Mg, and C as limiting factors and *G. ferruginea* necromass as C-source. Geomicrobiol J 37(2):190–199. https://doi.org/10.1080/01490451.2019.1686667

Ehrenreich A, Widdel F (1994) Anaerobic oxidation of ferrous iron by purple bacteria, a new type of phototrophic metabolism. Appl Environ Microbiol 60(12):4517–4526. https://doi.org/10.1128/aem.60.12.4517-4526.1994

Emerson D, Moyer C (1997) Isolation and characterization of novel iron-oxidizing bacteria that grow at circumneutral pH. Appl Environ Microbiol 63(12):4784–4792. https://doi.org/10.1128/aem.63.12.4784-4792.1997

Emerson D, Fleming EJ, McBeth JM (2010) Iron-oxidizing bacteria: an environmental and genomic perspective. Annu Rev Microbiol 64:561–583. https://doi.org/10.1146/annurev.micro.112408.134208

Finneran KT, Anderson RT, Nevin KP, Lovley DR (2002) Potential for bioremediation of uranium-contaminated aquifers with microbial U (VI) reduction. Soil Sediment Contam 11(3):339–357. https://doi.org/10.1080/20025891106781

Gemaque T, Costa DPD, Pereira LV (2019) Evaluation of iron toxicity in the tropical fish *Leporinus friderici*. Biomed J Sci Tech Res 18(2):13436–13441. https://doi.org/10.26717/BJSTR.2019.18.003127

Hallbeck L, Pedersen K (1991) Autotrophic and mixotrophic growth of *Gallionella ferruginea*. Microbiology 137(11):2657–2661. https://doi.org/10.1099/00221287-137-11-2657

Hallberg KB, González-Toril E, Johnson DB (2010) *Acidithiobacillus ferrivorans*, sp. nov.; facultatively anaerobic, psychrotolerant iron-, and sulfur-oxidizing acidophiles isolated from metal mine-impacted environments. Extremophiles 14(1):9–19. https://doi.org/10.1007/s00792-009-0282-y

He S, Tominski C, Kappler A, Behrens S, Roden EE (2016) Metagenomic analyses of the autotrophic Fe (II)-oxidizing, nitrate-reducing enrichment culture KS. Appl Environ Microbiol 82(9):2656–2668. https://doi.org/10.1128/AEM.03493-15

He S, Barco RA, Emerson D, Roden EE (2017) Comparative genomic analysis of neutrophilic iron (II) oxidizer genomes for candidate genes in extracellular electron transfer. Front Microbiol 8:1584. https://doi.org/10.3389/fmicb.2017.01584

Hedrich S, Schlömann M, Johnson DB (2011) The iron-oxidizing proteobacteria. Microbiology 157(6):1551–1564. https://doi.org/10.1099/mic.0.045344-0

Hegler F, Posth NR, Jiang J, Kappler A (2008) Physiology of phototrophic iron (II)-oxidizing bacteria: implications for modern and ancient environments. FEMS Microbiol Ecol 66(2):250–260. https://doi.org/10.1111/j.1574-6941.2008.00592.x

Heising S, Schink B (1998) Phototrophic oxidation of ferrous iron by a *Rhodomicrobium vannielii* strain. Microbiology 144(8):2263–2269. https://doi.org/10.1099/00221287-144-8-2263

Hohmann C, Winkler E, Morin G, Kappler A (2010) Anaerobic Fe (II)-oxidizing bacteria show As resistance and immobilize As during Fe (III) mineral precipitation. Environ Sci Technol 44(1):94–101. https://doi.org/10.1021/es900708s

Höhn R, Isenbeck-Schröter M, Kent DB, Davis JA, Jakobsen R, Jann S, Niedan V, Scholz C, Stadler S, Tretner A (2006) Tracer test with As (V) under variable redox conditions controlling arsenic transport in the presence of elevated ferrous iron concentrations. J Contam Hydrol 88(1–2):36–54. https://doi.org/10.1016/j.jconhyd.2006.06.001

Holmes DE, Risso C, Smith JA, Lovley DR (2011) Anaerobic oxidation of benzene by the hyperthermophilic archaeon *Ferroglobus placidus*. Appl Environ Microbiol 77(17):5926–5933. https://doi.org/10.1128/AEM.05452-11

Hundley GC, Navarro FKSP, Ribeiro Filho OP, Navarro RD (2018) Integration of Nile tilapia (*Oreochromis niloticus* L.) production *Origanum majorana* L. and *Ocimum basilicum* L. using aquaponics technology. Acta Sci Technol 40:e35460–e35460. https://doi.org/10.4025/actascitechnol.v40i1.35460

Hutchins DA, Hare CE, Weaver RS, Zhang Y, Firme GF, DiTullio GR, Alm MB, Riseman SF, Maucher JM, Geesey ME, Trick CG (2002) Phytoplankton iron limitation in the Humboldt Current and Peru Upwelling. Limnol Oceanogr 47(4):997–1011. https://doi.org/10.4319/lo.2002.47.4.0997

Imhoff J (2005) *Rhodomicrobium* Duchow and Douglas 1949, 415 AL emend. Imhoff, Trüper and Pfennig 1984, 341. Bergey's Manual® of Systematic Bacteriology, pp 543–545

Jiao Y, Newman DK (2007) The pio operon is essential for phototrophic Fe (II) oxidation in Rhodopseudomonas palustris TIE-1. J Bacteriol 189(5):1765–1773. https://doi.org/10.1128/JB. 00776-06

Jiao Y, Kappler A, Croal LR, Newman DK (2005) Isolation and characterization of a genetically tractable photoautotrophic Fe (II)-oxidizing bacterium, *Rhodopseudomonas palustris* strain TIE-1. Appl Environ Microbiol 71(8):4487–4496. https://doi.org/10.1128/AEM.71.8. 4487-4496.2005

Kappler A, Newman DK (2004) Formation of Fe (III)-minerals by Fe (II)-oxidizing photoautotrophic bacteria. Geochim Cosmochim Acta 68(6):1217–1226. https://doi.org/10.1016/j.gca.2003. 09.006

Kashefi K, Lovley DR (2003) Extending the upper temperature limit for life. Science 301(5635): 934–934. https://doi.org/10.1126/science.1086823

Katsoyiannis IA, Zouboulis AI (2004) Biological treatment of Mn (II) and Fe (II) containing groundwater: kinetic considerations and product characterization. Water Res 38(7): 1922–1932. https://doi.org/10.1016/j.watres.2004.01.014

Klueglein N, Zeitvogel F, Stierhof YD, Floetenmeyer M, Konhauser KO, Kappler A, Obst M (2014) Potential role of nitrite for abiotic Fe (II) oxidation and cell encrustation during nitrate reduction by denitrifying bacteria. Appl Environ Microbiol 80(3):1051–1061. https://doi.org/10.1128/AEM.03277-13

Kostka JE, Wu J, Nealson KH, Stucki JW (1999) The impact of structural Fe (III) reduction by bacteria on the surface chemistry of smectite clay minerals. Geochim Cosmochim Acta 63(22): 3705–3713. https://doi.org/10.1016/S0016-7037(99)00199-4

Li B, Tian C, Zhang D, Pan X (2014) Anaerobic nitrate-dependent iron (II) oxidation by a novel autotrophic bacterium, Citrobacter freundii strain PXL1. Geomicrobiol J 31(2):138–144. https://doi.org/10.1080/01490451.2013.816393

Li B, Pan X, Zhang D, Lee DJ, Al-Misned FA, Mortuza MG (2015) Anaerobic nitrate reduction with oxidation of Fe (II) by Citrobacter Freundii strain PXL1–a potential candidate for simultaneous removal of As and nitrate from groundwater. Ecol Eng 77:196–201. https://doi.org/10.1016/j.ecoleng.2015.01.027

Li Q, Zhang M, Yang J, Liu Q, Zhang G, Liao Q et al (2020a) Formation and stability of biogenic tooeleite during Fe (II) oxidation by *Acidithiobacillus ferrooxidans*. Mater Sci Eng C 111: 110755. https://doi.org/10.1016/j.msec.2020.110755

Li X, Qiao J, Li S, Häggblom MM, Li F, Hu M (2020b) Bacterial communities and functional genes stimulated during anaerobic arsenite oxidation and nitrate reduction in a paddy soil. Environ Sci Tech 54(4):2172–2181. https://doi.org/10.1021/acs.est.9b04308

Liu T, Chen D, Li X, Li F (2019) Microbially mediated coupling of nitrate reduction and Fe (II) oxidation under anoxic conditions. FEMS Microbiol Ecol 95(4):fiz030. https://doi.org/10.1093/femsec/fiz030

Lonergan DJ, Jenter HL, Coates JD, Phillips EJ, Schmidt TM, Lovley DR (1996) Phylogenetic analysis of dissimilatory Fe (III)-reducing bacteria. J Bacteriol 178(8):2402–2408. https://doi.org/10.1128/jb.178.8.2402-2408.1996

Lovley DR (2003) Cleaning up with genomics: applying molecular biology to bioremediation. Nat Rev Microbiol 1(1):35–44. https://doi.org/10.1038/nrmicro731

Lovley DR, Phillips EJ (1986) Organic matter mineralization with reduction of ferric iron in anaerobic sediments. Appl Environ Microbiol 51(4):683–689. https://doi.org/10.1128/aem.51. 4.683-689.1986

Magot M (2005) Indigenous microbial communities in oil fields. In: Petroleum microbiology. ASM Press, Washington, DC, pp 21–33. https://doi.org/10.1128/9781555817589.ch2

Means JL, Crerar DA, Duguid JO (1978a) Migration of radioactive wastes: radionuclide mobilization by complexing agents. Science 200(4349):1477–1481. https://doi.org/10.1126/science.200. 4349.1477

Means JL, Crerar DA, Borcsik MP, Duguid JO (1978b) Adsorption of Co and selected actinides by Mn and Fe oxides in soils and sediments. Geochim Cosmochim Acta 42(12):1763–1773. https://doi.org/10.1016/0016-7037(78)90233-8

Mishra M, Singh SK, Kumar A (2021) Role of omics approaches in microbial bioremediation. In: Microbe mediated remediation of environmental contaminants. Woodhead Publishing, Sawston, pp 435–445. https://doi.org/10.1016/B978-0-12-821199-1.00036-5

Noffke A, Hensen C, Sommer S, Scholz F, Bohlen L, Mosch T, Graco M, Wallmann K (2012) Benthic iron and phosphorus fluxes across the Peruvian oxygen minimum zone. Limnol Oceanogr 57(3):851–867. https://doi.org/10.4319/lo.2012.57.3.0851

Pettijohn FJ (1963) Chemical composition of sandstones, excluding carbonate and volcanic sands: representative analyses. US Government Printing Office. https://books.google.co.in/books?hl=en&lr=&id=NCQlAQAAIAAJ&oi=fnd&pg=PA1&dq=Pettijohn,+F.+J.,+1963.+Chemical+composition+of+sandstones%3B+excluding+carbonate+and+and+volcanic++sands.+United+States+Geological+Survey+Professional+Paper+440-S,+21.&ots=g7AhbkE-w8&sig=JdO9iFuF7rtc-gFTwlkv0zd5fes#v=onepage&q&f=false

Picardal F (2012) Abiotic and microbial interactions during anaerobic transformations of Fe (II) and NOx. Front Microbiol 3:112. https://doi.org/10.3389/fmicb.2012.00112

Prince RC, Walters CC (2016) Biodegradation of oil hydrocarbons and its implications for source identification. In: Standard handbook oil spill environmental forensics. Academic Press, London, pp 869–916. https://doi.org/10.1016/B978-0-12-803832-1.00019-2

Raiswell R, Canfield DE (2012) The iron biogeochemical cycle past and present. Geochem Perspect 1(1):1–2. https://pubs.geoscienceworld.org/perspectives/article-abstract/1/1/1/251624/The-Iron-Biogeochemical-Cycle-Past-and-Present?redirectedFrom=fulltext

Rickard D (2012) Sedimentary iron biogeochemistry. In: Developments in sedimentology, vol 65. Elsevier, Amsterdam, pp 85–119. https://doi.org/10.1016/B978-0-444-52989-3.00003-9

Scholz F, Severmann S, McManus J, Hensen C (2014) Beyond the Black Sea paradigm: the sedimentary fingerprint of an open-marine iron shuttle. Geochim Cosmochim Acta 127:368–380. https://doi.org/10.1016/j.gca.2013.11.041

Scholz F, Löscher CR, Fiskal A, Sommer S, Hensen C, Lomnitz U, Wuttig K, Göttlicher J, Kossel E, Steininger R, Canfield DE (2016) Nitrate-dependent iron oxidation limits iron transport in anoxic ocean regions. Earth Planet Sci Lett 454:272–281. https://doi.org/10.1016/j.epsl.2016.09.025

Senn DB, Hemond HF (2002) Nitrate controls on iron and arsenic in an urban lake. Science 296(5577):2373–2376. https://doi.org/10.1126/science.1072402

Singh VK, Chaudhari AK, Notarte KIR, Kumar A, Singh R, Bhadouria R (2021) Metal-oxidizing microbes and potential application in bioremediation. In: Microbe mediated remediation of environmental contaminants. Woodhead Publishing, Sawston, pp 107–114. https://doi.org/10.1016/B978-0-12-821199-1.00011-0

Straub KL, Benz M, Schink B, Widdel F (1996) Anaerobic, nitrate-dependent microbial oxidation of ferrous iron. Appl Environ Microbiol 62(4):1458–1460. https://doi.org/10.1128/aem.62.4.1458-1460.1996

Su J, Luo X, Huang T, Ma F, Zheng S, Lu J (2017) Effect of nitrite on autotrophic denitrification efficiency by *Pseudomonas* sp. SZF15. Environ Eng Sci 34(8):577–584. https://doi.org/10.1089/ees.2016.0173

Sudek LA, Templeton AS, Tebo BM, Staudigel H (2009) Microbial ecology of Fe (hydr)oxide mats and basaltic rock from Vailulu'u Seamount, American Samoa. Geomicrobiol J 26(8):581–596. https://doi.org/10.1080/01490450903263400

Sun W, Sierra-Alvarez R, Milner L, Oremland R, Field JA (2009) Arsenite and ferrous iron oxidation linked to chemolithotrophic denitrification for the immobilization of arsenic in anoxic environments. Environ Sci Technol 43(17):6585–6591. https://doi.org/10.1021/es900978h

Tong H, Liu C, Hao L, Swanner ED, Chen M, Li F, Xia Y, Liu Y, Liu Y (2019) Biological Fe (II) and As (III) oxidation immobilizes arsenic in micro-oxic environments. Geochim Cosmochim Acta 265:96–108. https://doi.org/10.1016/j.gca.2019.09.002

Waikhom D, Ngasotter S, Soniya Devi L, Devi S, Singh AS (2020) Role of microbes in petroleum hydrocarbon degradation in the aquatic environment: a review. Int J Curr Microbiol Appl Sci 9: 2990–2903. https://doi.org/10.20546/ijcmas.2020.905.342

Wang X, Liu T, Li F, Li B, Liu C (2018) Effects of simultaneous application of ferrous iron and nitrate on arsenic accumulation in rice grown in contaminated paddy soil. ACS Earth Space Chem 2(2):103–111. https://doi.org/10.1021/acsearthspacechem.7b00115

Weber KA, Achenbach LA, Coates JD (2006) Microorganisms pumping iron: anaerobic microbial iron oxidation and reduction. Nat Rev Microbiol 4(10):752–764. https://doi.org/10.1038/nrmicro1490

Weber KA, Hedrick DB, Peacock AD, Thrash JC, White DC, Achenbach LA, Coates JD (2009) Physiological and taxonomic description of the novel autotrophic, metal oxidizing bacterium, *Pseudogulbenkiania* sp. strain 2002. Appl Microbiol Biotechnol 83(3):555–565. https://doi.org/10.1007/s00253-009-1934-7

Widdel F, Schnell S, Heising S, Ehrenreich A, Assmus B, Schink B (1993) Ferrous iron oxidation by anoxygenic phototrophic bacteria. Nature 362(6423):834–836. https://doi.org/10.1038/362834a0

Wu P, Zhang LJ, Lin CB, Xie XX, Yong XY, Wu XY, Zhou J, Jia HH, Wei P (2020) Extracting heavy metals from electroplating sludge by acid and bioelectrical leaching using *Acidithiobacillus ferrooxidans*. Hydrometallurgy 191:105225. https://doi.org/10.1016/j.hydromet.2019.105225

Xiang L, Chan LC, Wong JWC (2000) Removal of heavy metals from anaerobically digested sewage sludge by isolated indigenous iron-oxidizing bacteria. Chemosphere 41(1-2):283–287. https://doi.org/10.1016/S0045-6535(99)00422-1

Zagury GJ, Narasiah KS, Tyagi RD (1994) Adaptation of indigenous iron-oxidizing bacteria for bioleaching of heavy metals in contaminated soils. Environ Technol 15(6):517–530. https://doi.org/10.1080/09593339409385458

Zahedi S, Vaezzade H, Rafati M, Zarei Dangesaraki M (2014) Acute toxicity and accumulation of iron, manganese and, aluminum in Caspian kutum Fish (*Rutilus kutum*). Iran J Toxicol 8(24): 1028–1033. http://ijt.arakmu.ac.ir/article-1-299-en.html

Zhang H, Wang H, Yang K, Chang Q, Sun Y, Tian J, Long C (2015) Autotrophic denitrification with anaerobic Fe2+ oxidation by a novel *Pseudomonas* sp. W1. Water Sci Technol 71(7): 1081–1087. https://doi.org/10.2166/wst.2015.071

Zhu-Barker X, Cavazos AR, Ostrom NE, Horwath WR, Glass JB (2015) The importance of abiotic reactions for nitrous oxide production. Biogeochemistry 126(3):251–267. https://doi.org/10.1007/s10533-015-0166-4

Bioactive Compounds from Aquatic Ecosystem

18

Surendra Puri, Rohit Mahar, and Gunjan Goswami

Abstract

Aquatic ecosystems, especially marine, which cover nearly 67% of the earth and 50% of the whole globe's biodiversity, offer scientists and researchers a promising sustainable source of many bioactive substances and their potential applications in pharmaceuticals, cosmeceuticals, and nutraceuticals industries. They contain different specific habitats characterized by a wide range of temperatures, hydrostatic pressures, and salinity. To survive in such harsh and challenging conditions, marine living beings have the potential to produce unique and valuable bioactive compounds. In the recent decades, the course of investigations of bioactive compounds from the aquatic ecosystems has greatly expanded and gained much importance because of their therapeutic uses for numerous diseases. Compounds isolated from aquatic or marine biotopes possess a broad spectrum of biological activities such as antitumor, anticancer, anti-human immunodeficiency virus (HIV), antidiabetic, antioxidant, antimicrotubule, antiproliferative, antibacterial, anti-inflammatory, cytotoxic, photoprotective, antibiotic, and antifouling. The huge chemical complexity, unique skeleton diversity, broad spectrum of biological activities of the compounds isolated from aquatic beings, and their therapeutic applications have engrossed the attention of scientists and researchers in the recent years.

S. Puri (✉) · R. Mahar
Department of Chemistry, H.N.B. Garhwal University (A Central University), Srinagar, Uttarakhand, India
e-mail: surendrapuri@hnbgu.ac.in

G. Goswami
Government Inter College (GIC) Pitradhar, Augustmuni, Rudraprayag, Uttarakhand, India

© The Author(s), under exclusive license to Springer Nature Singapore Pte Ltd. 2023
R. Soni et al. (eds.), *Current Status of Fresh Water Microbiology*,
https://doi.org/10.1007/978-981-99-5018-8_18

Keywords

Aquatic · Bioactive compounds · Ecosystem · Anti-inflammatory · Antioxidant

18.1 Introduction

Natural products have been used for the treatment of human ailments since the ancient time, and the remedial properties of these natural products are well described in Rigveda (2500–1800 BC), Charak Samhita, and Sushruta Samhita. Oceans cover nearly 67% of the earth and 50% of the whole globe's biodiversity and remain as one such storehouse for natural products. The importance of natural product discovery from microorganisms started only after the large-scale production of penicillin during World War II, although penicillin was discovered in 1928 by the Scottish bacteriologist Alexander Fleming from the *Penicillium* mold (Landau et al. 1999), and at that time, it was believed that soil microorganisms are the major source of novel drugs. The aquatic ecosystems are less explored than their terrestrial counterparts (Lordan et al. 2011; Rasmussen and Morrissey 2007) due to the time, cost, and geographical constraints, yet they possess infinite hidden sources for many useful chemical substances (Schröder 2010). The aquatic biotopes undergo harsh climate changes; live in stressful habitats; experience a cold, lightless, and high-pressure conditions; and biosynthesize a wide variety of secondary metabolites (SMs) for their defense against other microorganisms. These fascinating and structurally complex SMs serve as a source of bioactive compounds for use in human therapies (Bhatnagar and Kim 2010). Secondary metabolites (SMs) are generally defined as organic compounds formed as bioproducts in organisms and do not directly involve in the growth, development, and reproduction of thereof (i.e., hormones, antibiotics, and vaccines). The metabolites derived from marine bacteria, proteobacteria, cyanobacteria, microalgae, seaweed, fungi, marine invertebrates, fishes, and actinomycetes showed antibacterial (Lu et al. 2009), antiviral, antifungal (Rajanbabu and Chen 2011), antihypertensive (angiotensin-converting enzyme [ACE] inhibitor) (Ono et al. 2003), antioxidant (Mendis et al. 2005; Nalinanon et al. 2010), antimicrobial (Zhang et al. 2004), antitumor (Guha et al. 2013; Li et al. 2005), antithrombin (Salte et al. 1995), antidiabetic (Zhu et al. 2010), antilarvicidal (Kirst 2010), antineoplastic (Smith et al. 2000), antituberculosis (Barry et al. 2000; Jensen et al. 2007), and antimalarial (Jensen et al. 2007) activities.

Having a broad range of bioactivities, aquatic organisms and microorganisms have not been fully explored yet, and limited research and discoveries are available in literature to date. To fill this gap, there is still a scope for a collaborative research to explore the therapeutic potential of novel molecules of an aquatic biotope. In 2004, Blunt et al. reported that among the marine organisms and microorganisms, sponges (37%), coelenterates (21%), microorganisms (18%), algae (9%), echinoderms (6%), tunicates (6%), molluscs (2%), bryozoans (1%), etc. are the major contributors to the production of bioactive compounds (Leal et al. 2012). Table 18.1 summarizes the medicinal applications and activities of the molecules isolated from the marine ecosystems.

Table 18.1 Structures and biological activities of secondary metabolites isolated from marine sources

Sources	Order/species	Bioactive (secondary metabolites)	Biological activity	Structures
Marine sponge	*Streptomyces* sp. SBT345	Ageloline A	Antioxidant and antichlamydial activities	
	Hamigera tarangaensis	Hamigeran B	Antiviral activity against polio virus	
	Jaspis stellifera	Stellettin B (Wu et al. 2016)	Significant cytotoxic effects against the human glioblastoma cell line SF295	
	Xestospongia carbonaria	Halenaquinone (Budke et al. 2013)	Inhibits the secondary DNA binding of RAD51	
Marine fungi	*Penicillium oxalicum*	Secalonic acid D	Antitumor activity and antimicrobial activity (against the gram-positive bacteria and gram-negative bacteria); antifungal and antialgal activities	

(continued)

Table 18.1 (continued)

Sources	Order/species	Bioactive (secondary metabolites)	Biological activity	Structures
	Phoma herbarum	Bromo-chloro-gentisyl-quinones A and B	Antioxidant activity	
	Dysidea avara	Avarol and avarone (Loya and Hizi 1990)	Antiviral activity against HIV and anti-inflammatory activity	
Marine macroalgae	*Pelvetia siliquosa*	Fucosterol	Prevent the degradation of blood glucose	
	Ecklonia kurome	Phlorotannins	Inhibition of amylase precipitate proteins	
Marine microalgae	*Haematococcus pluvialis*	Astaxanthin	Antioxidant, anticancer, and anti-inflammatory activities	
	Chlorella vulgaris	Lutein	Antioxidant and anticancer activities	
	Chlorella ellipsoidea	Violaxanthin	Anti-inflammatory and anticancer activities	

Marine bacteria	*Streptomyces aureoverticillatus*	Aureoverticillactam	Cytotoxicity against various types of cell tumors	
	Salinispora tropica	Salinosporamide A (NPI-0052) (Beer and Moore 2007)	Potential anti-cancer agents	
	Streptomyces sp.	Caprolactones	Anticancer activity	
	Streptomyces sp.	Marmycin A and marmycin B	Marmycin A displays potent cytotoxicity against numerous cancer cell lines	
	Streptomyces sp.	Marineosins	Significant inhibition of human colon carcinoma and selective activities in diverse cancer cell types	
	Streptomyces sp.	Chinikomycins	Antitumor activity	

(continued)

Table 18.1 (continued)

Sources	Order/species	Bioactive (secondary metabolites)	Biological activity	Structures
	Streptomyces griseus	Streptomycin	Potent antibiotic used to treat numerous bacterial infections	
	Pseudoalteromonas	MC21-B	Antibacterial activity	
	Bacillus sp.	Macrolactin S	Antibacterial activity against *E. coli* and *S. aureus*	
	Streptomyces fradiae	Urdamycin A	Antibacterial and anticancer activities	
	Streptomyces strain CNQ-085	Daryamide D	Cytotoxic activity against human colon carcinoma and antifungal activity against *Candida albicans*	

Streptomyces sp. M045	Manumycin	Antitumor activity	
Streptomyces sioyaensis	Altemicidin	Antitumor and acaricidal activities	
Streptomyces sp.	Streptochlorin	Potent chemotherapeutic agent	
Janibacter limosus	Helquinoline	Antibacterial, antifungal, and antimicroalgal activities	
Streptomyces sp.	Himalomycins	Antimicrobial activity against gram-positive bacteria	
Micromonospora sp. GMKU326	Maklamicin	Antimicrobial activity against gram-positive bacteria	

(continued)

Table 18.1 (continued)

Sources	Order/species	Bioactive (secondary metabolites)	Biological activity	Structures
	Micromonospora sp.	Diazepinomicin (ECO-4601)	Antitumor, antioxidant, and antiprotease activities	
	Agelasidae, Axinellidae, and Halichondriidae	Hymenialdisine (Meijer et al. 2000)	Anti-inflammatory activity	
	Amycolatopsis rifamycinica	Rifampicin	Potent antibiotic used to treat various bacterial infections	
	Sporosarcina newyorkensis	Amikacin	Powerful activity against antibiotic-resistant clinical isolates	
	Streptomyces sp.	Cycloserine	Antitubercular activity	

Organism	Compound	Activity	Structure
Streptomyces sp.	Komodoquinone A (Itoh et al. 2003)	Induces neuronal cell differentiation in the neuroblastoma cell line	41
Streptomyces kanamyceticus	Kanamycin	Antibiotic used to treat severe bacterial infections and tuberculosis	42
Streptomyces sp.	Capreomycin	It is a second-line anti-TB drug used to treat drug-resistant tuberculosis.	43
Streptomyces puniceus	Viomycin	Viomycin is a peptide antibiotic exhibiting anti-TB activity	44
Streptomyces griseus	Frigocyclinone	Antibacterial activities against gram-positive bacteria	45
Chromobacterium violaceum and *Janthinobacterium lividum*	Violacein	Antiprotozoal activity	46

(continued)

Table 18.1 (continued)

Sources	Order/species	Bioactive (secondary metabolites)	Biological activity	Structures
Marine actinomycetes	*Marinispora* sp. strain CNQ-140	Marinomycin	Inhibits cancer cell proliferation	
Marine invertebrates	Mollusca (*Dolabella auricularia*)	Dolastatin 10	Anticancer activity, promising antiproliferative activity	
	Mollusca (*Dolabella auricularia*)	Dolastatin 15	Anticancer activity	
	Ascidians and tunicates (*Trididemnum* sp.)	Didemnin	Exhibits antitumor activity, strong antiviral agent against both DNA and RNA viruses, exhibits strong activity against murine leukemia cells	R = *N*-Me-L-Leu : Didemnin A **50** R = Lac-Pro-*N*-Me-L-Leu : Didemnin B **51** R = Lac-*N*-Me-L-Leu : Didemnin C **52**

Marine fish	Giant squid (*Dosidicus gigas*)	Xanthommatin	Antioxidant activity	
Marine fungus	*Penicillium* sp.	Isochromophilone	Anti-inflammatory activity	
	Penicillium sp.	(+)-Brefeldin A, (+)-brefeldin C, and 7-oxobrefeldin A	Antifungal activity against *Microsporum gypseum* SH-MU-4	
	Chaetomium sp.	Chaetocyclinones A–C	Antifungal activity	
	Streptomyces sp.	Toyocamycin	Antifungal activity	

(continued)

Table 18.1 (continued)

Sources	Order/species	Bioactive (secondary metabolites)	Biological activity	Structures
	Shewanella sp.	Eicosapentaenoic acid (EPA) (Braune et al. 2021)	Used to treat heart disease, anti-inflammatory activity	
	Chaetoceros sp.	Calothrixin	Antimalarial and anticancer activities	

18.2 Bioactivity of Secondary Metabolites Isolated from Marine Ecosystems

18.2.1 Anticancer Activity

In general, human cells grow and proliferate through the cell division process and form new cells. When cells become old, get damaged, or die, new cells take their place. Sometimes cell division process breaks down, and the abnormal or damaged cells grow up and multiply themselves uncontrollably. These abnormal cells have the ability to invade and destroy normal body tissue or spread to other parts of the body and cause cancer. The probable signs and symptoms of cancer are abnormal bleeding, prolonged cough, unexplained weight loss, and a change in bowel movements. There are more than hundred types of cancer, and the most common types are lung cancer, prostate cancer, breast cancer, stomach cancer, cervical cancer, and colorectal cancer (Cakir et al. 2012). Worldwide the number of cancer patients and the cost of treatments are increasing simultaneously, which remain a huge socioeconomic burden. Therefore, there is an urgent need to discover new anticancer drugs. A few natural compounds such as taxol, camptothecin, vincristine, and vinblastine were isolated from plants and used to treat cancer (Sung et al. 2017). In the recent decades, secondary metabolites isolated from marine biotopes have shown a promising activity when administered at different cancer stages and opened another dimension for the development of new anticancer drugs. Secalonic acid D (SAD, **5**) obtained from *Penicillium oxalicum* and marine-derived fungus shows promising anticancer properties and acts as a deoxyribonucleic acid (DNA) topoisomerase I inhibitor (Cherigo et al. 2015; Steyn 1970). It demonstrates a potent cytotoxic activity and induces apoptosis in K562 and HL60 myeloid leukemia cell lines by blocking the G1 phase of the cell cycle in the GSK-3β/β-catenin/c-Myc pathway. It also shows a potent antimicrobial activity against gram-positive bacteria such as *Bacillus megaterium* and the gram-negative bacteria *Escherichia coli*, an antifungal activity against *Microbotryum violaceum*, and an antialgal activity against *Chlorella fusca* (Zhang et al. 2008). Salinosporamide A (**17**) is a potent anticancer agent obtained from marine bacteria *Salinispora tropica* and *Salinispora arenicola*. In preliminary screening, a high percentage of the organic extracts of Salinispora strains displays antibiotic and anticancer activities. Compound **17** shows potent in vitro cytotoxicity against HCT-116 human colon carcinoma with an half-maximal inhibitory concentration (IC_{50}) value of 11 ng mL^{-1} and inhibits proteasome activity by covalently modifying the active site of threonine residues of the 20S proteasome (Groll et al. 2006). Caprolactones (R-10-methyl-6-undecanolide (**18**) and 6R, 10S-10-methyl-6-dodecanolide (**19**)) are isolated from the extract of a marine *Streptomyces* sp. B6007. These two compounds are found active against human cancer cell lines and show concentration-dependent inhibition of the cell growth of HM02 (gastric adenocarcinoma), MCF-7 (breast adenocarcinoma), and HepG2 (hepatocellular carcinoma). Cell cycle analysis was carried out in HepG2 cells, which showed an increased number of cells in the G1 phase than the S phase on the treatment of 5 μg mL^{-1} for 2–24 h (Stritzke et al. 2004). Marmycins A and B

(20, 21) isolated from the genus *Streptomyces* display significant cytotoxicity against numerous cancer cell lines at nanomolar concentrations. It is noteworthy that Compound 21 was found less potent than Compound 20. Compound 20 induces apoptosis and arrests the G1 phase of the cell cycle (Martin et al. 2007).

The secondary metabolites marineosins A (22) and marineosins B (23) are obtained from actinomycete of the genus *Streptomyces*. Marineosins A (22) shows significant cytotoxicity against HCT-116 human colon tumor cell line with an IC_{50} value of 0.5 µM. Marineosins B (23) displays significantly weaker cytotoxicity (IC_{50} = 46 µM). Compounds 22 and 23 show very weak antifungal activity against *Candida albicans* with a minimum inhibitory concentration (MIC) value of 100 µg/mL (Boonlarppradab et al. 2008). Similarly, Aureoverticillactam (16) (Mitchell et al. 2004), Chinikomycins (24) (Li et al. 2005), Urdamycin A (28) (Henkel et al. 1989), Daryamide D (29) (Asolkar et al. 2006), Manumycin (30) (Li et al. 2005), Altemicidin (31) (Takahashi et al. 1989), Streptochlorin (32) (Shin et al. 2007), Diazepinomicin (ECO-4601) (36) (Sung et al. 2017), Cycloserine (40) (Barry et al. 2000), Marinomycin (47) (Liu and Jiang 2017), Dolastatin 10 (48) (Pettit et al. 2017), Dolastatin 15 (49) (Pettit et al. 2017), and Didemnin (50, 51 and 52) (Rinehart et al. 1981), etc. are isolated from marine ecosystems and show a promising activity against a panel of cancer cell lines. Table 18.1 summarizes the potential applications of these compounds.

18.2.2 Antimicrobial Activity

Natural products are being widely and frequently utilized for antimicrobial therapy and other related diseases since the ancient times (Suleria et al. 2016). SMs secluded from marine ecosystem show a broad range of antimicrobial activities against various human pathogens. Fucosterol (11), isolated from *Fucus vesiculosus* (macroalgae), inhibits the germination of macroconidia in *Fusarium culmorum* (Meinita et al. 2021; Tyśkiewicz et al. 2019). In the recent years, the outbreak of drug-resistant tuberculosis (TB) increases the mortality rate across the globe and drew the attention of researchers to search for new potent anti-TB agents. A few marine-derived compounds such as Rifampicin (38) (Barry et al. 2000), Streptomycin (25) (Barry et al. 2000), Capreomycin, Viomycin, and Kanamycin (Barry et al. 2000) show promising anti-TB activity and are used as second-line anti-TB drugs. Compound 25 is also found active against several aerobic gram-negative bacteria. Macrolactin S (27) isolated from a culture broth of marine *Bacillus* sp. inhibits the bacterial growth against *E. coli* with an MIC of 0.2 mg mL^{-1}. Macrolactin S also shows an antifungal activity and inhibits the hyphal growth of *Pyricularia oryzae* (Lu et al. 2009). MC21-B is isolated from marine bacterium Pseudoalteromonas phenolica O-BC30^T and shows an antibacterial activity against various isolates of methicillin-resistant *Staphylococcus aureus* (MRSA) (Isnansetyo and Kamei 2009). Helquinoline (33), another secondary metabolite obtained from the cultures of *Janibacter limosus*, shows significant antibacterial and antifungal activities against *Streptomyces viridochromogenes*, *Staphylococcus aureus*, *Mucor miehei*, and

18 Bioactive Compounds from Aquatic Ecosystem 401

Chlorella vulgaris (Maskey et al. 2004). Similarly, Himalomycins **(34)** ("Himalomycin A and cycloheximide-producing marine actinomycete from Lagos Lagoon soil sediment" 2015), Frigocyclinone **(45)** (Bruntner et al. 2005), Amikacin **(39)** (Barry et al. 2000), and Maklamicin **(35)** (Igarashi et al. 2017) show a potent antimicrobial activity against gram-positive bacteria.

18.2.3 Anti-inflammatory Activity

Anti-inflammatory drugs reduce inflammation or swelling by binding to cortisol receptors in case of steroidal anti-inflammatory drugs (SAIDs) or inhibit cyclooxygenase (COX) enzymes in case of nonsteroidal anti-inflammatory drugs (NAISDs). The frequent use of anti-inflammatory drugs causes gastrointestinal ulcer, bleeding, heart attack, stroke, and hepatotoxicity. Because of severe side effects, natural products–based drugs are preferred over synthetic drugs for the treatment of inflammation. Azaphilone derivative, isochromophilone IX **(54)**, and sclerketide C isolated from coral-derived fungus *Penicillium sclerotiorin* exhibit potent anti-inflammatory activity (Liu et al. 2019). Fucoidans are polysaccharides containing a significant fraction of L-fucose and sulfate ester groups obtained from brown algae *Fucus vesiculosus*, which possess potent anti-inflammatory activity via inhibiting pro-inflammatory cytokine production (Vo et al. 2015). Marine lectins (protein) also show an anti-inflammatory activity due to their carbohydrate-binding site (Cheung et al. 2015).

18.2.4 Antioxidant Activity

Janibacter melonis and *Pseudomonas Stutzeri* are marine bacteria that produce antioxidant compounds (Shahidi and Santhiravel 2022). The extract of actinobacterium *Streptomyces* sp. *S2A* shows antioxidant and cytotoxic activities against different cell lines (Siddharth and Vittal 2018). Bromo-chloro-gentisylquinones A **(6)** and B **(7)**, obtained from a marine fungus *Phoma herbarum* strain, show a significant radical scavenging activity against 2,2-diphenyl-1-picrylhydrazyl (DPPH) (Trisuwan et al. 2009). Xanthommatin **(53)** is an ommochrome obtained from jumbo squid skin (JSS) and shows a potent antioxidant activity against 2,2-diphenyl-1-picrylhydrazyl (DPPH) and 2,2′-azino-bis-(3-ethylbenzothiazoline-6-sulfonic acid) (ABTS) (Swain et al. 2017; Williams et al. 2019).

18.2.5 Antifungal Activity

An antifungal medication, also known as an antimycotic medication, is a fungicide used to treat and prevent mycosis such as ringworm, athlete's foot, and candidiasis (thrush). Three compounds named (+)-brefeldin A **(55)**, (+)-brefeldin C **(56)**, and 7-oxobrefeldin A **(57)** are isolated from the marine fungi of the genus *Penicillium* sp. Compound **55** shows an excellent antifungal activity against *Microsporum gypseum* with an MIC value of 228.57 μM, while the rest of the compounds are found inactive (Nenkep et al. 2010). Chaetocyclinones A–C **(58, 59,** and **60)** are the marine SMs obtained from the cultures of *Chaetomium* sp. Compounds **59** and **60** are found inactive, while Compound **58** shows a potent antifungal activity against *Phytophthora infestans* at a concentration of 89.1 μM. Toyocamycin **(61)** is an adenosine analog produced by *Streptomyces diastatochromogenes*, which shows an antifungal activity (Burja et al. 2001; Chan-Higuera et al. 2019). Isochromophilone **(54)** derivatives isolated from the culture broth of *Penicillium* sp. show a broad range of activities, such as inhibition of acyl-CoA cholesterol acyltransferase (ACAT) and diacylglycerol acyltransferase (DGAT) (Arai et al. 1995).

18.2.6 Miscellaneous Activity of Marine-Derived Secondary Metabolites

Violacein **(46)** is a bisindole, water-insoluble purple pigment, produced by *Chromobacterium violaceum* and *Janthinobacterium lividum.* This compound displays antiprotozoal and antiparasitic activities and also shows a cytotoxic effect against several tumor cell lines (Almeida et al. 2021; Matz et al. 2008). Toyocamycin **(61)** isolated from culture broth of an *Actinomycete* strain, which belongs to a class nucleosides, shows significant antifungal, antibiotic, and antitumor activities (Katritzky and Rees 1984). Calothrix obtained from cyanobacteria shows a potent antimalarial activity and inhibits ribonucleic acid (RNA) polymerase and DNA topoisomerase-I poisoning activity. Compound **61** also shows a significant antiproliferative activity against human HeLa cancer cells (Ramkumar and Nagarajan 2013; Rickards et al. 1999). Eicosapentaenoic acid (EPA, **62**), also known as omega 3-fatty acid, obtained from microalgae is used to treat heart and inflammatory diseases. EPA is also effective against rheumatoid arthritis and immunodeficiency disease (Braune et al. 2021) (Table 18.1).

18.3 Conclusion

The various structurally diverse compounds and extracts obtained from marine biotopes possess a broad range of biological activities including anticancer, antibacterial, antiviral, fungicidal, cytotoxic, antimalarial, and antioxidant and have been utilized for the treatment of numerous human ailments. The well-documented medicinal applications of these marine compounds as evidenced from the past and current research make oceans so special and remain a large producer of many bioactive compounds. The hidden unexplored marine diversity would be a great treasure for scientists and researchers to discover the novel drugs. Under this paradigm, a growing and sustainable research and development programs must be fostered by the scientists, researchers, academicians, and industries under collaborative partnerships. It is also encouraging that few molecules obtained from marine sources are under clinical and preclinical trials, and many of them are on the market for the treatment of many diseases.

Acknowledgments Dr. Surendra Puri is thankful to Hemvati Nandan Bahuguna Garhwal University (A Central University), Srinagar, Garhwal, Uttarakhand, and University Grants Commission–Dr. D.S. Kothari Post-Doctoral Fellowships in Sciences for providing financial assistance (No.F.4-2/2006 (BSR)/CH/20-21/0206). Dr. Rohit Mahar is thankful to Hemvati Nandan Bahuguna Garhwal University (A Central University), Srinagar, Garhwal, Uttarakhand, for providing basic facilities. Dr. Gunjan Goswami is thankful to Government Inter College Pitradhar, Augustmuni, Rudraprayag, Uttarakhand, for providing basic facilities.

References

Almeida MC, Resende DISP, da Costa PM, Pinto MMM, Sousa E (2021) Tryptophan derived natural marine alkaloids and synthetic derivatives as promising antimicrobial agents. Eur J Med Chem 209:112945. https://doi.org/10.1016/j.ejmech.2020.112945

Arai N, Shiomi K, Tomoda H, Tabata N, Yang DJ, Masuma R, Kawakubo T, Omura S (1995) Isochromophilones III -VI, inhibitors of acyl-CoA: cholesterol acyltransferase produced by Penicillium multicolor FO-3216. J Antibiot 48:696–702. https://doi.org/10.7164/antibiotics. 48.696

Asolkar RN, Jensen PR, Kauffman CA, Fenical W (2006) Daryamides A–C, weakly cytotoxic polyketides from a marine-derived actinomycete of the genus *Streptomyces* strain CNQ-085. J Nat Prod 69:1756–1759. https://doi.org/10.1021/np0603828

Barry CE, Slayden RA, Sampson AE, Lee RE (2000) Use of genomics and combinatorial chemistry in the development of new antimycobacterial drugs. Biochem Pharmacol 59:221–231. https://doi.org/10.1016/S0006-2952(99)00253-1

Beer LL, Moore BS (2007) Biosynthetic convergence of Salinosporamides A and B in the marine Actinomycete *Salinispora tropica*. Org Lett 9:845–848. https://doi.org/10.1021/ol063102o

Bhatnagar I, Kim S-K (2010) Immense essence of excellence: marine microbial bioactive compounds. Mar Drugs 8:2673–2701. https://doi.org/10.3390/md8102673

Boonlarppradab C, Kauffman CA, Jensen PR, Fenical W (2008) Marineosins A and B, cytotoxic Spiroaminals from a marine-derived actinomycete. Org Lett 10:5505–5508. https://doi.org/10. 1021/ol8020644

Braune S, Krüger-Genge A, Kammerer S, Jung F, Küpper J-H (2021) Phycocyanin from Arthrospira platensis as potential anti-cancer drug: review of in vitro and in vivo studies. Life 11:91. https://doi.org/10.3390/life11020091

Bruntner C, Binder T, Pathom-aree W, Goodfellow M, Bull AT, Potterat O, Puder C, Hörer S, Schmid A, Bolek W, Wagner K, Mihm G, Fiedler H-P (2005) Frigocyclinone, a novel angucyclinone antibiotic produced by a Streptomyces griseus strain from Antarctica. J Antibiot 58:346–349. https://doi.org/10.1038/ja.2005.43

Budke B, Kalin JH, Pawlowski M, Zelivianskaia AS, Wu M, Kozikowski AP, Connell PP (2013) An optimized RAD51 inhibitor that disrupts homologous recombination without requiring Michael acceptor reactivity. J Med Chem 56:254–263. https://doi.org/10.1021/jm301565b

Burja AM, Banaigs B, Abou-Mansour E, Grant Burgess J, Wright PC (2001) Marine cyanobacteria—a prolific source of natural products. Tetrahedron 57:9347–9377. https://doi.org/10.1016/S0040-4020(01)00931-0

Cakir BÖ, Adamson P, Cingi C (2012) Epidemiology and economic burden of nonmelanoma skin cancer. Facial Plast Surg Clin North Am 20:419–422. https://doi.org/10.1016/j.fsc.2012.07.004

Chan-Higuera JE, Santacruz-Ortega HDC, Carbonell-Barrachina ÁA, Burgos-Hernández A, Robles-Sánchez RM, Cruz-Ramírez SG, Ezquerra-Brauer JM (2019) Xanthommatin is behind the antioxidant activity of the skin of Dosidicus gigas. Molecules 24:3420. https://doi.org/10.3390/molecules24193420

Cherigo L, Lopez D, Martinez-Luis S (2015) Marine natural products as breast cancer resistance protein inhibitors. Mar Drugs 13:2010–2029. https://doi.org/10.3390/md13042010

Cheung RCF, Wong JH, Pan W, Chan YS, Yin C, Dan X, Ng TB (2015) Marine lectins and their medicinal applications. Appl Microbiol Biotechnol 99:3755–3773. https://doi.org/10.1007/s00253-015-6518-0

Groll M, Huber R, Potts BCM (2006) Crystal structures of Salinosporamide A (NPI-0052) and B (NPI-0047) in complex with the 20S proteasome reveal important consequences of β-lactone ring opening and a mechanism for irreversible binding. J Am Chem Soc 128:5136–5141. https://doi.org/10.1021/ja058320b

Guha P, Kaptan E, Bandyopadhyaya G, Kaczanowska S, Davila E, Thompson K, Martin SS, Kalvakolanu DV, Vasta GR, Ahmed H (2013) Cod glycopeptide with picomolar affinity to galectin-3 suppresses T-cell apoptosis and prostate cancer metastasis. Proc Natl Acad Sci U S A 110:5052–5057. https://doi.org/10.1073/pnas.1202653110

Henkel T, Ciesiolka T, Rohr J, Zeeck A (1989) Urdamycins, new angucycline antibiotics from Streptomyces fradiae. V. Derivatives of urdamycin A. J Antibiot 42:299–311. https://doi.org/10.7164/antibiotics.42.299

(2015) Himalomycin A and cycloheximide-producing marine actinomycete from Lagos Lagoon soil sediment. J Coast Life Med 3. https://doi.org/10.12980/JCLM.3.201514J82

Igarashi M, Sawa R, Yamasaki M, Hayashi C, Umekita M, Hatano M, Fujiwara T, Mizumoto K, Nomoto A (2017) Kribellosides, novel RNA 5′-triphosphatase inhibitors from the rare actinomycete Kribbella sp. MI481-42F6. J Antibiot 70:582–589. https://doi.org/10.1038/ja.2016.161

Isnansetyo A, Kamei Y (2009) Anti-methicillin-resistant Staphylococcus aureus (MRSA) activity of MC21-B, an antibacterial compound produced by the marine bacterium Pseudoalteromonas phenolica O-BC30T. Int J Antimicrob Agents 34:131–135. https://doi.org/10.1016/j.ijantimicag.2009.02.009

Itoh T, Kinoshita M, Aoki S, Kobayashi M (2003) Komodoquinone A, a novel neuritogenic anthracycline, from marine Streptomyces sp. KS3. J Nat Prod 66:1373–1377. https://doi.org/10.1021/np030212k

Jensen PR, Williams PG, Oh D-C, Zeigler L, Fenical W (2007) Species-specific secondary metabolite production in marine actinomycetes of the genus Salinispora. Appl Environ Microbiol 73:1146–1152. https://doi.org/10.1128/AEM.01891-06

Katritzky AR, Rees CW (eds) (1984) Comprehensive heterocyclic chemistry: the structure, reactions, synthesis, and uses of heterocyclic compounds, 1st edn. Pergamon Press, Oxford

Kirst HA (2010) The spinosyn family of insecticides: realizing the potential of natural products research. J Antibiot 63:101–111. https://doi.org/10.1038/ja.2010.5

Landau R, Achilladelis B, Scriabine A (eds) (1999) Pharmaceutical innovation: revolutionizing human health. Chemical Heritage Press, Philadelphia

Leal MC, Madeira C, Brandão CA, Puga J, Calado R (2012) Bioprospecting of marine invertebrates for new natural products — a chemical and zoogeographical perspective. Molecules 17:9842–9854. https://doi.org/10.3390/molecules17089842

Li F, Maskey RP, Qin S, Sattler I, Fiebig HH, Maier A, Zeeck A, Laatsch H (2005) Chinikomycins A and B: isolation, structure elucidation, and biological activity of novel antibiotics from a marine Streptomyces sp. isolate M045. J Nat Prod 68:349–353. https://doi.org/10.1021/np030518r

Liu J, Jiang SP (eds) (2017) Mesoporous materials for advanced energy storage and conversion technologies. CRC Press, Boca Raton

Liu Z, Qiu P, Liu H, Li J, Shao C, Yan T, Cao W, She Z (2019) Identification of anti-inflammatory polyketides from the coral-derived fungus Penicillium sclerotiorin: in vitro approaches and molecular-modeling. Bioorg Chem 88:102973. https://doi.org/10.1016/j.bioorg.2019.102973

Lordan S, Ross RP, Stanton C (2011) Marine bioactives as functional food ingredients: potential to reduce the incidence of chronic diseases. Mar Drugs 9:1056–1100. https://doi.org/10.3390/md9061056

Loya S, Hizi A (1990) The inhibition of human immunodeficiency virus type 1 reverse transcriptase by avarol and avarone derivatives. FEBS Lett 269:131–134. https://doi.org/10.1016/0014-5793(90)81137-D

Lu Y, Dong X, Liu S, Bie X (2009) Characterization and identification of a novel marine Streptomyces sp. produced antibacterial substance. Mar Biotechnol 11:717–724. https://doi.org/10.1007/s10126-009-9186-1

Martin GDA, Tan LT, Jensen PR, Dimayuga RE, Fairchild CR, Raventos-Suarez C, Fenical W (2007) Marmycins A and B, cytotoxic Pentacyclic C-glycosides from a marine sediment-derived actinomycete related to the genus Streptomyces. J Nat Prod 70:1406–1409. https://doi.org/10.1021/np060621r

Maskey RP, Helmke E, Kayser O, Fiebig HH, Maier A, Busche A, Laatsch H (2004) Anti-cancer and antibacterial trioxacarcins with high anti-malaria activity from a marine Streptomycete and their absolute stereochemistry. J Antibiot 57:771–779. https://doi.org/10.7164/antibiotics.57.771

Matz C, Webb JS, Schupp PJ, Phang SY, Penesyan A, Egan S, Steinberg P, Kjelleberg S (2008) Marine biofilm bacteria evade eukaryotic predation by targeted chemical defense. PLoS One 3: e2744. https://doi.org/10.1371/journal.pone.0002744

Meijer L, Thunnissen A-M, White A, Garnier M, Nikolic M, Tsai L-H, Walter J, Cleverley K, Salinas P, Wu Y-Z, Biernat J, Mandelkow E-M, Kim S-H, Pettit G (2000) Inhibition of cyclin-dependent kinases, GSK-3β and CK1 by hymenialdisine, a marine sponge constituent. Chem Biol 7:51–63. https://doi.org/10.1016/S1074-5521(00)00063-6

Meinita MDN, Harwanto D, Tirtawijaya G, Negara BFSP, Sohn J-H, Kim J-S, Choi J-S (2021) Fucosterol of marine macroalgae: bioactivity, safety and toxicity on organism. Mar Drugs 19: 545. https://doi.org/10.3390/md19100545

Mendis E, Rajapakse N, Kim S-K (2005) Antioxidant properties of a radical-scavenging peptide purified from enzymatically prepared fish skin gelatin hydrolysate. J Agric Food Chem 53:581–587. https://doi.org/10.1021/jf048877v

Mitchell SS, Nicholson B, Teisan S, Lam KS, Potts BCM (2004) Aureoverticillactam, a novel 22-atom macrocyclic lactam from the marine actinomycete Streptomyces aureoverticillatus. J Nat Prod 67:1400–1402. https://doi.org/10.1021/np049970g

Nalinanon S, Benjakul S, Kishimura H (2010) Purification and biochemical properties of pepsins from the stomach of skipjack tuna (Katsuwonus pelamis). Eur Food Res Technol 231:259–269. https://doi.org/10.1007/s00217-010-1275-x

Nenkep VN, Yun K, Li Y, Choi HD, Kang JS, Son BW (2010) New production of haloquinones, bromochlorogentisylquinones A and B, by a halide salt from a marine isolate of the fungus Phoma herbarum. J Antibiot 63:199–201. https://doi.org/10.1038/ja.2010.15

Ono S, Hosokawa M, Miyashita K, Takahashi K (2003) Isolation of peptides with angiotensin I-converting enzyme inhibitory effect derived from hydrolysate of upstream chum Salmon muscle. J Food Sci 68:1611–1614. https://doi.org/10.1111/j.1365-2621.2003.tb12300.x

Pettit GR, Melody N, Chapuis J-C (2017) Antineoplastic agents. 604. The path of quinstatin derivatives to antibody drug conjugates. J Nat Prod 80:2447–2452. https://doi.org/10.1021/acs.jnatprod.7b00237

Rajanbabu V, Chen J-Y (2011) Applications of antimicrobial peptides from fish and perspectives for the future. Peptides 32:415–420. https://doi.org/10.1016/j.peptides.2010.11.005

Ramkumar N, Nagarajan R (2013) Total synthesis of calothrixin A and B via C–H activation. J Org Chem 78:2802–2807. https://doi.org/10.1021/jo302821v

Rasmussen RS, Morrissey MT (2007) Marine biotechnology for production of food ingredients, in: advances in food and nutrition research. Elsevier, Amsterdam, pp 237–292. https://doi.org/10.1016/S1043-4526(06)52005-4

Rickards RW, Rothschild JM, Willis AC, de Chazal NM, Kirk J, Kirk K, Saliba KJ, Smith GD (1999) Calothrixins A and B, novel pentacyclic metabolites from Calothrix cyanobacteria with potent activity against malaria parasites and human cancer cells. Tetrahedron 55:13513–13520. https://doi.org/10.1016/S0040-4020(99)00833-9

Rinehart KL, Gloer JB, Hughes RG, Renis HE, McGovren JP, Swynenberg EB, Stringfellow DA, Kuentzel SL, Li LH (1981) Didemnins: antiviral and antitumor depsipeptides from a Caribbean tunicate. Science 212:933–935. https://doi.org/10.1126/science.7233187

Salte R, Norberg K, Ole Rasmus Ø (1995) Some functional properties of teleost antithrombin. Thromb Res 80:193–200. https://doi.org/10.1016/0049-3848(95)00167-P

Schröder T (ed) (2010) Living with the oceans, World Ocean review. Maribus gGmbH, Hamburg

Shahidi F, Santhiravel S (2022) Novel marine bioactives: application in functional foods, nutraceuticals, and pharmaceuticals. J Food Bioactives 19. https://doi.org/10.31665/JFB.2022. 18316

Shin HJ, Jeong HS, Lee H-S, Park S-K, Kim HM, Kwon HJ (2007) Isolation and structure determination of streptochlorin, an antiproliferative agent from a marine-derived Streptomyces sp. 04DH110. J Microbiol Biotechnol 17:1403–1406

Siddharth S, Vittal R (2018) Evaluation of antimicrobial, enzyme inhibitory, antioxidant and cytotoxic activities of partially purified volatile metabolites of marine Streptomyces sp.S2A. Microorganisms 6:72. https://doi.org/10.3390/microorganisms6030072

Smith WC, Xiang L, Shen B (2000) Genetic localization and molecular characterization of the nonS gene required for macrotetrolide biosynthesis in *Streptomyces griseus* DSM40695. Antimicrob Agents Chemother 44:1809–1817. https://doi.org/10.1128/AAC.44.7.1809-1817.2000

Steyn PS (1970) The isolation, structure and absolute configuration of secalonic acid D, the toxic metabolite of Penicillium oxalicum. Tetrahedron 26:51–57. https://doi.org/10.1016/0040-4020 (70)85006-2

Stritzke K, Schulz S, Laatsch H, Helmke E, Beil W (2004) Novel caprolactones from a marine Streptomycete. J Nat Prod 67:395–401. https://doi.org/10.1021/np030321z

Suleria HAR, Gobe G, Masci P, Osborne SA (2016) Marine bioactive compounds and health promoting perspectives; innovation pathways for drug discovery. Trends Food Sci Technol 50: 44–55. https://doi.org/10.1016/j.tifs.2016.01.019

Sung JE, Choi JY, Kim JE, Lee HA, Yun WB, Park JJ, Kim HR, Song BR, Kim DS, Lee CY, Lee HS, Lim Y, Hwang DY (2017) Hepatotoxicity and nephrotoxicity of saponin-enriched extract of *Asparagus cochinchinensis* in ICR mice. Lab Anim Res 33:57. https://doi.org/10.5625/lar.2017. 33.2.57

Swain SS, Paidesetty SK, Padhy RN (2017) Antibacterial, antifungal and antimycobacterial compounds from cyanobacteria. Biomed Pharmacother 90:760–776. https://doi.org/10.1016/j. biopha.2017.04.030

Takahashi A, Kurasawa S, Ikeda D, Okami Y, Takeuchi T (1989) Altemicidin, a new acaricidal and antitumor substance. I. Taxonomy, fermentation, isolation and physico-chemical and biological properties. J Antibiot 42:1556–1561. https://doi.org/10.7164/antibiotics.42.1556

Trisuwan K, Rukachaisirikul V, Sukpondma Y, Phongpaichit S, Preedanon S, Sakayaroj J (2009) Lactone derivatives from the marine-derived fungus Penicillium sp. PSU-F44. Chem Pharm Bull 57:1100–1102. https://doi.org/10.1248/cpb.57.1100

Tyśkiewicz K, Tyśkiewicz R, Konkol M, Rój E, Jaroszuk-Ściseł J, Skalicka-Woźniak K (2019) Antifungal properties of Fucus vesiculosus L. supercritical fluid extract against Fusarium culmorum and Fusarium oxysporum. Molecules 24:3518. https://doi.org/10.3390/ molecules24193518

Vo T-S, Ngo D-H, Kang K-H, Jung W-K, Kim S-K (2015) The beneficial properties of marine polysaccharides in alleviation of allergic responses. Mol Nutr Food Res 59:129–138. https://doi. org/10.1002/mnfr.201400412

Williams TL, Lopez SA, Deravi LF (2019) A sustainable route to synthesize the Xanthommatin biochrome via an electro-catalyzed oxidation of tryptophan metabolites. ACS Sustain Chem Eng 7:8979–8985. https://doi.org/10.1021/acssuschemeng.9b01144

Wu Z, Su X, Xu Y, Kong B, Sun W, Mi S (2016) Bioprinting three-dimensional cell-laden tissue constructs with controllable degradation. Sci Rep 6:24474. https://doi.org/10.1038/srep24474

Zhang Y-A, Zou J, Chang C-I, Secombes CJ (2004) Discovery and characterization of two types of liver-expressed antimicrobial peptide 2 (LEAP-2) genes in rainbow trout. Vet Immunol Immunopathol 101:259–269. https://doi.org/10.1016/j.vetimm.2004.05.005

Zhang W, Krohn K, Zia-Ullah, Flörke U, Pescitelli G, Di Bari L, Antus S, Kurtán T, Rheinheimer J, Draeger S, Schulz B (2008) New mono- and dimeric members of the secalonic acid family: blennolides A–G isolated from the fungus Blennoria sp. Chem Eur J 14:4913–4923. https://doi.org/10.1002/chem.200800035

Zhu C-F, Li G-Z, Peng H-B, Zhang F, Chen Y, Li Y (2010) Treatment with marine collagen peptides modulates glucose and lipid metabolism in Chinese patients with type 2 diabetes mellitus. Appl Physiol Nutr Metab 35:797–804. https://doi.org/10.1139/H10-075

19

Freshwater Blue–Green Algae: A Potential Candidate for Sustainable Agriculture and Environment for the Welfare of Future Planet Earth

Arun Kumar Rai, Binu Gogoi, and Rabina Gurung

Abstract

Different types of microorganisms are known to exist in freshwater habitats. These microbes function similarly to the microorganisms found in soil and air. Freshwater, brackish, marine and terrestrial cyanobacteria (blue–green algae [BGA]) are a diverse group of prokaryotes and are also the most successful and oldest life forms on the planet. They play an important role in maintaining and improving soil fertility, increasing plant growth and yield as a natural biofertilisers, nutrient cycling, nitrogen (N_2) fixation and environmental protection. Cyanobacteria demonstrate the potential for effectively converting light energy into chemical energy. The aim of this chapter is to provide valuable information about the potential role of freshwater cyanobacteria in solving the agricultural and environmental problems on the planet earth.

Keywords

BGA · Nitrogen fixation · Plant growth · Bioremediation · Biofuels

19.1 Introduction

Crop production has mostly relied on conventional agriculture for thousands of years. However, the world is evolving (Mutale-Joan et al. 2022). By 2050, there will be roughly nine billion people on the planet, according to the current growth rates. Abiotic stresses, global climate change and instances of land degradation will all rise in tandem with this (Gr et al. 2021). Producing adequate food for an

---

A. K. Rai (✉) · B. Gogoi · R. Gurung
Department of Botany, School of Life Sciences, Sikkim University, Tadong, Sikkim, India
e-mail: akrai@cus.ac.in

© The Author(s), under exclusive license to Springer Nature Singapore Pte Ltd. 2023
R. Soni et al. (eds.), *Current Status of Fresh Water Microbiology*,
https://doi.org/10.1007/978-981-99-5018-8_19

409

expanding population is one of the issues faced by agriculture today (Iniesta-Pallares et al. 2021). The development of chemicals like synthetic fertilisers and pesticides has significantly increased crop yield over time. However, the widespread use of such chemicals has an adverse effect on the ecology and ecosystem (Mutale-Joan et al. 2022). The continuous increase of human population and growing concerns about energy crisis, food security, disease, global warming and other environmental problems require a sustainable solution from nature, and one of the promising resources is cyanobacteria which need simple materials to grow and have relatively simple genomes (Zahra et al. 2020). Blue–green bacteria, sometimes known as cyanobacteria, are Gram-negative prokaryotes that may grow in photoautotrophic environments. They occupy a unique place in the earth's evolutionary process. In addition to living in other harsh environments, cyanobacteria may survive in deserts and soil that has been impacted by salt. Based on their physical characteristics and molecular approaches, around 150 genera and 2000 species have been discovered (Li et al. 2019). Cyanobacteria are classified as members of the kingdom Monera and the division Cyanophyta (Singh et al. 2005). They are also known as blue–green algae, found in a variety of environments including freshwater, oceans, soil and bare rock (Mehdizadeh Allaf and Peerhossaini 2022). They frequently form vast colonies, are tiny and are typically unicellular. Cyanobacteria are made up of a wide variety of bacteria in various sizes and forms. They can be found in 150 of the currently recognised genera. They resemble the earliest fossils, which date back to more than 3.5 billion years ago (Chittora et al. 2020) and also have an evolutionary significance, as they are necessary for today's oxygenic climate (Awasthi and Singh 2021). Being prokaryotic oxygenic phototrophs, cyanobacteria are resilient to adverse environmental conditions, such as temperature extremes, ultraviolet radiation, drought, salt and quick hydration/dehydration cycles. As a result, they may colonise a wide range of environments (Chamizo et al. 2018). Their cellular structure is simple prokaryotic, and they perform photosynthesis like plants, but without plant cell walls like primitive bacteria. These are also similar to animals in that they have complex sugars on their cell membranes, such as glycogen (Singh et al. 2005). Cyanobacteria are one of the most important organisms to have ever evolved on our planet. Their ancestors developed the ability to oxidize photosynthesis, making it possible to oxidise the earth's atmosphere and oceans (Sánchez-Baracaldo et al. 2022).

19.2 Diversity in Freshwater Ecosystem

Algae predominantly are a group of photosynthetic organisms, having a nucleus like plants but lacking true roots, stems and leaves. Algae are divided into two classes based on cellularity: microalgae and macroalgae. Microalgae are unicellular species and comprises red algae, green algae, brown algae and blue–green algae. Macroalgae are multicellular and commonly known as seaweeds. Based on their pigments, macroalgae are divided as red algae (Rhodophyta), green algae (Chlorophyta) and brown algae (Phaeophyta).

19 Freshwater Blue–Green Algae: A Potential Candidate for... 411

Blue–green algae are also known as cyanobacteria. They are Gram-negative prokaryotes. Because of the presence of the blue pigment phycocyanin, they are able to fix atmospheric nitrogen and help in carbon assimilation. Their structural organisation comprises simple unicellular to complex multicellular forms. They have highly differentiated cell types. The cells of cyanobacteria are differentiated into special cell types that are known as heterocysts. With the aid of this structure, cyanobacteria are able to carry out nitrogen fixation (Waterbury 2006). Photosynthesis is carried out by vegetative cells present in them. Cyanobacteria are present in paddy fields, soil, saline habitats and brackish, fresh and marine waters (Thajuddin and Subramanian 2005). Numerous studies have been conducted on the diversity of blue–green algae from various sources of their existence. Blue–green algae consist of 2000 species in 150 genera.

19.2.1 Classification

Blue–green algae (cyanobacteria) are divided into five orders (Rippka et al. 1979).

1. Chroococcales.
2. Oscillatoriales.
3. Pleurocapsales.
4. Nostocales.
5. Stigonematales.

This classification was based on the observation of morphological characteristics. These orders are further divided into families, subfamilies, genera and species. Blue–green algae are made up of the genera *Anabaena, Aphanizomenon, Calothrix, Cylindrospermopsis, Gloeotrichia, Microcystis, Nostoc, Nodularia, Oscillatoria planktothrix, Phormidium, Scytonema*, etc.

19.3 Role in Agriculture

19.3.1 Biofertilisers (Nitrogen Fixation)

The ongoing increase in human population and the depletion of energy resources pose a threat to the environment's requirements and to the sustainable production of food and energy. Anthropogenic activities can lead to soil barrenness, degradation and salinity, which can reduce the area of arable land and threaten food security. The demand for food like rice increases as the world's population increases, necessitating a greater use of nitrogen fertilisers to boost crop yields (Song et al. 2021). The most environmentally friendly method, or 'green technology', has been used to prepare biofertilisers. The prokaryotic organisms that have evolved and survived the longest are cyanobacteria (Chittora et al. 2020). Biofertilisers are organisms that improve soil nutrient availability for crops (Awasthi and Singh 2021). Cyanobacteria, since they have the ability to fix atmospheric nitrogen, might be utilised as a biofertiliser for the growth of commercially significant crops like rice and beans (Chittora et al. 2020). Cyanobacteria have unique properties of nitrogen fixation in paddy fields, resulting in enhanced rice growth and productivity. The biological nitrogen fixation process is aided by the use of cyanobacteria as biofertilisers, which also increases the organic matter level of the soil. They improve the grain weight, yield, fresh and dry weight, root and shoot length and uptake of micronutrients by the crop (Kalyanasundaram et al. 2020). The agricultural importance of cyanobacteria in rice cultivation is based on their ability to fix nitrogen and other beneficial effects on soil and plants (Zahra et al. 2020). In rice fields, green manure, particularly nitrogen-fixing cyanobacteria (NFC), has been shown to be an environmentally benign N_2 source. NFC inoculation has been proved in studies to greatly boost the nitrogen-fixing capacity in rice fields, hence improving the rice yield (Song et al. 2021). Cyanobacteria called diazotrophs are helpful in producing affordable, conveniently accessible biofertilisers that are friendly to the environment. They can reduce plant nitrogen shortage, aerate the soil, increase soil water retention and supply vitamin B12 (Awasthi and Singh 2021; Chittora et al. 2020). Rice crop cultivation area contains the most effective nitrogen-fixing cyanobacteria, including *Nostoc linckia, Anabaena variabilis, Aulosira fertilisima, Calothrix* sp., *Tolypothrix* sp. and *Scytonema* sp. These species act as biofertilisers, sequestering atmospheric nitrous oxide and allowing it to be converted to ammonium, helping to maintain soil fertility and sustainability over the long term (Zahra et al. 2020). In addition to being more affordable and less noxious to the environment than chemical fertilisers, biofertilisers are also a renewable energy source. Therefore, if we want to boost the productive capability of land on the earth, we must utilise biofertilisers (Tiwari et al. 2022). The use of consortia or biofilms of green algae and cyanobacteria as biofertilisers has recently been shown to be effective (Jemilakshmi 2021).

19.3.2 Soil Fertility Improvement

The soil microbiota are essential for preserving the soil's health, carbon sink and nutritional balance (Abinandan et al. 2019). Cyanobacteria have a wide range of contaminant-digesting abilities and serve a number of roles in the soil ecosystem to maintain soil fertility (Koul and Kumar 2022). Cyanobacteria are not dominant microorganisms in the soil, but they have a significant impact on soil fertility, microbial community and soil productivity (Li et al. 2019). It is undeniably true that cyanobacteria and microalgae play a significant part in preserving the fertility and health of the soil required for long-term agriculture. The balance in the dynamics of nutrients like carbon, nitrogen and phosphate for the yield of agricultural crops is the main benefit of utilising microalgae and cyanobacteria (Abinandan et al. 2019). So, by utilising microbes, researchers have used 'green technology' to create eco-friendly environments. The use of cyanobacteria to boost soil fertility and agricultural yield is covered in a number of green technology articles (Chittora et al. 2020). By storing significant volumes of water, cyanobacteria promote soil stability, decrease water and wind erosion, increase surface moisture and boost soil fertility by fixing carbon (C) and elemental nitrogen (N) (Chamizo et al. 2018). They are advantageous over rivals in numerous ecological niches because they can withstand a variety of stressors, including salt, pH, high or low temperatures and drought. They bind soil particles together, causing soil aggregation, an increase in organic matter and an improvement in the soil's top layer's ability to store water (Mutale-Joan et al. 2022). Through atmospheric nitrogen fixation, phosphate solubilisation and nutrient release, cyanobacteria increase soil fertility and crop yield (Gr et al. 2021). The improvement and reclamation of soil infertility can be facilitated by plant growth promoting rhizobacteria (PGPRs) and cyanobacteria that produce exopolysaccharides (EPS) (Awasthi and Singh 2021). The ability of *Nostoc* sp. to absorb nitrogen in the environment improves soil fertility and plant growth (Maltseva and Maltsev 2021).

19.3.3 Wasteland Reclamation

A sizeable amount of land on the earth is classified as wastelands. They are deteriorated and underutilised lands as a result of many factors. They include places impacted by waterlogging, riverine lands, shifting cultivation, degraded forest land, sandy area and salt-affected land. One of the most pressing issues today is the recovery of natural ecosystems in areas disturbed by human activities (Maltseva and Maltsev 2021). Global warming is contributing to an increase in the desertification of soil, and water shortage makes traditional restoration techniques ineffective (Roncero-Ramos et al. 2022). Due to the full loss of topsoil and the ecological instability of these wastelands, they are unsuited for farming. Salinity is a significant issue among the causes of wastelands and may be brought on by improper agricultural practises, waterlogging and mining (Devi et al. 2021). Soil deterioration and salinisation are caused by both natural and human-made processes. The security of our food supply is threatened by the loss of cropland. Around one billion acres of salt-affected soils can be restored using chemical, physical and biological methods. Exceptional abilities of *Nostoc, Anabaena* and other cyanobacterial species include the capacity to fix nitrogen from the atmosphere, build an extracellular matrix and synthesise suitable solutes (Li et al. 2019). Therefore, a substitute technique based on cyanobacteria that create biocrusts on deteriorated soils has arisen. Communities of mosses, lichens, cyanobacteria or fungi called biocrusts invade the soil's surface and create a permanent and fertile layer (Roncero-Ramos et al. 2022). Since native cyanobacterial strains have developed an innate resilience to salt and osmotic stress from repeated exposure to such stress over time, using them as biofertilisers for the restoration of salt-affected soil is anticipated to be more effective. Utilising native cyanobacterial strains that are osmo- and halotolerant and have high levels of stress tolerance and adaptations as biofertilisers will help add organic matter, promote the development of other microbial communities, maintain nutrient cycles, reduce soil erosion by improving soil aggregation and structural stability and facilitate increased crop yield (wheat and pearl millet) under water scarcity (Nisha et al. 2018). Species from the genus *Nostoc* are frequently listed as dominants among cyanobacteria. It is proof that the closest steppe phytocenoses and agrocenoses have a substantial impact on group formation in revegetation habitats, where cyanobacteria typically play a key role (Maltseva and Maltsev 2021). There are large expanses of unproductive usar lands in many Indian states, such as Uttar Pradesh (12.95 lakh hectares) and Rajasthan (12.14 lakh hectares). Salinity-affected or alkaline lands are considered to be usar soils. Thus, the need to reclaim such barren terrain has been recognised in order to meet the growing need for food from a growing world population. Usar soil can be reclaimed using a variety of techniques that lower its salt content, including irrigation with freshwater, gypsum treatment, the growing of crops that can withstand salt and the use of cyanobacteria (Rai et al. 2019). By promoting plant growth and development and boosting soil fertility in unfavourable environmental conditions, algae are crucial to the rehabilitation of soil. Use of algae in soil reclamation is not only economical, but it will also improve the soil fertility status and plant growth conditions under adverse environmental stresses (Kalyanasundaram et al. 2020).

19.3.4 Biocontrol

The majority of chemical pesticides are extremely hazardous to plants and should not be used. These chemical pesticides are dangerous for soil and plants to use because they cause crops to accumulate a variety of harmful toxins. By producing hydrolytic enzymes such as carboxymethyl cellulase (CMCase), chitosanase and 1,3-endoglucanase, cyanobacteria inhibit the growth of phytopathogens (Kalyanasundaram et al. 2020). The development of hydrolytic chemicals and biocidal mixtures, such as benzoic acid and majusculonic acid (Renuka et al. 2018), has resulted in microalgae, particularly cyanobacteria, being considered as potential biocontrol agents because of their antagonistic effects against a variety of plant pathogens, such as microscopic organisms, parasites and nematodes. Microalgae and cyanobacteria contain a wide range of bioactive chemicals; therefore, using them (or their extracts) can help crops be adequately protected from both biotic and abiotic influences (Gonçalves 2021). The existence of organisms like insects, nematodes, bacteria and fungi, among others, can have a detrimental effect on the productivity of crops (Pan et al. 2019). Polysaccharides are frequently used to combat these pathogenic organisms because they recognise the signalling molecules found in the pathogenic organism cell wall and trigger a number of defensive reactions (Chanda et al. 2019). Regulation of signalling pathways, gene expression and induction of particular biosynthetic pathways are examples of common defence reactions. These actions typically produce secondary metabolites with antioxidant, antimicrobial and fungicidal properties (such as phenolic compounds, terpenoids and other compounds) (Renuka et al. 2018; Farid et al. 2019). Microalgae and cyanobacteria, which are both used in agricultural crops and are a major source of polysaccharides, may help to strengthen these defence mechanisms (Chanda et al. 2019). Indeed, the induction of plant defence mechanisms in the presence of microalgae and cyanobacteria has been described previously. For example, inoculating spice crop seeds with the cyanobacteria *Anabaena laxa* and *Calothrix elenkinii* significantly increased the activity of the enzyme 1,3-endoglucanase in shoots and roots, which is responsible for the breakdown of pathogen cell wall components. The authors also reported an increase in fungicidal activity, plant dry weight and shoot and root length (Kumar et al. 2013). Increased activity of plant defence enzymes is linked to the activation of metabolic pathways that result in the production of thousands of metabolites, such as phenolic compounds, which are toxic to pathogenic organisms. For example, phenylalanine ammonia-lyase causes an increase in the phenylpropanoid pathway, which produces phytoalexins, phenolic substances with antimicrobial and antioxidant properties (Chanda et al. 2019). The seaweeds *Spatoglossum variabile, Stokeyia indica* and *Melanothamnus afaqhusainii* were found to have a significant suppressive effect against eggplant and watermelon pathogens (e.g. the root-rotting fungi *Fusarium solani, Fusarium oxysporum* and *Macrophomina phaseolina* and the root-knot nematode *Meloidogyne* spp.). Apart from the inhibitory effect on these organisms, the authors reported improved plant growth (increased fresh weight of shoots, vine length in watermelon and shoot lengths in eggplant) (Baloch et al. 2013). In addition, the cyanobacterium *Anabaena minutissima* provides bioactive substances such polysaccharides and phycobiliproteins (PBPs). Pre-harvest treatments with polysaccharides on straw-berry fruits decreased sporulation and symptoms of *Botrytis cinerea* by 50% and 67%, respectively (Righini et al. 2022).

19.3.5 Crop Productivity Enhancement

The physiologically active chemicals produced by microalgae and cyanobacteria are a key source of molecules that can greatly increase agricultural productivity. It is possible to use microalgae and cyanobacteria as pure extracts or as a crude algal compost due to the large variety of substances they contain and the several roles they play in crop production (Gonçalves 2021). It has already been established that cyanobacteria have advantageous impacts in agricultural soils for the growth of various crops. For instance, applying a cyanobacterial inoculum in a rice paddle boosted the yields of grain and straw while also increasing the soil's ability to hold onto nitrogen (Jha and Prasad 2006). The strains of *Anabaena variabilis* encouraged an overall improvement in the rice crops by using them in a rice crop field (e.g. increase in plant height and leaf length, improvement of seeds, grain and straw productivity) (Singh and Datta 2007). Despite the fact that the majority of studies focus on the utilisation of cyanobacteria in rice fields, more recent research has also highlighted the benefits of cyanobacteria in other cultures. After introducing the cyanobacterial species *Nostoc entophytum* and *Oscillatoria angustissima* to the soil used to cultivate pea plants, there was a considerable improvement in the germination rate, photosynthetic pigments and growth characteristics of this plant. The authors also noted that both cyanobacteria had greater N_2-fixing activity in correlation with this increase (Osman et al. 2010). Using a biofilm made by the cyanobacterium *Anabaena torulosa* in a wheat crop considerably enhanced the amount of nitrogen that was available in the soil (Swarnalakshmi et al. 2013). Consequently, the addition of microalgae and cyanobacteria to soils can increase the availability of macro- and micronutrients for plant growth (Gonçalves 2021).

19.4 Role in Environment

19.4.1 Bioremediation

The health of the ecosystem has been severely harmed by the release of untreated sewage effluents into the environment, which has led to an alarming rise in pollution. A novel and efficient method for removing both organic and inorganic toxins is to use cyanobacteria for bioremediation. Cyanobacteria are effective at combating with various pollutants due to their capacity to fix atmospheric nitrogen and their special ability to adapt to adverse environmental circumstances (Cui et al. 2022). Because cyanobacteria exopolysaccharides are anionic, they have the ability to remediate heavy metals (Potnis et al. 2021). A study showed how *Spirulina* sp. was highly tolerant and helpful in biosorption of heavy metals like cobalt and chromium (Murali et al. 2014). Bioremediation of phenol, which is another harmful organic pollutant discharged by industries and is known to cause severe health issues, was carried out using *Leptolyngbya* sp. (Guha Thakurta et al. 2018). Bacteria and cyanobacteria are capable of cleaning up hydrocarbon-contaminated environment (Kumar et al. 2016). A study on heavy metal removal by *Cyanothece* and *Cyanospira capsulata* strains showed high metal adsorption (De Philippis et al. 2001). Some cases have demonstrated how a consortium of green algae such as *Ankistrodesmus* sp. and cyanobacteria such as *Nostoc* and *Anabaena* were used to remove heavy metals such as iron (Fe), copper (Cu), lead (Pb), chromium (Cr) and zinc (Zn) through biosorption and bioaccumulation (Iqbal et al. 2022).

Cyanobacteria also have the capability for soil bioremediation. They are known to increase the soil organic content. When cyanobacteria are added to the soil, they bind to the soil particles, resulting in soil impermeability and improved aeration. Cyanobacteria and microalgae, when compared to other species, are more environment-friendly, have large cell sizes and generate extra biomass, making them the greatest and most sustainable solution to the problems with soil fertility and accessible water resources. Remediation by microalgae and cyanobacteria may be a harmless and long-term solution for removing heavy metals found in municipal sewage water and for assisting in the breakdown of contaminants in industrial waste water (Koul et al. 2022).

19.4.2 Carbon Dioxide (CO_2) Sequestration

The rise in the earth's temperature, because of the increase of CO_2 gas in the atmosphere, is a major cause of global warming. The disruption in the amount of

Table 19.1 Blue–green algae species contributing to CO_2 sequestration and fixation

Microalgae/cyanobacteria	Application	References
Limnothrix redekei, Planktolyngbya crassa	CO_2 sequestration	Manjre and Deodhar (2013)
Leptolyngbya tenuis, Chlorella ellipsoidea	CO_2 fixation	Satpati and Pal (2021)
Thermosynechococcus sp. *Nannochloropsis* sp.	CO_2 fixation	Hsueh et al. (2009)
Desmodesmus sp.	CO_2 tolerance and assimilation	Xie et al. (2014)
Oscillatoria sp.	CO_2 capture	Anguselvi et al. (2019)

CO_2 gas majorly affects all the biogeochemical cycles. Higher amount of CO_2 emitted in the atmosphere due to the burning of fossil fuels must be removed. This can be achieved by carbon sequestration that is the method of holding and storing carbon dioxide present in air (Gayathri et al. 2021). This is carried out through biological, chemical and physical processes. Carbon sequestration helps in maintaining the ecological balance by placing carbon from its origin. The amount of fossil fuel emissions is related to carbon sequestration. If the amount of fossil fuels is limited, there will not be any need to go for carbon sequestration. An efficient way by which carbon sequestration is accomplished is through photosynthetic organisms. Such microorganisms that can carry out carbon sequestration are able to withstand a high temperature. Microalgae have the ability to convert CO_2 present in the atmosphere and the emitted CO_2 from flue gas, thus reducing greenhouse gases. The application of microalgae has elevated the CO_2 sequestration in a sustainable way (Cheah et al. 2015) (Table 19.1).

Literature studies have shown that microalga-based carbon sequestration is quite beneficial over other methods as it involves quick biomass production, a high rate of photosynthetic conversion for bioremediation and the ability to produce various value-added products. Other conventional methods of carbon sequestration are high power driven and are of high cost. As microalgae grow in aquatic habitats, they are able to capture CO_2 gas passing through them. Microalgae under light are able to take in solar energy and assimilate CO_2 based on the air and nutrients present in the aquatic habitat (Anguselvi et al. 2019). Microalgae utilise the carbon dioxide present in the atmosphere for their growth by converting it into glucose (Ho et al. 2014). Photosynthesis involves two reactions: light and dark reactions. Microalgae in the first stage of photosynthesis, which is a light-dependent reaction, captures the energy so obtained from the sunlight, thereby transforming it into adenosine triphosphate (ATP) and nicotinamide adenine dinucleotide phosphate (NADPH) to energy carriers ATP and NADPH. Besides their role in ATP and NADPH formation, microalgae also capture CO_2-producing organic compounds in the light-independent stage. This ability of microalgae in capturing and storing CO_2 for bioconversion makes them an excellent choice for carbon sequestration (Zhao and Su 2014).

By using microalgae for carbon sequestration, greenhouse gases can be moderated. In addition to this, microalgae are also known to produce a variety of products like biofuels. There is an urgent need to reduce carbon emissions so that carbon sequestration does not have to be considered.

19.4.3 Biofuels

Algae are an alluring source for various bioproducts in agriculture, pharmaceuticals and other industries, particularly as biofuels. Blue–green algae (cyanobacteria) can convert solar energy and carbon dioxide into biofuels. They are also able to produce biofuels because of the high fatty acid and lipid content (Quintana et al. 2011). Among the members of cyanobacteria, the highest lipid content is present in *Oscillatoriales.* Cyanobacteria produce biofuels in the presence of alcohol as an acyl acceptor and a catalyst. According to Mata et al. (2010), the two different ways in which cyanobacteria produce biofuels are excess production and transesterification of fatty acids to produce fatty acid methyl esters.

The ability of microalgae in growing fast either on brackish or on saline water as compared to other plants makes them a good option to be considered as a renewable biofuel (Dismukes et al. 2008). Due to their numerous beneficial characteristics, cyanobacteria have a high potential for producing cyanodiesel. Some of the advantages are listed below: (Sarsekeyeva et al. 2015).

1. Cyanobacteria do not claim farmlands.
2. Cyanobacteria growth gathers biomass quickly.
3. They have the ability to fix atmospheric CO_2.
4. They have less requirement of elements like sunlight, water and a few inorganic trace elements.
5. They serve as a medium for genetic modification of metabolic pathways.

By utilising cyanobacteria for the production of biofuels, we can mitigate issues concerning the dependence, depletion and overuse of oil and gas reserves.

19.4.4 Methane (CH_4) Oxidation Augmentation

Methane (CH_4) is one of the chief constituents of natural gas, and it also accounts for the most potent greenhouse gas. It is the simplest among the members of the paraffin series of hydrocarbons. The major sources of methane are from geological deposits and coal bed methane extraction. Rivers, lakes and gas hydrates are some other sources of atmospheric methane gas. Another contributor of CH_4 is the methanogenesis in anaerobic flooded paddy soils. The amount of CH_4 increases also from the burning of fossil fuels and excessive waste generation. The application of cyanobacteria is a sustainable approach for limiting the amount of CH_4 generated (Cuellar-Bermudez et al. 2017; Singh et al. 2016).

19 Freshwater Blue–Green Algae: A Potential Candidate for...

Cyanobacteria are known to increase the amount of oxygen in the rhizosphere region of paddy. This increases the methanotrophs to take in the methane. These microbes carry out N_2 fixation, which aids in combating global warming. By releasing oxygen into the flooded soils that are unsuitable for methanogenesis, cyanobacteria generate an aerobic environment during this process. By boosting methanotroph activity and population, the oxygen supplied also promotes methane oxidation (Singh et al. 2011; Singh and Pandey 2013).

19.4.5 Food Supplements

The demand for food supplements made from microbes has increased due to the increase in global population. Many blue–green algae species like *Arthrospira platensis*, *Aphanizomenon flosaquae*, *Oscillatoria funiformis*, *Limnospira maxima* and *Nostoc sphaeroides* have been consumed by humans since a long time (Mutoti et al. 2022). Due to their great nutritional value, such blue–green algae species are available as food supplements in the form of pills, capsules and liquids. Blue–green algae are considered to be both food and feed supplements. Some species of blue–green algae like *Spirulina* are excellent food sources. *Spirulina* is a microscopic, filamentous cyanobacterium that grows in alkaline water bodies. It was described as the best food for the future by the United Nations World Food Conference in 1974. The food and feed supplement of *Spirulina* are made mainly from two species of cyanobacteria: *Arthrospira platensis* and *Arthrospira maxima*. Their dried biomass is produced on a large scale due to the abundant nutrient content (Shinde et al. 2022).

The biological components of blue–green algae include minerals, like sodium, potassium, calcium, iron, manganese, selenium, magnesium and zinc, and roughly 70–80% protein content, polysaccharides, polyunsaturated fatty acids, carotenoid colour and the presence of vitamins. The other constituents of spirulina include lipid, protein and chlorophyll, which have high antioxidant properties. Another important constituent found in *Spirulina* sp. is gamma-linoleic acid, which is known for its various health benefits like boosting hair growth, strengthening bone and maintaining the reproductive system (Raja et al. 2016).

Blue–green algae's anti-inflammatory and antioxidant properties have been found to stimulate neurological development, prevent nutrient deficiencies and protect against an overactive immune system. In addition to this, blue–green algae have also been known to maintain cholesterol level and are known to possess numerous dietary benefits boosting the immune system (Al-Thawadi 2018).

In a study by Appel et al. (2018), two commercial extract products of spirulina, Immulina® and immunLoges®, showed anti-inflammatory properties by inhibiting the immunoglobulin E (IgE)–antigen complex-induced production of TNFα, histamine, interleukin-4 (IL-4) and leukotrienes by also releasing histamine from mast cells.

19.5 Climate Change and Its Impacts on Freshwater Cyanobacteria Species

Globally, the effects of climate change have profoundly changed ecosystems, with freshwater lakes being especially sensitive because of their propensity to absorb atmospheric and terrestrial reactions. Northern latitudes have seen the fastest rates of change as several locations have warmed to previously unheard of levels. In reaction to increasing temperatures, northern latitude lakes are now experiencing 'longer summers' (warmer and longer ice-free situations) and hydrological intensification (wetter climates) (Erratt et al. 2022). Among the physical responses of lakes to climate change include changes in mixing regimes, loss of ice cover, water budget changes and increasing surface water temperatures. Fish deaths brought on by hypoxia in shallow lakes, the growth of cyanobacteria and modifications to the energy and carbon cycling are only a few of the effects of climate change (Cremona et al. 2022). A greater proportion of cyanobacteria was found in the mat-forming algae specifically as a result of warming and even more so as a result of warmth plus acidity. Only around 25% of the benthic cover was made up of cyanobacteria under control conditions; however, heat and acidity increased this proportion to over 75% while decreasing the amount of edible algae (Hofer 2018). The two most common climatic variables are temperature and precipitation, which have been investigated as potential drivers of cyanobacteria bloom production (Erratt et al. 2022). Toxic cyanobacteria (cyanobacterial harmful algal blooms [cyanoHABs]), which favour warmer temperatures and thrive under increasingly intense oscillations in the wet/dry cycle, have been particularly affected by global warming and have seen an increase in their species that are associated with harmful algal blooms (Zeppernick et al. 2022).

19.6 Future Prospects and Constraints

Cyanobacteria are a resource with potential for use in many different fields. They are an appealing renewable source of bioactive substances and other fine chemicals because their production has a generally minimal negative influence on the environment. However, due to the high degree of variety seen in cyanobacteria in natural ecosystems, this field of microorganisms is currently understudied. In particular, there are no genetic improvement studies on cyanobacteria with a view to eco-sustainable agriculture. Large-scale production and commercialisation of blue–green algae are required as they have high applicability and they can be used in the development of various products. Therefore, more attempts are required to study and screen blue–green algae species that show major potential traits of bioremediation as it has a high market potential in the coming few years.

References

Abinandan S, Subashchandrabose SR, Venkateswarlu K, Megharaj M (2019) Soil microalgae and cyanobacteria: the biotechnological potential in the maintenance of soil fertility and health. Crit Rev Biotechnol 39(8):981–998

Al-Thawadi S (2018) Public perception of algal consumption as an alternative food in the Kingdom of Bahrain. Arab J Basic Appl Sci 25(1):1–12

Anguselvi V, Masto RE, Mukherjee A, Singh PK (2019) CO2 capture for industries by algae. In: Algae. IntechOpen, London

Appel K, Munoz E, Navarrete C, Cruz-Teno C, Biller A, Thiemann E (2018) Immunomodulatory and inhibitory effect of Immulina®, and Immunloges® in the Ig-E mediated activation of RBL-2H3 cells. A new role in allergic inflammatory responses. Plan Theory 7(1):13

Awasthi A, Singh MDP (2021) Cyanobacteria: a source of bio fertilizers for sustainable agriculture. Int J Mod Agric 10(2):4117–4122

Baloch GN, Tariq S, Ehteshamul-Haque S, Athar M, Sultana V, Ara J (2013) Management of root diseases of eggplant and watermelon with the application of asafoetida and seaweeds. J Appl Bot Food Qual 86(1):138–142

Chamizo S, Rodríguez-Caballero E, Cantón Y, De Philippis R (2018) Soil inoculation with cyanobacteria: reviewing its' potential for agriculture sustainability in drylands. Agric Res Technol 18(556046):10–19080

Chanda MJ, Merghoub N, El Arroussi H (2019) Microalgae polysaccharides: the new sustainable bioactive products for the development of plant bio-stimulants? World J Microbiol Biotechnol 35(11):1–10

Cheah WY, Show PL, Chang JS, Ling TC, Juan JC (2015) Biosequestration of atmospheric CO2 and flue gas-containing CO2 by microalgae. Bioresources 184:190

Chittora D, Meena M, Barupal T, Swapnil P, Sharma K (2020) Cyanobacteria as a source of biofertilizers for sustainable agriculture. Biochem Biophys Rep 22:100737

Cremona F, Öglü B, McCarthy MJ, Newell SE, Nõges P, Nõges T (2022) Nitrate as a predictor of cyanobacteria biomass in eutrophic lakes in a climate change context. Sci Total Environ 818: 151807

Cuellar-Bermudez SP, Aleman-Nava GS, Chandra R, Garcia-Perez JS, Contreras-Angulo JR, Markou G et al (2017) Nutrients utilization and contaminants removal. A review of two approaches of algae and cyanobacteria in wastewater. Algal Res 24:438–449

Cui H, Zhu X, Zhu Y, Huang Y, Chen B (2022) Ecotoxicological effects of DBPs on freshwater phytoplankton communities in co-culture systems. J Hazard Mater 421:126679

De Philippis R, Sili C, Paperi R, Vincenzini M (2001) Exopolysaccharide-producing cyanobacteria and their possible exploitation: a review. J Appl Phycol 13(4):293–299

Devi S, Rani N, Sagar A (2021) Bioreclamatory studies on salt affected soil by using cyanobacterial biofertilizers. Plant Arch 21(2):416–422

Dismukes GC, Carrieri D, Bennette N, Ananyev GM, Posewitz MC (2008) Aquatic phototrophs: efficient alternatives to land-based crops for biofuels. Curr Opin Biotechnol 19(3):235–240

Erratt K, Creed IF, Favot EJ, Todoran I, Tai V, Smol JP, Trick CG (2022) Paleolimnological evidence reveals climate-related preeminence of cyanobacteria in a temperate meromictic lake. Can J Fish Aquat Sci 79(4):558–565

Farid R, Mutale-Joan C, Redouane B, Mernissi Najib EL, Abderahime A, Laila S, Hicham ELA (2019) Effect of microalgae polysaccharides on biochemical and metabolomics pathways related to plant defense in Solanum lycopersicum. Appl Biochem Biotechnol 188(1):225–240

Gayathri R, Mahboob S, Govindarajan M, Al-Ghanim KA, Ahmed Z, Al-Mulhm N et al (2021) A review on biological carbon sequestration: a sustainable solution for a cleaner air environment, less pollution and lower health risks. J King Saud Univ Sci 33(2):101282

Gonçalves AL (2021) The use of microalgae and cyanobacteria in the improvement of agricultural practices: a review on their biofertilising, biostimulating and biopesticide roles. Appl Sci 11(2): 871

Gr S, Yadav RK, Chatrath A, Gerard M, Tripathi K, Govindsamy V, Abraham G (2021) Perspectives on the potential application of cyanobacteria in the alleviation of drought and salinity stress in crop plants. J Appl Phycol:1–18

Guha Thakurta S, Aakula M, Chakrabarty J, Dutta S (2018) Bioremediation of phenol from synthetic and real wastewater using *Leptolyngbya* sp.: a comparison and assessment of lipid production. *3*. Biotech 8(4):1–10

Ho SH, Chan MC, Liu CC, Chen CY, Lee WL, Lee DJ, Chang JS (2014) Enhancing lutein productivity of an indigenous microalga Scenedesmus obliquus FSP-3 using light-related strategies. Bioresour Technol 152:275–282

Hofer U (2018) Climate change boosts cyanobacteria. Nat Rev Microbiol 16:122–123

Hsueh HT, Li WJ, Chen HH, Chu H (2009) Carbon bio-fixation by photosynthesis of *Thermo synechococcus* sp. CL-1 and *Nannochloropsis oculta*. J Photochem Photobiol B Biol 95(1): 33–39

Iniesta-Pallarés M, Álvarez C, Gordillo-Cantón FM, Ramírez-Moncayo C, Alves-Martínez P, Molina-Heredia FP, Mariscal V (2021) Sustaining rice production through biofertilization with N2-fixing cyanobacteria. Appl Sci 11(10):4628

Iqbal J, Javed A, Baig MA (2022) Heavy metals removal from dumpsite leachate by algae and cyanobacteria. Biorem J 26(1):31–40

Jemilakshmi TV (2021) The inherentpotential of algae for forthcoming future: a comprehensive review. Ann Romanian Soc Cell Biol 5:12452–12462

Jha MN, Prasad AN (2006) Efficacy of new inexpensive cyanobacterial biofertilizer including its shelf-life. World J Microbiol Biotechnol 22(1):73–79

Kalyanasundaram GT, Ramasamy A, Rakesh S, Subburamu K (2020) Microalgae and cyanobacteria: role and applications in agriculture. In: Applied algal biotechnology. Nova Science, USA, Hauppauge

Koul A, Kumar R (2022) A review on diatom flora and cyanobacteria from fresh water. Int J Sci Res (IJSR) 11:34–36

Koul B, Yakoob M, Shah MP (2022) Agricultural waste management strategies for environmental sustainability. Environ Res 206:112285

Kumar M, Prasanna R, Bidyarani N, Babu S, Mishra BK, Kumar A et al (2013) Evaluating the plant growth promoting ability of thermotolerant bacteria and cyanobacteria and their interactions with seed spice crops. Sci Hortic 164:94–101

Kumar BNP, Mahaboobi S, Satyam S (2016) Cyanobacteria: a potential natural source for drug discovery and bioremediation. J Ind Pollut Control 32:508–517

Li H, Zhao Q, Huang H (2019) Current states and challenges of salt-affected soil remediation by cyanobacteria. Sci Total Environ 669:258–272

Maltseva IA, Maltsev YI (2021) Diversity of cyanobacteria and algae in dependence to forest-forming tree species and properties rocks of dump. Int J Environ Sci Technol 18(3):545–560

Manjre SD, Deodhar MA (2013) Screening of thermotolerant microalgal species isolated from Western Ghats of Maharashtra, India for CO2 sequestration. J Sustain Energy Environ 4:61–67

Mata TM, Martins AA, Caetano NS (2010) Microalgae for biodiesel production and other applications: a review. Renew Sust Energ Rev 14:217–232

Mehdizadeh Allaf M, Peerhossaini H (2022) Cyanobacteria: model microorganisms and beyond. Microorganisms 10(4):696

Murali O, Shaik G, Mehar SK (2014) Assessment of bioremediation of cobalt and chromium using cyanobacteria. Ind J Fund Appl Life Sci 4(1):252–255

Mutale-Joan C, Sbabou L, Hicham EA (2022) Microalgae and cyanobacteria: how exploiting these microbial resources can address the underlying challenges related to food sources and sustainable agriculture: a review. J Plant Growth Regul 42:1–20

Mutoti M, Gumbo J, Jideani AIO (2022) Occurrence of cyanobacteria in water used for food production: a review. Phys Chem Earth Parts A/B/C 125:103101

Nisha R, Kiran B, Kaushik A, Kaushik CP (2018) Bioremediation of salt affected soils using cyanobacteria in terms of physical structure, nutrient status and microbial activity. Int J Environ Sci Technol 15(3):571–580

Osman MEH, El-Sheekh MM, El-Naggar AH, Gheda SF (2010) Effect of two species of cyanobacteria as biofertilizers on some metabolic activities, growth, and yield of pea plant. Biol Fertil Soils 46(8):861–875

Pan S, Jeevanandam J, Danquah MK (2019) Benefits of algal extracts in sustainable agriculture. In: Grand challenges in algae biotechnology. Springer, Cham, pp 501–534

Potnis AA, Raghavan PS, Rajaram H (2021) Overview on cyanobacterial exopolysaccharides and biofilms: role in bioremediation. Rev Environ Sci Biotechnol 20(3):781-794.0

Quintana N, Van der Kooy F, Van de Rhee MD, Voshol GP, Verpoorte R (2011) Renewable energy from cyanobacteria: energy production optimization by metabolic pathway engineering. Appl Microbiol Biotechnol 91(3):471–490

Rai AN, Singh AK, Syiem MB (2019) Plant growth-promoting abilities in cyanobacteria. In: Cyanobacteria. Academic Press, London, pp 459–476

Raja R, Hemaiswarya S, Ganesan V, Carvalho IS (2016) Recent developments in therapeutic applications of cyanobacteria. Crit Rev Microbiol 42(3):394–405

Renuka N, Guldhe A, Prasanna R, Singh P, Bux F (2018) Microalgae as multi-functional options in modern agriculture: current trends, prospects and challenges. Biotechnol Adv 36(4):1255–1273

Righini H, Francioso O, Martel Quintana A, Roberti R (2022) Cyanobacteria: a natural source for controlling agricultural plant diseases caused by fungi and oomycetes and improving plant growth. Horticulturae 8(1):58

Rippka R, Deruelles J, Waterbury JB, Herdman M, Stanier RY (1979) Generic assignments, strain histories and properties of pure cultures of cyanobacteria. Microbiology 111(1):1–61

Roncero-Ramos B, Román JR, Acién G, Cantón Y (2022) Towards large scale biocrust restoration: producing an efficient and low-cost inoculum of N-fixing cyanobacteria. Sci Total Environ 848: 157704

Sánchez-Baracaldo P, Bianchini G, Wilson JD, Knoll AH (2022) Cyanobacteria and biogeochemical cycles through Earth history. Trends Microbiol 30(2):143–157

Sarsekeyeva F, Zayadan BK, Usserbaeva A, Bedbenov VS, Sinetova MA, Los DA (2015) Cyanofuels: biofuels from cyanobacteria. Reality and perspectives. Photosynth Res 125(1): 329–340

Satpati GG, Pal R (2021) Co-cultivation of Leptolyngbya tenuis (cyanobacteria) and Chlorella ellipsoidea (green alga) for biodiesel production, carbon sequestration, and cadmium accumulation. Curr Microbiol 78(4):1466–1481

Shinde S, Bhosale M, Tambe T, Sonawane P (2022) Overall review on therapeutic effects of spirulina supplement. Res J Sci Technol 14(2):115–120

Singh S, Datta P (2007) Outdoor evaluation of herbicide resistant strains of Anabaena variabilis as biofertilizer for rice plants. Plant Soil 296(1):95–102

Singh JS, Pandey VC (2013) Fly ash application in nutrient poor agriculture soils: impact on methanotrophs population dynamics and paddy yields. Ecotoxicol Environ Saf 89:43–51

Singh S, Kate BN, Banerjee UC (2005) Bioactive compounds from cyanobacteria and microalgae: an overview. Crit Rev Biotechnol 25(3):73–95. https://doi.org/10.1080/07388550500248498

Singh JS, Singh DP, Dixit S (2011) Cyanobacteria: an agent of heavy metal removal. Bioremediation of pollutants. IK International Publisher, New Delhi, pp 223–243

Singh JS, Kumar A, Rai AN, Singh DP (2016) Cyanobacteria: a precious bio-resource in agriculture, ecosystem, and environmental sustainability. Front Microbiol 7:529

Song X, Zhang J, Peng C, Li D (2021) Replacing nitrogen fertilizer with nitrogen-fixing cyanobacteria reduced nitrogen leaching in red soil paddy fields. Agric Ecosyst Environ 312: 107320

Swarnalakshmi K, Prasanna R, Kumar A, Pattnaik S, Chakravarty K, Shivay YS, Singh R, Saxena AK (2013) Evaluating the influence of novel cyanobacterial biofilmed biofertilizers on soil fertility and plant nutrition in wheat. Eur J Soil Biol 55:107–116

Thajuddin N, Subramanian G (2005) Cyanobacterial biodiversity and potential applications in biotechnology. Curr Sci 89:47–57

Tiwari R, Para P, Sharma A, Singh R, Upadhyay S (2022) Biofertilizer as prospective input for sustainable agriculture in India: a review. Pharma Innov J 11(3):1811–1816

Waterbury JB (2006) The cyanobacteria—isolation, purification and identification. Prokaryotes 4: 1053–1073

Xie YP, Ho SH, Chen CY, Chen CNN, Liu CC, Ng IS et al (2014) Simultaneous enhancement of CO2 fixation and lutein production with thermo-tolerant Desmodesmus sp. F51 using a repeated fed-batch cultivation strategy. Biochem Eng J 86:33–40

Zahra Z, Choo DH, Lee H, Parveen A (2020) Cyanobacteria: review of current potentials and applications. Environments 7(2):13

Zeppernick BN, Wilhelm SW, Bullerjahn GS, Paerl HW (2022) Climate change and the aquatic continuum: a cyanobacterial comeback story: mini review. Environ Microbiol Rep. https://doi.org/10.1111/1758-2229.13122

Zhao B, Su Y (2014) Process effect of microalgal-carbon dioxide fixation and biomass production: a review. Renew Sust Energ Rev 31:121–132

Factors Affecting Fish Migration

20

Uddesh Ramesh Wanjari, Anirban Goutam Mukherjee, Abilash Valsala Gopalakrishnan, Reshma Murali, Sandra Kannampuzha, and D. S. Prabakaran

Abstract

Multiple environmental and endogenous elements work together to cause fish migrations. Abiotic and biotic environmental conditions, such as seasonal variations in temperature and photoperiod, changes in the number of food sources, and the behavior of conspecifics, are all known to impact migration timing. The size, age, and sex of fish and energy reserves are endogenous variables affecting the time of migration. For instance, temperature and photoperiod influence juvenile salmonid movements, which are further altered by endogenous variables such as size and metabolic rate. In order to survive, thrive, and depend on the river flow regime, particularly the time and predictability of flows for spawning and raising their young, fish have evolved features and life history characteristics. It is unknown how much environmental factors affect migratory behavior, such as pulsed flows from hydropower plants. Research on some fish species' exceptional migrations is still ongoing. The adaptive benefit of adopting higher productive habitats for growth is probably how diadromy arose. This chapter explains all the possible aspects and the factors affecting fish migration.

U. R. Wanjari · A. G. Mukherjee · A. V. Gopalakrishnan (✉) · R. Murali · S. Kannampuzha
Department of Biomedical Sciences, School of Bio-Sciences and Technology, Vellore Institute of Technology, Vellore, Tamil Nadu, India
e-mail: abilash.vg@vit.ac.in

D. S. Prabakaran (✉)
Department of Radiation Oncology, College of Medicine, Chungbuk National University, Cheongju, Republic of Korea

Department of Biotechnology, Ayya Nadar Janaki Ammal College (Autonomous), Sivakasi, India
e-mail: prababio@chungbuk.ac.kr

© The Author(s), under exclusive license to Springer Nature Singapore Pte Ltd. 2023
R. Soni et al. (eds.), *Current Status of Fresh Water Microbiology*,
https://doi.org/10.1007/978-981-99-5018-8_20

Keywords

Ecological barrier · Factors · Fish · Migration

20.1 Introduction

Many organisms exhibit migration behaviors (Dingle 2014). Animals migrate for the same reason humans do: to maximize an individual's chances of growth and survival (Dingle and Drake 2007). Individuals must constantly adjust their metabolic demands for somatic growth, maturation, and reproduction with the energy available throughout their lives (Zera and Harshman 2001). Differences in food availability between marine and freshwater habitats are hypothesized to have led to the evolution of diadromy, migration between these two types of environments (Gross et al. 1988). All species in the Salmonidae family spawn in freshwater, but many are anadromous, meaning that individuals move to the sea at some point to take advantage of the more abundant food supplies (Jonsson and Jonsson 1993). It is common for salmonid population to include both permanent freshwater residents and seasonal marine migrants (Chapman et al. 2012). Salmonids' body mass positively correlates with their reproductive success (Elliott 1995). If the fitness gains from a larger body outweigh the costs of migrating, such as an increase in mortality, disease, and the inability to reach spawning grounds, then marine migration is beneficial (Klemetsen et al. 2003; Thorstad et al. 2016; Eldøy et al. 2021).

The unidirectional flow of river and narrow channels give migrating fishes definite reference points. In a dendritic river system, physical and chemical flow properties may vary spatially and temporally, which may confuse migrants but also allow branch detection. River migration requires upstream and downstream components. Early life history stages usually have the former and vice versa. However, life history stage migration may entail downstream and upstream components. Upstream migration must be active. Juveniles and adults migrate downstream nocturnally but diurnally in murky rivers. Some riverine and anadromous species spawn downstream, while growing adults usually spawn upstream. The timing of an upstream migration can vary from night to day (Northcote 1984; Jungwirth et al. 1998; Smith 2012; Tsukamoto et al. 2009; Morais and Daverat 2016).

20.2 Fish Migration

Sometimes, a single habitat cannot meet the needs of a population of fish (e.g., foraging and reproduction) (Wang et al. 2020). There are several reasons for this, including variations in habitat conditions (e.g., temperature) or changing needs of the population (e.g., foraging habitat vs. spawning habitat) (Zorn and Kramer 2022). An alternate habitat can benefit an individual's fitness in such a situation. This has led many fishes to develop a life history that involves coordinated movement across habitats (Cooke et al. 2022). The movement between discrete habitats of part or all of a population is what is known as "migration" (Dingle and Drake 2007). Migratory fish make up about 2.5% of all fish species. Some species of coastal- and stream-dwelling fish migrate hundreds of meters, while others migrate thousands of kilometers, such as eels (e.g., *Anguilla* species) and tunas (e.g., *Thunnus* species) (Kumari 2020; Cooke et al. 2011). Most species migrate on a seasonal basis, although some show coordinated daily movements (e.g., vertical migrations or tidal migrations) (Secor 2015). Many authors consider this to be migratory activity, while others view it as a specialized form of foraging.

20.3 Energetics of Migration

The long-distance migrations are energy-demanding, and feeding occurs rarely during long-distance migrations. Due to energy reallocation for foraging and digestion, feeding reduces the metabolic scope available for migratory activity (Naisbett-Jones et al. 2017; Eliason et al. 2011). Most species accumulate energy reserves prior to migration, so they rely heavily on those reserves for survival (Churova et al. 2021; Olsen et al. 2021; Paton et al. 2020). The most energy-dense form of energy storage is lipids, which have an energy density of 9.45 kcal per gram that is roughly twice as high as that of proteins and carbohydrates, which have 4.8 and 4.1 kcal per gram, respectively (Cooke et al. 2011). Fishes store their lipid reserves throughout a range of organs, including adipose-like tissue in the viscera, the muscle tissue, and the liver, in contrast to higher vertebrates, which mostly store lipids in adipose tissue (Azeez et al. 2014; Sheridan 1988). Except during spawning migrations, where the energy is used for gonadal growth, the majority of energy generated during migration is utilized by the muscles (da Silva Souza et al. 2020; Yin et al. 2020). Fishes use their energy reserves to varying degrees depending on the species and their associated population. Energy depletion can reach up to 85% of total energy stored during various anadromous spawning migrations (Birnie-Gauvin et al. 2021; Baktoft et al. 2020). The frequency of semelparity in fish population is correlated with the amount of energy depleted during migration; the more the energy depleted, the less likely they are to be a repeat spawner. In semelparous communities, mortality is most likely influenced by the depletion of energy supplies, although it is not the only reason (Tentelier et al. 2021; Kern and Gems 2022). It is clear that there are additional genetic and/or environmental factors at work because certain semelparous population travel just a moderate distance, which is significantly less than what would deplete their energy stores (Birnie-Gauvin et al. 2021).

20.4 Marine Migration Behavior

Among sea trout individuals, marine migration behavior varies widely. Prior to migration in the spring, the majority of sea trout had poor nutritional conditions (Jönsson et al. 2010). Low plasma triglyceride levels and body condition parameters in sea trout were associated with earlier sea entrance and increased likelihood of migration to sea. When compared to fish in better health, fishes with poor physical condition are more likely to stay in the sea longer and migrate farther offshore. Males were less likely than females to move to the sea. More often than the smaller fish, the larger fish scattered farther from the river and were also more likely to migrate to the sea rather than staying in freshwater and estuaries (Eldøy et al. 2021).

Fish nutritional status can be assessed using a blood plasma sample-derived nutritional correlation in addition to the body condition component. Low levels of plasma triglycerides, total protein, and calcium may be signs of inadequate nutrition, which has been linked to an increase in brown trout and Atlantic salmon migration in the ocean (Halttunen et al. 2018; Congleton and Wagner 2006). Cortisol levels that are elevated, perhaps as a result of scarce food sources, have been linked to earlier seaward migration (Birnie-Gauvin et al. 2019). Sex and body size have been observed to affect sea trout migration behavior. Consequently, genetically determined sex and physical size were also included. Body size, sex, and nutritional status all have an impact (Ferguson et al. 2019). Sea trout population from two different fjord systems in Northern Norway were studied for their maritime migration patterns (Eldøy et al. 2021; Berg and Berg 1989). Fish who were in poorer physical condition before migrating spend more time at sea and move further from the fjord than fish that were in better condition. When migrating, sea trout with low body condition parameters tended to spend more time at sea and migrate further from shore, presumably indicating a larger need for reconditioning compared to those with higher body condition scores (Davidsen et al. 2014; Eldøy et al. 2015; Bordeleau et al. 2018; Pemberton 1976; Jensen et al. 2019). Due to the close association between a female's body size and the number of eggs they may lay, females likely gain more from additional feeding chances than do males (Elliott 1995). A well-known phenomenon called sexual bias in migratory behavior has been documented in the range of salmonid species (Dodson et al. 2013).

20.5 Prediction of Fish Migration

The significant rise in global ocean temperatures has caused several issues, including ecological repercussions, the relocation of some traditional fisheries, increased competition among fisheries, and effects on biological reproduction. Investing in the consequences of ocean temperatures on fisheries is vital, which accounts for a significant amount of some nations' economies and affects people's ability to support themselves. Therefore, models for predicting fish migration to address the issue was explored and presented. The impacts of seawater temperature and salinity was studies and an ArcGIS-based model for predicting fish school migration was developed. Using this ArcGIS software, kriging interpolation, and nuclear density analysis, a thermodynamic map of fish school dispersion was generated from these data. In addition, linear interpolation and fitting are utilized to determine the link between the yearly fish movement distance and the year to anticipate fish migration paths over the following 50 years (Yipeng et al. 2020; Yalçin 2019).

The explosion of weather data has made weather forecasting more difficult (Belavadi et al. 2020). Artificial intelligence algorithms based on machine learning have steadily given meteorological research new ideas and ways to handle such massive data. Research now uses support vector machine (SVM), convolutional neural network (CNN), back-propagation (BP), etc. methods. These approaches have good temperature prediction results but cannot analyze the time correlation of time series data, such as forgetting premature data and reflecting periodic changes. However, ocean temperature data restrict prediction accuracy in a typical time series and influence migratory fish prediction. Researchers developed a recurrent neural network (RNN) (Ma et al. 2021) to include meteorological data time series information into the prediction model and improve forecast outcomes (Olivetti et al. 2021). The classic RNN model can process time series data, but its deep memory is short-term. Because numerous Jacobian matrix multiplication is required to calculate connections between distant nodes, it will struggle with long-term dependencies. The gradient disappears and explodes. RNN cannot handle long-term timing concerns due to these drawbacks. The distorted long short-term memory (LSTM) network may use previous data to help make current decisions and overcome the problem that existing approaches are hard to learn vast data and do not account for temperature time series data's time correlation (Jiao 2021; Zhang et al. 2021).

20.6 Ecological Barrier

Dams are obvious vertical barriers in fixed locations along river courses. Fish cannot swim upstream due to a dam, but they can swim downstream using the dam's structures, including the spillways or turbines (in hydroelectric dams). It is not uncommon for fish to congregate below the dam wall, particularly during spawning season (Loures and Pompeu 2012). Because of this problem, which can be seen at any dam, fishways have been proposed by scientists, managers, and regular people for a long time. Ladders, elevators, bypass channels, locks, and trap-truck systems are examples of technologies that help migratory fish get through dams and fulfil their life cycle needs by reuniting once-separate river sections (Agostinho et al. 2007; Clay 2017; Larinier 2002). However, most fish passes are insufficient (Brown et al. 2013; Noonan et al. 2012), and fishways may only assist many upstream passages depending on the area and the species. For a selected number of species, fishways in South America allow for the passage of just a tiny percentage of migratory individuals (Fernandez et al. 2004). From the time the first ladders were set up in Brazil until now, there has been much room for improvement in terms of both planning and engineering and hydraulics. Increased understanding of the swimming skills of local species and the development of more efficient fish passages have contributed to improved conservation efforts (Agostinho et al. 2007; Pelicice et al. 2015).

First and foremost, when a river is dammed, the hydrophysical and morphological structure of the water flow is altered, producing a distinct biotopic pattern that influences the life activities of fish, including their migratory behavior. Ecological barriers, such as the water reservoir and the dam, are established in a dammed river, fundamentally altering the conditions that determine the features of the downstream migration from the higher reaches to the lower reaches of a natural river. Both the physical complexity of a body of water and the intensity of water exchange play crucial roles in the establishment of these barriers and in regulating downstream migration. These influences operate not just at the scale of the entire water reservoir but also at small biotopes, where there is a downstream migration (Pavlov et al. 2019; Larinier, 2000).

20.7 Abiotic Factors

20.7.1 Climatic Factors

Migration is an important, expanding, and an extremely complex phenomenon (Black et al. 2011). Changes in temperature were primarily linked to variations in migration and reproduction timing, age at maturity, age at juvenile migration, growth, survival, and fecundity (Crozier and Hutchings 2014). Migration is one option among the many that might be used to address the effects of climate change, and it has drawn more and more interest from policymakers and researchers. The underlying hypothesis of most of these reports is that climate-related migration essentially reflects a failure to slow down climate change and/or a failure to adapt. Together, they have generated some concerning projections for the overall magnitude of migration that such failures might cause (Douglas et al. 2008; Warner et al. 2009; Renaud et al. 2007; Conisbee and Simms 2003). However, migration can also be considered to be a legitimate coping strategy for the increased stresses and shocks that may be brought on by climate change and a number of other causes (Tacoli 2009). The claim that climate change is already affecting marine ecosystems is supported by scientific research. The structure of marine communities, changes in the intensity and timing of coastal upwelling, and increases in ocean temperature all have an impact on the primary and secondary productivity (Leonard et al. 2015; Sissener and Bjørndal 2005).

20.7.2 Anthropogenic Factors

A significant anthropogenic contributor to the reduction in biodiversity, particularly in aquatic ecosystems, is habitat fragmentation. The effects of habitat fragmentation in rivers are influenced by the interactions of biota migration characteristics, temporal hydrological change that mediates connection, and the location of anthropogenic obstacles (Rolls et al. 2014). Anthropogenic factors are typically believed to have significant impacts on migration rates. The anadromous fish Hilsa exhibits the same migratory patterns and breeding habits as the Atlantic salmon (*Salmo* sp.) (Shohidullah Miah 2015).

20.7.3 Global Warming

Ocean temperatures have been gradually rising in the recent years as a result of global warming. Fish population (such as those of herring and mackerel) will migrate in order to survive and reproduce (Xu et al. 2021). The research by Nicolas et al. found that the distribution of several typical estuarine species has shifted northward, which is consistent with the idea that these species will migrate north in response to climate change (Nicolas et al. 2011). Global warming may make fish in streams in the southern Great Plains and southwest North America particularly susceptible to extirpation or extinction. The streams in this area already have some of the warmest free-flowing water on the planet, and fish continue to survive occasionally extremely close to their lethal thermal limits. In the case of global warming, fish in these prairie stream systems cannot migrate northward to colder temperatures, unlike many terrestrial and marine animals or fish of some rivers (Matthews and Zimmerman 1990).

20.7.4 Effects of the Photoperiod

Numerous physiological and biochemical systems in fish experience daily variations (circadian rhythms), and the endocrine system regulates these processes (Björnsson 1997). Different photoperiod modes cause the parr–smolt transformation and smoltification, which are linked to increased endocrine and Na^+ and K^+-ATPase activity, an increase in lipid and nitrogen metabolism, modifications in the bioconversion of physiologically important polyunsaturated fatty acids (PUFAs) like eicosapentaenoic acid, docosahexaenoic acid, and arachidonic acid from essential fatty acids and linoleic acid (Nemova et al. 2020). The physiological state of the organism, including age, gut fullness, fitness, reproductive status, maturity, and water temperature, determines how the fish react to different light types (Sautin 1989). Light conditions (not less than 16,000 and not more than 80,500 lx) limit the salmon parr's ability to move around and hunt (Nesterov 1985).

20.8 Bioaccumulation of Organic Compounds

Due to habitat loss and shrinking biodiversity, persistent organic pollutants (POPs) and metals can have a negative impact on an entire aquatic ecosystem. They can also be toxic to aquatic organisms, either chronically or acutely (Teunen et al. 2021a). Pollutants that result from a variety of anthropogenic activities (such as industry, agriculture, and combustion by-products) can enter the environment through discharge, leaching, erosion, and atmospheric deposition (Schweitzer and Noblet 2018). The Stockholm Convention has significantly reduced the use and production of numerous pollutants over the past few decades, but previous contamination is still pervasive in the aquatic environment (Maes et al. 2008). As hydrophobic organic compounds (HOCs) do not dissolve easily in water, they are not easily measured in the water phase (Belpaire and Goemans 2007; Jürgens et al. 2013). Pollutant bioavailability can be influenced by sediment properties such as clay concentration and total organic carbon (TOC) and water qualities such as oxygen content, pH, conductivity, and dissolved organic carbon (DOC). The bioavailability of lipophilic substances may be decreased by greater organic complexation caused by high DOC or TOC levels (Dittman and Driscoll 2009; Li et al. 2015; Moeckel et al. 2014). Consequently, it is advised to measure polycyclic aromatic hydrocarbons (PAHs) in bivalves or crustaceans rather. Active biomonitoring has frequently been used to monitor bioaccumulative pollutants using translocated individuals (Babut et al. 2020; Catteau et al. 2021). This standardized sampling method is based on exposing a specific species to various sampling locations that reflect the level of local pollution under controlled low background concentrations, sizes, or circumstances. For this, the bivalve genus Dreissena has been employed frequently (Bashnin et al. 2019; Potet et al. 2018; Teunen et al. 2021b).

20.9 Environmental Cues

The study of migratory biology frequently centers on how environmental factors affect migration. The fact that environmental variables are frequently highly associated with one another makes it challenging to determine which variables are the most relevant for affecting a migration, which presents the biggest obstacle in investigating how environmental factors influence migration (Gahagan et al. 2010; Davidsen 2010). The local features of the habitat where the migration is taking place may affect the relative impact of each environmental element. Nevertheless, despite these difficulties, it has consistently been demonstrated that a variety of environmental conditions affect fish migration (Ahn et al. 2020). Discharge-related occurrences can significantly affect migratory behavior in fluvial habitats. Changing discharge often correlates with changes in other environmental factors (Sykes et al. 2009; Bhat and Qayoom 2021; Schwevers and Adam 2020). A significant environmental factor that stimulates migratory movement is changing light levels. Changes in photoperiod on a seasonal scale offer calendar information that is used to start and synchronize migratory behavior among individuals within a group. The photoperiod appears to be particularly important for coordinating migratory activity during long-distance migrations, as the spawning migrations of Pacific salmon and Pacific lampreys (*Lampetra tridentata*) (Cooke et al. 2011). Fish migratory activity can be synchronized and triggered by temperature. Migration can be seen as a type of behavioral thermoregulation when temperature serves as a trigger (Haesemeyer 2020; Archer et al. 2020). The temperature may fluctuate in thermally diverse situations beyond what a specific population can tolerate. The fish are forced to look for a new thermal environment as a result. The population's thermal needs could alter. The ideal temperature, for instance, may not be the same for development and reproduction (Haesemeyer 2020; Wheeler et al. 2020; Volkoff and Rønnestad 2020).

20.10 Conclusion

The term "migration" refers to the periodic changes in habitat that are an integral component of the life cycle of many species and are used for survival, development, and reproduction. An expansion or relocation of a migration loop may cause diadromy or speciation in diadromous fishes, respectively; hence, migration loop models might be helpful for conceptualizing the process of speciation in these organisms. According to their phylogenetic relationships, salmonids originated in freshwater. They became increasingly dependent on the sea through anadromous migrations, while anguillid eels originated in the ocean and appeared to be facultatively catadromous in temperate regions, with some individuals not entering freshwater for growth (Tsukamoto et al. 2009). Changes in productivity between marine and freshwater ecosystems at high and low latitudes appear to cause these variations. Using the upstream migration of amphidromous Ayu as a behavioral model to try to figure out what makes fish move, a behavioral approach suggests that both internal and external factors, like an increase in water temperature, fish density, or hunger level, can raise the drive level and release behavior associated with upstream migration.

Acknowledgments The authors thank the Vellore Institute of Technology, Vellore, Tamil Nadu, India, for supporting this work.

References

Agostinho AA, Gomes LC, Pelicice FM (2007) Ecologia e manejo de recursos pesqueiros em reservatórios do Brasil. UEM

Ahn H et al (2020) Evaluation of fish biodiversity in estuaries using environmental DNA metabarcoding. PLoS One 15(10):e0231127

Archer LC et al (2020) Food and temperature stressors have opposing effects in determining flexible migration decisions in brown trout (Salmo trutta). Glob Chang Biol 26(5):2878–2896. (in eng)

Azeez OI, Meintjes R, Chamunorwa JP (2014) Fat body, fat pad and adipose tissues in invertebrates and vertebrates: the nexus. Lipids Health Dis 13:71. (in eng)

Babut M et al (2020) Monitoring priority substances in biota under the Water Framework Directive: how effective is a tiered approach based on caged invertebrates? A proof-of-concept study targeting PFOS in French rivers. Environ Sci Eur 32(1):1–13

Baktoft H et al (2020) Can energy depletion of wild Atlantic salmon kelts negotiating hydropower facilities lead to reduced survival? Sustainability 12(18):7341

Bashnin T, Verhaert V, De Jonge M, Vanhaecke L, Teuchies J, Bervoets L (2019) Relationship between pesticide accumulation in transplanted zebra mussel (Dreissena polymorpha) and community structure of aquatic macroinvertebrates. Environ Pollut 252(pt A):591–598. (in eng)

Belavadi SV, Rajagopal S, Ranjani R, Mohan R (2020) Air quality forecasting using LSTM RNN and wireless sensor networks. Proc Comput Sci 170:241–248

Belpaire C, Goemans G (2007) The European eel Anguilla anguilla, a rapporteur of the chemical status for the water framework directive? Vie Milieu 57:235–252

Berg OK, Berg M (1989) The duration of sea and freshwater residence of the sea trout, Salmo trutta, from the Vardnes River in northern Norway. Environ Biol Fishes 24(1):23–32

Bhat SU, Qayoom U (2021) Implications of sewage discharge on freshwater ecosystems. IntechOpen, London

Birnie-Gauvin K et al (2019) Cortisol predicts migration timing and success in both Atlantic salmon and sea trout kelts. Sci Rep 9(1):2422. (in eng)

Birnie-Gauvin K et al (2021) Life-history strategies in salmonids: the role of physiology and its consequences. Biol Rev 96(5):2304–2320

Björnsson BT (1997) The biology of salmon growth hormone: from daylight to dominance. Fish Physiol Biochem 17(1):9–24

Black R, Kniveton D, Schmidt-Verkerk K (2011) Migration and climate change: towards an integrated assessment of sensitivity. Environ Plann A Econ Space 43(2):431–450

Bordeleau X, Davidsen JG, Eldøy SH, Sjursen AD, Whoriskey FG, Crossin GT (2018) Nutritional correlates of spatiotemporal variations in the marine habitat use of brown trout (Salmo trutta) veteran migrants. Can J Fish Aquat Sci 75(10):1744–1754

Brown JJ, Limburg KE, Waldman JR, Stephenson K, Glenn EP, Juanes F, Jordaan A (2013) Fish and hydropower on the U.S. Atlantic coast: failed fisheries policies from half-way technologies. Conserv Lett 6(4):280–286. https://doi.org/10.1111/conl.2013.6.issue-4. 10.1111/conl.12000

Catteau A et al (2021) Water quality of the Meuse watershed: assessment using a multi-biomarker approach with caged three-spined stickleback (Gasterosteus aculeatus L.). Ecotoxicol Environ Saf 208:111407. (in eng)

Chapman BB et al (2012) Partial migration in fishes: causes and consequences. J Fish Biol 81(2): 456–478. (in eng)

Churova M, Shulgina N, Krupnova MY, Efremov D, Nemova N (2021) Activity of energy and carbohydrate metabolism enzymes in the juvenile Pink Salmon Oncorhynchus gorbuscha (Walb.) during the transition from freshwater to a marine environment. Biol Bull 48(5):546–554

Clay CH (2017) Design of fishways and other fish facilities. CRC Press

Congleton J, Wagner T (2006) Blood-chemistry indicators of nutritional status in juvenile salmonids. J Fish Biol 69(2):473–490

Conisbee M, Simms A (2003) Environmental refugees: the case for recognition. New Economics Foundation, London

Cooke SJ, Binder TR, Hinch SG (2011) The biology of fish migration. In: Encyclopedia of fish physiology: from genome to environment. Academic Press, San Diego

Cooke SJ et al (2022) The movement ecology of fishes. J Fish Biol 101(4):756–779

Crozier LG, Hutchings JA (2014) Plastic and evolutionary responses to climate change in fish. Evol Appl 7(1):68–87

Davidsen JG (2010) Effects of environmental factors on migratory behaviour of northern Atlantic salmon. Universitetet i Tromsø. https://hdl.handle.net/10037/2480

Davidsen J, Daverdin M, Sjursen A, Rønning L, Arnekleiv J, Koksvik J (2014) Does reduced feeding prior to release improve the marine migration of hatchery brown trout Salmo trutta smolts? J Fish Biol 85(6):1992–2002

Dingle H (2014) Migration: the biology of life on the move. Oxford University Press, Oxford

Dingle H, Drake VA (2007) What is migration? Bioscience 57(2):113–121

Dittman JA, Driscoll CT (2009) Factors influencing changes in mercury concentrations in lake water and yellow perch (Perca flavescens) in Adirondack lakes. Biogeochemistry 93(3): 179–196

Dodson JJ, Aubin-Horth N, Thériault V, Páez DJ (2013) The evolutionary ecology of alternative migratory tactics in salmonid fishes. Biol Rev 88(3):602–625

Douglas I, Alam K, Maghenda M, Mcdonnell Y, McLean L, Campbell J (2008) Unjust waters: climate change, flooding and the urban poor in Africa. Environ Urban 20(1):187–205

Eldøy SH et al (2015) Marine migration and habitat use of anadromous brown trout (Salmo trutta). Can J Fish Aquat Sci 72(9):1366–1378

Eldøy SH et al (2021) The effects of nutritional state, sex and body size on the marine migration behaviour of sea trout. Mar Ecol Prog Ser 665:185–200

Eliason EJ et al (2011) Differences in thermal tolerance among sockeye salmon populations. Science 332(6025):109–112. (in eng)

Elliott J (1995) Fecundity and egg density in the redd for sea trout. J Fish Biol 47(5):893–901

Ferguson A, Reed TE, Cross TF, McGinnity P, Prodöhl PA (2019) Anadromy, potamodromy and residency in brown trout Salmo trutta: the role of genes and the environment. J Fish Biol 95(3): 692–718. (in eng)

Fernandez DR, Agostinho AA, Bini LM (2004) Selection of an experimental fish ladder located at the dam of the Itaipu Binacional Paraná River Brazil. Braz Arch Biol Technol 47(4):579–586. https://doi.org/10.1590/S1516-89132004000400012

Gahagan BI, Gherard KE, Schultz ET (2010) Environmental and endogenous factors influencing emigration in juvenile anadromous alewives. Trans Am Fish Soc 139(4):1069–1082

Gross MR, Coleman RM, McDowall RM (1988) Aquatic productivity and the evolution of diadromous fish migration. Science 239(4845):1291–1293. (in eng)

Haesemeyer M (2020) Thermoregulation in fish. Mol Cell Endocrinol 518:110986. (in eng)

Halttunen E et al (2018) Sea trout adapt their migratory behaviour in response to high salmon lice concentrations. J Fish Dis 41(6):953–967. (in eng)

Jensen AJ, Finstad B, Fiske P (2019) The cost of anadromy: marine and freshwater mortality rates in anadromous Arctic char and brown trout in the Arctic region of Norway. Can J Fish Aquat Sci 76(12):2408–2417

Jiao Y (2021) Prediction of fish migration based on LSTM model. In: 2021 IEEE International Conference on Power Electronics, Computer Applications (ICPECA), pp 697–701. IEEE

Jonsson B, Jonsson N (1993) Partial migration: niche shift versus sexual maturation in fishes. Rev Fish Biol Fish 3(4):348–365

Jönsson E, Kaiya H, Björnsson BT (2010) Ghrelin decreases food intake in juvenile rainbow trout (Oncorhynchus mykiss) through the central anorexigenic corticotropin-releasing factor system. Gen Comp Endocrinol 166(1):39–46. (in eng)

Jungwirth M, Schmutz S, Weiss S (1998) Fish migration and fish bypasses. Fishing News Books, Oxford

Jürgens MD, Johnson AC, Jones KC, Hughes D, Lawlor AJ (2013) The presence of EU priority substances mercury, hexachlorobenzene, hexachlorobutadiene and PBDEs in wild fish from four English rivers. Sci Total Environ 461–462:441–452. (in eng)

Kern CC, Gems D (2022) Semelparous death as one element of iteroparous aging gone large. Front Genet 13:880343. (in eng)

Klemetsen A et al (2003) Atlantic salmon Salmo salar L., brown trout Salmo trutta L. and Arctic charr Salvelinus alpinus (L.): a review of aspects of their life histories. Ecol Freshw Fish 12(1): 1–59

Kumari P (2020) Migration of fishes-a spectacular event of nature. Int Res J Eng Technol 07(04): 3321–3324

Larinier M (2000) Dams and fish migration. World Commission on Dams, Toulouse, France

Larinier M (2002) Fishways-general considerations. Bulletin Français de la Pêche et de la Pisciculture (364):21–27

Leonard N, Fritsch M, Ruff J, Fazio J, Harrison J, Grover T (2015) The challenge of managing the Columbia River Basin for energy and fish. Fish Manag Ecol 22(1):88–98

Li YL et al (2015) Influences of binding to dissolved organic matter on hydrophobic organic compounds in a multi-contaminant system: coefficients, mechanisms and ecological risks. Environ Pollut 206:461–468. (in eng)

Loures RC, Pompeu PS (2012) Temporal variation in fish community in the tailrace at Três Marias Hydroelectric Dam São Francisco River Brazil. Neotrop Ichthyol 10(4):731–740. https://doi.org/10.1590/S1679-62252012000400006

Ma S et al (2021) Improved seagull optimization algorithm to optimize neural networks with gated recurrent units for network intrusion detection. In: 2021 11th IEEE international conference on Intelligent Data Acquisition and Advanced Computing Systems: Technology and Applications (IDAACS), Cracow, vol 1. IEEE, pp 100–104

Maes J, Belpaire C, Goemans G (2008) Spatial variations and temporal trends between 1994 and 2005 in polychlorinated biphenyls, organochlorine pesticides and heavy metals in European eel (Anguilla anguilla L) in Flanders, Belgium. Environ Pollut 153(1):223–237. (in eng)

Matthews WJ, Zimmerman EG (1990) Potential effects of global warming on native fishes of the Southern Great Plains and the Southwest. Fisheries 15(6):26–32

Moeckel C, Monteith DT, Llewellyn NR, Henrys PA, Pereira MG (2014) Relationship between the concentrations of dissolved organic matter and polycyclic aromatic hydrocarbons in a typical U.K. upland stream. Environ Sci Technol 48(1):130–138. (in eng)

Morais P, Daverat F (2016) An introduction to fish migration. CRC Press, Boca Raton

Naisbett-Jones LC, Putman NF, Stephenson JF, Ladak S, Young KA (2017) A magnetic map leads juvenile European eels to the Gulf stream. Curr Biol 27(8):1236–1240. (in eng)

Nemova NN et al (2020) The effect of the photoperiod on the fatty acid profile and weight in hatchery-reared underyearlings and yearlings of Atlantic Salmon Salmo salar L. Biomolecules 10(6):845. (in eng)

Nesterov VD (1985) The behavior of juvenile Atlantic Salmon Salmo Salar during downhill migration. PhD, Russian Federal Research Institute of Fisheries and Oceanography, Specialty 03.00.10 (Ichthyology), Moskow

Nicolas D et al (2011) Impact of global warming on European tidal estuaries: some evidence of northward migration of estuarine fish species. Reg Environ Change 11(3):639–649

Noonan MJ, Grant JWA, Jackson CD (2012) A quantitative assessment of fish passage efficiency. Fish Fish 13(4):450–464. 10.1111/faf.2012.13.issue-4. https://doi.org/10.1111/j.1467-2979.2011.00445.x

Northcote T (1984) Mechanisms of fish migration in rivers. In: Mechanisms of migration in fishes. Springer, Boston, pp 317–355

Olivetti S, Gil MA, Sridharan VK, Hein AM (2021) Merging computational fluid dynamics and machine learning to reveal animal migration strategies. Methods Ecol Evol 12(7):1186–1200

Olsen L, Thum E, Rohner N (2021) Lipid metabolism in adaptation to extreme nutritional challenges. Dev Cell 56(10):1417–1429

Paton KR, Cake MH, Potter IC (2020) Lipid and protein catabolism contribute to aerobic metabolic responses to exhaustive exercise during the protracted spawning run of the lamprey Geotria australis. J Comp Physiol B 190(1):35–47

Pavlov DS, Mikheev VN, Kostin VV (2019) Migrations of fish juveniles in dammed rivers: the role of ecological barriers. J Ichthyol 59(2):234–245. https://doi.org/10.1134/S0032945219020140

Pelicice FM, Pompeu PS, Agostinho AA (2015) Large reservoirs as ecological barriers to downstream movements of Neotropical migratory fish. Fish Fish 16(4):697–715. https://doi.org/10.1111/faf.2015.16.issue-4. 10.1111/faf.12089

Pemberton R (1976) Sea trout in North Argyll sea lochs, population, distribution and movements. J Fish Biol 9(2):157–179

Potet M, Giambérini L, Pain-Devin S, Louis F, Bertrand C, Devin S (2018) Differential tolerance to nickel between Dreissena polymorpha and Dreissena rostriformis bugensis populations. Sci Rep 8(1):700. (in eng)

Renaud FG, Bogardi JJ, Dun O, Warner K (2007) Control, adapt or flee: how to face environmental migration? UNU-EHS, Bonn

Rolls RJ, Stewart-Koster B, Ellison T, Faggotter S, Roberts DT (2014) Multiple factors determine the effect of anthropogenic barriers to connectivity on riverine fish. Biodivers Conserv 23(9):2201–2220

Sautin E (1989) The problem of regulation of adaptive changes in lipogenesis, lipolysis and lipid transport in fish. Biol Bull Rev 107:131–149

Schweitzer L, Noblet J (2018) Water contamination and pollution. In: Dransfield TE, Török B (eds) Green chemistry. Elsevier, Amsterdam, pp 261–290

Schwevers U, Adam B (2020) Fish protection technologies and fish ways for downstream migration. Springer, Cham

Secor DH (2015) Migration ecology of marine fishes. JHU Press, Baltimore

Sheridan MA (1988) Lipid dynamics in fish: aspects of absorption, transportation, deposition and mobilization. Comp Biochem Physiol B 90(4):679–690. (in eng)

Shohidullah Miah M (2015) Climatic and anthropogenic factors changing spawning pattern and production zone of Hilsa fishery in the Bay of Bengal. Weather Clim Extremes 7:109–115

da Silva Souza JG et al (2020) A method to analyze the relationship between locomotor activity and feeding behaviour in larvae of Betta splendens. Aquac Int 28(3):1141–1152

Sissener EH, Bjørndal T (2005) Climate change and the migratory pattern for Norwegian spring-spawning herring—implications for management. Mar Policy 29(4):299–309

Smith RJF (2012) The control of fish migration. Springer, Berlin

Sykes GE, Johnson CJ, Shrimpton JM (2009) Temperature and flow effects on migration timing of Chinook salmon smolts. Trans Am Fish Soc 138(6):1252–1265

Tacoli C (2009) Crisis or adaptation? Migration and climate change in a context of high mobility. Environ Urban 21(2):513–525

Tentelier C et al (2021) The dynamics of spawning acts by a semelparous fish and its associated energetic costs. Peer Community J 1:e49

Teunen L et al (2021a) Effect of abiotic factors and environmental concentrations on the bioaccumulation of persistent organic and inorganic compounds to freshwater fish and mussels. Sci Total Environ 799:149448. (in eng)

Teunen L, Bervoets L, Belpaire C, De Jonge M, Groffen T (2021b) PFAS accumulation in indigenous and translocated aquatic organisms from Belgium, with translation to human and ecological health risk. Environ Sci Eur 33(1):1–19

Thorstad EB et al (2016) Marine life of the sea trout. Mar Biol 163(3):1–19

Tsukamoto K, Miller MJ, Kotake A, Aoyama J, Uchida K (2009) The origin of fish migration: the random escapement hypothesis. In: American Fisheries Society symposium, vol 69, pp 45–61

Volkoff H, Rønnestad I (2020) Effects of temperature on feeding and digestive processes in fish. Temperature 7(4):307–320

Wang L, Chen Q, Zhang J, Xia J, Mo K, Wang J (2020) Incorporating fish habitat requirements of the complete life cycle into ecological flow regime estimation of rivers. Ecohydrology 13(4):e2204

Warner K, Ehrhart C, Sherbinin AD, Adamo S, Chai-Onn T (2009) In search of shelter: mapping the effects of climate change on human migration and displacement. Climate Change CARE International, London, UK

Wheeler CR et al (2020) Anthropogenic stressors influence reproduction and development in elasmobranch fishes. Rev Fish Biol Fish 30(2):373–386

Xu F, Du Y-A, Chen H, Zhu J-M (2021) Prediction of fish migration caused by ocean warming based on SARIMA model. Complexity 2021:5553935

Yalçin E (2019) The behaviour of Turkish bluefin tuna (Thunnus thynnus) fishing fleet in the Mediterranean Sea. Mar Sci Technol Bull 8(2):64–68

Yin D et al (2020) Metabolic mechanisms of Coilia nasus in the natural food intake state during migration. Genomics 112(5):3294–3305

Yipeng Z, Dichao N, Wentao H (2020) Global warming prediction model of fish migration based on ArcGIS. In: E3S web of conferences, vol 204. EDP Sciences, p 01004

Zera AJ, Harshman LG (2001) The physiology of life history trade-offs in animals. Annu Rev Ecol Syst 32:95–126

Zhang X, Zhu C, He M, Dong M, Zhang G, Zhang F (2021) Failure mechanism and long short-term memory neural network model for landslide risk prediction. Remote Sens (Basel) 14(1):166

Zorn TG, Kramer DR (2022) Changes in habitat conditions, fish populations, and the fishery in northern Green Bay, Lake Michigan, 1989–2019. N Am J Fish Manag 42(3):549–571